Halide Perovskite Semiconductors

Halide Perovskite Semiconductors

Structures, Characterization, Properties, and Phenomena

Edited by Yuanyuan Zhou and Iván Mora-Seró

WILEY-VCH

Editors

Prof. Yuanyuan Zhou
The Hong Kong University of Science and Technology
Department of Chemical and Biological Engineering
Clear Water Bay, Hong Kong
SAR 999077
China

Prof. Iván Mora-Seró
Universitat Jaume I (UJI)
Institute of Advanced Materials (INAM)
Avenida de Vicent Sos Baynat
s/n, 12071 Castelló de la Plana
Spain

Cover Image: © NANOCLUSTERING/ SCIENCE PHOTO LIBRARY/ Getty Images

Library of Congress Card No.: applied for

British Library Cataloguing-in-Publication Data
A catalogue record for this book is available from the British Library.

Bibliographic information published by the Deutsche Nationalbibliothek
The Deutsche Nationalbibliothek lists this publication in the Deutsche Nationalbibliografie; detailed bibliographic data are available on the Internet at <http://dnb.d-nb.de>.

Print ISBN: 978-3-527-34809-1
ePDF ISBN: 978-3-527-82901-9
ePub ISBN: 978-3-527-82903-3
oBook ISBN: 978-3-527-82902-6

Typesetting Straive, Chennai, India
Printing and Binding CPI Group (UK) Ltd, Croydon, CR0 4YY

C9783527348091_030724

Contents

Preface

Despite that fact that metal halide perovskites have been known for several decades, they are receiving significant attention, especially in the last one, as a new type of semiconductor material family that has injected excitement in both material and device research. Structurally, halide perovskite semiconductors are like conventional inorganic semiconductors, exhibiting long-range ordering at the atomic scale and crystalline grain characteristics at the microscopic scale. However, the predominantly ionic character of their bonds provides unusual ionic conductivity, which can limit the performance of some device configurations but promote other properties and even self-healing processes. In addition, some halide perovskites also present benign defect physics. This favors the interactions of the materials themselves with photons, charge carriers, and phonons for excellent semiconducting properties even in polycrystalline films with numerous crystallographic defects. From the perspective of processing, perovskite semiconductors are similar to soft organic semiconductors, which can be fabricated using high-throughput, low-temperature solution printing at low cost. This unleashes the potential of flexible, massive device integration into our future intelligent world. In fact, intense investigations aimed at creating better and cheaper semiconductors have been performed for many decades, but the emergence of a semiconductor type that can combine all the above merits has been extremely rare. Therefore, perovskite semiconductors have triggered the interest of the science community to devote significant research efforts to probing the fundamental sciences that underpin their unprecedented materials and device behaviors. With the continuous blooming of perovskite research for more than a decade, as well as the substantially rising interest from the industry in scaling-up perovskite technologies, a high level of new, overarching fundamental knowledge on perovskite semiconductors may be synthesized by the field based on the existing literature studies. We thus consider it urgent to summarize these new semiconductor sciences in a systematic manner. In this context, we have invited world-reputed researchers to contribute chapters on near-full range of fundamental topics covering from versatile crystal structures to characterization methods and from various properties to device implications. All the chapter authors have not only presented excellent summaries of reported findings but also incorporated their forward-looking thinking to guide future research. We deeply appreciate their dedicated efforts that ensure the high quality and lasting impacts of this book. We sincerely hope that this book will create inspiration for the readers to discover more fascinating sciences and to invent more frontier technologies in future research and development of perovskite semiconductors.

The Hong Kong University of Science and Technology, China *Yuanyuan Zhou*
Universitat Jaume I, Spain *Iván Mora-Seró*
1 October 2023

1

Introduction to Perovskite

Tianwei Duan[1], *Iván Mora-Seró*[2], *and Yuanyuan Zhou*[1,3]

[1] *Hong Kong Baptist University, Department of Physics, Kowloon Tang, Hong Kong, SAR 999077, China*
[2] *Universitat Jaume I (UJI), Institute of Advanced Materials (INAM), Avenida de Vicent Sos Baynat, s/n, 12071 Castelló de la Plana, Spain*
[3] *The Hong Kong University of Science and Technology, Department of Chemical and Biological Engineering, Clear Water Bay, Hong Kong, SAR 999077, China*

1.1 Evolution of Perovskite

Perovskite refers to a crystalline structure and extends to all the materials sharing this structure, despite the fact that it can present very different nature and properties. Initially, perovskites just denoted metal oxide minerals with a crystallography family of ABO_3 stoichiometry. The beginning of perovskite dates back to the discovery of chlorite-rich skarn at the Ural Mountains by the mineralogist German Gustav Rose in 1839. The component $CaTiO_3$ was found in this mineral and named after the notable Count Lev A. Perovski (1792–1856), president of the Russian Geological Society. Thereafter, many metal oxides with perovskite structures, such as $BaTiO_3$, $PbTiO_3$ and $SrTiO_3$, were widely studied. Many of the oxide perovskites were found to exhibit ferroelectric or piezoelectric properties [1–3].

More than 50 years after the discovery of oxide perovskite, a series of lead halide compounds with the general formula $CsPbX_3$ (X = Cl, Br, I) were synthesized by Wells [4]. These metal halides were later proved to have a perovskite structure, ABX_3, which is cubic at high temperatures and transforms from a tetragonally distorted structure at a lower temperature. The tunable photoconductivity of $CsPbX_3$ has drawn much attention to the electronic property study, and also evolved the idea of organic molecules addition [5, 6]. Weber discovered that the organic cation methylammonium ($CH_3NH_3^+$) substitutes for Cs^+ form $CH_3NH_3MX_3$ (M = Pb, Sn, X = I, Br) and published the first crystallographic study on organic lead halide perovskites [7, 8].

At the end of the twentieth century, abundant organic–inorganic halide perovskites were synthesized by Mitzi et al. [9–11]. Organic molecules, such as small and large organic cations, breathe new life into halide perovskite, embracing more diverse structures and physical properties in optoelectronic, photovoltaic, ferro- and antiferromagnetic, and non-linear optical fields. In addition to flexible components and versatile functionality, the low-forming energy makes halide perovskites facile

Halide Perovskite Semiconductors: Structures, Characterization, Properties, and Phenomena, First Edition. Edited by Yuanyuan Zhou and Iván Mora-Seró.

to be fabricated into films, which makes them a promising material for commercialization in next-generation semiconductors, and their interest in the development of light-emitting diodes (LEDs) and transistors was demonstrated.

1.2 Structure of Perovskite

In a traditional view, perovskite represents a crystallographic family with the chemical formula ABX_3, in which A and B are cations and X is an anion. The ideal perovskite is a cubic structure, having B cations as sixfold coordination surrounded by an octahedron of X anions, and A cations as 12fold cuboctahedral coordination, see Figure 1.1. Taking inorganic perovskite $CsPbI_3$ as an example, the Cs^+ cations are shown at the corners of the cube, and Pb^{2+} cations are in the center with I^- anions in the face-centered positions. In three-dimensional (3D) perovskites, all six anions at the corners of the octahedra, with Pb at the center, are shared with the six nearest octahedra, see Figure 1.1. When large cations are included in the structure, not all the six halides can be shared with other octahedra, forming 2D, 1D, or 0D perovskite-inspired materials. Many composition types of perovskites have been reported, involving lead halide perovskite, all-inorganic cesium/rubidium lead halide perovskite, lead-free or lead-low halide perovskite, and halide double perovskite, as it will be extensively discussed in this book.

In the case of organic–inorganic perovskite, at least one of the ions in ABX_3 is organic, e.g. $MAPbI_3$ and $FAPbI_3$ (MA is methylammonium, $CH_3NH_3^+$; FA is formamidinium, $HC(NH_2)_2^+$). Recently, metal-free perovskite has also been synthesized with the chemical formula ANH_4X_3, where A is a divalent organic cation, and X is halogen ions, e.g. MDABCO–NH_4I_3 (MDABCO is *N*-methyl-*N*′-diazabicyclo[2.2.2]octonium).

Several conditions must be satisfied in order for perovskite structure to be formed. Generally, the valences of A and B cations must total to three times those of the X anion to preserve charge balance. Furthermore, the perovskite structure can only

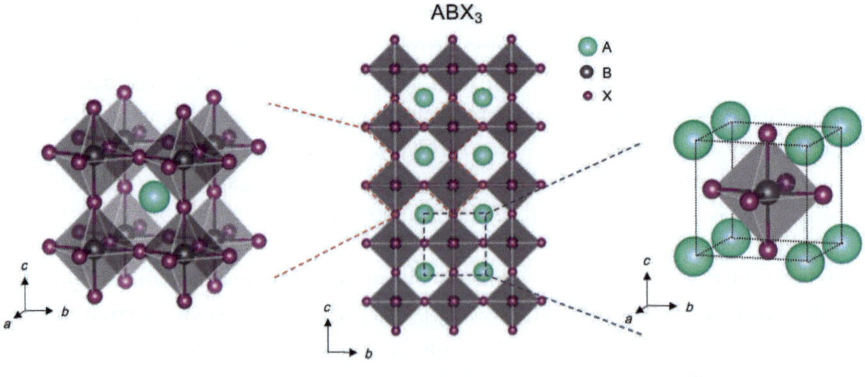

Figure 1.1 Crystal structure of 3D cubic perovskite. Cation A is located in the void between the BX_6 octahedra. In a crystal unit cell, A is located in the corners, B is in a body-centered position, and X is in a face-centered position.

tolerate particular ion combinations because of the size restrictions between ions in order to preserve the anion-corner-sharing structure. This ionic size relationship is expressed in terms of the Goldschmidt tolerance factor τ, which is correlated to the ionic radii r_A, r_B, and r_X [12]:

$$\tau = \frac{r_A + r_X}{\sqrt{2}(r_B + r_X)}$$

where τ is an empirical index to predict the different structures of ABX_3. When $0.9 < \tau < 1$, the perfect cubic perovskite structure is formed; when $0.8 < \tau < 0.9$, the distorted perovskite structure with tilted octahedra is preferred; when $\tau < 0.8$ or $\tau > 1$, the structure is non-perovskite [13]. Another factor is the octahedral factor,

$$\mu = {}^{r_B}/r_X$$

which determines whether the B atoms will favor the octahedral coordination of X atoms over greater or lower coordination numbers; this criterion is met for values between 0.4 and 0.9 [14]. In addition to size and charge, the coordination preference of metal ions is also taken into consideration.

Nowadays, many structural variants of perovskite have been synthesized, and they are all derived from the original 3D perovskite structure based on the corner-sharing BX_6 structure. Although the ABX_3 perovskite structure has rigid constraints, the low-dimensional perovskite allows for broader structural and compositional tunability. When the 3D perovskite is conceptually excised into slices, the size restrictions for the A′, which is the interlayer cation, are lifted for low-dimensional derivatives. According to the connectivity, the segregated component made of BX_6 octahedra is usually present 2D, 1D, or even separately 0D types, see Figure 1.2. From the perspective of the dimensions of morphologies, perovskite materials can be categorized into 3D bulk, 2D nanoplatelets, 1D nanowires, and 0D nanocrystals.

In the case of 2D perovskite, such structures are made up of a cation monolayer or bilayers alternating with sheets of the corner-sharing BX_6 octahedra. The 2D perovskite features mono- or diammonium cations A′, showing the chemical formulas of A'_2BX_4 and $A'BX_4$, which are frequently referred to as Dion–Jacobson (divalent A′) or Ruddlesden–Popper (monovalent A′) phases [15]. In A'_2BX_4 formed by monovalent cations, such as PEA^+ (phenethylammonium, $C_6H_5(CH_2)_2NH_3^+$) and BA^+ (butylammonium, $C_4H_9NH_3^+$), a van der Waals gap was generated by a bilayer of monovalent cations from two neighboring lead halide sheets. Instead, in the $A'BX_3$ system, each pair of cations can be substituted by a single divalent cation with tethering groups at each end to attach to neighboring halide sheets. The other low-dimensional perovskite derivatives feature much more separated BX_6 links, including 1D "pillar"-like BX_6 octahedra connected chains and 0D isolated "dot"-like octahedra, respectively; see Figure 1.2. Especially, the BX_6 connectivity can also be separated by the different compositions, which form the mixed perovskites known as *pseudo* members.

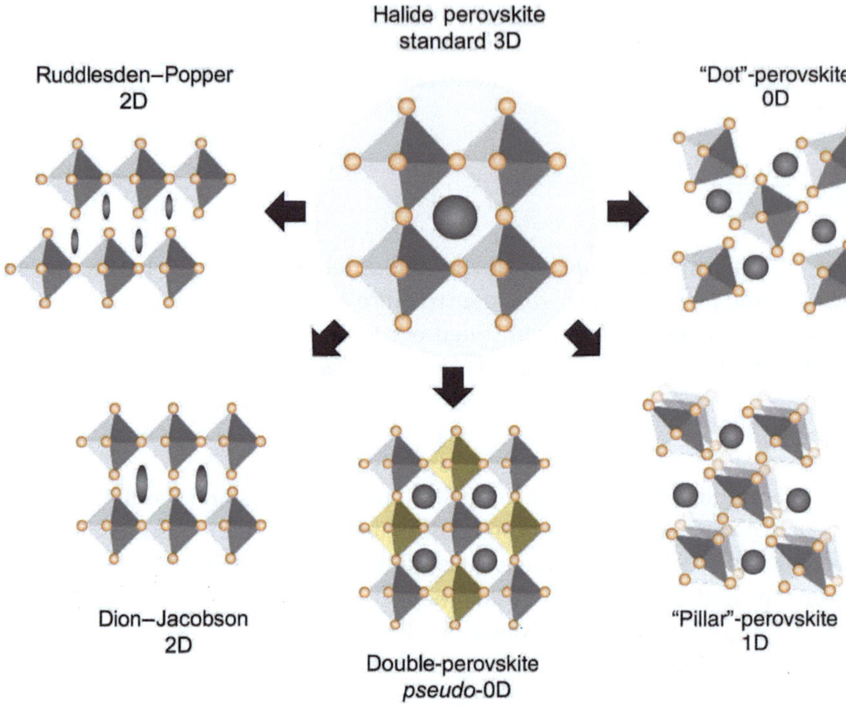

Figure 1.2 Halide perovskite family tree. Schematic illustration of standard 3D perovskite and the low-dimensional derivates, including Ruddlesden–Popper 2D, Dion–Jacobson 2D, "Pillar"-perovskite 1D, "Dot"-perovskite 0D, and the double perovskite *pseudo*-0D perovskite.

1.3 Property and Application of Perovskite

Early research on perovskite oxides focused on the biaxial optical properties and ferroelectric properties. In contrast, halide perovskites open the door to studying the optoelectrical properties because of their unique electronic structures, including direct tunable bandgap, strong absorption, small and balanced electron-hole effective masses, and defect resistance, thereby improving their photoluminescence quantum yield. Moreover, the unprecedented flexibility of perovskite composition can be brought about by organic or inorganic components with optical or electronic functionalities. The most important advantage of halide perovskites is their facile, accessible, high-quality crystals and films, enabling structure–property correlation exploration and prototype device optimization. Thus, halide perovskite semiconductors will hold promise for a variety of fascinating applications, including photovoltaics (PVs), LEDs, photodetectors, memristors and lasers, see Figure 1.3, just to cite the ones that probably receive more attention. In addition to this versatility, it is important to highlight the enormous potential for the development of high-performance devices on flexible substrates, extending the application range of high-performance rigid photovoltaics.

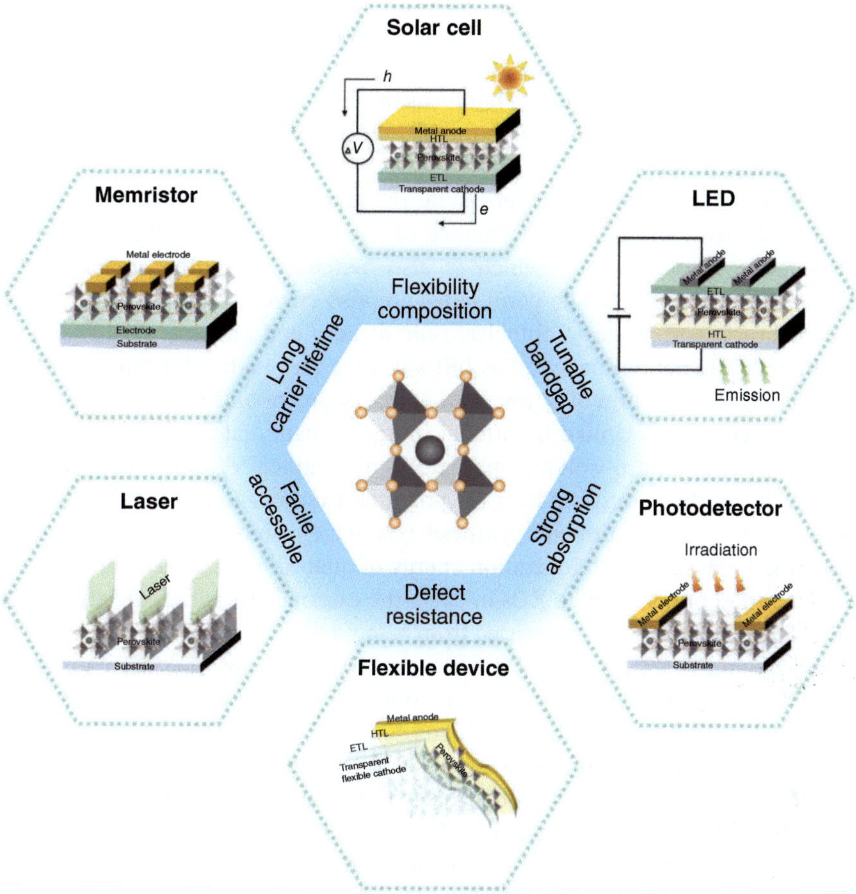

Figure 1.3 Example of properties and applications of halide perovskite. Halide perovskites are promising semiconductors with excellent properties, including flexible composition, facile accessibility, tunable bandgap, strong absorption, long carrier lifetimes, and defects resistance. These merits enable halide perovskite to be competent in solar cells, LEDs, photodetectors, memristors, and lasers. Interestingly, halide perovskite devices can also be developed on flexible substrates due to the good performance of polycrystalline films and low-temperature growth conditions.

The organic–inorganic halide perovskites have received wide attention, mostly due to their high efficiency and low cost in next-generation PVs, which have made these materials start to compete with commercial thin-film cells. Such materials harvest the energy of sunlight efferently because of their high absorption coefficients in both visible and near-infrared light. The first report of the perovskite solar cell was in 2009 [16], and now the record laboratory-scale power conversion efficiency of perovskite film is certified at 25.7% (https://www.nrel.gov/pv/cell-efficiency). Perovskite film-based solar cells are easy to be fabricated at low temperature, more energy-saving, and environmentally friendly than the conventional silicon wafer with a lower payback time [17], but with a lower contrasted long- term stability.

Consequently, further commercialization process of perovskite-based solar cells has been hindered by the stability problem, including degradation due to moisture, oxygen, heat, light, mechanical stress, and reverse bias. These failings do not detract from the overall excellence but focus the research effort on increasing device stability. Halide perovskite materials remain a cost-effective solution to address vast electrical energy supplies.

While initial boost of halide perovskite research was the development of photovoltaic devices, the good performance of these solar cells is founded on low nonradiative recombination, which is also beneficial for other optoelectronic devices. Consequently, halide perovskites are also currently impacting the development of a broad range of optoelectronic devices and systems. One significant benefit of halide perovskites used as LEDs is their very high color purity, with full width at a half-maximum of 20 nm for the blue or green–blue electroluminescence spectrum peaks [18]. Unlike traditional inorganic nanomaterials, the exceptional color purity of quantum-well nanoparticles is maintained regardless of crystal size. As a result, halide perovskites have the potential to solve some of the drawbacks of existing LEDs, such as difficult synthesis challenges, high cost, poor color purity of organic LEDs, and high ionization energy of quantum dot LEDs. Since the first demonstration of perovskite LEDs in 2014, the external quantum efficiency (EQE) of these devices has rapidly increased from below 1% to 25.8% for red [19], 28.1% for green [20], and 14.8% for blue [21]. The new LED technology has seen a meteoric rise in device efficiencies, but many scientific and technical obstacles, such as the unsatisfied stability and efficiency of blue LEDs, still stand in the way of perovskite LEDs further advancement into real-world applications.

Halide perovskite-based photodetectors exhibit comparable performance to commercially available photodetectors based on crystalline Si and III–V, offering significant potential for the technology of light-signal detection. The outstanding intrinsic optoelectronic properties of halide perovskites, such as photoinduced polarization, high drift mobilities, and effective charge collection, have contributed to the current growth of cutting-edge material studies in the field of light-signal detection. Halide perovskite semiconductors feature effective light absorption, enabling the detection of a wide range of electromagnetic waves from ultraviolet and visible to near-infrared and even radiations (X-ray, γ-ray, etc.), with low-cost solution processability and high photon yield. This class of semiconductors may empower ground-breaking photodetector technology in the areas of imaging, optical communications, and biomedical sensing; in this last case, further stability in polar solvent media, such as water, could increase enormously the range of applications of these systems.

Moreover, halide perovskites present a high ionic bonding character and ionic conductivity, causing the coexistence and coupling of ionic and electronic components of current and capacitance. This fact is at the base of nonconventional effects on optoelectronic systems, which could be a source of instabilities but can also be exploited. In this sense, halide perovskites exhibited good memristive properties supported by their electronic–ionic conductivity properties [22]. Memristor, that is memory resistor, is a leading candidate with robust capabilities in information storage and neuromorphic computing applications to address the growing challenge of approaching the end of Moore's law and the von Neumann bottleneck. The memristive property of halide perovskite is achieved through the synergistic coupling of photonic, electronic, and ionic processes, which enable perovskite to demonstrate novel functions such as optical-erasing memory, optogenetically inspired synaptic functions, and light-accelerated learning with multifunctionalization and novel photonic, logical, multilevel, and flexible functions.

Aside from the above-mentioned, halide perovskites have much more extensive applications due to their outstanding attributes, such as lasers, X-ray detectors, waveguides, scintillators, gas sensors, spintronics, and photocatalysis.

1.4 Summary and Outlook

The last two decades have seen the rapid development of halide perovskite materials, with researchers in particular pioneering systematic structural and property correlation studies based on halide perovskite composition and phases. The structure of perovskites has also been derived from ABX_3 to various derivatives that are important for changing chemical properties, controlling energy bands, and granting new physical properties. Perovskite materials, as an excellent new generation of semiconductor materials, have been demonstrated to be useful in a wide range of application scenarios, including photovoltaics, displays, and sensing, storage.

However, the environmental and thermodynamic stability of halide perovskite-based applications are two major challenges impeding their development, and some related phenomena and mechanisms should be thoroughly investigated to address perovskite device long-term use problems. Scientists have begun to use multimodal characterization to study the structural changes of halide perovskites, to monitor structural changes related to physical processes using *in situ* technology, and to conduct large-scale studies to establish correlations between components and properties using AI technology. The halide perovskite is like a treasure trove, and more interdisciplinary collaboration will lead to even more unexpected discoveries. This book will overview the intriguing properties of halide perovskites, making this system significantly different from other optoelectronic materials. Properties of materials and devices will be overviewed as well as the perspective on material and device development, always focusing on the fundamental properties.

References

1 Wainer, E. (1946). High titania dielectrics. *Transactions of the Electrochemical Society* 89 (1): 331.

2 Miyake, S. and Ueda, R. (1932). On polymorphic change of $BaTiO_3$. *Journal of the Physical Society of Japan* 1 (1): 32.

3 Cross, L. and Newnham, R. (1987). History of ferroelectrics. *Ceramics and Civilization* 3: 289.

4 Wells, H.L. (1893). Über die cäsium-und kalium-bleihalogenide. *Zeitschrift für anorganische Chemie* 3 (1): 195.

5 Møller, C. (1957). A phase transition in cæsium plumbochloride. *Nature* 180 (4593): 981.

6 Møller, C. (1958). Crystal structure and photoconductivity of caesium plumbohalides. *Nature* 182 (4647): 1436.

7 Weber, D. (1978). $CH_3NH_3PbX_3$, ein Pb(II)-system mit kubischer perowskitstruktur/$CH_3NH_3PbX_3$, a Pb(II)-system with cubic perovskite structure. *Zeitschrift für Naturforschung B* 33 (12): 1443.

8 Weber, D. (1978). $CH_3NH_3SnBr_xI_{3-x}$ (x = 0–3), ein Sn(II)-system mit kubischer perowskitstruktur/$CH_3NH_3SnBr_xI_{3-x}$ (x = 0–3), a Sn(II)-system with cubic perovskite structure. *Zeitschrift für Naturforschung B* 33 (8): 862.

9 Mitzi, D., Wang, S., Feild, C. et al. (1995). Conducting layered organic-inorganic halides containing⟨110⟩-oriented perovskite sheets. *Science* 267 (5203): 1473.

10 Mitzi, D.B. (1999). Synthesis, structure, and properties of organic-inorganic perovskites and related materials. In: *Progress in Inorganic Chemistry* (ed. K.D. Karlin).

11 Mitzi, D.B. (2000). Organic-inorganic perovskites containing trivalent metal halide layers: the templating influence of the organic cation layer. *Inorganic Chemistry* 39 (26): 6107.

12 Goldschmidt, V.M. (1926). Die Gesetze der Krystallochemie. *Naturwissenschaften* 14 (21): 477.

13 Han, G., Hadi, H.D., Bruno, A. et al. (2018). Additive selection strategy for high performance perovskite photovoltaics. *The Journal of Physical Chemistry C* 122 (25): 13884.

14 Li, C., Lu, X., Ding, W. et al. (2008). Formability of ABX_3 (X = F, Cl, Br, I) halide perovskites. *Acta Crystallographica Section B* 64 (6): 702.

15 Mao, L., Ke, W., Pedesseau, L. et al. (2018). Hybrid Dion–Jacobson 2D lead iodide perovskites. *Journal of the American Chemical Society* 140 (10): 3775.

16 Kojima, A., Teshima, K., Shirai, Y., and Miyasaka, T. (2009). Organometal halide perovskites as visible-light sensitizers for photovoltaic cells. *Journal of the American Chemical Society* 131 (17): 6050.

17 Vidal, R., Alberola-Borràs, J.-A., Sánchez-Pantoja, N., and Mora-Seró, I. (2021). Comparison of perovskite solar cells with other photovoltaics technologies from the point of view of life cycle assessment. *Advanced Energy and Sustainability Research* 2 (5): 2000088.

18 Adjokatse, S., Fang, H.-H., and Loi, M.A. (2017). Broadly tunable metal halide perovskites for solid-state light-emission applications. *Materials Today* 20 (8): 413.

19 Jiang, J., Chu, Z., Yin, Z. et al. (2022). Red perovskite light-emitting diodes with efficiency exceeding 25% realized by co-spacer cations. *Advanced Materials* 34 (36): 2204460.

20 Liu, Z., Qiu, W., Peng, X. et al. (2021). Perovskite light-emitting diodes with EQE exceeding 28% through a synergetic dual-additive strategy for defect passivation and nanostructure regulation. *Advanced Materials* 33 (43): 2103268.

21 Shen, Y., Li, Y.-Q., Zhang, K. et al. (2022). Multifunctional crystal regulation enables efficient and stable sky-blue perovskite light-emitting diodes. *Advanced Functional Materials* 32 (41): 2206574.

22 John, R.A., Shah, N., Vishwanath, S.K. et al. (2021). Halide perovskite memristors as flexible and reconfigurable physical unclonable functions. *Nature Communications* 12 (1): 3681.

2

Halide Perovskite Single Crystals

Clara Aranda-Alonso[1,2] *and Michael Saliba*[1,2]

[1] *Institute for Photovoltaics (IPV), University of Stuttgart, 70569 Stuttgart, Germany*
[2] *IEK5-Photovoltaics, Forschungszentrum Jülich, 52425 Jülich, Germany*

2.1 Introduction

To further explore the potential of hybrid halide perovskite materials, including solar cells and beyond, their optoelectronic properties still need to be thoroughly investigated. In this respect, single crystals (SC), with fewer defects, are the ideal candidates to further analyze these properties without the detriment of the instability under atmospheric conditions that polycrystalline perovskites exhibit.

In this chapter, we address the main characteristics and applications of SC based on perovskite materials. The chapter is organized as follows: first, the three principal perovskite crystal structures are described, distinguishing between lead-based, lead-free, and all inorganic perovskite single crystals. Then, the different synthesis methods are described, highlighting the ones with significant impact. To further understand the potential of this material, the optoelectronic properties of the most commonly used SC are discussed in depth. Finally, the main applications of these materials in different technologies are described, including their abilities as photodetectors, scintillators, solar cells, light emitting diodes, and memristors.

2.2 Crystal Structure

In the general perovskite oxide structure, ABX_3, A and B are cations of different sizes. *A* used to be larger than *B*, being B six-coordinated by an X-site anion to form a BX_6 octahedron complex. To arrange a three-dimensional (3D) system, the octahedron must share the corners, locating A-cations in the cavities of the framework. The charge of each cation and anion must be such as to preserve electroneutrality. In the hybrid perovskite materials, the A-cation is a monovalent organic amine, B is a divalent metal (Pb^{2+}, Sn^{2+}) and X is the halide element (I^-, Br^-, and Cl^-), or an extended version associating molecular linkers (azides N^{3-}, cyanides CN^-, or even borohydrides BH^{4-}). The divalent metal could also be replaced by mixed monovalent and trivalent metals, forming double perovskites $A_2BB'X_6$ structures. As well as

Halide Perovskite Semiconductors: Structures, Characterization, Properties, and Phenomena, First Edition. Edited by Yuanyuan Zhou and Iván Mora-Seró.

for their polycrystalline counterparts, the capability of tuning the chemistry of perovskite single crystals is governed by the ionic ratio sizes of their components. The Goldschmidt tolerance factor (t) and the octahedral factor (μ) should be considered to design a stable and useful perovskite material. In Sections 2.2.1, 2.2.2, and 2.2.3, we will summarize the structure of the most commonly used perovskite formulations for SC growth, including lead-based, lead-free, and all inorganics.

2.2.1 Lead-Based Perovskite Single Crystals

In Pb-based halide perovskites, crystallization occurs in the ABX_3 structure, which is isostructural to the initial perovskite oxide $CaTiO_3$. The A and B cations coordinate with 12 and 6 X anions, leading to the 3D corner-sharing cuboctahedral (12-fold coordinated) and octahedral (sixfold coordinated) structures. In the $MAPbX_3$ structure, the methylammonium position is disordered in the tetragonal phase from 160 K to room temperature, whereas it is ordered below 160 K in the orthorhombic phase (Figure 2.1) [1].

Lead-based perovskite single crystals are the most studied to date. They correspond to the perovskite structures that have given record efficiency results in thin film solar cells. Thus, the information that can be extracted from the bulk of these materials will also be valuable for research on thin films. On the other hand, they can be synthesized by a wide range of methods, some of which are fully described in Section 2.3. Examples of different lead-base perovskite single crystals already grown are shown in Figure 2.2 [2].

It is well known that temperature influences the structural transitions of perovskite materials, having a direct impact on their optoelectronic properties. It is then essential to clarify the relationship between both phenomena in perovskite single crystals to understand their polycrystalline counterparts further as well. Ding and coworkers reported a depth analysis of the different crystal facets of $MAPbI_3$ single crystal. They confirm the different atom densities for the different facets and how this influences ionic migration [3]. In Figure 2.3a, the X-ray diffraction (XRD) pattern of facet (100) of both single crystal and powder diffraction is shown.

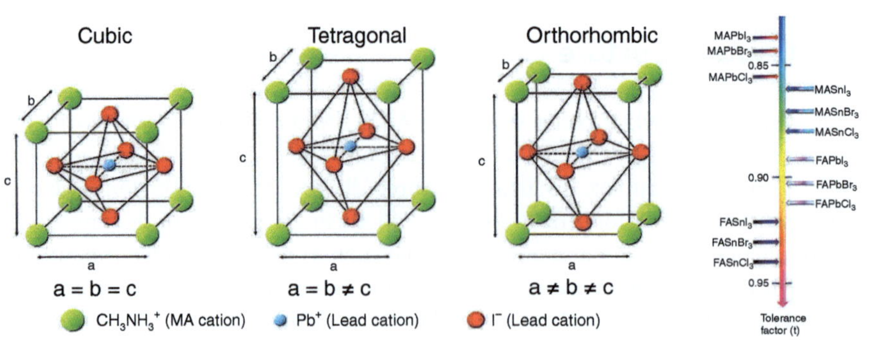

Figure 2.1 (a) Crystal structures for $MAPbX_3$ single crystals. Cubic, tetragonal, and orthorhombic groups. (b) Tolerance factors for the different perovskite compositions, Pb and Sn containing. Source: Reproduced with permission from Murali et al. [1]/American Chemical Society.

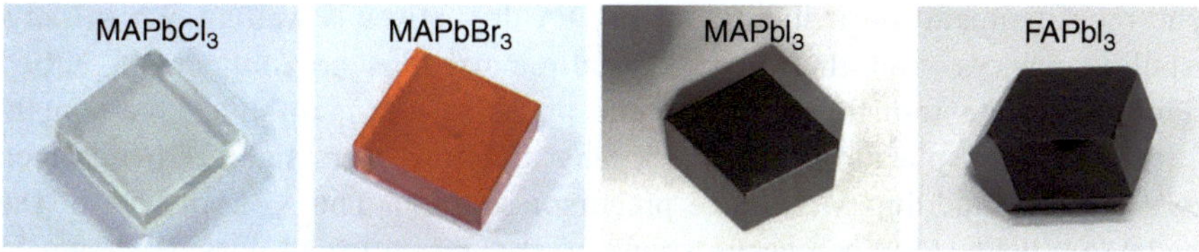

Figure 2.2 Photographs of Pb-based perovskite SCs: $MAPbCl_3$, $MAPbBr_3$, $MAPbI_3$, and $FAPbI_3$ Source: Reproduced with permission from Liu et al. [2]/John Wiley & Sons.

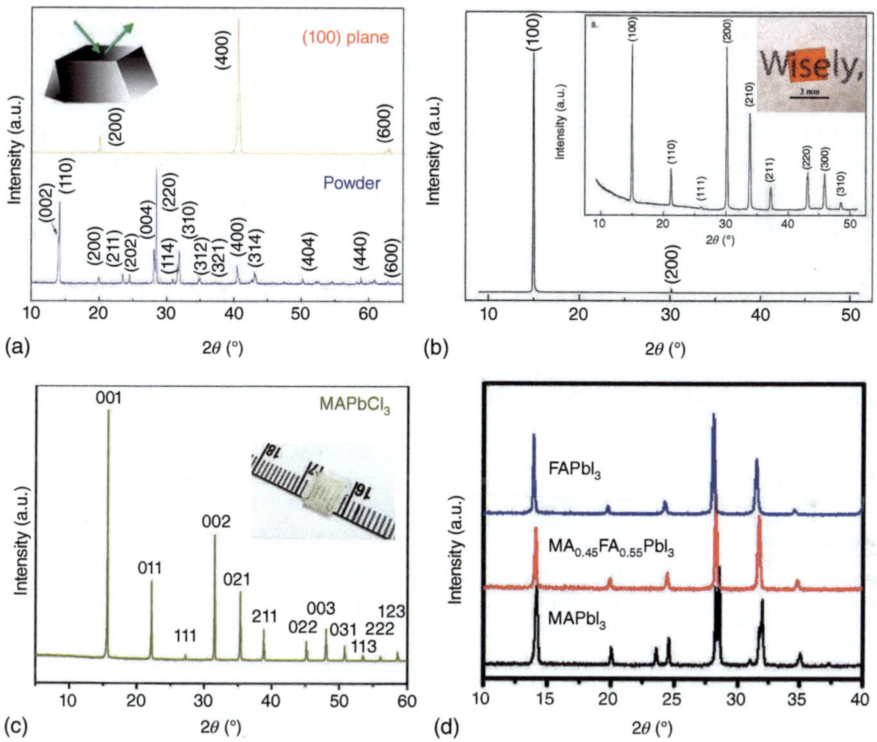

Figure 2.3 X-Ray diffraction patterns from the most commonly used lead-based perovskite structures. (a) $MAPbI_3$. Source: Reproduced with permission Ding et al. [3]/American Chemical Society. (b) $MAPbBr_3$ single crystal. Source: Reproduced from Wang et al. [4]/Springer Nature/CC BY 4.0. (c) $MAPbCl_3$ single crystal. Source: Reproduced from Lee et al. [5]/MDPI/CC BY 4.0. (d) Mixed MA and FA lead iodide single crystal stabilized. Source: Reproduced with permission from Li et al. [6]/Royal Society of Chemistry.

In the case of $MAPbBr_3$, the crystal belongs to the cubic $Pm\bar{3}m$ space group at room temperature, as seen in Figure 2.3b from both single crystal and powder diffraction methods. However, the crystal experiences several structural transitions with lowering temperatures, from cubic to tetragonal, tetragonal to orthorhombic I, and orthorhombic I to orthorhombic II [4]. But this composition is not the only one suffering from this temperature instability; $MAPbCl_3$ (Figure 2.3c) also shows two structural phase transitions: (i) from the cubic-to-tetragonal phase at around −95 °C and (ii) from the tetragonal to the orthorhombic when lowering to −116 °C [5].

The room-temperature instability of the FA derivatives is well-known in polycrystalline devices, and their monocrystalline partners are not spared either. $FAPbI_3$ perovskite transforms from cubic phase to non-perovskite phase, making its practical application difficult. Kuang and coworkers stabilized the black phase (cubic) by adding MA to the precursor solution. The XRD pattern of the mixed $MA_{0.45}FA_{0.55}PbI_3$ perovskite single crystal, stable for 14 days, is shown in Figure 2.3d [6].

2.2.2 Lead-Free Perovskite Single Crystals

Pb-free SCs presume a variety of crystal structures, which are not necessarily limited to the typical ABX_3 architecture. The replacement of Pb^{2+} ions leads to a nanoscale deformation and a depth change in the optoelectronic properties due to the differences in chemical valence and ionic size. In this case, the crystal structure depends mainly on substitutions, including group-14 elements (Sn and Ge), adjacent elements such as Sb or Bi, and double elements (i.e. Bi combined with Ag) [7]. These substitutions extend from orthorhombic to trigonal lattice systems, generating quaternary structure $A_2B^+B^{3+}X_6$. A *vacancy-ordered* double perovskite can also be synthesized by removing part of the B atoms from the octahedron center. Isolated clusters can be obtained using transition or post-transition elements to form two face-sharing $[M_2X_9]^{3-}$ octahedra. Below is a summary of the possible substitutes for lead to develop lead-free perovskite single crystals (Figure 2.4).

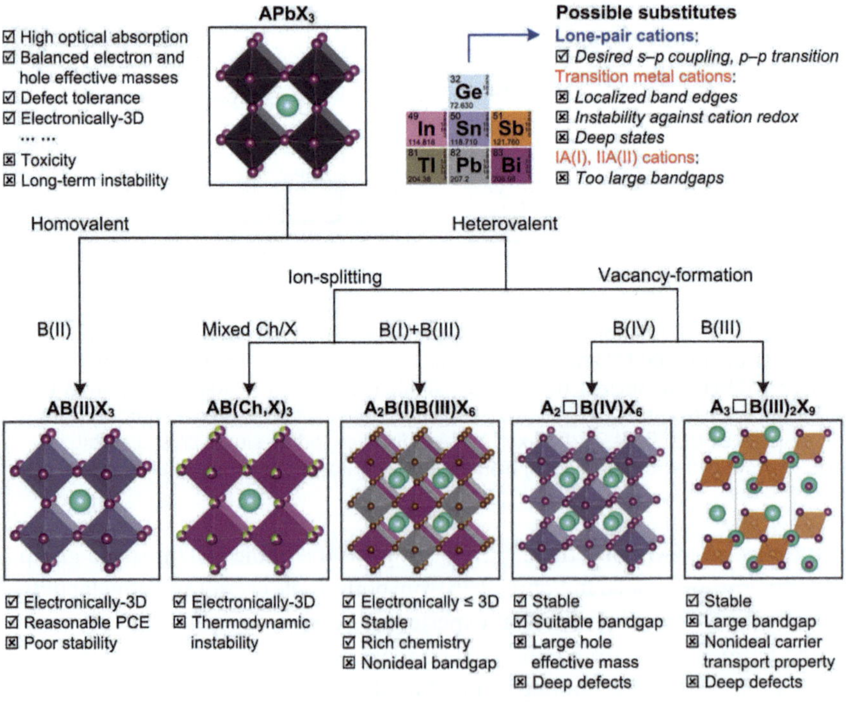

Figure 2.4 Diagram of the main routes to replace Pb in perovskite crystals. Source: Reproduced with permission from Zhang et al. [7]/John Wiley & Sons.

2.2.3 All-Inorganic Perovskite Single Crystals

If perovskite does not contain any organic component, it can be categorized as all-inorganic perovskite. The main advantage of this inorganic variant is the absence of organic cations due to their associated intrinsic instability. Therefore, the all-inorganic perovskites show impressive thermal and environmental stability, opening new ways to further applications with limited restrictions.

Regarding the structure, all-inorganic perovskite single crystals can be divided into three main categories: ABX_3 type, special structures, and doped perovskites (Figure 2.5) [8].

Regarding the traditional ABX_3 type, the principal protagonist lacking lead is the $CsPbBr_3$ formulation. This inorganic single crystal is grown by the Bridgman method (described in Section 2.3) and the solution-grown method.

He et al. optimized the growth conditions using the Bridgman method, considering the relation between the crystal's defects and an excessive temperature gradient [9]. After thoughtful optimization, $CsPbBr_3$ with high purity was obtained,

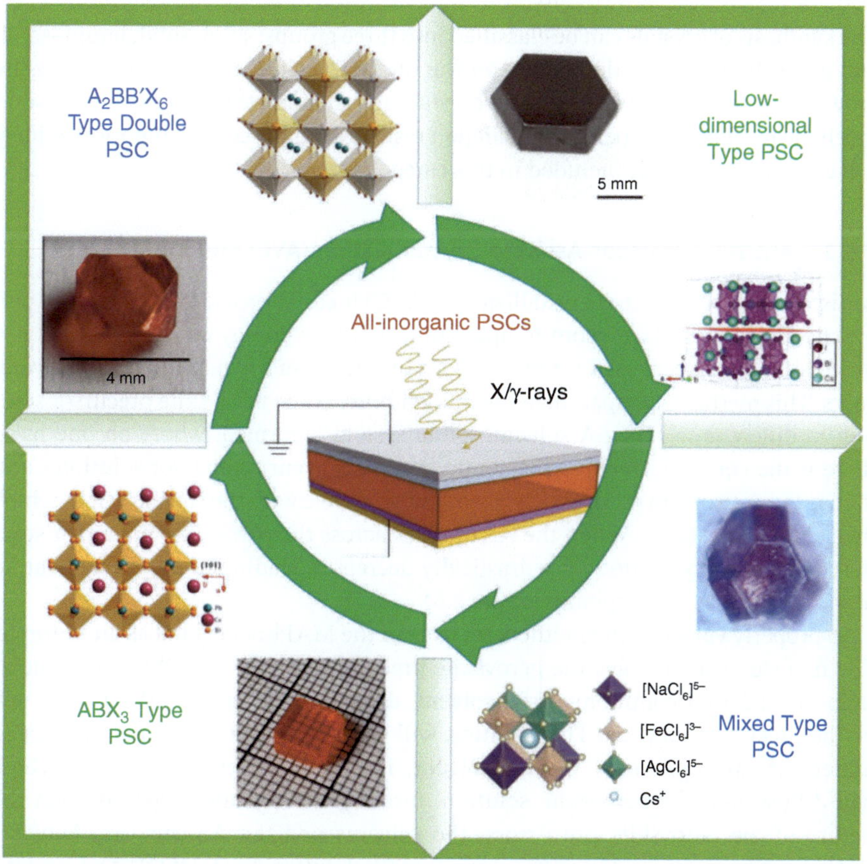

Figure 2.5 Summary of all-inorganic PSC: $AB^IB^{III}X_6$ type, low-dimensional type, mixed type, and ABX_3 traditional type PSC. Source: Reproduced from Wu et al. [8]/MDPI/CC BY 4.0.

performing as a great gamma-ray detector. Their work reported that $CsPbBr_3$ exhibits two nondestructive phase transitions at low temperatures. The transition first occurs around 130 °C, from a cubic to tetragonal system. This is followed by a second-order transition at around 88 °C to the orthorhombic phase, which is stable at room temperature.

Another structure that shows high stability in air and excellent optoelectronic properties is $TlPbI_3$. This material has a low melting temperature, which reduces the thermally activated defects during growth of the process, which is typically the Bridgman method. Several improvements have been made to exploit this material's full potential as a gamma-ray detector [10].

$CsSnI_3$ SCs have shown p-type metallic behavior, presenting the highest hole mobility among the p-type semiconductors. Chung et al. applied doped $CsSnI_3$ to solar cells, developing all-solid-state dye-sensitized solar cells with outstanding PCE of up to 10.2% [11].

2.3 Synthesis Methods

The synthesis of crystals can be classified into three groups: solid–solid, liquid–solid, and gas–solid methods, depending on the phase transition that occurs during the growing process. Most technological crystals are obtained through the liquid–solid mechanism, including perovskite single crystals. In this section, the most widely used synthesis methods included in this category are summarized.

2.3.1 Antisolvent Vapor-Assisted Crystallization (AVC) Method

Antisolvent vapor-assisted crystallization (AVC) method provides high-quality perovskite single crystals at room temperature, reducing energy consumption. It is a process that has been used for preparation of micro- and nanoparticles for several years. This method's key parameter is the solubility of the perovskite precursor solution in different solvents. A solvent–solute system is formed, where on one hand there is the couple formed by the solute and the solvent (precursor solution), and then is the antisolvent (AS), which is perfectly miscible with the solvent but in which the solute is insoluble. When the AS diffuses across the perovskite precursor solution, the solubility of the solute drastically decreases, leading to the crystallization of the material as a precipitate (Figure 2.6).

To properly visualize the method, let us take the $MAPbBr_3$ crystal as an example. The first step is to dissolve the perovskite precursors, MABr and $PbBr_2$, in a molar ratio of 1 : 1 in a commonly used solvent, dimethylformamide (DMF). This will be the precursor solution. This solution will then be placed in a small vial sealed under the atmosphere of the antisolvent, i.e. dichloromethane (DCM), which will diffuse into the perovskite solution, forming the crystals. Depending on the nature of the perovskite precursors, the solvents and the AS can vary between DMF, gamma-butyrolactone (GBL), or dimethyl sulfoxide (DMSO). The options for choosing AS are quite wide if the insolubility of the solute is ensured. As might

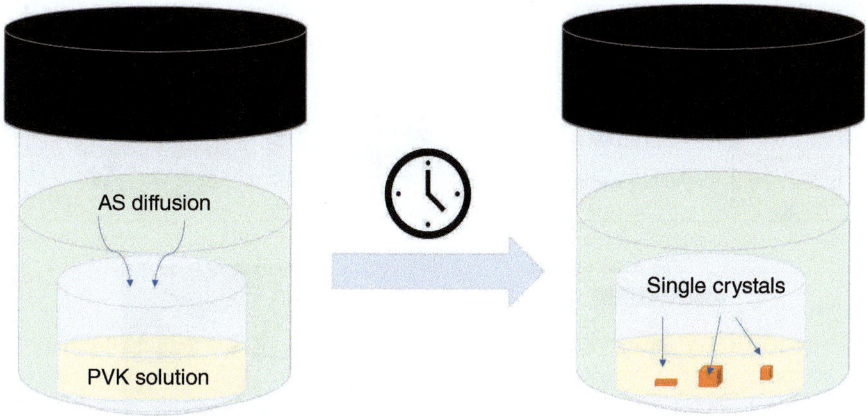

Figure 2.6 Schematic illustration of the antisolvent vapor-assisted crystallization (AVC) method. Source: Clara Aranda-Alonso.

be expected, the speed of the crystal growth is proportional to the diffusion rate of the AS into the solution. However, the reproducibility of this method can be challenging, constituting a constraint for large-scale processing.

2.3.2 Solution Temperature Lowering (STL) Method

In the case of solution temperature lowering (STL), the change in the solubility of perovskite precursors in acid halide solvents (e.g. HI, HBr, and HCl) due to different temperatures is the protagonist. The first step is to dissolve the perovskite precursors at high temperature, and then the solution is progressively cooled down. This process leads to a supersaturated solution, which promotes the crystal's growth. The STL method has two variants as a seed-assisted method: (i) crystal seed at the bottom (BSSG) of a vial containing the stock solution, or (ii) on the top (TSSG), suitably fixed as illustrated in Figure 2.7.

In the case of BSSG-type growth, immersing the seed in the solution shall be repeated as many times as necessary until the desired crystal size is achieved. For TSSG, small seeds need to be fixed on a substrate and immersed in the top half of the

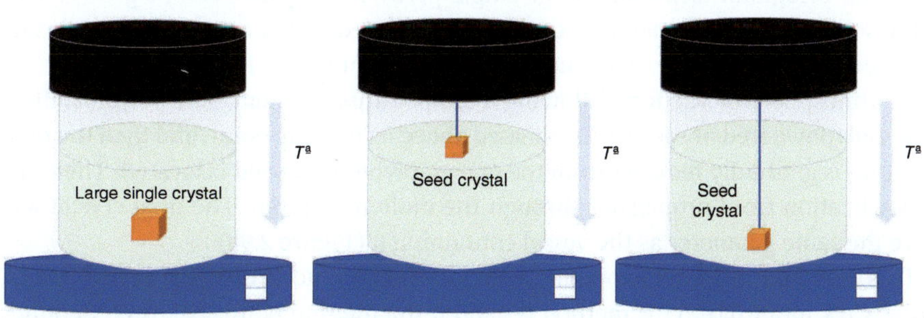

Figure 2.7 Schematic representation of the solution temperature lowering (STL) method. Source: Clara Aranda-Alonso.

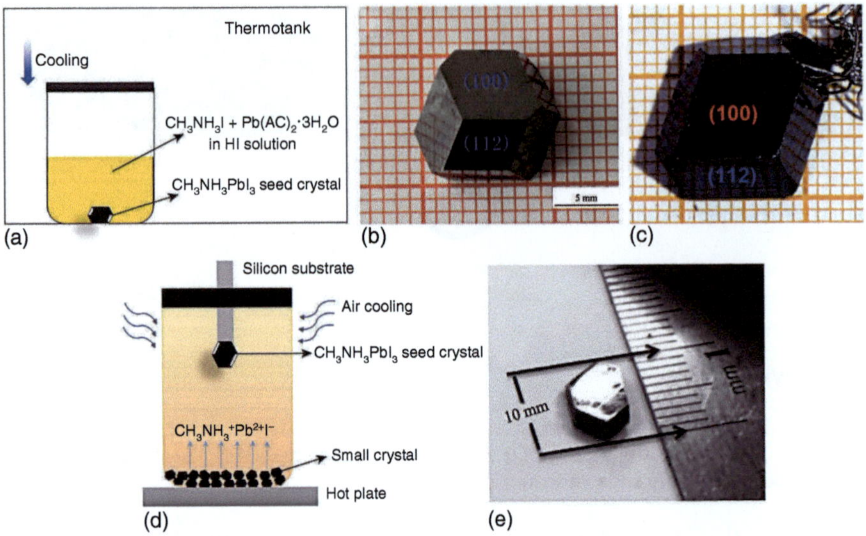

Figure 2.8 Schematic diagram of STL crystallization. (a–c) Crystallization process of BSSG and images of as-prepared $CH_3NH_3PbI_3$ single crystal; (d, e) TSSG method and the obtained crystal [12]. Source: (b) Dang et al. (2015)/Royal Society of Chemistry. (c) Lian et al. (2015)/ Springer Nature/CC BY 4.0.

precursor. The lower half of the solution containing the seed is heated in an oil bath. The upper half is cooled with air to create a temperature gradient, inducing crystallization. At the same time, a convection current providing ions is produced in the lower half. With this method, large single crystals can be obtained by strictly controlling the temperature. An example of this growing method is shown in Figure 2.8 [12].

However, it is not suitable for crystals with low solubility at elevated temperatures. The prolonged rate of growth is also a disadvantage, together with the formation of by-products such as $MA_4PbX_6 \cdot 2H_2O$ and irregular crystals if the temperature drops too fast.

2.3.3 Bridgman Method

With the Bridgman technique, large single perovskite crystals can be grown. This solid–solid technique's operation mechanism is based on the crystal's growth inside a sealed quartz ampoule. This ampoule is placed inside a furnace filled with an inert atmosphere or vacuum, allowing a temperature gradient. The polycrystalline material (powdered or seed) is first heated above its melting point, and then the temperature is gradually reduced from the one end where the seed is located. Then, the crystallization front propagates through the molten material. The final crystal will have the same geometry as the vessel containing it (Figure 2.9a).

This method has been historically used for semiconductor crystals, such as GaAs, ZnSe, CdS, and CdTe. This method is valid if the melting points of the compounds to be crystallized are defined. In fact, it is not possible to use organic compounds, which are chemically unstable at their melting point.

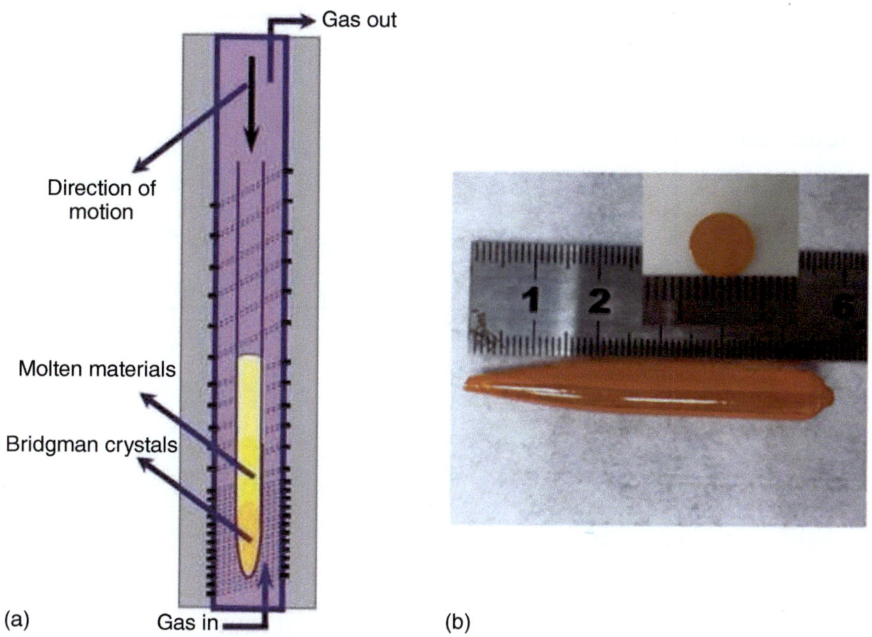

Figure 2.9 (a) Schematic explanation of the Bridgman crystallization method. (b) $CsPbBr_3$ perovskite single crystal grown by the BM. Source: Reproduced with permission from Zhang et al. [13]/John Wiley & Sons.

Another limitation of this method is that the crystal grows along the walls of the ampoule, which can lead to several defects due to the mechanical stress. The defects that may appear include small grain boundaries in the crystals, reducing their purity and their potential for technological applications. In addition, the crystallization parameters, such as the temperature gradient, dropping speed, and cooling rate, need to be well controlled to avoid cracks in the ingots.

One of the examples of single perovskite crystals grown through this method was reported by Zhang et al. [13]. They developed a simplified version of the method using a homemade vertical two-zone furnace. The large crystal of $CsPbBr_3$ is shown in Figure 2.9b.

2.3.4 Slow Evaporation Method

The slow evaporation method is a liquid–liquid technique, which uses a controlled rate of solvent evaporation from a solution precursor close to the saturated state. As we said before, this saturated state represents the driving force and can be achieved by a change in temperature or by the solvent evaporating (Figure 2.10A).

The steady-state nucleation rate on the crystal surface is defined as:

$$j_0 = \omega^* \Gamma N_0 \exp\left(-\frac{\Delta G^*}{k_B T}\right)$$

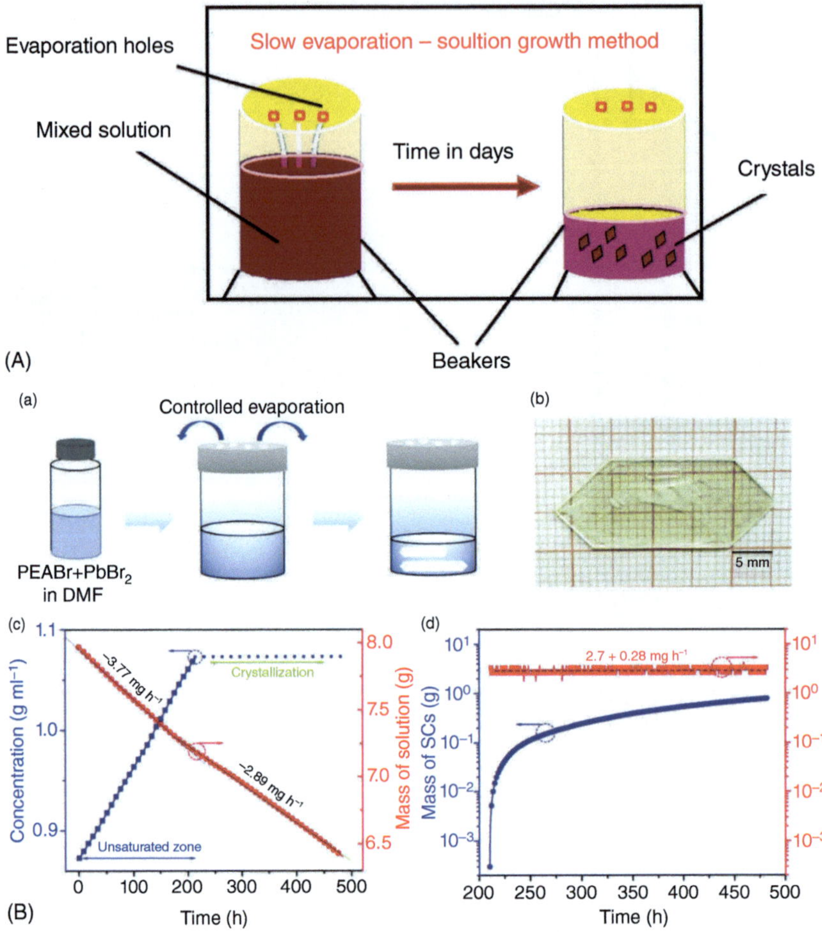

Figure 2.10 Top: (A) Schematic representation of the slow evaporation method. Bottom: (B) Example of the method applied to the growth process of $(PEA)_2PbBr_4$ single crystals. (b) Photograph of a transparent, ~27 × 11 mm^2 $(PEA)_2PbBr_4$ single crystal. (c) Mass and concentration of $(PEA)_2PbBr_4$ precursor solution as a function of evaporation time. (d) Mass and growth rate of $(PEA)_2PbBr_4$ as a function of time. Source: Reproduced with permission from Zhang et al. [14]/Royal Society of Chemistry.

where ω^* is the frequency at which the solute molecules diffuse from the solution to the critical nucleus, Γ is the nonequilibrium Zeldovich factor, N_0 is the density of adsorption sites, ΔG^* is the Gibbs free energy change for the nucleus formation, k_B is the Boltzmann constant, and T is the temperature. This equation defines the existence of a supersaturation level below which the nucleation rate becomes zero and increases rapidly above it. Attending to the exponential term, to achieve only the required high-quality single crystal, the growth conditions need to be controlled by regulating the temperature.

This crystallization technique represents one of the simplest methods to grow organic crystals, both in 3D and 2D architectures. Zhang and coworkers reported an example of 2D perovskite single crystal grown by this method in 2019 (Figure 2.10B) [14]. They developed an optimized growing process leading to a well-defined large-size (>200 mm^2) 2D $(PEA)_2PbBr_4$ ($C_6H_5CH_2CH_2NH_3^+$, PEA^+) single crystals. The excellent performance of the synthesized crystals made them very ideal for ultrafast optical computing and optical communications, reaching the quality of the state-of-the-art materials such as ZnO, TiO_2, and GaN in the field.

2.3.5 Inverse Temperature Crystallization (ITC) Method

As the name suggests, this method is based on the phenomenon that solubility is inversely proportional to temperature. Also known as *retrograde solubility,* this process describes the change in a solute concentration above the eutectic temperature (Figure 2.11a) [15]. The effects that this phenomenon can cause are clear: high enthalpy, variable distribution coefficient, and maximum chemical potential. However, A. L. McKelvey in 1996 described the origin of this mechanism from a thermodynamic point of view. That work concluded that retrograde solubility could occur if the ΔH_{mix} is large enough to cause a miscibility gap in ΔG_{mix} at temperatures larger than the eutectic temperature (the subindex *mix* refers to the

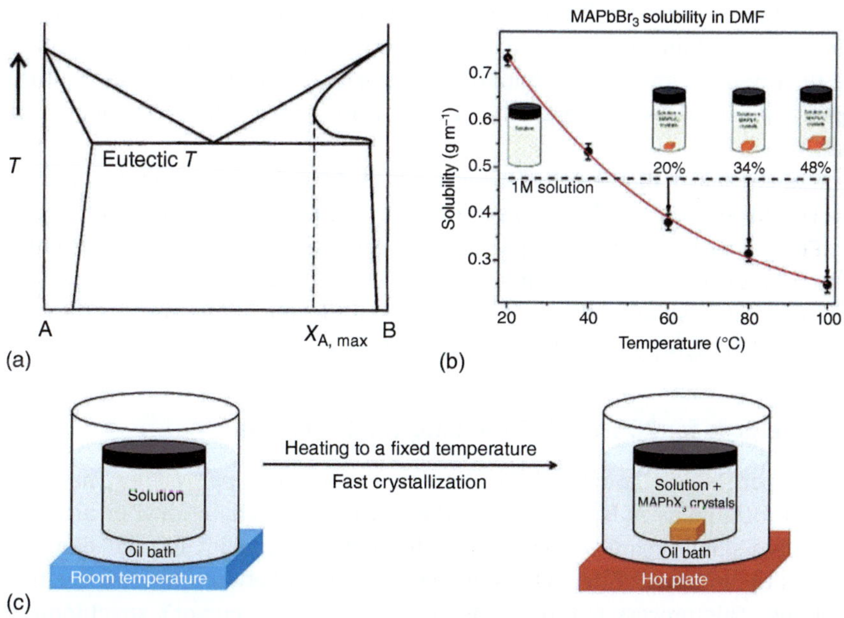

Figure 2.11 (a) Schematic phase diagram showing the phenomenon of retrograde solubility and the solubility maximum above the eutectic temperature. Source: Reproduced with permission from Mckelvey [15]/Springer Nature. (b) Temperature-dependent solubility of $MAPbBr_3$ in DMF shows yield at different temperatures. Source: Reproduced with permission from Saidaminov et al. [16]/Royal Society of Chemistry. (c) ITC setup in which the crystallization vial is heated from room temperature and kept at an elevated temperature. Source: Reproduced with permission from Saidaminov et al. [17]/Springer Nature/CC BY 4.0.

mixture of components in the solution; ΔH_{mix} and ΔG_{mix} are the enthalpies and the free energy of the mixing). It was established that this effect can occur in any binary system where the enthalpy dominates the free energy of mixing above the given temperature invariant. Few cases of materials presenting this effect have been reported so far; luckily, perovskite is one of them.

Bakr and coworkers demonstrated this when they observed an apparent solubility decrease of $MAPbX_3$ perovskites in the habitual solvents (DMF, GBL, etc.) with increasing temperature [17]. The inverse solubility does not occur with all solvents. For instance, in the case of $MAPbBr_3$, this phenomenon is clear for DMF (Figure 2.11b), but does not happen using DMSO [16]. The reason behind this is the different coordination capabilities of both within the lead halide precursor. In the case of DMSO, this capability is much higher, leading to a stronger binding with the lead. Once the solvent is chosen considering this premise, the second thing to do is control the temperature and concentration of the solution. The result will depend strongly on that.

The setup to execute the method is easy: just a vial containing the precursor solution and a hotplate coupled with a precise temperature controller (Figure 2.11c) [17]. This method allows the growth of high-quality perovskite single crystals and drastically reduces the synthesis times. If the temperature is higher than the needed one, the process's speed will increase, leading to multiple small crystals. Then, it is crucial to maintain a balance between the ramp temperature and concentration. Once the crystal is grown, it is important to isolate it outside the solution; if the temperature decreases to room temperature while the crystal is inside, it will redissolve.

A variant of this method involves using a seed to obtain larger crystals with further control of shape and size. This works by carefully taking the crystal and placing it in a fresh precursor solution. A comparison between seeded and unseeded $MAPbBr_3$ single crystals using this ITC method is shown in Figure 2.12.

This method is suitable for different perovskite formulations such as $FAPbI_3$, $MAPbCl_3$, $MAPbI_xBr_{1-x}$, and $MA_xFA_{1-x}PbI_3$ crystals.

2.3.6 Methods for 2D and 1D Perovskite Single Crystals

Quasi-2D perovskites known as Ruddlesden–Popper (RP) perovskites, have the structural formula as $A_2B_{n-1}M_nX_{3n+1}$, where A is a long-chain alkyl or aromatic group, B is a small organic cation, M is a divalent metal, and X is a halogen. The index n is the number of the 3D system layers inserted between the two layers of A^+ cations. This process is tunable by controlling the precursor's stoichiometry during the synthesis. In this low-dimensional system, the layers of A^+ cations have the role of "barriers" for carriers due to the large bandgap. On the other hand, in the 3D system, there are the B^+ cations, which serve as "wells." The wells' thickness in these quantum well structures is determined by the index n, which can be easily tailored. ITC and AVC methods widely used for 3D structures are not feasible to grow 2D perovskite SC. Instead, cooling temperature appears to be an effective method to reduce the solubility of a saturated solution.

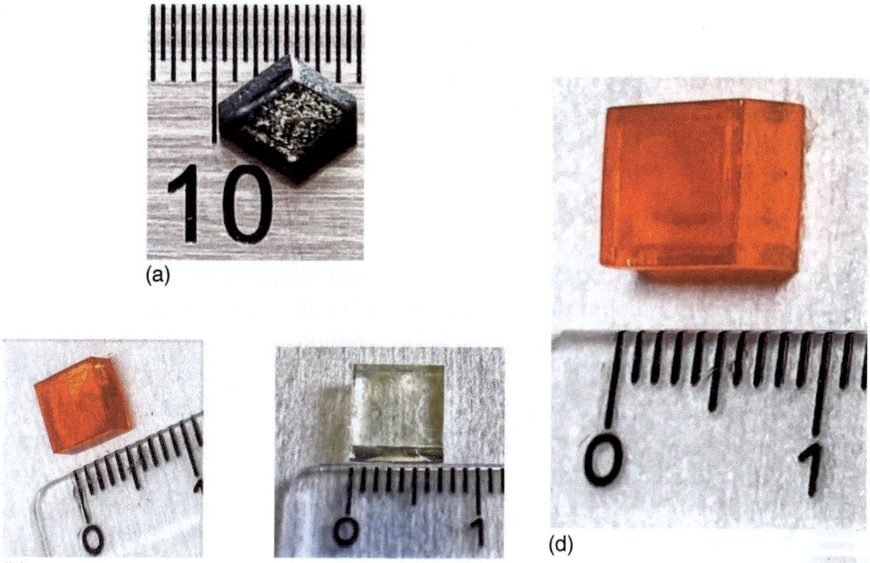

Figure 2.12 (a–c) Real pictures of $MAPbI_3$, $MAPbBr_3$, and $MAPbCl_3$ single crystals grown through unseeded ITC method. (d) $MAPbBr_3$ single crystal grown by the seeded assisted ITC method. Notice the boundaries of the seeded one demonstrating the dimensions control. Courtesy of Waqas Zia.

An example can be found in the work reported by Kanatzidis and coworkers. With this method, they successfully obtained a 2D $(BA)_2(MA)_{n-1}PbnI_{3n+1}$ (n = 1–4) SCs [18].

2.4 Optoelectronic Properties of Halide Perovskite Single Crystals

Perovskite single crystals have demonstrated their potential to be applied in various technologies such as photodetectors, X-ray or γ-ray detectors, high-performance solar cells, or light emitting devices. This is possible due to the optoelectronic properties of these materials in their monocrystalline phase, which are mainly characterized by the absence of grain boundaries. These properties: large carrier diffusion length, long carrier lifetime, extended absorption range, and low carrier defects. In this section, we will summarize these properties, highlighting the instrumental techniques used to understand them.

2.4.1 UV–Vis Absorption, Photoluminescence (PL), and Transient Decays: TRPL and TPV

As mentioned above, the main characteristic of single crystals is that they lack of grain boundaries. It is precisely this feature that gives them optoelectronic properties that are in several aspects different from those of their polycrystalline counterparts. One of the main consequences of this is a decrease in charge recombination. On the other hand, the highly ordered pattern in perovskite single crystal lattice significantly reduces the charge carrier trapping, favoring a long carrier diffusion length which is defined as follows:

$$L_D = \frac{\sqrt{kT}\,\mu\tau}{q}$$

where k is the Boltzmann constant, T the absolute temperature, q the elementary charge, μ the carrier mobility, and τ the carrier lifetime. This property will define the architecture and composition of the device containing the single crystal.

The absorption range of single perovskite crystals is redshifted compared with the polycrystalline films (Figure 2.13) [19]. This is caused by enhanced below-bandgap absorption and a clear band edge cutoff with no excitonic signature. This is also a signal of low defect states. In the case of $MAPbI_3$ single crystals; below-bandgap absorption is detected, originating in a weakly indirect bandgap of ≈60 meV below

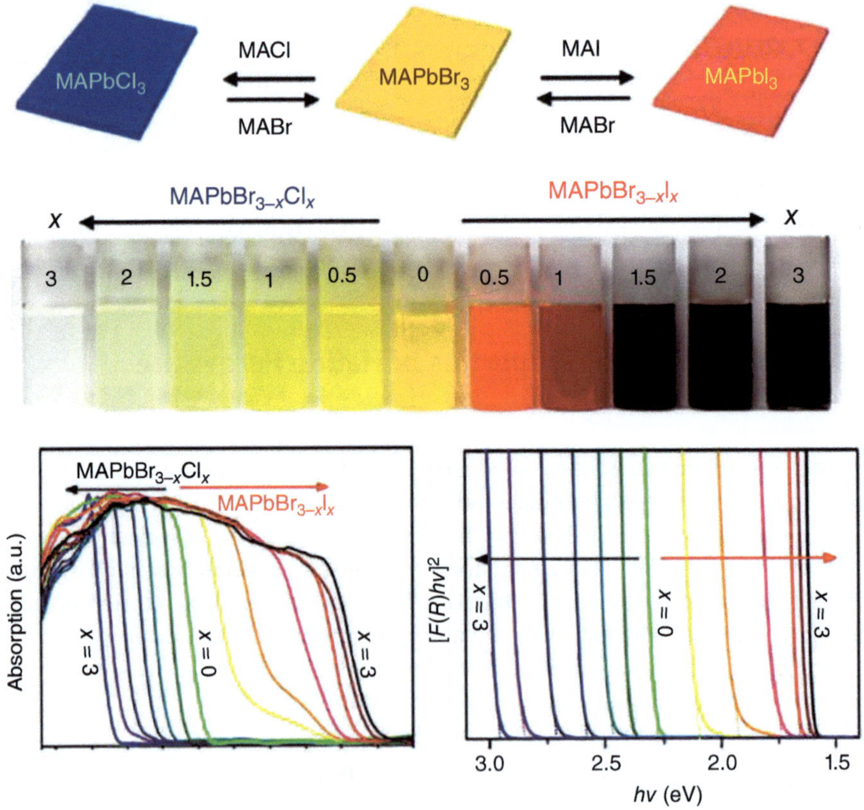

Figure 2.13 Tunable bandgaps in perovskite single crystals. The absorption spectra redshifted is also shown. Source: Reproduced with permission from Jang et al. [19]/American Chemical Society.

the direct bandgap transition. The absorption coefficient of the below-bandgap transition is several orders of magnitude smaller than that of the above-bandgap transition, resulting in negligible below-bandgap absorption in polycrystalline perovskite films but obvious in SCs due to the thicker thickness and larger carrier diffusion length. The extended absorption range makes more photons absorbed and increases the short-circuit current density when applied to solar cells, for instance.

Regarding photoluminescence (PL), it has been demonstrated that single perovskite crystals show weaker PL in vacuum than in ambient conditions. This was attributed to the oxygen and water passivation of the crystal's surface. Temperature-dependent XRD and PL have also been performed to analyze the phase transitions and trap states. In the case of $MAPbBr_3$, it was reported that a phase transition occurs from the orthogonal phase (Pna21) at <140 K to the tetragonal phase (*I4*/mcm/or *P4*/mmm) ranging from 140 to 230 K, and to cubic phase (*Pm3m*) of >230 K. PL measurement and time-resolved photoluminescence (TRPL) measurement performed in single perovskite crystals, show that free excitons (FEs) have a short lifetime of nanoseconds and bound excitons (BEs) have a long lifetime of a few microseconds (Figure 2.13). Therefore, the TRPL spectra are visualized to verify that the highest energy level peaks originate from FEs, and the low energy level peaks originate from Bes, which are attributed to trap states [20].

2.4.2 Electronic Properties

Like their polycrystalline counterparts, single perovskite crystals also stand out for their notable electrical properties. In this section, the electronic properties of perovskite monocrystalline devices are summarized through two main techniques: space charge limited current (SCLC) and impedance spectroscopy (IS).

2.4.2.1 Space-Charge-Limited Current (SCLC)

The SCLC is a steady-state technique to analyze the charge transport properties of a semiconductor [21]. The carrier mobility and trap density of single perovskite crystals are the two main properties that can be determined with this technique. The procedure is to record the dark current as a function of the applied voltage [22]. However, the data obtained from the SCLC model can easily be misinterpreted if the correct model is not selected considering the nature of the device [23, 24].

The most widely used SCLC method is based on the *Mott–Gurney (MG)* law (Eq. (2.1)) [25].

$$J = \frac{9}{8}\varepsilon\mu\frac{V^2}{L^3}$$

where μ, ε, and L are the charge carrier mobility, permittivity, and thickness of the perovskite semiconductor. To apply this model accurately, the device must satisfy some conditions: (i) The semiconductor material under study has to be undoped and trap-free, (ii) the configuration should be the sandwiched one, where the semiconductor is placed between two ohmic contacts, and (iii) the diffusion contributions to the current need to be insignificant, which can occur only for specific voltage ranges, also in the case of devices satisfying the two first conditions. If our system meets at

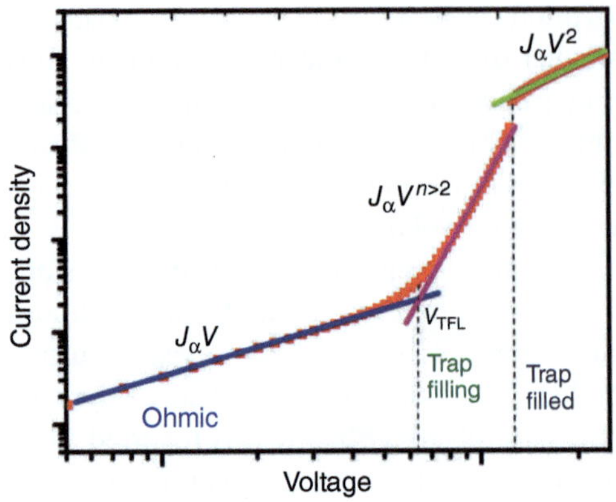

Figure 2.14 The jV characteristics visualize different transport regions. A linear ohmic regime ($J\alpha V$, blue line) is followed by the trap-filling regime, marked by a steep increase in current ($J\alpha V^{n>2}$, purple line). The $J-V$ curve shows a trap-free child's regime ($J\alpha V^2$, green line). Source: Reproduced with permission from Khan et al. [26]/John Wiley & Sons.

least conditions (i) and (ii), when plotted in a logarithmic scale, the current–voltage curve (J–V) can be easily divided into three main regions (Figure 2.14): (i) Ohmic region ($J \propto V$), where the conductivity can be determined, (ii) Trap-filling region, where the trap states are filled by the charge carriers and a sharp increase in the current is found at the trap-filled limit voltage (V_{TFL}), and (iii) Child's region at higher voltages where the current shows a quadratic dependence on voltage ($J \propto V^2$) [26, 27].

From the ohmic region of the jV curve, the free carrier concentration can be calculated as follows (Eq. (2.2)):

$$n_C = \frac{\sigma}{e\mu}$$

where σ is the conductivity extracted from the first region.

The density of deep trap states can also be determined using the following relation (Eq. (2.3)):

$$n_{\text{trap}} = \frac{2\varepsilon\varepsilon_0 V_{\text{TFL}}}{eL^2}$$

where ε_0 is the vacuum permittivity, ε is the relative dielectric constant, L is the crystal thickness, and e is the electron charge, and V_{TFL} is the transition voltage from the ohmic to the trap filling region.

As mentioned at the beginning of this section, if the model used is not well adjusted to the nature of the device, several discrepancies can appear in the final results [28–30]. One of the factors that also need to be considered is the geometry of the device. The MG model adjusts for a sandwiched configuration, but what happens if the electrodes' geometry is different? It must be considered that three types of geometry are possible when contacting the semiconductor under study (Figure 2.15).

For the gap-type geometry, or lateral configuration, the carrier mobility follows Geurst's SCLC model (Eq. (2.4)) [31]:

$$J = \frac{2\mu\varepsilon\varepsilon_0 W}{\pi}\left(\frac{V}{L}\right)^2$$

Figure 2.15 Various electrode configurations for a thin semiconductor layer. (a) Sandwich-type structures, (b) and (c) gap-type structures.

where W and L are the width and the interelectrode distance. According to the Geurst theory, the threshold voltage for a gap-type structure is defined as (Eq. (2.5)):

$$V_{\mathrm{TFL}} = \frac{\pi \sigma_0 L}{4\varepsilon\varepsilon_0}$$

where μ is the carrier mobility, σ_0 is the surface charge density per unit area, ε_0 is the vacuum permittivity, and ε is the relative dielectric constant of perovskite.

But not only the geometry of the electrodes can affect the final results; the growth method used to synthesize the single crystals may also modify the final device's electronic properties as well [32, 33]. In Figure 2.16, the shape of the jV curves can be appreciated for the three main cases of $MAPbI_3$ single crystals other electrode

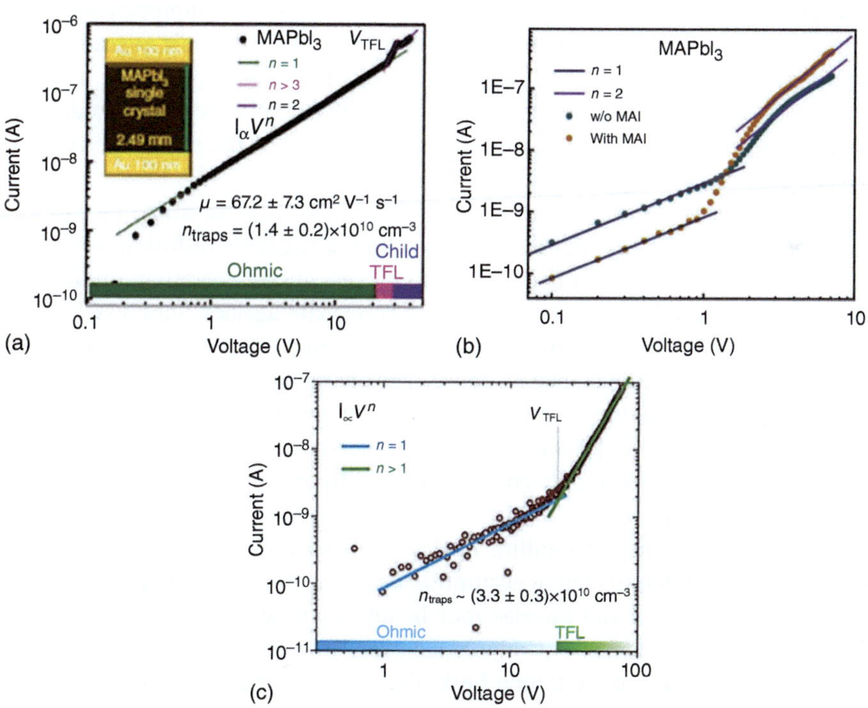

Figure 2.16 jV characteristic showing three different regimes for (a) $MAPbI_3$ single crystal grown by rapid ITC method and fabricated in sandwiched configuration. Source: Reproduced with permission from Saidaminov et al. [17]/Springer Nature/CC BY 4.0. (b) $MAPbI_3$ single crystal with gap-type geometry. Source: Reproduced with permission from Song et al. [32]/Springer Nature/CC BY 4.0. (c) $MAPbI_3$ single crystal grown via AVC method. Source: Reproduced with permission from Peng et al. [33]/Royal Society of Chemistry/CC BY 3.0.

Table 2.1 Comparison of the trap density and mobility values for the three different devices and methods.

	$MAPbI_3$ (a)	$MAPbI_3$ (b)	$MAPbI_3$ (c)
Trap density (n_{trap})	$1.4 \pm 0.2 \times 10^{10}$ cm^{-3}	4.51×10^9 cm^{-2}	$3.3 \pm 0.3 \times 10^{10}$ cm^{-3}
Charge mobility (μ)	67.2 ± 7.3 cm^2 V^{-1} s^{-1}	1.16 cm^2 V^{-1} s^{-1}	410 cm^2 V^{-1} s^{-1}

configurations and with varying methods of growth. Table 2.1 summarizes the values obtained in each case.

2.4.2.2 Impedance Spectroscopy (IS)

As it is known from previous chapters, the mixed ionic-electronic nature of perovskite material is responsible for increasing the complexity of the physico-chemical processes that govern it, making their understanding challenging. The main difference between the polycrystalline and monocrystalline perovskite devices is the absence of grain boundaries in the latter. This facilitates the comprehension of the different processes occurring in this material, shedding light on ionic migration, conductivity and dielectric responses, carrier lifetime, or interfacial processes. In fact, in monocrystalline perovskite devices, a Debye-type dielectric relaxation is predictable, accompanied by a single relaxation time corresponding to the unique grain of the crystals and the lack of grain boundaries.

The ionic migration constitutes one of the main bottlenecks in the race to fully stable perovskite systems, and the quantification of the ionic diffusion is a crucial step to minimizing the harmful effects of this process during operating conditions. It can easily be determined in monocrystalline perovskite devices. Suppression of the electronic contribution is the key factor needed to make it possible, together with depth analysis of the impedance response. The characteristic signature of the ionic diffusion in the impedance spectra is a Warburg element and a transmission line in the equivalent circuit [34]. Figure 2.17a shows a device based on $MAPbBr_3$ material with a TiO_2/perovskite/gold configuration, with the perovskite synthesized as a monocrystalline bulk.

In the case of the monocrystalline device, a Warburg-like coefficient can appear at the HF domain due to charge accumulation at the interfaces, with the interface between the perovskite and the electrode being very reflective. On the other hand, at LF domain, it becomes entirely capacitive. However, in the case of similar device made with a polycrystalline bulk, the charge carriers can diffuse over the perovskite-metal contact interface with an additional arc at the LF region. The contact then becomes partially absorptive (Figure 2.17b) [35].

Another typical feature that can appear in the impedance spectra of a single crystal perovskite device is an inductive behavior traduced in a negative capacitance (Figure 2.18a). The clear sign of this process is a loop in the LF domain. Several works pointed out the vacancy-assisted ionic diffusion responsible for this performance.

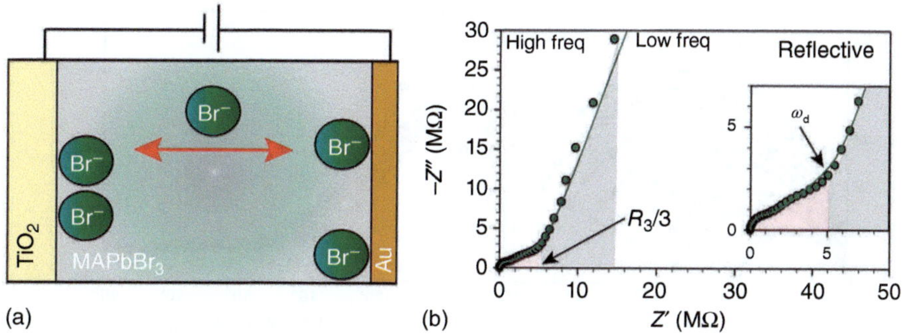

Figure 2.17 (a) Configuration of the monocrystalline perovskite device and its corresponding impedance spectra showing the Warburg behavior. (b) The reflective character of the contact in this case is shown in the inset of the impedance plot, differentiating between ionic transport (up to 5 MΩ) and charge accumulation (from 5 to 50 MΩ). Source: Reproduced with permission from Peng et al. [34]/American Chemical Society.

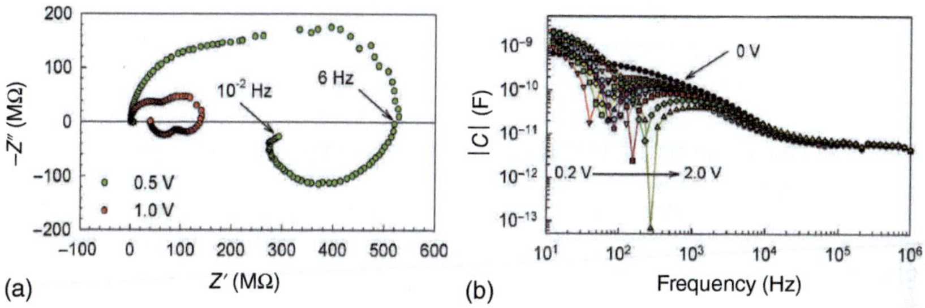

Figure 2.18 (a) Impedance plot of perovskite single crystals exhibiting the LF inductive behavior with their negative capacitance, more pronounced the higher the applied potential. (b) The collapse of the geometrical capacitance at HF is also shown. Source: Reproduced with permission from Kovalenko et al. [35]/AIP Publishing.

The interfaces govern the carrier injection into perovskite layers through an inductive process with a response time of 1–100 seconds under dark conditions and under an inert atmosphere.

Another substantial difference in the impedance responses of monocrystalline and polycrystalline perovskite devices is the time constant in the HF domain. In the case of the monocrystalline systems, these values are found to be 10^2–10^3 Hz (in the case of $MAPbI_3$), while being in the frequency range of 10^5–10^6 Hz in the polycrystalline counterpart [36]. This difference is ascribed to the recombination resistance and geometrical capacitance affected by the absence of grain boundaries and symmetrical contacts in the single perovskite crystals. The mobile ions mostly govern the net capacitance below 10^3 Hz, and the associated resistance is related to the ionic conductivity of the SC (Figure 2.18b). In addition, the low-frequency capacitance values are influenced by the ion density and mobility, which at the same time are influenced by the applied voltage and temperature (Figure 2.19).

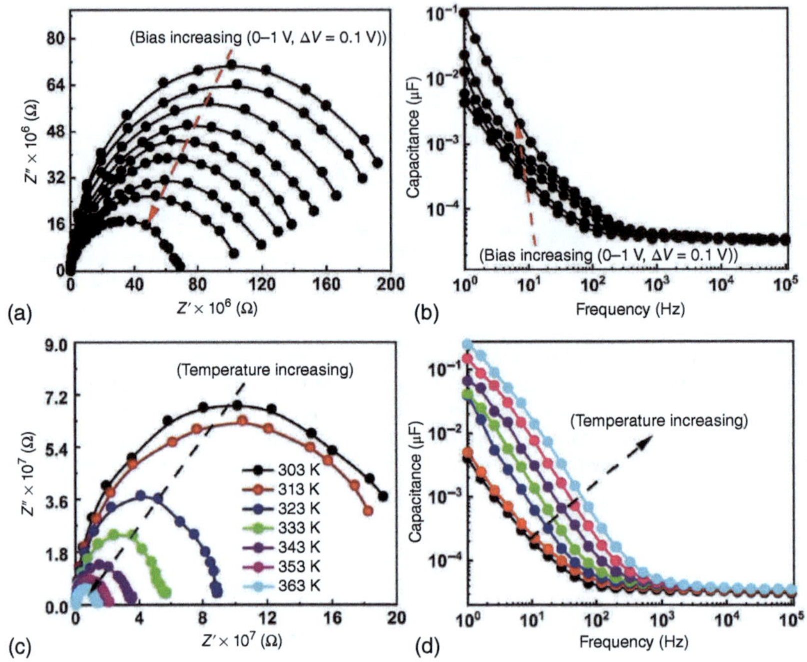

Figure 2.19 Impedance spectra of $MAPbI_3$ SC at ambient temperature (303 K) as a function of applied bias measured under dark conditions (a) and its corresponding capacitance response (b). IS under dark conditions of the $MAPbI_3$ SC at 0 V DC as a function of temperature (313–363 K) (c and d). Source: Reproduced with permission from Kalam et al. [36]/American Chemical Society.

The activation energy of ion migration can be highly modified with surface passivation treatments. In the case of $MAPbBr_3$, a treatment with $PbSO_4$ as a passivation layer can increase from 0.28 to 0.36 eV, suggesting a reduction in the ionic migration [37]. This effect is also shown in the impedance spectra. Instead of the two typical arcs performed for the non-passivated crystals, a single arc resulted in the case of the passivated one at the LF domain. The electronic conductivity of the single perovskite crystals is also affected by the moving ions, which can cause dynamic doping and vary the carrier density. This effect appears as a linear increment in the LF region and an increment in the capacitive response (Figure 2.20).

The carrier lifetime determines the extraction of charge carriers before recombination happens, and it is defined by the product between the recombination resistance (R_{rec}) and the chemical capacitance (C_{μ}). Therefore, it can be determined by IS under different light intensities. As an example, the carrier lifetime of a $MAPbI_3$ SC with a configuration of Au/$MAPbI_3$ SC/Au is around 95 ± 8 ms, which is more than 10 times the carrier lifetime detected for polycrystalline thin-film devices after surface passivation treatments.

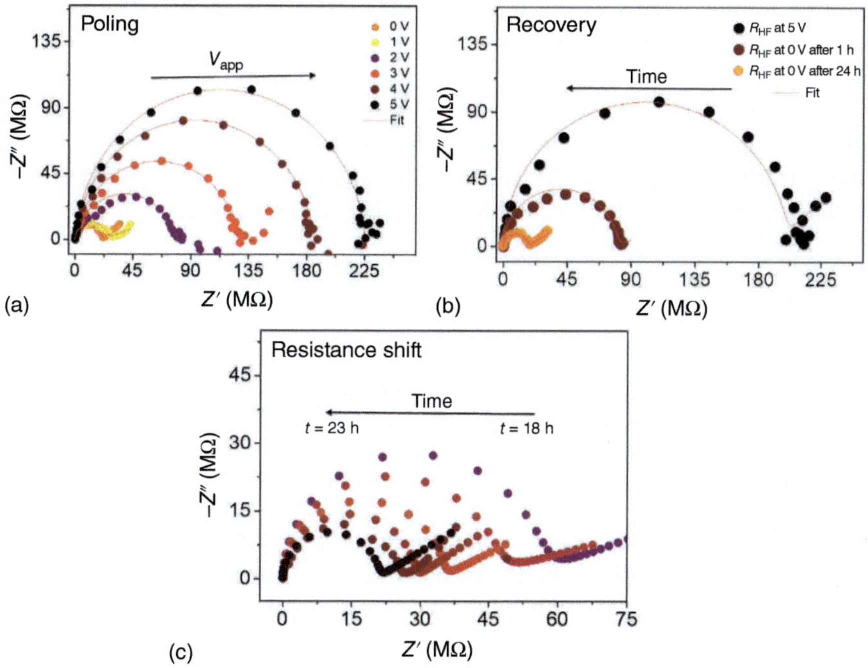

Figure 2.20 Variation of the impedance (a) as a function of the increasing bias with total measuring time ~20 minutes at each voltage, (b) after removing 5 V bias and evolution with time, and (c) a detailed view of the variation of the impedance after removing 5 V bias and evolution with time for smaller impedances. Source: Reproduced with permission from García-Batlle et al. [37]/John Wiley & Sons.

2.5 Applications

2.5.1 Photodetectors

Photodetection is perhaps the most direct application of perovskite single crystals. The working principle of a photodetector is to convert the incident radiation into an electrical output characterized by several properties. These properties of the electrical signal serve to evaluate the photodetector's performance. The main ones include responsivity (R), gain (G), detectivity (D^*), linear dynamic range (LDR), noise equivalent power (NEP), and response time (rise/decay time).

The responsivity is defined as the ratio of photocurrent to incident-light intensity and is proportional to the external quantum efficiency as follows (Eq. (2.6)):

$$R = EQE\frac{\lambda e}{hv}$$

where h is the Planck constant, c refers to the speed of light in vacuum, λ is the light wavelength, and e refers to the absolute value of electron charge.

The photoconductive gain G is defined as the average number of circuit electrons generated per photocarrier pair (Eq. (2.7)):

$$G = \frac{\tau}{L/\mu E}$$

where τ is the excess of minority carrier lifetime, L is the channel length, μ is the carrier mobility, and E is the electric field intensity provided by the applied voltage. The detectivity refers to the weakest level of light that the photodetector can detect, and it is defined as (Eq. (2.8)):

$$D^* = \frac{R}{(2qJ_d)^{1/2}}$$

where J_d is the dark current. The LDR labels the linearity of the current response of the photodetector with respect to the incident radiation occurring in a certain range of illumination intensity. It is defined as follows:

$$\text{LDR} = 20\log\frac{J_{\text{upper}} - J_d}{J_{\text{lower}} - J_d}$$

where J_{upper} is the current value of the device's response deviating from linearity, J_{lower} is the lower resolution limit, and J_d is the dark current. The NEP refers to the noise output produced by the photodetector when no light is applied, and it is correlated with the sensitivity of the photodetector. It is defined as the inverse of the detectivity (Eq. (2.9)):

$$\text{NEP} = \frac{(A\Delta f)^{1/2}}{D^*}$$

where A is the area of the device, and Δf is the bandwidth.

2.5.2 X-Ray Detection

The ability of perovskite single crystals as photodetectors includes radiation from a wide range of energies. The high atomic weight of the metallic cation (Pb) and the halides (I, Br), together with the high carrier mobility and carrier lifetime, allow the possibility to detect high-energy radiations as X-Rays. A single crystal of $MAPbBr_3$ synthesized by Wei et al. was the first perovskite X-ray photodetector ever reported [38]. A record high mobility–lifetime ($\mu\tau$) product of $1.2 \times 10^{-2}\ cm^2\ V^{-1}$ was obtained using a non-stoichiometry precursor for crystal growth, ensuring effective carrier transport, and suppressing the undesired dark current by lowering the applied bias. The carrier lifetime and extraction efficiency were increased by decreasing the surface charge recombination velocity with UV-O_3 treatment. Finally, the as-fabricated single crystal devices showed detectable X-ray dose rates as low as 0.5 μ Gyair s^{-1} with a sensitivity of 80 μC $Gyair^{-1}\ cm^{-2}$, four times higher than the sensitivity achieved with α-Se X-ray detectors (Figure 2.21).

2.5.3 γ-Ray Detection and Scintillators

The emission of gamma (γ) photons is due to the decay of radioactive isotopes. This emission has energies in the range of ~50 keV to 10 MeV. Part of this high energy

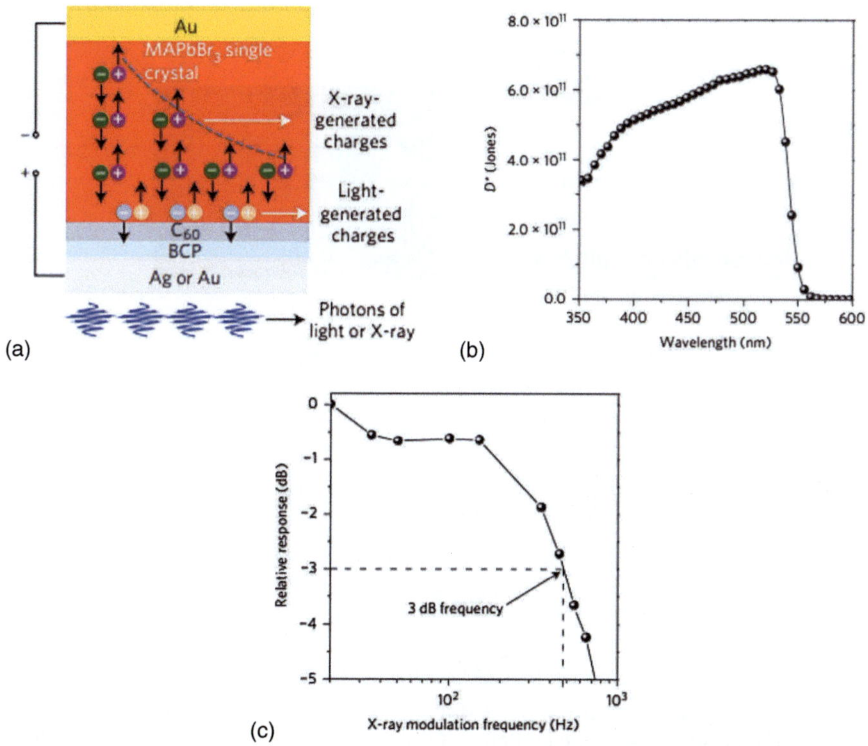

Figure 2.21 (a) Single-crystal radiation detector structure. (b) Specific detectivity of the $MAPbBr_3$ single-crystal device calculated from the directly measured IQE and NEP. (c) Normalized response as a function of input X-ray frequency, showing that the 3 dB cutoff frequency is 480 Hz. Source: Reproduced with permission from Wei et al. [38]/Springer Nature.

can penetrate the matter directly, creating an electron–positron pair. The result is an electric current that can be quantified to evaluate the energy and direction of the original γ-ray. For such hard energy, the detectors need to have high charge-carrier mobility–lifetime ($\mu\tau$) product, which is the case with perovskite materials. There are two main groups of γ-ray detectors, depending on their nature and functionality: (i) spectrometers, formed by scintillators or solid-state materials, to transform the γ-ray into optical or electronic signals, and (ii) based on the physical principle of electron-positron production or *Compton scattering* to perform γ-ray imaging. In any case, the γ-ray detectors need to work in weak radiation field pulse mode. An example of a γ-photodetection using single perovskite crystals was performed by Kovalenko and co-workers. They reported the properties as detectors of large (3–12 mm in size) SCs based on $MAPbI_3$, $MAPbBr_3$, $MAPbCl_3$, $FAPbI_3$, $FAPbBr_3$, and I-treated $MAPbBr_3$, with $MAPbI_3$ having highest sensitivity [39]. They used $MAPbI_3$ SC for single photon counting, considering the current generated by a single γ photon must be sufficiently above the noise level.

Scintillators convert ionized photon radiation into UV-visible photons. The detection of this radiation is used in a broad range of fields, including diagnostic

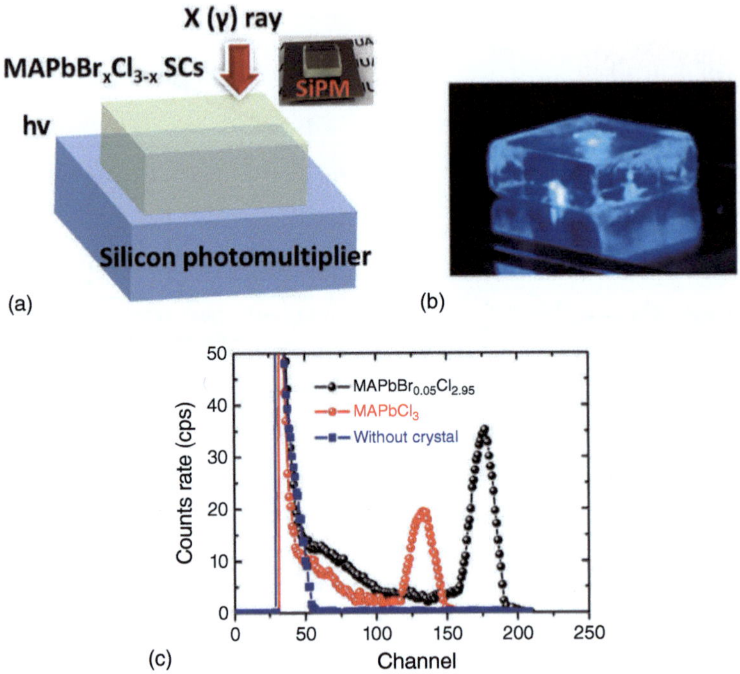

Figure 2.22 (a) Schematic illustration of $MAPbBr_xCl_{3-x}$ SCs integrating on the window of SiPM. Inset is the side view of the $MAPbBr_{0.05}Cl_{2.95}$ SC scintillator integrated on SiPM MPPC. (b) Scintillation of $MAPbBr_xCl_{3-x}$ SCs excited with a pulsed X-ray, generated by an accelerator. (c) Pulse height spectra acquired with $MAPbBr_xCl_{3-x}$ SCs excited with a 1.4 μCi button-sized ^{137}Cs source. Source: Reproduced with permission from Xu et al. [40]/American Chemical Society.

nuclear medicine, astrophysics, and homeland security. Ouyang and coworkers demonstrated that perovskite single crystals with a composition of $MAPbBr_{0.05}Cl_{2.95}$ coupled with a silicon photomultiplier (SiPM) fits the requirements of low cost and good time resolution [40]. They used a source of ^{137}Cs radioactive isotope at room temperature and obtained a successful X (γ)-ray excited luminescence (XEL) spectra in the visible range (Figure 2.22).

2.5.4 Solar Cells

As described in previous chapters, the PCE of solar cells based on polycrystalline perovskites is astonishing. However, this polycrystalline nature is also responsible for the vulnerability of these devices to environmental conditions. The grain boundaries are one of the loci responsible for allowing water and oxygen to pass through them, weakening the photovoltaic performance of the devices. In addition, the ionic migration, responsible for the hysteretic behavior and the consequent instability, also occurs to a greater extent, due to high trap densities. It has been proven that this factor is much stronger in polycrystalline perovskites than in their monocrystalline counterparts [41]. In fact, single perovskite crystals displayed a higher carrier diffusion length, above tens of micrometer, due to the absence of these grain boundaries, which significantly reduces the defect density in the bulk.

It seems clear, then, that if we remove these grain boundaries and reduce the trap densities, the stability of PSCs must inevitably increase together with the photovoltaic performance. What is less clear is whether efficiency can also be maintained [42]. Monocrystalline materials typically offer better photovoltaic properties than their polycrystalline counterparts due to the low density of structural and chemical defects (Figure 2.23a). However, single-crystalline perovskites are, at least for the time being, far from this rule. Although the bulk trap density is exceptionally low, the surface trap density is significantly high on perovskite single crystals. In fact, one of the main limitations of solar cells based on perovskite single crystals is that their surface recombination velocity is higher than that of polycrystalline counterparts by a factor of 6 (Figure 2.23b) [42].

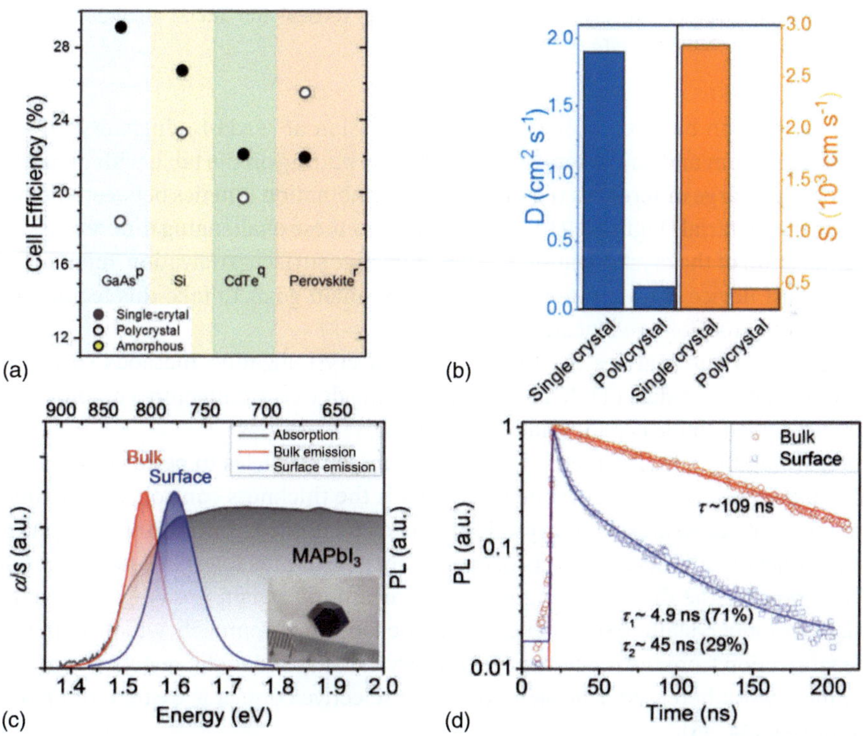

Figure 2.23 (a) Best-efficiency comparison of polycrystalline and single-crystalline solar cells of GaAs, Si, CdTe, and perovskite. (b) Diffusion coefficient (*D*) and surface recombination velocity (*S*) of single-crystal and polycrystalline $MAPbI_3$. Source: Reproduced with permission from Turedi et al. [42]/American Chemical Society. (c) $MAPbI_3$ α/s (absorption/scattering), bulk emission ($\lambda_{ex} = 1200$ nm), and surface emission ($\lambda_{ex} = 600$ nm) profiles (inset: $MAPbI_3$ single-crystal image). (d) $MAPbI_3$ bulk emission and surface transient photoluminescence (TRPL) dynamics at the same wavelengths. Source: Reproduced with permission from Wu et al. [43]/John Wiley & Sons.

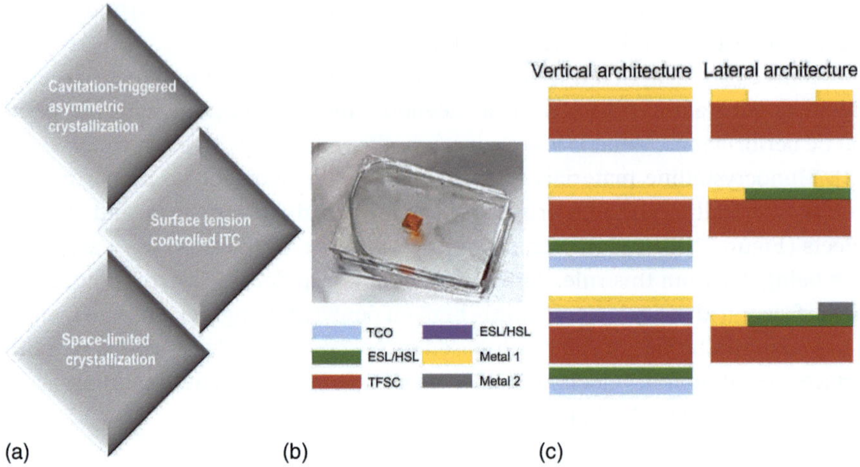

Figure 2.24 (a) Crystallization techniques for the TFSC's growth. (b) TFSC of $MAPbBr_3$ grown by the space-limited crystallization method. (c) Vertical and lateral architectures for TFSC solar cells. Source: Clara Aranda-Alonso.

A comparison between bulk and surface emission of $MAPbI_3$ single crystal was reported by Sum and coworkers (Figure 2.23c) by putting on the table with quantitative data the large differences regarding the recombination kinetics between the two loci of the material (Figure 2.23d) [43]. To address these challenging differences, the optimization of the crystallization method, effective surface passivation approaches, and robust device configurations appear as the main goals to take this technology beyond the proof-of-concept stage.

Figure 2.24 summarizes the most important crystallization methods so far for thin film single crystals (TFSC) together with the device architectures employed to achieve a good single crystal solar cell response.

Regarding the growing method, one of the major obstacles to performing highly efficient single crystals perovskite solar cells is the thickness control. The intrinsic properties of perovskite single crystals have been focused on the thick bulk materials in the order of millimeters, which, of course, are not suitable for solar cell application if we consider the diffusion length. In this regard, two main growing approaches exist for monocrystalline solar cells. First is the top-down approach, which addresses the fabrication from the ready-made single crystal attached to the selective contacts. We must work here under pressure once the selective contact is treated for greater adherence [44, 45].

But this adherence requirement of the bottom layer is a considerable limitation. The layer needs to perform the charge extraction process at the same time as it must mechanically fix the crystal. To address this challenge, Müller-Buschbaum and coworkers reported an advanced top-down approach in 2018. They analyzed two different structures: (i) a p–i–n architecture with ITO\PEDOT:PSS\perovskite\ PCBM(spray)\silver paste or\Al and (ii) an approach using a polymer bezel which enables a n–i–p architecture with FTO\PCBM\perovskite\spiro-OMeTAD\Au

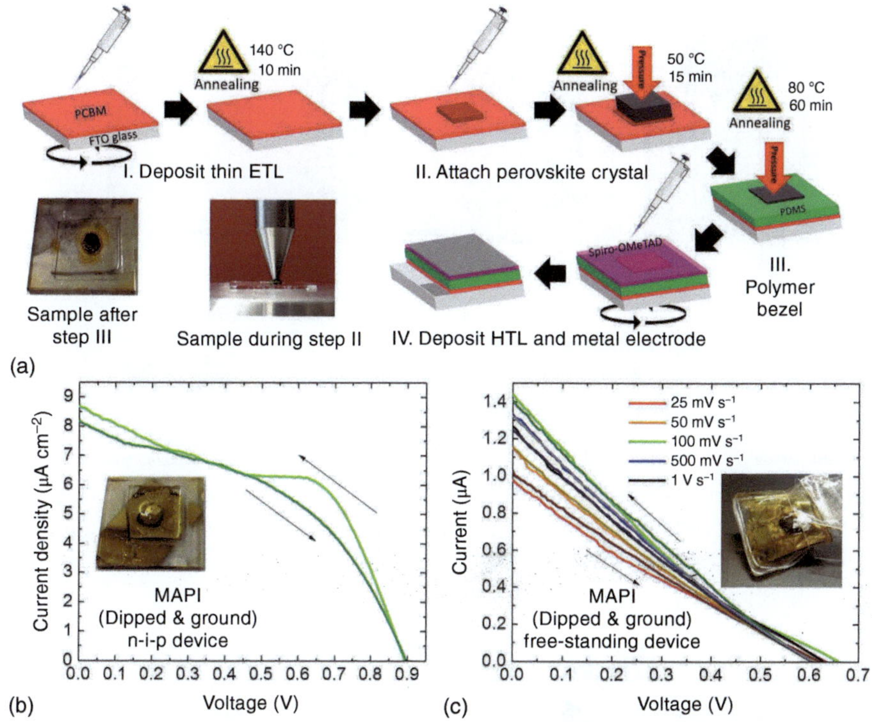

Figure 2.25 (a) Fabrication process of the n-i-p single crystal perovskite solar cell with the polymer bezel made from PDMS. (b) *jV* curves showing negligible hysteresis at 100 mV s^{-1}. Inset: photograph of the device. (c) Performance of the free-standing device incorporated in the polymer bezel. Inset: photograph of the device. Source: Reproduced from Schlipf et al. [46]/Springer Nature/CC BY 4.0.

or even a substrate-free architecture with silver paste\PCBM\perovskite\spiro-OMeTAD\Au [46]. In Figure 2.25a is shown the n-i-p single crystal perovskite solar cell made with the polymer bezel from a polydimethylsiloxane (PDMS) which solves the problem of needing a material that both fixes the crystal, favoring at the same time the charge extraction. Crystal is placed during the PCBM annealing and subsequent deposition of the PDMS precursor. First, the PCBM solution is used to benefit good attachment of the single crystal. Then the crystal is placed during the annealing time of PCBM, followed by the deposition of the PDMS precursor. Finally, spiro-OMeTAD is used as HSL, and Au is thermally evaporated as back contact. The enhanced photovoltaic performance of the final device is presented in Figure 2.25b, showing reduced hysteresis.

On the other hand, there is the *bottom-up* method, typically used for polycrystalline devices, which is performed layer by layer from the bottom substrate to the top. In the case of single crystals, this method is modified as an in-situ growing method, in which a monocrystalline film crystallizes (i.e. on the substrate) during device fabrication or in an already built device. This presents some restrictions regarding the choice of materials for selective contacts [47]. One of the first

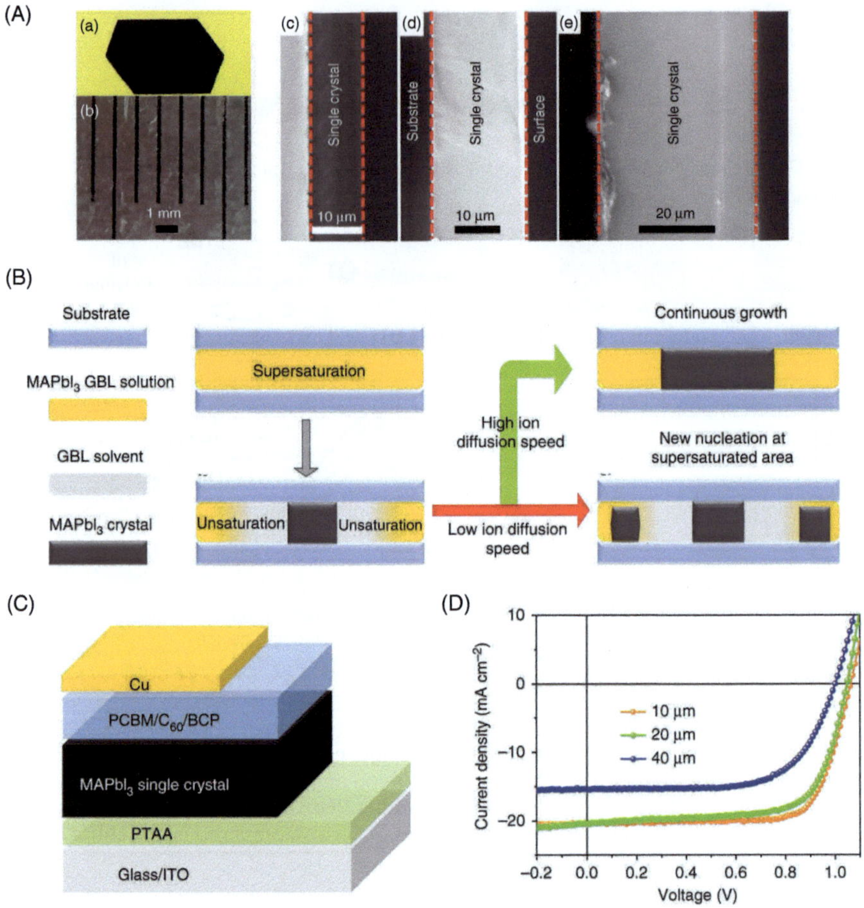

Figure 2.26 (A) Methylammonium lead triiodide ($MAPbI_3$) thin single crystal with the cross-section SEM images showing different thicknesses. (B) Growth mechanism of perovskite thin single crystals showing correlation between ion diffusion and thin single-crystal growth. (C) Schematic representation of device architecture. (D) *jV* curves of the monocrystalline perovskite solar cell corresponding with the different single crystal thicknesses. Inset: photograph of device. Source: Reproduced from Chen et al. [48]/Springer Nature/CC BY 4.0.

successful approaches to address the challenge of the in-situ growing process was reported by Jinsong Huang and coworkers in 2017. For the first time, a confined growing method that can handle a balance between micrometer-thick (20 μm) iodide-based single perovskite crystals with a large area was presented (Figure 2.26a) [48]. The reported work shows how the space confinement facilitates ion diffusion using the well-known ITC method. The mechanism lies in the interaction between the substrate surface and solvent in precursor solution (Figure 2.26b).

The gap between the substrates used to perform the confinement becomes key for the ion diffusion in the solvent. This diffusion rate of precursor ions is determined by how fast the solvent molecules can carry them through the substrate.

The other key factor is the surface tension, which determines the speed of solvent molecules moving forward in-plane direction near the substrate. When the surface tension increases due to a more hydrophobic material (i.e. PTAA, which can also act as hole selective contact), the solvent-substrate interaction becomes weak, promoting the diffusion of the solvent and avoiding the ions from being dragged during the process. In Figure 2.26c is shown the schematic configuration of the final device composed as follows: ITO/PTAA/$MAPbI_3$/phenyl-C61-butyric acid methyl ester (PCBM)/C60/bathocuproine (BCP)/copper (Cu). It showed a PCE of 16.1%, as presented in Figure 2.26d, and the *jV* curve shape variation with respect to the crystal thickness. As we mentioned before, the surfaces of single perovskite crystals represent the greatest difficulty to be solved because of their higher density of traps than the bulk. Hence, a treatment with MAI was done to passivate those defects and control the surface trap population, enhancing the final performance.

In the following, we will summarize the crystallization methods, surface passivation techniques, and device configurations with which the best efficiency results have been obtained to date.

The bottom-up methodology coupled with the ITC technique used in a *half-made* device, completing the device once the crystal is grown, has proven to be the optimal combination to date, allowing PCE of around 23%. In Figure 2.27, we can see the progress of the outstanding results performed by Turedi et al. They reported 20-μm thick single crystals made from $MAPbI_3$ perovskites. It was grown directly on a PTAA film (semi-made device) [49]. They used the solution-limited inverse-temperature crystal growth method, yielding power conversion efficiencies

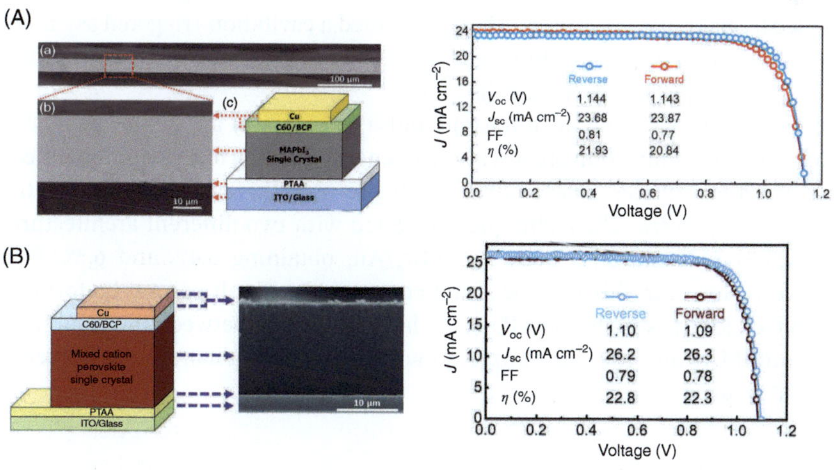

Figure 2.27 (A) Cross-sectional SEM image of the $MAPbI_3$ SC-PSCs with their corresponding *jV* curves of the champion cell showing the photovoltaic parameters under 1 sun illumination. Source: Reproduced with permission from Alsalloum et al. [49]/American Chemical Society. (B) Device architecture of the single-crystal device and its cross-sectional SEM image of the single-crystal device. Corresponding *jV* curves of the best device under reverse (blue) and forward (red) scans. Source: Reproduced with permission from Alsalloum et al. [50]/American Chemical Society.

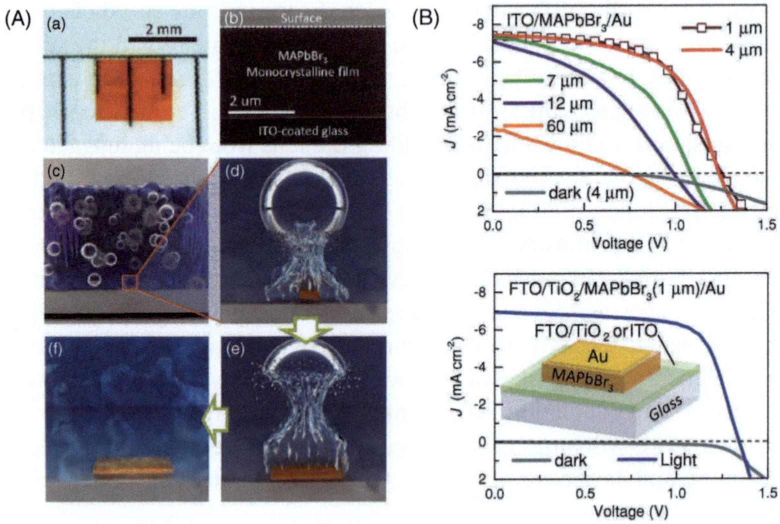

Figure 2.28 (A) Optical image of the TFSC, cross-section SEM image of it, and the schematic representation of the CTAC mechanism on a microscopic scale. (B) Dark and light *jV* curves of the two different architectures. Source: Reproduced with permission from Peng et al. [47]/John Wiley & Sons.

up to 21.09% (Figure 2.27a). In 2021, they pushed forward their results by using a mixed-cation formulation. By varying the passivation of the crystal surface, they finally obtained the record PCE to date, up to 22.8% (Figure 2.27b) [50].

For other perovskite formulations, such as $MAPbBr_3$, an original method was developed by Bakr and coworkers. They performed a cavitation-triggered asymmetrical crystallization (CTAC) technique (Figure 2.28) to grow two different thin-film single-crystalline perovskite solar cells [47].

This method is based on an ultrasonic pulse (<1 second) applied to perovskite solution. The ultrasonication creates a shot of energy, leading to a cavitation process accompanied by bubbles. These bubbles collapse near the substrate, accelerating the nucleation process. The technique was tested with two different architectures: ITO/$MAPbBr_3$/Au and FTO/TiO_2/$MAPbBr_3$/Au, obtaining 5.4% and 6.5% PCE, respectively. However, this method is not compatible with all perovskite formulations. In fact, for the simplest $MAPbI_3$, the large difference between the solubilities of PbI_2 and MAI and the anisotropic growth in the tetragonal system prevent a successful result.

2.5.5 Light Emitting Diodes

In addition to the extraordinary properties of perovskites in the field of photovoltaics, their high color purity, low Auger recombination, and exceptional PL should be highlighted. These properties make them ideal candidates for application as light-emitting diodes (LEDs). In polycrystalline perovskites, coherent light emission is relatively inefficient due to the additional recombination caused by the presence of grain boundaries (loci of nonradiative recombination pathways) and the different orientations of the crystalline domains. To produce good band-to-band or bimolecular recombination and enhanced electroluminescence, it is necessary to confine the generated electron-hole pairs spatially. To use single-crystal perovskites in LED technology, it is necessary to be able to control the size and dimensionality of the synthesized crystals and reduce as much as possible the defects responsible for the radiative losses. Thus, the properties of emitters will depend on whether the crystals are a few centimeters or go down to the nanometer scale. Similarly, the performance of LEDs will change depending on whether we have microdisks, nanosheets, nanoplatelets, or nanowires. But controlling the size, dimension, and the number of defects in the final crystal is insufficient to develop an efficient and long-lasting perovskite single crystals LEDs.

Despite the advantages of having no grain boundaries, the instability of perovskites is still there, making it difficult to achieve long operating times. This fact distances them from their organic competitors (OLEDs) and slows the road to industrialization. On the other hand, we must not forget the lead content. Its toxicity creates an additional obstacle that needs to be overcome with Pb-free perovskites without compromising their PLQY. To this respect, the main crystallization growth techniques for LEDs application are the following: solution growth with top seed ($MAPbI_3$, $MASnI_3$, and $FASnI_3$) and bottom seed ($MASnCl_3$), antisolvent vapor crystallization ($MAPbI_3$ and $MAPbBr_3$), inverse temperature crystallization ($MAPbX_3$, $FAPbX_3$, and $FA_{1-x}MA_xPbX_3$) and chemical vapor deposition ($CsPbBr_3$).

Below are some of the most recent works that have reported remarkable results. In Figure 2.29a, a single bulk crystal is shown contacting carbon nanotubes in a sandwich configuration. This work reported by Horvàth and coworkers shows an in-situ growth of the crystals between the two electrodes [51]. This technique leads to a p–i–n junction, decreasing the interface energy barrier, which enhances the charge injection and produces a green emission up to 1800 cd m^{-2} at room temperature. In Figure 2.29b, we find another exciting configuration for a perovskite single-crystal LED. To avoid lead toxicity, Huang and coworkers developed a successful white emitter using a $CsCu_2I_3$ perovskite [52]. The 1D structure is formed by the Cu–I octahedron separated by Cs atoms.

In Figure 2.29c, another lead-free perovskite single crystal is shown to perform with red photoluminescence. In this case, a centimeter-size, 2D-layered $(CH_3NH_3)_2MnCl_4$ single crystal was grown using the ITC method [53]. They demonstrated that a good red emission was possible with a QE of 2.4% upon excitation at 417 nm.

Beyond lead substitution, the stability issue is still the main target for the next work presented in Figure 2.29d. Developing a liquid-to-liquid self-encapsulation inkjet printing technique, perovskite inks are placed into a PDMS precursor to in-situ grow $MAPbBr_3$ single crystals [54]. The liquid performs a space-confinement effect, slowing down the crystallization rate, and acts as a protection material against environmental damage. In Figure 2.29e, Chen et al. played with the different layer thicknesses of 2D RP phase single crystals to obtain different emission wavelengths [55]. The LED device was developed using Cs–Pb–Br-based perovskites with

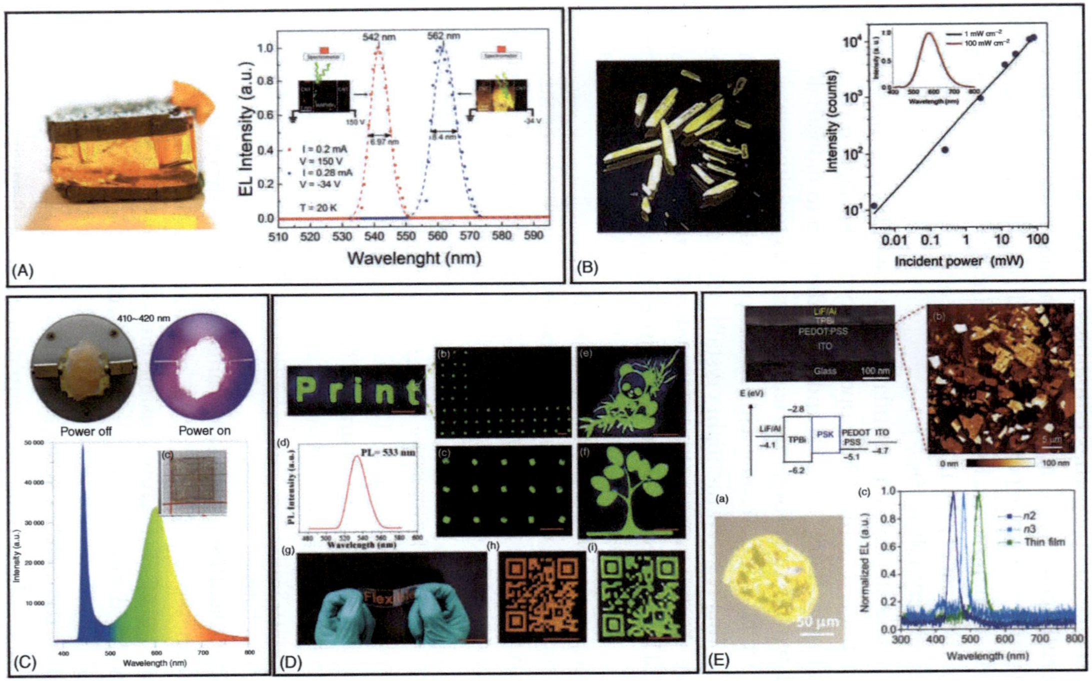

Figure 2.29 (A) Optical image of a $MAPbBr_3$ single-crystal sandwiched between two VACNT electrodes. Normalized EL spectra recorded at 20 K in vacuum, at positive (red, +150 V) and negative (blue, −34 V) bias voltages. Source: Reproduced with permission from Andričević et al. [51]/American Chemical Society. (B) Optical image of a 1D $CsCu_2I_3$ perovskite crystal and its PL intensity under various light intensities. Source: Reproduced with permission from Lin et al. [52]/John Wiley & Sons. (C) LED device and the effects of powering off/on together with the corresponding spectrum of the LED emission. Source: Reproduced with permission from Cheng et al. [53]/Royal Society of Chemistry. (D) Wafer-scale fabrication of perovskite single-crystal array embedded in PDMS by inkjet printing. Fluorescence images of complicated patterns in the flexible substrate. Scale bars are 5 mm. Source: Reproduced with permission from Gu et al. [54]/American Chemical Society. (E) A cross-sectional SEM image and energy band alignment of the LED device consisting of a $(BA)_2Cs_{n-1}Pb_nBr_{3n+1}$ single crystal (*n*1, *n*2, and *n*3 packing along the long crystallographic axis). AFM image of the perovskite nanostructures. Real image of the crystal *n*3 and the corresponding EL spectra of *n*2 (blue), *n*3 (sky blue), and thin film (green)-based LED. Source: Reproduced with permission from Chen et al. [55]/American Association for the Advancement of Science.

different n layers. Synthesized by structural confinement engineering with the help of organic layers, the RP single crystal perovskite showed a successful blue-emission response.

2.5.6 Memristors

The mixed nature of perovskites in terms of their ionic-electronic conductivity is responsible for instabilities during operation times, traduced in hysteretic responses. However, this same mixture is relevant in their application as memory devices. Merging these combined conductive capabilities with the sensitivity to light radiation, optical-erase memory devices, neuromorphic computing, and light-accelerated learning capability are new possibilities in the perovskite field.

In a memristor (memory + resistor), the resistance can be regulated reversibly by applying an electrical signal. They are considered passive devices and usually consist of two terminals. The main characteristic of a memristor property is the presence of a hysteresis loop during current-voltage (jV) measurement when driven by a voltage or current modulation. It was in the 1970s when Leon Chua categorized this device as one of the fourth main circuit elements (Figure 2.30a) [56]. Memristors are classified into three categories: ideal, generic, and extended; all with their corresponding

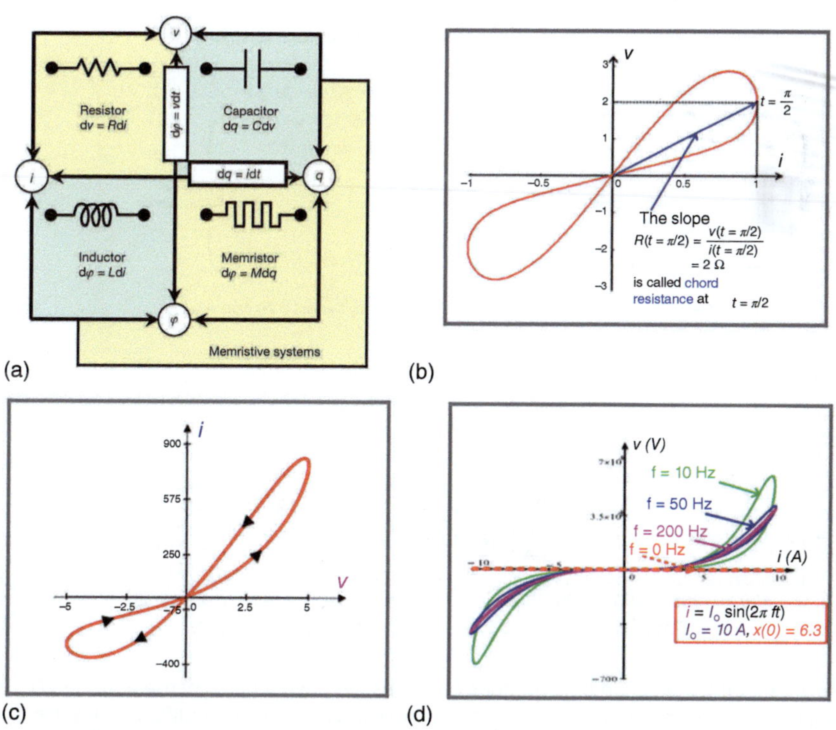

Figure 2.30 Basic electronic circuits include a memristor as the fourth element. Typical IV hysteresis loops of (b) ideal, (c) generic, and (d) extended memristors. Source: Reproduced with permission from Zhao et al. [56]/John Wiley & Sons/CC BY 4.0.

hysteretic behavior (Figure 2.30b). They can be of biological origin or developed by semiconductor technology, existing both analogically or digitally, depending on whether the resistive switching is continuous or discrete, respectively.

Although the application of perovskites as memristors is still in its infancy, several relevant works have highlighted the incredible potential of these materials in this field. Most of the perovskite-based memristors are configured as metal-insulator-metal (MIM), and their memristor responses are explained through two different mechanisms: ionic migration or charge trapping and detrapping. These mechanisms often depend on the contacts used in their architecture.

Regarding the application of perovskites single crystals as memristors, it was in 2017 when one of the first works in this respect was published by Rand and coworkers [57]. They used a 2D exfoliated $MAPbBr_3$ single crystal as a memory device for neuromorphic computing. Neuromorphic computing requires minimal energy consumption to massively process parallel information. To achieve this ability, memory devices need to have considerable ionic transport and extremely low operation current. 3D perovskites and polycrystalline films present a large intrinsic electronic current and parasitic leakage current due to grain boundaries. This fact leads to a leakage current three orders of magnitude higher than that of state-of-the-art resistive memory devices. Synapses from human brain process

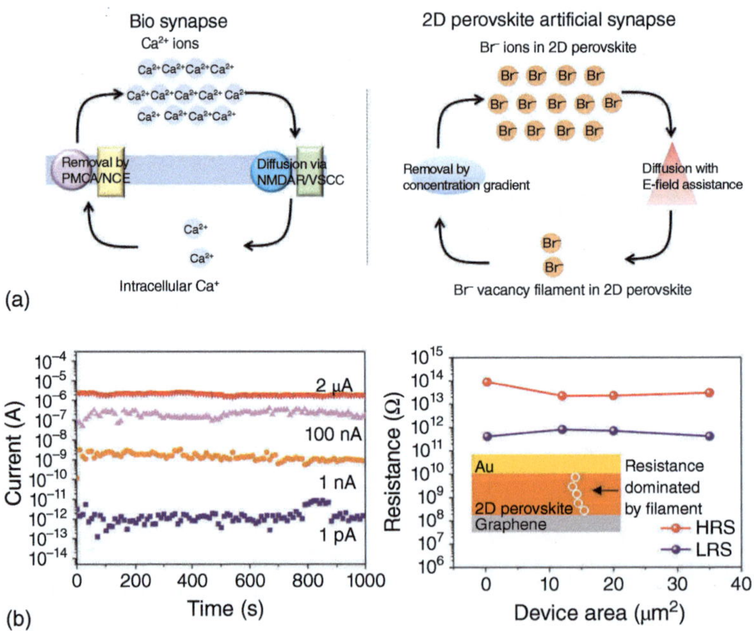

Figure 2.31 (a) Comparison of analogous functions between a bio-synapse and the 2D perovskite-based artificial synapse. The Br^- ion movement in 2D perovskite mimics the Ca^{2+} ion movement in a biosynapse. (b) Demonstration of multilevel storage at different current levels showing data stability for up to 1000 seconds. HRS and LRS data as function of the device area. Source: Reproduced with permission from Tian et al. [57]/American Chemical Society.

massively parallel information with an energy consumption of around ~1–100 fJ per synaptic event, which means an operating current in the pA range. On the other hand, artificial synapses perform set/reset programs in the mA and μA range. The set program is defined as the resistive switching from a high-resistance state (HRS) to a low-resistance state (LRS); reset program is described as the opposite (from LRS to HRS).

To reach pA, the electronic transport should be minimized, and the ionic contribution must be enough. To enhance the ionic transport while minimizing the electronic one, defects are usually introduced in the memory material. However, these defects can provoke leakage currents, causing a counterproductive effect.

The work from Randal and coworkers explored the potential of 2D exfoliated wide bandgap perovskite single crystals, avoiding leakage currents due to the absence of grain boundaries. The wide bandgap enables high ionic migration due to the low carrier density, and the layered 2D single crystal's exfoliation allows for a thin film, which causes a favored bromine migration. In Figure 2.31a, it is shown the comparison between a biosynapse and the artificial synapse with the 2D perovskite SC developed through the bromide ionic migration. This migration is followed by the accumulation of the ions at the anode interface, leading to a conducting channel under an electric field. Finally, an off current as low as 0.1 pA was obtained, leading to an operation current of 10 pA with an on/off ratio of 10. This made it possible to suppress the charge transport in the out-of-plane direction of the 2D material.

Acknowledgments

C. A. A thanks the Helmholtz Young Investigator Group FRONTRUNNER. M. S. thanks the German Research Foundation (DFG) for funding (SPP2196, GRK 2642). M. S. acknowledges funding from ProperPhotoMile. Project ProperPhotoMile is supported under the umbrella of SOLAR-ERA.NET Cofound 2 by the Spanish Ministry of Science and Education and the AEI under the project PCI2020-112185 and CDTI project number IDI-20210171; the Federal Ministry for Economic Affairs and Energy based on a decision by the German Bundestag project numbers FKZ 03EE1070B and FKZ 03EE1070A; and the Israel Ministry of Energy with project number 220-11-031. SOLAR-ERA.NET is supported by the European Commission within the EU Framework Programme for Research and Innovation HORIZON 2020 (Cofound ERA-NET Action, No. 786483).

References

1 Murali, B., Kolli, H.K., Yin, J. et al. (2020). Single crystals: the next big wave of perovskite optoelectronics. *ACS Materials Letters* 2 (2): 184–214.

2 Liu, Y., Zhang, Y., Zhu, X. et al. (2021). Triple-cation and mixed-halide perovskite single crystal for high-performance X-ray imaging. *Advanced Materials* 33 (8): 2006010.

3 Ding, J., Jing, L., Cheng, X. et al. (2018). Design growth of $MAPbI_3$ single crystal with (220) facets exposed and its superior optoelectronic properties. *Journal of Physical Chemistry Letters* 9 (1): 216–221.

4 Wang, K.-H., Li, L.-C., Shellaiah, M., and Wen, S.K. (2017). Structural and photophysical properties of methylammonium lead tribromide ($MAPbBr_3$) single crystals. *Scientific Reports* 7 (1): 13643.

5 Lee, J.W., Naqvi, F.H., Ko, J.-H. et al. (2022). Acoustic anomalies and the critical slowing-down behavior of $MAPbCl_3$ single crystals studied by brillouin light scattering. *Materials* 15: 3692.

6 Li, W.-G., Rao, H.-S., Chen, B.-X. et al. (2017). A formamidinium–methylammonium lead iodide perovskite single crystal exhibiting exceptional optoelectronic properties and long-term stability. *Journal of Materials Chemistry A* 5 (36): 19431–19438.

7 Zhang, L., Wang, K., and Zou, B. (2019). Bismuth halide perovskite-like materials: current opportunities and challenges. *ChemSusChem* 12 (8): 1612–1630.

8 Wu, X., Li, P., Wei, X., and Liu, J. (2022). All-inorganic perovskite single crystals for optoelectronic detection. *Crystals* 12 (6): 792.

9 He, Y., Matei, L., Jung, H.J. et al. (2018). High spectral resolution of gamma-rays at room temperature by perovskite $CsPbBr_3$ single crystals. *Nature Communications* 9 (1): 1609.

10 Hany, I., Yang, G., Phan, Q.V., and Kim, H.J. (2021). Thallium lead iodide ($TlPbI_3$) single crystal inorganic perovskite: electrical and optical characterization for gamma radiation detection. *Materials Science in Semiconductor Processing* 121: 105392.

11 Chung, I., Lee, B., He, J. et al. (2012). All-solid-state dye-sensitized solar cells with high efficiency. *Nature* 485 (7399): 486–489.

12 Zeng-Yong, C., Zhen-Hua, J., Chun-Hua, W., and Lig-L (2018). Recent progress in high-quality perovskite $CH_3NH_3PbI_3$ single crystal. *Journal of Inorganic Materials* 33: 1035–1045.

13 Zhang, C., Liu, X., Chen, J., and Lin, J. (2021). Solution and solid-phase growth of bulk halide perovskite single crystals. *Chinese Journal of Chemistry* 39 (5): 1353–1363.

14 Zhang, Y., Liu, Y., Xu, Z. et al. (2019). Two-dimensional $(PEA)_2PbBr_4$ perovskite single crystals for a high performance UV-detector. *Journal of Materials Chemistry C* 7 (6): 1584–1591.

15 Mckelvey, A.L. (1996). Retrograde solubility in semiconductors. *Metallurgical and Materials Transactions A: Physical Metallurgy and Materials Science* 27 (9): 2704–2707.

16 Saidaminov, M.I., Abdelhady, A.L., Maculan, G., and Bakr, O.M. (2015). Retrograde solubility of formamidinium and methylammonium lead halide perovskites enabling rapid single crystal growth. *Chemical Communications* 51 (100): 17658–17661.

17 Saidaminov, M.I., Abdelhady, A.L., Murali, B. et al. (2015). High-quality bulk hybrid perovskite single crystals within minutes by inverse temperature crystallization. *Nature Communications* 6.

18 Stoumpos, C.C., Cao, D.H., Clark, D.J. et al. (2016). Ruddlesden–popper hybrid lead iodide perovskite 2D homologous semiconductors. *Chemistry of Materials* 28 (8): 2852–2867.

19 Jang, D.M., Park, K., Kim, D.H. et al. (2015). Reversible halide exchange reaction of organometal trihalide perovskite colloidal nanocrystals for full-range band gap tuning. *Nano Letters* 15 (8): 5191–5199.

20 Shi, D., Adinolfi, V., Comin, R. et al. (2015). Low trap-state density and long carrier diffusion in organolead trihalide perovskite single crystals. *Science (80-)* 347 (6221): 519–522.

21 Mark, P. and Helfrich, W. (1962). Space-charge-limited currents in organic crystals. *Journal of Applied Physics* 33 (1): 205–215.

22 Zhang, X.G. and Pantelides, S.T. (2012). Theory of space charge limited currents. *Physical Review Letters* 108 (26): 266602.

23 Röhr, J.A., Moia, D., Haque, S.A. et al. (2018). Exploring the validity and limitations of the Mott–Gurney law for charge-carrier mobility determination of semiconducting thin-films. *Journal of Physics. Condensed Matter* 30 (10): 105901.

24 Röhr, J.A., Kirchartz, T., and Nelson, J. (2017). On the correct interpretation of the low voltage regime in intrinsic single-carrier devices. *Journal of Physics. Condensed Matter* 29 (20): 205901.

25 Leighton, P.A. (1941). Electronic processes in ionic crystals (Mott, N. F.; Gurney, R. W.). *Journal of Chemical Education* 18: 249.

26 Khan, M.T., Almohammedi, A.R., Kazim, S., and Ahmad, S. (2019). Electrical methods to elucidate charge transport in hybrid perovskites thin films and devices. *Chemical Record* 20: 452.

27 Zhang, P., Valfells, Á., Ang, L.K. et al. (2017). 100 years of the physics of diodes. *Applied Physics Reviews* 4 (1): 11304.

28 Duijnstee, E.A., Ball, J.M., Le Corre, V.M. et al. (2020). Toward understanding space-charge limited current measurements on metal halide perovskites. *ACS Energy Letters* 5 (2): 376–384.

29 Varo, P.L., Tejada, J.A.J., Villanueva, J.A.L. et al. (2012). Modeling the transition from ohmic to space charge limited current in organic semiconductors. *Organic Electronics* 13: 1700–1709.

30 Sajedi Alvar, M., Blom, P.W.M., and Wetzelaer, G.-J.A.H. (2020). Space-charge-limited electron and hole currents in hybrid organic-inorganic perovskites. *Nature Communications* 11 (1): 4023.

31 Geurst, J.A. (1966). Theory of space-charge-limited currents in thin semiconductor layers. *Physica Status Solidi* 15 (1): 107–118.

32 Song, Y., Bi, W., Wang, A. et al. (2020). Efficient lateral-structure perovskite single crystal solar cells with high operational stability. *Nature Communications* 11 (1): 274.

33 Peng, J., Chen, Y., Zheng, K. et al. (2017). Insights into charge carrier dynamics in organo-metal halide perovskites: from neat films to solar cells. *Chemical Society Reviews* 46 (19): 5714–5729.

34 Peng, W., Aranda, C., Bakr, O.M. et al. (2018). Quantification of ionic diffusion in lead halide perovskite single crystals. *ACS Energy Letters* 3 (7): 1477–1481.

35 Kovalenko, A., Pospisil, J., Krajcovic, J. et al. (2017). Interface inductive currents and carrier injection in hybrid perovskite single crystals. *Applied Physics Letters* 111 (16): 163504.

36 Kalam, A., Runjhun, R., Mahapatra, A. et al. (2020). Interpretation of resistance, capacitance, defect density, and activation energy levels in single-crystalline $MAPbI_3$. *Journal of Physical Chemistry C* 124 (6): 3496–3502.

37 García-Batlle, M., Baussens, O., Amari, S. et al. (2020). Moving ions vary electronic conductivity in lead bromide perovskite single crystals through dynamic doping. *Advanced Electronic Materials* 6 (10): 2000485.

38 Wei, H., Fang, Y., Mulligan, P. et al. (2016). Sensitive X-ray detectors made of methylammonium lead tribromide perovskite single crystals. *Nature Photonics* 10 (5): 333–339.

39 Yakunin, S., Dirin, D.N., Shynkarenko, Y. et al. (2016). Detection of gamma photons using solution-grown single crystals of hybrid lead halide perovskites. *Nature Photonics* 10 (9): 585–589.

40 Xu, Q., Shao, W., Liu, J. et al. (2019). Bulk organic–inorganic methylammonium lead halide perovskite single crystals for indirect gamma ray detection. *ACS Applied Materials & Interfaces* 11 (50): 47485–47490.

41 Shao, Y., Fang, Y., Li, T. et al. (2016). Grain boundary dominated ion migration in polycrystalline organic–inorganic halide perovskite films. *Energy & Environmental Science* 9 (5): 1752–1759.

42 Turedi, B., Yeddu, V., Zheng, X. et al. (2021). Perovskite single-crystal solar cells: going forward. *ACS Energy Letters* 6 (2): 631–642.

43 Wu, B., Nguyen, H.T., Ku, Z. et al. (2016). Discerning the surface and bulk recombination kinetics of organic–inorganic halide perovskite single crystals. *Advanced Energy Materials* 6 (14): 1600551.

44 Meier, R., Birkenstock, C., Palumbiny, C.M., and Müller-Buschbaum, P. (2012). Efficiency-improved organic solar cells based on plasticizer assisted soft embossed PEDOT:PSS layers. *Physical Chemistry Chemical Physics* 14 (43): 15088–15098.

45 Palumbiny, C.M., Heller, C., Schaffer, C.J. et al. (2014). Molecular reorientation and structural changes in cosolvent-treated highly conductive PEDOT:PSS electrodes for flexible indium tin oxide-free organic electronics. *Journal of Physical Chemistry C* 118 (25): 13598–13606.

46 Schlipf, J., Askar, A.M., Pantle, F. et al. (2018). Top-down approaches towards single crystal perovskite solar cells. *Scientific Reports* 8 (1): 4906.

47 Peng, W., Wang, L., Murali, B. et al. (2016). Solution-grown monocrystalline hybrid perovskite films for hole-transporter-free solar cells. *Advanced Materials* 28 (17): 3383–3390.

48 Chen, Z., Dong, Q., Liu, Y. et al. (2017). Thin single crystal perovskite solar cells to harvest below-bandgap light absorption. *Nature Communications* 8 (1): 1890.

49 Alsalloum, A.Y., Turedi, B., Zheng, X. et al. (2020). Low-temperature crystallization enables 21.9% efficient single-crystal $MAPbI_3$ inverted perovskite solar cells. *ACS Energy Letters* 5 (2): 657–662.

50 Alsalloum, A.Y., Turedi, B., Almasabi, K. et al. (2021). 22.8%-Efficient single-crystal mixed-cation inverted perovskite solar cells with a near-optimal bandgap. *Energy & Environmental Science* 14 (4): 2263–2268.

51 Andričević, P., Mettan, X., Kollár, M. et al. (2019). Light-emitting electrochemical cells of single crystal hybrid halide perovskite with vertically aligned carbon nanotubes contacts. *ACS Photonics* 6 (4): 967–975.

52 Lin, R., Guo, Q., Zhu, Q. et al. (2019). All-Inorganic $CsCu_2I_3$ single crystal with high-PLQY (≈15.7%) intrinsic white-light emission via strongly localized 1D excitonic recombination. *Advanced Materials* 31 (46): 1905079.

53 Cheng, X., Jing, L., Yuan, Y. et al. (2019). Centimeter-size square 2D layered Pb-free hybrid perovskite single crystal $(CH_3NH_3)_2MnCl_4$ for red photoluminescence. *CrystEngComm* 21 (27): 4085–4091.

54 Gu, Z., Huang, Z., Hu, X. et al. (2020). In situ inkjet printing of the perovskite single-crystal array-embedded polydimethylsiloxane film for wearable light-emitting devices. *ACS Applied Materials & Interfaces* 12 (19): 22157–22162.

55 Chen, H., Lin, J., Kang, J. et al. (2022). Structural and spectral dynamics of single-crystalline Ruddlesden-Popper phase halide perovskite blue light-emitting diodes. *Science Advances* 6 (4): eaay4045.

56 Zhao, X., Xu, H., Wang, Z. et al. (2019). Memristors with organic-inorganic halide perovskites. *InfoMat* 1 (2): 183–210.

57 Tian, H., Zhao, L., Wang, X. et al. (2017). Extremely low operating current resistive memory based on exfoliated 2D perovskite single crystals for neuromorphic computing. *ACS Nano* 11 (12): 12247–12256.

3

Halide Perovskite Nanocrystals

Samrat Das Adhikari[1], *Andrés F. Gualdrón-Reyes*[1,2], *and Iván Mora-Seró*[1]

[1] *Universitat Jaume I (UJI), Institute of Advanced Materials (INAM), Avenida de Vicent Sos Baynat, s/n, 12071 Castelló de la Plana, Spain*
[2] *Universidad Austral de Chile, Instituto de Ciencias Químicas, Facultad de Ciencias, Isla Teja, 5090000 Valdivia, Chile*

3.1 Introduction

Halide perovskites are, without doubt, one of the most promising and suitable candidates for the future technologies based on energy conversion and production, due to their low-cost preparation and mesmerizing intrinsic properties, such as their high absorption coefficient, low exciton binding energy, improved carrier transport ability, and tuneable optical features [1, 2]. The term "perovskite" denotes a crystalline structure shared by a broad range of materials. The first material for which this term was employed is calcium titanate, $CaTiO_3$, which was discovered by the German minerologist Gustav Rose in 1839 and named by the Russian minerologist Lev Perovsky, followed by the first report in the year 1893 [3]. Beyond materials using O^{2-} as anion, as $CaTiO_3$, halide perovskites present a halide or combination of halides as anions. Halide perovskites were just an academic interest until the last decade of the past century [4], and beyond this period, this material gained interest among the scientific research and engineering communities. In the past decade, this material has proficiently gained interest due to the tremendous enhancement of photovoltaic technology, expanding its potentiality to analogous hot-topic areas such as light-emitting diodes, photonics, photodetectors, and so on [1]. Important studies about defect tolerance, ion migration phenomena, and carrier dynamics have triggered the rapid advance in design of a photoactive material with the desired requirements to fabricate efficient devices [5, 6].

Similar to the ABO_3 metal oxide structure, where the A and B-site cations show an oxidation state of "+2" and "+4," and oxygen exhibits a value of "−2," respectively, halide perovskites depict the ABX_3 formula. Here, A- and B-site cations hold oxidation states of "+1" and "+2" while anions (halides, X = Cl, Br and I) are "−1," respectively. Attending to an ideal cubic structure, metal into the B-site is bonded to six halides forming corner-sharing octahedra type coordination [PbX_6], leaving a cavity for the presence of the A-site cation (see Figure 3.1) [5].

Halide Perovskite Semiconductors: Structures, Characterization, Properties, and Phenomena, First Edition. Edited by Yuanyuan Zhou and Iván Mora-Seró.

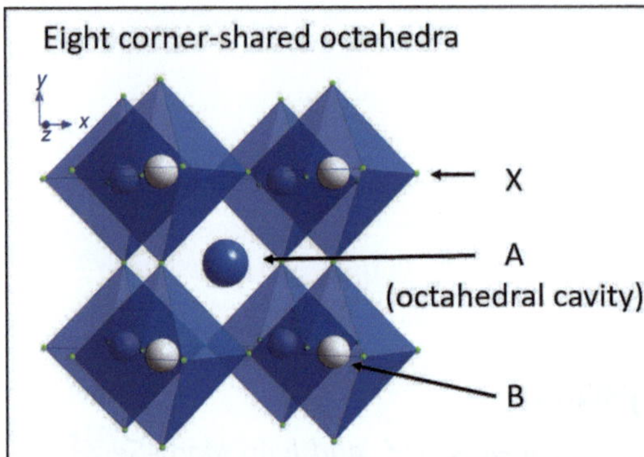

Element	Co-ordinates	Occupancy
A (e.g. Cs/Rb)	0,0,0	8*1/8 = 1
B (e.g. Pb/Sn)	½, ½, ½	1*1 = 1
X (e.g. Cl/Br/I)	½, ½, 0	6*½ = 3

Figure 3.1 Atomic model of typical ABX_3 perovskite unit cell. Table provides the co-ordinates and occupancy of individual atoms of ABX_3.

The synthesis of colloidal halide perovskite nanocrystals (PNCs) has also represented a second major breakthrough after the demonstration of the outstanding photovoltaic performance, as in a relatively easy way nanoparticles with very high photoluminescence quantum yield (PLQY) can be obtained without complex core/shell structures [7]. This synthesis was first demonstrated for hybrid halide perovskites, $MAPbBr_3$, with methylammonium (MA) as A-site cation by Perez-Prieto and coworkers in 2014 [8] and for all-inorganic $CsPbX_3$ perovskites by Kovalenko and coworkers in 2015 [9]. In case of PNCs, the surface atoms are coordinated with organic ligands, where the halides are usually bonded with long-chained alkylamines, while the A- and B-site cations are coordinated with the long-chained carboxylic acids [10]. The surface coordination, for example with the organic acids and amines likely the most extended capping molecules, controls the morphology, helps to provide the particle size distribution in nanometer range, and creates the hydrophobicity over the nanocrystals, which make them suitable for the suspension in nonpolar solvents [1, 11]. In addition, through acid-base reactions, carboxylic acids mediate the protonation–deprotonation of the alkylamines, generating the corresponding alkylammonium carboxylate. This species is the key factor to mediate the PbX_2 dissolution and final PNC stabilization [12]. Alongside, the surface-ligand coordination at the perovskite surface makes these nanocrystals smaller and even lower than their corresponding excitonic Bohr radius (discussed in Section 3.3), providing in this case the quantum confinement effect to the PNCs [13], which in this regime are also known as quantum dots (QDs). In this chapter, most of the topics are discussed with the properties of lead halide PNCs due to simplicity, broad physical phenomena, and wide applicability. First, the synthesis of PNCs following the possible colloidal methods is discussed, followed by the quantum size effects of as-synthesized size-tuneable

PNCs. The soft and dynamic crystallinity of PNCs offers fast halide exchanges, and even the compensation of halide deficiency in the PNCs by post-synthetic treatments is discussed in successive sections. Another important physical phenomenon of PNCs is the formation of different morphologies with the control of reaction parameters, which are discussed accordingly. B-site doping with foreign metals to substitute Pb to obtain dopant emission and enhance the excitonic photoemissions is discussed herein. On the other hand, Pb-free PNCs are also materials of interest due to their identical physical properties to their homologous Pb-based ones, which have been summarized along with the challenges in the last section of this chapter.

3.2 Methodology

The synthesis of PNCs is adapted from the concept of colloidal QD synthesis methodology, where the ligands are introduced for a surface capping. This fact provides them with stability and hydrophobicity over the surface to harvest in nonpolar solvent. A variety of colloidal synthesis methodologies are conducted to prepare PNCs. While most of the methodologies are the bottom-up approaches, top-down ones have also been explored for the synthesis of these materials. Details of the synthesis methodologies are discussed in the following section.

3.2.1 Hot-injection (HI) Method

Hot-injection synthesis method is one of the most convenient procedures for the preparation of PNCs, where the injection of the precursors into a supersaturated solution at high temperatures is achieved. In this context, PNCs can be synthesized following HI approach in two ways: (i) Type I, where metal halide precursor solution is processed in the reaction pot and A-site cation precursor is hot injected. This method is suitable for all-inorganic and hybrid organic–inorganic lead halide perovskite PNCs [9, 14], and (ii) Type II, A-site cation precursors and metal precursors (excluding the halide salts) are processed in the reaction pot, and halide precursors are injected [15]. Herein, $APbX_3$ (A = Cs^+; FA^+, formamidinium) is the most synthesized material for a wide range of applications [16], chosen for the discussion of the reaction procedure and the mechanism. However, this synthetic procedure has also been addressed for carrying out the growth of Pb-free perovskites in NCs as well [17]. The schematic presentation of typical HI synthesis procedure is shown in Figure 3.2a, where colloidal materials with diverse bandgaps (depending on the composition, mainly halide content) with high PLQY are observed. The multicolor emission under the UV-lamp and their corresponding PL spectra are provided in Figure 3.2b,c.

Following the type-I approach, the metal halide precursors have to be dissolved in presence of the suitable ligand mixtures in a stoichiometric ratio prior to the hot injection of the third element. For the preparation of $CsPbX_3$ through the hot-injection synthesis approach, PbX_2 is being introduced to the reaction pot along

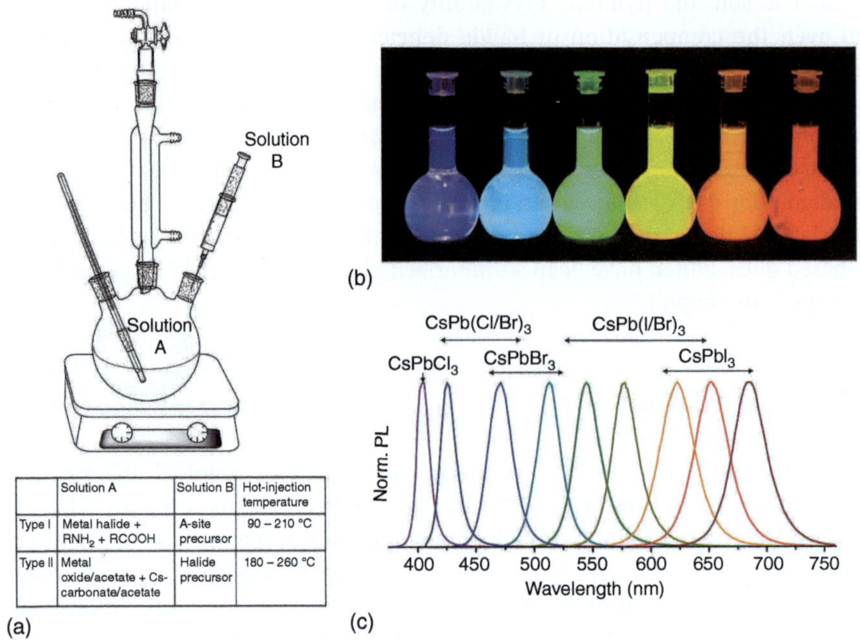

	Solution A	Solution B	Hot-injection temperature
Type I	Metal halide + RNH_2 + RCOOH	A-site precursor	90 – 210 °C
Type II	Metal oxide/acetate + Cs-carbonate/acetate	Halide precursor	180 – 260 °C

Figure 3.2 (a) Schematic representation of hot-injection synthesis for the preparation of PNCs. (b) Multicolor PNCs colloidal solutions and their (c) emission spectra. Source: Figures (b) and (c) are reproduced with permission from Protesescu et al. [9], American Chemical Society.

with oleic acid and oleylamine as the ligands. Upon the introduction of oleylamine (RNH_2) and oleic acid (RCOOH) into the reaction medium, acid-base reactions take place, where oleic acid releases one proton (H^+), which is captured by oleylamine, to form oleylammonium oleate ($RNH_3^+ \cdot RCOO^-$) as an adduct [12, 14, 18]. In the presence of this species, PbX_2 gets dissolute to produce Pb-oleate and oleylammonium halide (RNH_3X) at a suitable temperature (120–180 °C), which is considered the lead halide precursor solution. The ammonium species are able to intercalate into the [PbX_6] octahedral frameworks, transforming the face-share structure of the halide precursor into a corner-sharing structure [19]. When, the PbX_2 is totally dissolved, the third element (A-site cation, e.g. Cs-oleate) is swiftly injected at a desired temperature (which promotes the formation of PNCs of a certain size), followed by fast cooling to quench the reaction (ice cooling) [9, 14]. Noteworthy, the fast-cooling step is mandatory at this stage because of the faster growth of PNCs. A prolonged reaction annealing time might lead to the formation of larger and irregular-sized bulk materials, with high particle size distribution. Typical reaction steps during the formation of PNCs through type-I synthetic route are involved as follows (Eqs. (3.1) and (3.2)) [20]:

$$PbX_2 + RCOOH + RNH_2 \rightarrow Pb(RCOO)_2 + RNH_3X \quad (3.1)$$

$$Pb(RCOO)_2 + RNH_3X + Cs(RCOO) \rightarrow CsPbX_3 \quad (3.2)$$

The reaction parameters need to be considered for (i) Cs-precursor to PbX_2 ratio, (ii) hot-injection temperature, and (iii) role of acid-base ligands.

(i) **Cs-precursor to PbX_2 ratio**: In the synthesis, PbX_2 is used as the Pb and halide source. For the settlement of the stoichiometry of $CsPbX_3$, the Cs-to-PbX_2 ratio should be up to 1 : 1.5 or lower. Higher content of Cs-precursor leads to the formation of Cs-rich phase, Cs_4PbX_6, either exclusively or as a by-product alongside the desired product [21, 22]. In some cases, a low Cs^+ content can trigger the formation of a highly defective PNC structure, considering that oleylammonium cations are consumed to fill A-site cation empty positions. Therefore, a low density of oleylammonium halide species will not be part of the PNC's surface passivation, reducing the stability of the materials.

(ii) **Hot-injection temperature**: This synthetic parameter is controlled to stabilize the precursor solution. Basically, the PbX_2 precursor dissolves in presence of the ligands at 120–150 °C, and the desired reaction temperature varies between 90–200 °C [9, 23]. Reaction temperature is very important to modulate the shape and size of the PNCs (discussed in later section). Besides, reaction temperature should not exceed 200–210 °C to stabilize the mixture reaction; otherwise, lead halide is prone to precipitate (the explanation is discussed in the following subsection "role of acid-base equilibrium").

(iii) **Role of acid-base equilibrium**: The role of organic acids and amines is very crucial for the hot-injection synthesis of PNCs. In this case, oleylamine (RNH_2) and oleic acid (RCOOH) forms react to produce $RNH_3^+{\cdot}RCOO^-$ species in 1-octadecene, an aprotic solvent. As mentioned previously, the as-formed $RNH_3^+{\cdot}RCOO^-$ is pivotal to dissolve lead halides. The formation of $RNH_3^+{\cdot}RCOO^-$ adduct generates an exothermic reaction, and according to the Le Chatelier's principle, acid-base adduct formation equilibrium shifts toward the reactant side with increasing the temperature [12]. Hence, the optimum temperature to carry out such reaction is typically up to 200–210 °C. Beyond 210 °C, $RNH_3^+{\cdot}RCOO^-$ species transforms to free RNH_2 and RCOOH, destabilizing the precursor solution to precipitate ligand-free PbX_2.

Role of excess halide into the reaction medium: Excess halide in the reaction medium helps oleylammonium cation be protonated and mitigates the tendency of proton release even upon increasing the reaction temperature. Introduction of preformed oleylammonium halide into the system prohibits the salting-out of the lead halide precursor (taking into account that more halide species can be introduced into PbX_2 structure to decrease the amount of intrinsic defects) [24], and hence the reaction temperature can be increased up to 260 °C [25, 26]. Alongside, the presence of oleylammonium halide in the reaction medium helps to form smaller PNCs. Basically, negative charge of halide species filling X-sites at the PNCs surface is compensated by the presence of oleylammonium cations, preventing Ostwald ripening (abrupt growth of the PNCs) and decreasing the surface energy [27, 28]. This process is called instant capping, reducing the PNCs dimensions. Then, annealing at high reaction temperature (250 °C) causes the phase transformation of $CsPbX_3$ to $CsPb_2X_5$. The phase transformation can also be prevented by the addition of oleylammonium halide salt [29]. Herein, not only the excess of halides but also long-chained oleylammonium ions at the PNC surface prevent the phase transformation and retain mean size of PNCs even after long-time annealing (30–45 minutes).

Alternatively, another HI reaction methodology (type II) has been developed, where the A-site cation and lead precursors (excluding the halides) are introduced along with oleic acid (RCOOH) to prepare metal oleate solution [14]. This mixture reaction is stable up to 260 °C, and the halide precursor is injected into the metal oleate solution at desired temperature [15, 30]. Typical examples of the halide precursors are: oleylammonium halide, trimethylsilyl halide, benzoyl halide, etc.

3.2.2 Ligand-assisted Reprecipitation (LARP) Method

Alongside the HI synthesis method, Ligand-assisted reprecipitation (LARP) technique is also widely used to prepare PNCs [31]. In such method, the concept of antisolvent-mediated recrystallization is adopted. The key parameter for the antisolvent-mediated recrystallization procedure is the choice of solvents. In a typical method, two different solvents are used: one is denominated polar solvent (good solvent), which helps to dissolving the precursor salts to obtain ions, and another is a nonpolar solvent (bad solvent), which helps to rearrange the ions to recrystallize [32]. These methods are widely followed for the preparation of perovskite thin films. When ligands are introduced to the nonpolar solvent prior to recrystallization process, typical technique is known as LARP. Most of the time these reactions are carried out at room temperature under ambient conditions (sometimes in inert conditions as well, depending on the stability of the precursors and the as-prepared perovskite material). The ligands are used to coordinate with the ions and help with the formation of perovskites in the nanoscale regime. In absence of the ligands, this strategy is useful for forming bulk perovskite materials. In a typical method, AX (A=Cs^+/FA^+/MA^+) and PbX_2 are added to a certain volume of any of the polar solvents (typically DMF or DMSO) to prepare the precursor solution. In another reaction pot, a certain volume of antisolvent (typically hexane/octane or toluene) and ligands are added under continuous stirring conditions. Similar to HI method, long-chained fatty acids and fatty amines are used as the ligands, which coordinate with metals and halides, respectively [31, 33]. In the following step, the precursor solution is added dropwise into the antisolvent-ligand mixture, followed by the reprecipitation of PNCs (see Figure 3.3a). The main parameters are the controlling factors for LARP technique: (i) precipitation temperature, (ii) the choice of the solvent system, (iii) the nature of the capping ligands, (iv) precursor concentration, and (v) water concentration in the precursor solution. As in the case of HI method, the stoichiometry of the precursors is crucial to provide diverse morphologies. Mostly, by modifying the A-cation side/Pb molar ratio, shapes such as nanoplatelets or nanocubes can be produced; reaching different intrinsic properties in the nanocrystals (see Figure 3.3b). Interestingly, through LARP, PNCs synthesis can be scaled up to factor of 50, obtaining high shape control and monodispersity similar to the standard procedure [32].

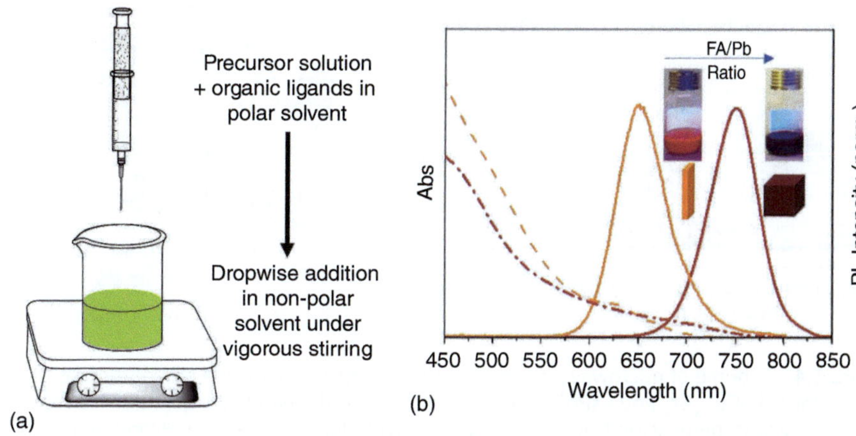

Figure 3.3 (a) Schematic presentation of LARP method for the preparation of PNCs. (b) UV–vis absorption and PL spectra of PNCs obtained by the presence of different FA/Pb content: low FA/Pb, nanoplatelets (orange color), while a high FA/Pb contents produces nanocubes (brown color). Insets exhibit the corresponding colloidal solutions under UV light. Source: Reproduced from Huang et al. [32], © 2019/CC BY 4.0, John Wiley & Sons.

3.2.3 Microwave-assisted Synthesis

Another synthetic route to scale up the formation of PNCs is through the use of microwave (MW)-assisted synthesis. The basic concept of the reaction mechanism is similar to the HI approach. The advantages of MW-assisted synthesis are: (i) initial loading of all precursors in one pot; and (ii) shorter reaction times. Nevertheless, one disadvantage is related to the presence of air during the reaction. Even though it is observed that PNC's growth has a tolerance for the presence of air [34, 35].

Considering the advantages of MW-assisted synthesis method, PNCs are synthesized by controlling the reaction parameters identically to the HI synthesis method. Figure 3.4 provides the schematic presentation of the MW-assisted synthesis of mixed halide $CsPbX_3$ PNCs. In this example, the controlled reaction parameters are the temperature of mixture reaction and the precursor dissolved prior to the synthesis, which yields different shape-selective (nanocubes, nanoplatelets, and nanorods) preparations of PNCs. When the reaction is carried out at 160 °C, nanocubes are formed; by lowering the reaction temperature to 80 °C, nanoplatelets are formed. When the precursors were dissolved previously, prior to the MW heating, nanorods were formed [34].

3.2.4 Ball-milling Process

Mechanochemical synthesis of PNCs is performed in order to benefit from their ionic nature and low crystallization and lattice energy. Mechanochemical synthesis of bulk perovskites can be conducted by grinding the precursors in a mortar and pestle. However, to prepare the colloidal nanocrystals mechanochemically, high-energy ball-milling is a promising method [36, 37]. Here, tiny rigid balls

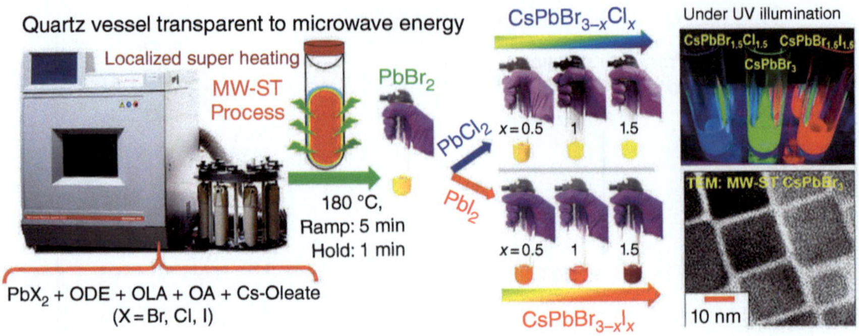

Figure 3.4 Scheme of microwave-assisted synthesis procedure for the preparation of multicolor $CsPbCl_{3-x}Br_x$ and $CsPbBr_{3-x}I_x$ PNCs by varying the halide composition. Source: Reproduced with permission from Thesika and Vadivel Murugan [34], American Chemical Society.

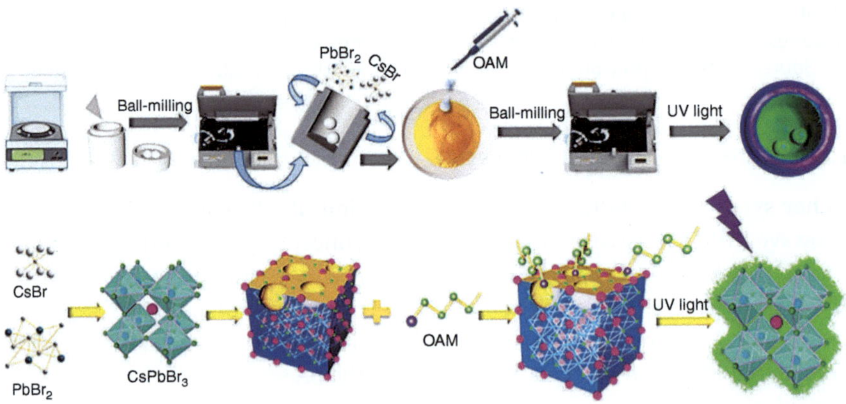

Figure 3.5 Schematic illustration of the mechanochemical synthesis of $CsPbBr_3$ PNCs by grinding the corresponding precursors, CsBr and $PbBr_2$, through ball-milling process. Source: Reproduced with permission from Wang et al. [38], © 2021, Elsevier.

(made of zirconia, corundum, or stainless steel) collide among each other, generating a high pressure to promote the grinding of the material(s) (see Figure 3.5). There two ways to prepare $APbX_3$ PNCs by ball-milling process: (i) mixing an equimolar quantity of A-site cation halide (e.g. CsX/methylammonium halide, etc.) and PbX_2 [38], and (ii) bulk crystal of $APbX_3$ as starting precursor to grind it into small pieces in presence of the organic capping ligands (e.g. oleylamine, oleic acid, oleylammonium halide, etc.) [39]. To obtain uniform PNCs, one of the important parameters is the ball-to-material weight ratio, which needs to be controlled along with the milling time. With a higher ball-to-material ratio, lower milling time is required. In case of $CsPbBr_3$ and $FAPbBr_3$, when the optimum milling time is exceeded, mixed-shaped nanocubes and nanoplatelets are formed. Another parameter is the choice of ligands to control the uniformity of the as-obtained PNCs [39].

3.3 Quantum Confinement Effect

Quantum confinement regime is reached in a nanocrystal when its radius is comparable with the exciton Bohr radius. In this regime, the electron wave function is not just defined by the crystalline structure of the material but also by the border between the nanocrystal and the external media. The most characteristic effect of this regime is the dependence of the bandgap on the nanocrystal size, increasing as the particle size decreases. Excitonic Bohr diameter is defined as the separation between the hole and electron of an exciton (electron-hole pair) [40]. The exciton binding energy remains unchanged until the size of the PNCs is lower than the excitonic Bohr diameter of the semiconductor. When the size of the PNCs is less than excitonic Bohr diameter, the excitonic energy needs to be increased to create the electron-hole pair [9]. Accordingly, PNCs are under quantum confinement regime, and the nanocrystals are referred to as "QDs." Importantly, smaller-sized PNCs cannot be conferred as QDs unless their sizes are smaller than the excitonic Bohr diameter. The quantum confinement effect is well studied for II–VI and III–V semiconductor NCs. The quantum confinement effect is mostly reported for $APbX_3$ ($A=Cs+/MA^+$, FA^+) PNCs [41]. In this context, the excitonic Bohr diameter of $CsPbCl_3$ is 5 nm, $CsPbBr_3$ is 7 nm, and $CsPbI_3$ is 10 nm [9]. While the chalcogenide QDs (e.g. CdSe QDs) form very small spherical dots (that's why these are referred to as "quantum dots"), $CsPbX_3$ PNCs form a variety of shapes, e.g. nanocubes, nanoplatelets, nanowires, etc. In case of the chalcogenide QDs, the diameter of the dots is considered to compare with excitonic Bohr diameter. In case of the PNCs, the edge length of the nanocubes is considered to compare, whereas the thickness and the width are considered for nanoplatelets and nanowires, respectively (Figure 3.6) [13]. In the next subsections, a few examples are discussed about synthetic control and the excitonic energy tailoring effects of different shaped $CsPbBr_3$ NCs.

3.3.1 Nanocubes

The synthetic control over the average size of the PNCs can be performed by the variation of X-to-Pb ratio during the synthesis. With the increase of the halide concentration into the solution, the average size of the PNCs reduces. For instance, the bromide concentration can be mediated using a foreign metal bromide salt (e.g. $ZnBr_2$) or oleylammonium bromide (OLAm-Br) salt along with the lead bromide precursor [42]. The size tailoring effect is justified by the thermodynamic equilibrium between the bromide ions in solution and the as-formed PNCs, which indeed controls their further growth [43]. Another argument in light of OLAm-Br, is that this species provides an excess of bulky-oleylammonium ions to replace the Cs^+ ions at the surface of the PNCs and hinders further growth of the PNCs (explained by the emergence of instant caping phenomenon, see above). An example of quantum size effect of $CsPbBr_3$ nanocubes is presented in Figure 3.7a, where the excitonic absorption and emission have a hypsochromic shift with the decrease of nanocube edge length [28]. Noteworthy, similar size control effect is also available for other halide counterparts (e.g. Cl and I), while $CsPbI_3$ nanocubes also

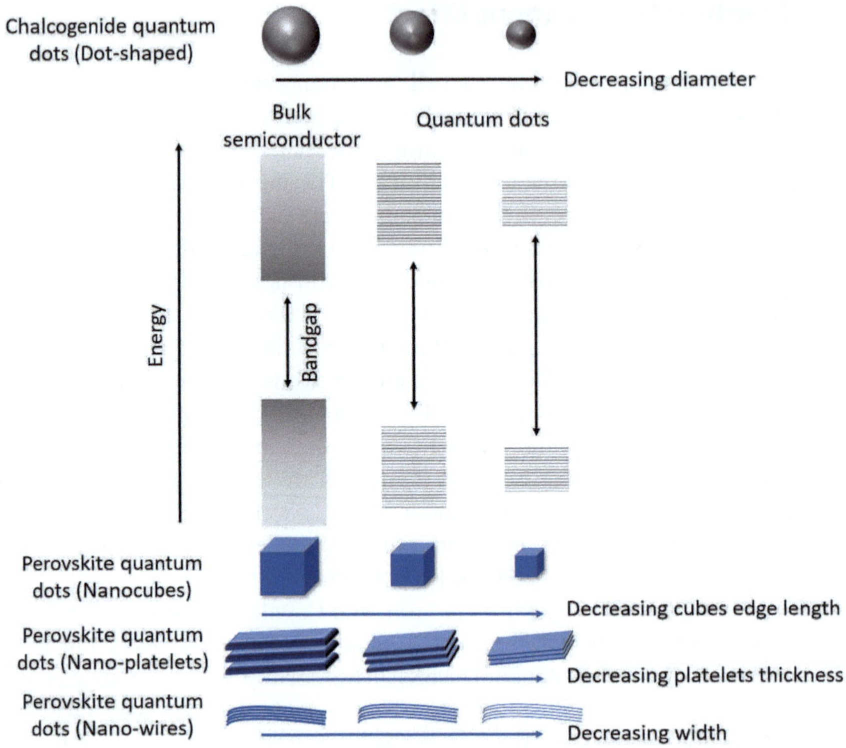

Figure 3.6 Schematic presentation of quantum confinement effect of chalcogenide semiconductors (top) and perovskite quantum dots with different shapes (bottom).

provides similar quantum confinement regime [46], it is very difficult to observe the quantum size effect in $CsPbCl_3$ PNCs because of their lower excitonic Bohr radius (5 nm). For the case of $CsPbI_3$ nanocubes, synthetic parameters such as temperature of reaction, mediated by a suitable amount of capping ligands is pivotal to control the PNCs size, and thereby, modulate the quantum confinement regime [18].

3.3.2 Nanoplatelets

Tuning of thickness of the nanoplatelets is the major concern to embellish the quantum confinement effect. The major synthetic parameter is the reaction temperature to control the thickness of the $CsPbX_3$ nanoplatelets. While $CsPbBr_3$ nanocubes are synthesized at 150–200 °C, a relatively lower temperature diapason (90–130 °C) is favorable to obtain thickness tuned nanoplatelets. Indeed, the low temperature reaction promotes the anisotropic growth producing two-dimensional (2D) geometry. Lowering the reaction temperature, one can obtain ultrathin $CsPbBr_3$ nanoplatelets (even only 1 atomic layer of corner-shared $[PbBr_6]$ octahedra). Figure 3.7b provides the typical example of quantum confinement effect of $CsPbBr_3$ nanoplatelets having a strong increase of the excitonic energy with the decrease of nanoplatelet thickness [44].

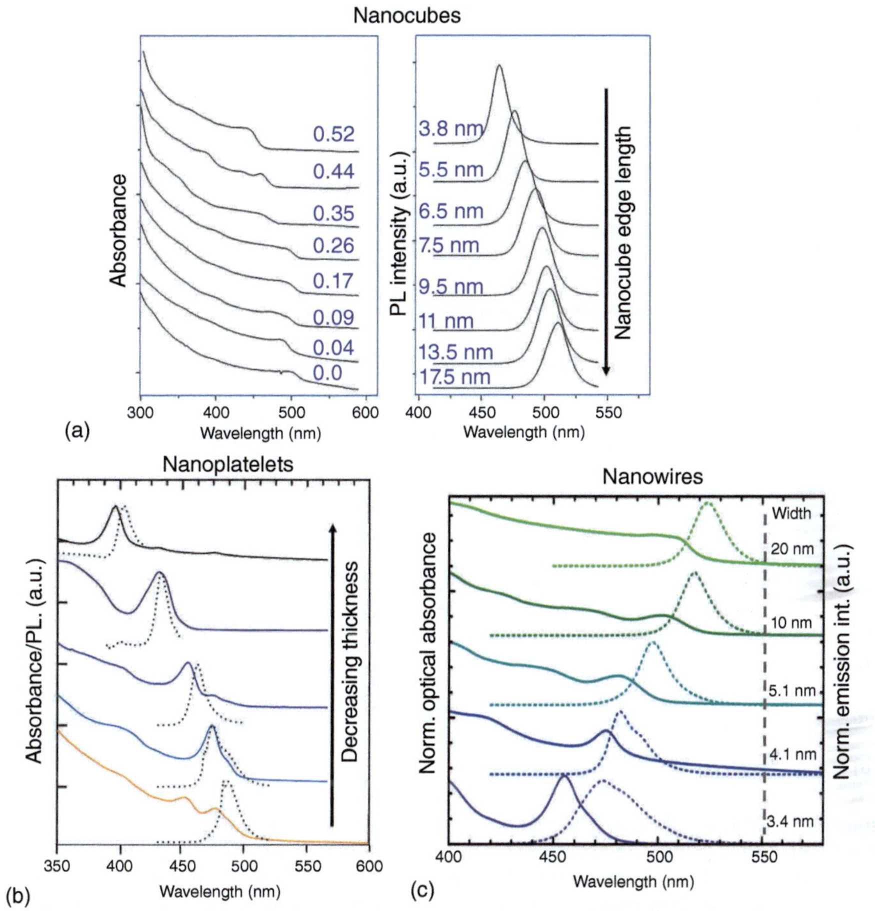

Figure 3.7 Quantum confinement effect of LHP NCs: absorbance and PL spectra of (a) $CsPbBr_3$ nanocubes, (b) $CsPbBr_3$ nanoplatelets, and (c) $CsPbBr_3$ nanowires. Source: Figure (a), (b), and (c) is adapted from Refs. [28, 44, 45] with permissions from ACS publications.

3.3.3 Nanowires

Similar to the nanoplatelets, quantum confinement of the nanowires can be achieved by tuning the width of the nanowires below the excitonic Bohr diameter. An experimental result of such effect in nanowires is provided in Figure 3.7c, where the blue shift of the excitonic absorbance and emission spectra is obtained with the shrinking of the nanowire width [45].

3.4 Solution-processed Halide Exchange

The halide ions of PNCs possess lower activation energy for the substitutional halide exchange reaction [47]. Thereby, with the addition of different halide

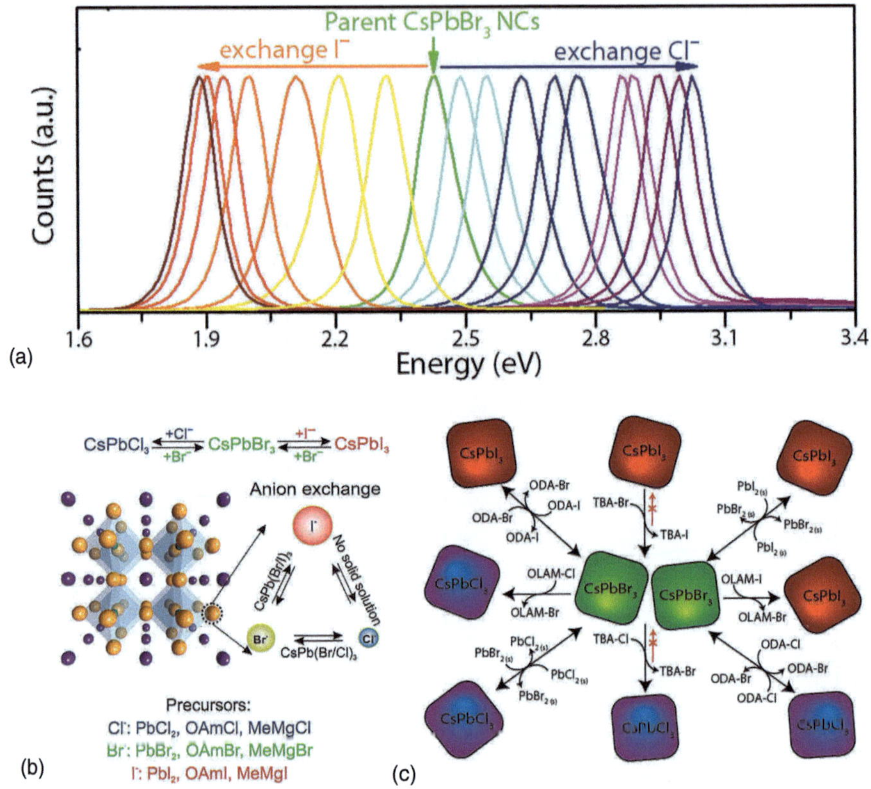

Figure 3.8 (a) A set of PL spectra of partially and completely halide-exchanged $CsPbX_3$ (X = Cl, Br, I) NCs. (b, c) Schematic presentations of solution-processed halide exchange steps with different halide precursors. Source: (a) and (c) are reproduced from Akkerman et al. [48]/American Chemical Society, and (b) is reproduced from Nedelcu et al. [23] with permission of American Chemical Society.

precursors, mutual fast exchanges occur among the halides at PNC surface, forming mixed halide perovskites, with emissions into a new spectral window. In this context, diverse luminescent PNCs can be produced by adding different halide precursor to an alternative PNCs solution. While only blue, green, and red emissions are obtained from the pure $CsPbCl_3$, $CsPbBr_3$, and $CsPbI_3$ PNCs, mixed halide perovskites possess whole PL features in the entire visible optical window (see Figure 3.8a) [23, 48]. Then, halide exchanges can also be performed by mixing individual perovskite PNCs solutions as well. As shown in Figure 3.8b, to obtain an emission from mixed halide perovskites, there must have to be presence of consecutive halides, e.g. Cl–Br or Br–I. Mixing of Cl–I system never produces any hybrid emission, which involves structural stress in the perovskite lattice during the exchanges between Cl and I [23]. A set of different halide precursors are mentioned in Figure 3.8b,c, which are useful for a successful halide exchange.

3.5 Post-synthesis Defect Recovery

The surface of the PNC is very sensitive toward the loss of halides. The surface halide vacancy creates structural defects in the crystal and hence, quenching their PLQY. Halide vacancies arise during synthesis and/or purification step. While the PLQY of red-emitting $CsPbI_3$ perovskite PNCs almost retains after the purification step, a noticeable loss is observed for $CsPbBr_3$ PNCs, and $CsPbCl_3$ PNCs providing poor PLQY even before the purification step especially in the case of $CsPbCl_3$ [49]. The main reason behind the low PLQY of $CsPbCl_3$ is the presence of deep trap states in its large bandgap, while for other PNCs with Br or I with narrower bandgap these defects just produce shallow traps. The deep trap states favor the nonradiative recombination dynamics [50], and consequently lowering of PLQY. In contrast, the halide deficiency in case of $CsPbBr_3$ and $CsPbI_3$ is related with the emergence of shallow-defect states, where the electrons are weakly localized and holes are delocalized in the bandgap due to their relatively narrower bandgap [50].

With the purpose of improving the photophysical properties of highly defective PNCs, as in the case of $CsPbCl_3$, several postsynthetic treatments to compensate for the density of halide defects are employed [7]. One of them is the use of metal halides or nonmetal halides. For instance, after adding metal chlorides or nonmetal chlorides to $CsPbCl_3$ PNCs, PL intensity increases sharply up to several multiplication order (see Figure 3.9) [49]. In this case, deep trap states are filled, hindering the nonradiative recombination channel and facilitating the radiative recombination pathway. On the other hand, Figure 3.10a provides such experiment of quantum confined $CsPbBr_3$ nanoplatelets, formed by different number of monolayers (ML), and their post synthesis defect recovery with the addition

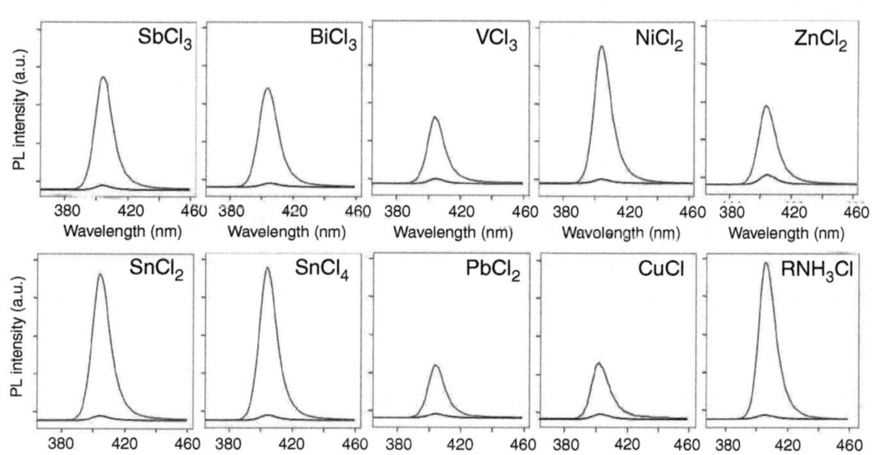

Figure 3.9 PL spectra of as-synthesized $CsPbCl_3$ and multifold intensity enhancement with the addition of several metal chloride salts post-synthetically. Source: Reproduced from Behera et al. [49] with permission of American Chemical Society.

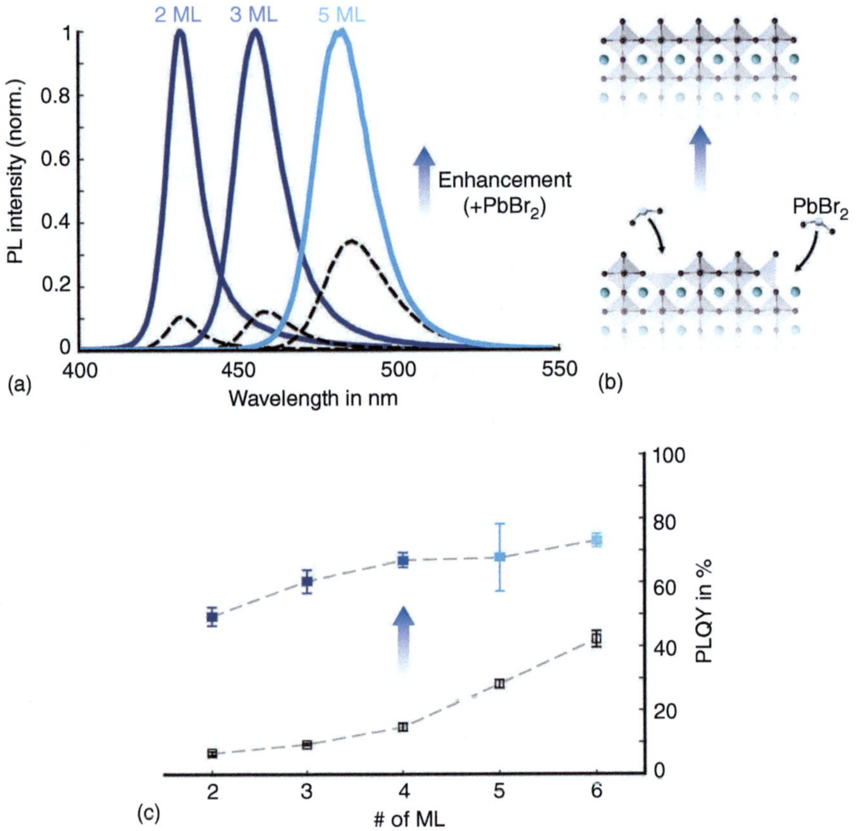

Figure 3.10 (a) PL spectra of quantum-confined $CsPbBr_3$ nanoplatelets and their post synthesis defect recovery with the addition of $PbBr_2$. (b) Schematic presentation of postsynthetic defect recovery process. (c) A plot of PLQY versus nanoplatelet thickness with before and after postsynthesis defect recovery processes. Source: Reproduced from Bohn et al. [51] with permission of American Chemical Society.

of $PbBr_2$ (see Figure 3.10b). In this example, a few ML thicknesses of $CsPbBr_3$ nanoplatelets were synthesized. A lowering of the PLQY is observed with the decrease of the nanoplatelet thickness (MLs), which is eventually enhanced with the addition of $PbBr_2$ postsynthetically (see Figure 3.10c) [51].

3.6 Different Shapes of the Nanocrystals

Ligand capped colloidal PNCs crystallize in diverse shapes following different growth kinetics by altering the reaction parameters (see Figure 3.11). $CsPbX_3$ PNCs typically forms 3D cubic morphology. Apart from 3D structures, they also form 2D platelets and 1D nanorods due to their anisotropic growths and 0D spherical dots. A synthetic control with the modulation of synthetic parameters such as

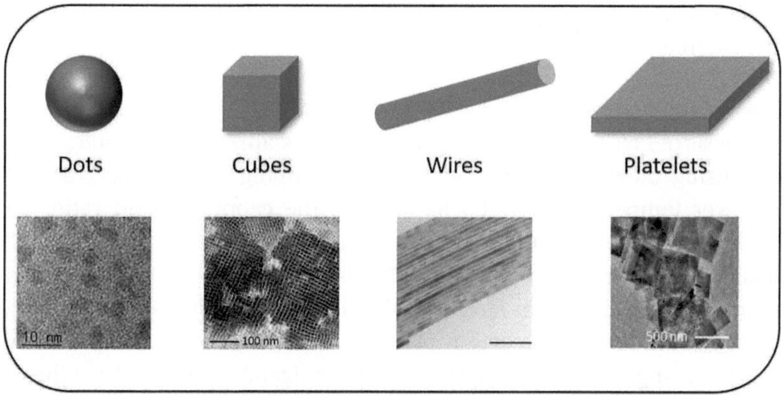

Figure 3.11 Different shapes of LHP NCs follow different growth kinetics. TEM images of dots, cubes, wires, and platelets. Source: Reproduced with permission from Refs. [28, 29, 52, 53], American Chemical Society.

capping ligands [52], B-site doping of metal ions [54], thermal annealing [53], and temperature tuning [44] triggers the formation of different shapes of the nanocrystals.

3.6.1 Shape-controlling Reaction Parameters

3.6.1.1 Temperature

Temperature is an important parameter for the preparation of PNCs following HI approach. Typical reaction temperature is in the range of 140–200 °C for the preparation of $CsPbBr_3$ nanocubes. Indeed, lowering the temperature from 200 to 140 °C, the mean size of the nanocubes decreases. Further lowering the temperature down to 90 °C, asymmetric growth of the PNCs leads to the formation of nanoplatelets instead of nanocubes. In the temperature range of 90–130 °C, thickness of the nanoplatelets tailors instead of lateral dimension. Thereby, following the same generic synthesis protocol, temperature tuning provides the control of edge length and thickness of nanoplatelets and nanocubes, respectively [14]. Temperature is a key parameter to obtain 3D nanocubes at higher range, and lower dimensional PNCs at the lower range [44].

3.6.1.2 Annealing Time

The term "Annealing" refers to high-temperature treatment to change the physical properties of a material. Annealing of $CsPbX_3$ NCs solution at 150–200 °C for 720–960 minutes yields high aspect ratio nanowires. Just after the hot injection of Cs-precursor into the Pb-halide solution, nanocubes are formed instantly. Further heat treatment leads to the attachment of smaller-sized cubes and the formation of nanowires with high aspect ratio. In general, annealing causes the Ostwald ripening of smaller-sized nanocrystals to bigger- sized ones. Then, for the preparation of nanowires, the use of oleylamine as solvent instead of 1-octadecene, leads to the growth of the PNCs only through one dimension [53]. Thereby, annealing time is not an exclusive reaction parameter to control the size, but also other parameters should be considered.

3.6.1.3 Role of Capping-ligand

Colloidal synthesis of PNCs follows ligand-mediated crystallization of smaller-sized nanoparticles using oleylamine and oleic acid as capping ligands. Synthesis of $CsPbBr_3$ at lower temperature (90 °C) by altering the capping-ligand ratio, the shape of the PNCs is varied, including 0D dots (single and lamellar structures), 2D nanoplatelets (stacked by self-assembly), and 2D flat-lying nanosheets [14, 52].

Furthermore, altering the chain length of the ligands promotes shape modulation, keeping constant all other reaction parameters. if oleylamine is replaced by octylamine, shortening the chain length of capping ligand, during HI synthesis, nanoplatelets are formed instead of nanocubes [28].

3.7 Doping in Perovskite Nanocrystals

Doping in perovskites is defined by the substitution of former element by a new foreign atom. In each of the three sites of halide perovskites (A, B, and X-site), doping is vastly studied with the exploration of new optoelectronic properties, defect recovery, and phase-stability (see Figure 3.12). In case of A-site doping, the only possible elements are the alkali metal ions (Li, Na, K, and Rb) and organic alkylammonium (e.g. methylammonium) or amidinium (e.g. formadinium), which retain their optoelectronic and structural identity [55]. The suitable element for X-site doping is thiocyanate (SCN), which is a pseudohalide to substitute corresponding halide ions, with a noticeable PL shifting and size tuning of PNCs [56]. Needless to say, in comparison to B-site doping, the influence of dopant ions is not so strong in A-site and X-site doping. In the following subsections, discussions are made of synthetic strategies of possible B-site doped PNCs, their influences on optoelectronic, phase-stability, etc., and their synthetic strategies.

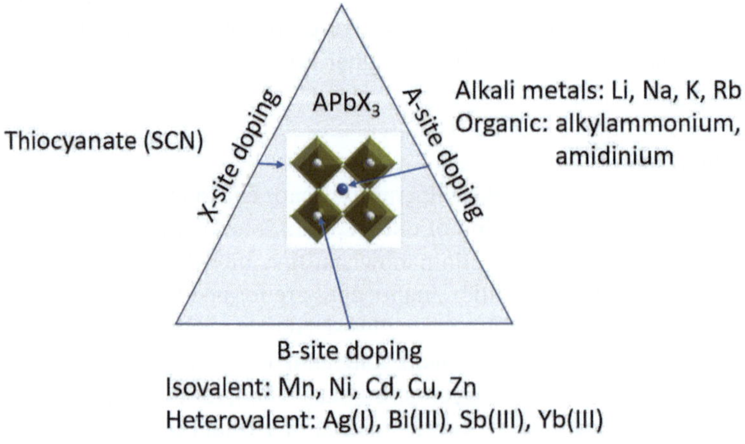

Figure 3.12 Sketch of dopant elements in each of the sites of $APbX_3$ perovskite.

3.7.1 Mn^{2+} Doping

A few percentages of Mn doping in high band gap semiconductors results from an excitonic energy transfer to the Mn(II) dopant state, which results from a broad PL (bandwidth = 75 nm) with millisecond-order lifetime and short-range tuneable (580–625 nm) and yellow–orange emission originated from the spin flipped d–d transition ($^4T_1$-$^6A_1$). The criteria to induce Mn(II) doping in perovskite semiconductors is high-bandgap of the host semiconductor to transfer efficient excitonic energy to the Mn(II) dopant [57]. For the same reason, Mn(II) dopant emission is only available efficiently from $CsPbCl_3$ [57], $CsPb(Cl/Br)_3$ [58], and quantum confined $CsPbBr_3$ NCs [59]. However, doping of Mn(II) can also be performed successfully in bulk-$CsPbBr_3$ and $CsPbI_3$ NCs as well, which only provides an extended stability without any Mn(II)-dopant related optical emission. The main reason behind the unfeasibility of Mn dopant emission from bulk-$CsPbBr_3$ and $CsPbI_3$ is their low excitonic energy to activate the Mn-dopant centers and reach the d–d transition. In summary, the influences of Mn(II) dopant in PNCs are: (i) new optical emission originated from the dopant centers [57], (ii) stability of the PNCs [60], and (iii) average particle size reduction [61]. Mn(II) coordinates with halides octahedrally (MnX_6) into the lattice identical to PbX_6 octahedra. The ionic radius of Pb^{2+} is 119 pm, and Mn^{2+} is 80 pm, thereby the substitution of smaller-sized [MnX_6] with [PbX_6] in PNCs influences the lattice shrinking and reduction of average particle size. Alongside, Mn(II) incorporation in $CsPbI_3$ reflects an enhancement of Goldsmith tolerance factor to prevent α-to-δ phase transformation and hence generates the stability of the black phase $CsPbI_3$. Mn dopant emission is tuneable and dependent on Mn^{2+} concentration in lattice. With the increasing of dopant ion concentration causes bathochromic shift of dopant emission. Under excess dopant content in $CsPb(Mn)Cl_3$ PNCs (more than 40% at the B-site), a little hypsochromic shift in excitonic emission is also observed [61]. The reason behind the hypsochromic shift of the excitonic emission can be explained by the weak confinement effect, where extreme Mn concentration in $CsPb(Mn)Cl_3$ causes the reduction of average particle size. The example of dual-emitting Mn(II) doped $CsPbCl_3$ NCs is provided in Figure 3.13a, synthesized at 150–210 °C [61]. A schematic sketch of excitonic energy transferred Mn(II) dopant emission transition is shown in Figure 3.13b [62].

3.7.2 Lanthanide Doping

Similar to Mn-dopant, lanthanides are another option to provide characteristic dopant emission from PNCs. Here, lanthanide halides can also offer octahedral coordination (LnX_6), which can successfully replace the PbX_6 octahedra. Herein, the origin of the emission is excitonic energy transferred f–f transition of the lanthanide dopants. Figure 3.13c,d provides a set of absorption and emission of a series of Ln^{3+} doped $CsPbCl_3$ PNCs, where the absorption is absolutely from the host excitonic transition, no trace of dopant related absorption, and the emission from the host excitonic transition as well as from the f–f transition from the

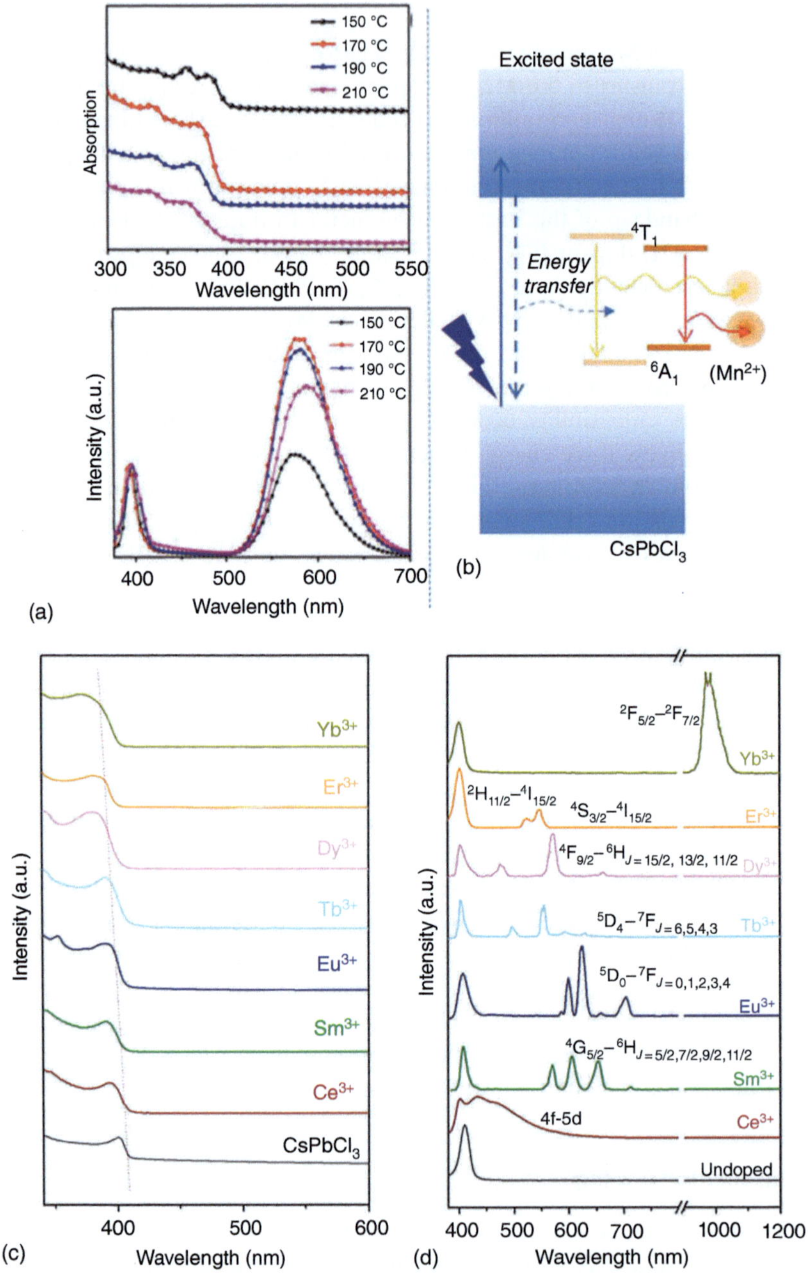

Figure 3.13 (a) Absorbance and PL of Mn(II) doped $CsPbCl_3$ PNCs following HI synthesis approach at different reaction temperatures. (b) Schematic presentation of excitonic energy transfer to the Mn(II) dopant state is followed by spin-flipped transition from $^4T_1$ to $^6A_1$ state. Source: Figures (a) and (b) are reproduced from Guria et al. [62]/with permission of American Chemical Society. (c) Absorbance and (d) PL of various lanthanide-doped $CsPbCl_3$ NCs. Source: Figures (c) and (d) are reproduced from Pan et al. [63]/with permission of ACS publications.

dopants. Indeed, this observation is evidence of host-sensitized lanthanide dopant emission from Ln^{3+} doped $CsPbCl_3$ PNCs [63]. Among the lanthanides, Yb^{3+} is the most suitable dopant for quantum cutting phenomenon, which emits near 990 nm (absorption onset 1.25 eV, more than twice of the host exciton, 2.95 eV), and provides more PLQY close to 200% [64].

3.7.3 Other B-site Dopants

Apart from the Mn(II) and lanthanide doping, other dopant ions are also proven to have successful Pb-site doping. Main effect of those dopant ions is to provide extended stability, which is very supportive to stabilize the α-phase of $CsPbI_3$. $CsPbBr_3$ is also shown some extended stability as per the laboratory reports under ambient conditions, and better photovoltaic performances. In case of $CsPbCl_3$, B-site doping with Cu(II), Ni(II), Cd(II) has proven to provide enhanced PLQY, which comparably very low in comparison to the undoped one [65–67]. Hence, alongside the post synthetically chloride salts treated defect recovery, B-site doping is also another useful stratagem to recover the deep-lying defects of high bandgap $CsPbCl_3$ NCs [68]. In case of the iodide-PNCs, the incorporation of Sr(II) as a doping level to replace Pb, can promote the decrease of Schottky defects into the perovskite structure, favoring the fill/replacement of iodide vacancies and thereby, the enhancement of radiative recombination dynamics [24, 69]. The different Pb-site dopants and their corresponding doping effect is summarized in Table 3.1.

Synthesis of doped perovskite PNCs is followed by typical synthesis procedures discussed in the previous section. Typically, dopant ions can be incorporated either in situ or postsynthetically. In situ dopant incorporation is carried out by taking dopant precursors along with the host precursor prior to the synthesis by controlling several reaction parameters such as temperature, reaction time, and supportive additives.

3.7.4 Postsynthesis Doping

Postsynthesis halide exchange of PNCs promotes facile reaction due to their soft ionic crystallinity and low formation energy and the exchange reaction completes within a few minutes. When a metal halide is added to the as-synthesized PNCs, halide exchanges take place followed by the substitution of Pb with corresponding foreign metal ions. In this typical procedure, desired dopant metals can be substituted postsynthetically. The main driving force behind this postsynthesis doping is the halide exchange, which indeed promotes the metal ions to be substituted with Pb. This procedure is known as halide exchange-driven cation exchange (HEDCE) [79]. This method has been studied extensively to dope Mn(II) postsynthetically. In such typical approach, green-emitting $CsPbBr_3$ is treated with $MnCl_2$, which leads to the blue-shifting of the excitonic emission instantly due to the fast formation of mixed halide perovskite. After a few minutes to a couple of hours, the Mn d–d emission arises and intensifies gradually. The postsynthesis doping is not only limited to Mn dopants, but also studied for Yb^{3+}, Cd^{2+}, and Ni^{2+} [68, 78, 80]. Noteworthy, it is

Table 3.1 Different Pb-site dopants and corresponding doping effect.

B-site doped NCs/QDs	PL (nm)	Effect of doping	References
$CsPbBr_3$: Mn[a)]	520	Stability PLQY enhancement	Chen and coworkers [70]
$CsPbBr_3$: Mn (quantum confined host)	<500–520/580	Mn d–d emission Stability PLQY enhancement	Son and coworkers [59]
$CsPbCl_3$: Mn	420/580–625	Mn d–d emission PLQY enhancement (exciton)	Pradhan and coworkers [25] Samanta and co-workers [71]
$CsPbI_3$: Mn[a)]	680	Stability PLQY enhancement	Chen and coworkers [60] Manna and coworkers [72]
$CsPbI_3$:Sb^{3+}[a)]	680	Stability PLQY enhancement	Pradhan and coworkers [73]
$CsPb_2Cl_xBr_{5-x}$: Mn	470/560	Mn d–d emission stability PLQY enhancement (exciton)	Meijerink and coworkers [74]
$CsPbCl_3$:Cd[a)]	410	PLQY enhancement	Samanta and coworkers [68]
$CsPb_{1-x}Sn_xBr_3$[a)]	517	Stability PLQY enhancement (exciton)	Donega and coworkers [75]
$CsPbI_3$: Ag[a)]	- ~780	Stability Enhanced photovoltaic performances	Rogach and coworkers [76]
$CsPbBr_3$: Ce^{3+}[a)]	515	Stability Enhanced photovoltaic performances	Zeng and coworkers [77]
$CsPbCl_3$: Ni^{2+}[a)]	405	PLQY enhancement	Sun and coworkers [66]
$CsPbX_3$: Yb^{3+}	986	Yb f–f emission	Nag and coworkers [78]
$FAPbI_3$: Sr^{2+}	750	Long-term stability Suppression of nonradiative traps PLQY enhancement	Mora-Sero and coworkers [24]

a) Do not provide dopant emission.

not clear or well-studied whether this procedure can be adopted for all other metal halides to incorporate at B-site, and an open question for the future prospects.

3.8 Lead-free Perovskite Nanocrystals

An ongoing demand for lead-free perovskite came to the forefront of current research due to the toxicity concern of Pb-based PNCs, and more importantly, efficient attributes of PNCs are another interest to look for alternatives to have similar characteristics. Sn(II), Sn(IV), Ge(II), Bi(III), In(III), Ag(I), Cu(I), and Te(IV) are the alternative low-toxic or nontoxic B-site substitutes to replace Pb(II) and forms all-inorganic or hybrid organic-inorganic metal halide perovskites [81, 82]. Classifications are made over the possibilities of stoichiometric formation and dimensionality of different perovskite and perovskite-inspired structures: these are 3D ABX_3, identical to the lead-based halide perovskites, $A_2BB'X_6$, double-halide perovskite, 0D A_2BX_6, vacancy-ordered double perovskite, and $A_3B_2X_9$. Discussions are made on considering the several B-site metals that forms these kinds of structural perovskites NCs in following subsections.

3.8.1 Classifications According to the Structure and Compositions

Figure 3.14 provides a few examples of different lead-free PNCs (LF-PNCs) having 3D (upper panel) and 0D (lower panel) structures. Noteworthy, these are not only the exclusive LF-PNCs members, but there are also more examples of these materials. $ASnX_3$ (A=Cs+/MA^+/FA^+) type of perovskites are direct bandgap and very poor band-to-band emitting materials, whereas lead-free double halide perovskites and 0D materials are highly emissive but the origin of the emissions is from the self-trapped excitons. Synthesis of these materials is similar to the Pb-PNCs following mostly HI and LARP techniques. For the preparation of $CsSnX_3$ PNCs, HI synthesis approach is followed by changing the reaction parameters. SnX_2 salts are needed to be solubilized in presence of trioctylphosphene, a mildly reducing agent instead of dissolving in long-chained acid and amines [90]. In case of the other PNCs, typically oleylamine and oleic acid is used as capping ligand. In terms of stability, Sn-based perovskites are very unstable in ambient atmosphere, where Sn(II) readily oxidizes to Sn(IV) and the perovskite compound degrades very fast. Double halide perovskites and vacancy-ordered double halide perovskites contain higher valence B-site metal ions, e.g. Bi(III), Sb(III), Zr(IV), Sn(IV), Te(IV), etc., provide superior stability under ambient condition.

3.8.2 Challenges of the Lead-free Perovskites

(i) Sn(II)-based perovskites issued instability in ambient conditions and hence limits potential applicability. This problem has partially solved by ligand engineering with perfluorooctanoic acid (PFOA). PFOA provides strong electron-withdrawing F- to have an interaction with Sn^{2+}, besides the steric

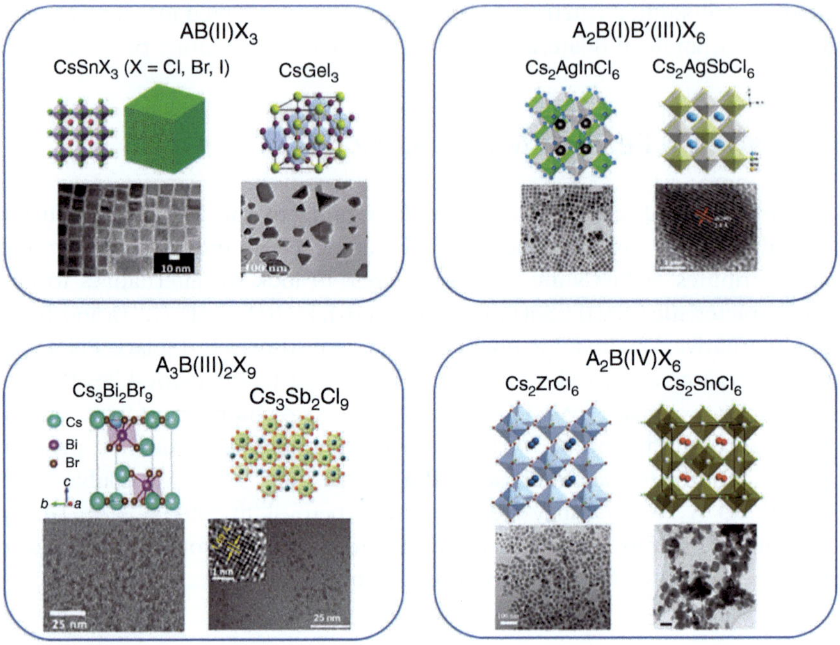

Figure 3.14 Classifications of LFPs in terms of structure and composition, and the example of LFP NCs. Source: Reproduced from Refs. [17, 83–89], American Chemical Society and John Wiley & Sons. Permissions are obtained from ACS publications [17, 83, 87], Wiley publishing group [84–86, 88], and RSC publications [89].

hindrance of PFOA group at the surface of the PNCs combinedly enhances the stability of the PNCs [91].

(ii) Among all listed possible LF-PNCs, theoretical studies predict a series of lead-free material screening with the combination of several A-site cations (both organic and inorganic), B-site metals with halides. However, some of those possibilities are ruled out during their PNCs formation due to the difficulties in formation mechanism in presence of the ligands.

(iii) Poor PLQY from LF-PNCs compared to the bulk perovskites is experienced due to the under-coordinated halide ions at the surface of the NCs, which enhances the number of nonradiative recombination. For the same reason, optoelectronic applications of LF-PNCs are still a few steps behind LHP NCs.

3.9 Summary

This chapter compiles relevant state of the art on synthesis pathways and emphasizes the key role of reaction parameters in the preparation of PNCs. Colloidal PNCs are the point of interest due to their diverse physical properties, chemistry of formation, generation of surface defects and the chemical way of their recovery, outstanding optical properties with high PLQY, etc. Temperature is an important parameter for the PNC synthesis, which controls size and shape of the PNCs to understand the quantum size effect and anisotropic growth of these materials. On the other hand, capping ligands and their ratio is very important to study morphology. Purification of the colloidal PNCs removes excess unreacted ligands and byproducts. However, purification of also leaves halide vacancy at the PNCs surface, creating defect states and quenching the emission PNCs proficiently. Postsynthesis halide treatment for the recovery of defect states is a broad area of interest in this field, discussed herein. A journey of the synthesis of various metal ions doped PNCs is extensively studied and discussed in this chapter. Interestingly, substitution of Pb(II) with other dopant metals enhances optoelectronic performances and reduces the toxic Pb content. Finally, a short outlook of Pb-free perovskites is summarized, which is one of the hot topics in halide perovskite to furnish an eco-friendly and alternative of Pb-based perovskites.

References

1 Dey, A., Ye, J., De, A. et al. (2021). State of the art and prospects for halide perovskite nanocrystals. *ACS Nano* 15 (7): 10775–10981.

2 Frost, J.M. and Walsh, A. (2016). Molecular motion and dynamic crystal structures of hybrid halide perovskites. In: *Organic-Inorganic Halide Perovskite Photovoltaics: From Fundamentals to Device Architectures* (ed. N.-G. Park, M. Grätzel, and T. Miyasaka), 1–17. Cham: Springer International Publishing.

3 Wells, H.L. (1893). Über die Cäsium- und Kalium-Bleihalogenide. *Zeitschrift für anorganische Chemie* 3 (1): 195–210.

4 Mitzi, D.B. (1999). Synthesis, structure, and properties of organic-inorganic perovskites and related materials. *Progress in Inorganic Chemistry* 1–121.

5 Jena, A.K., Kulkarni, A., and Miyasaka, T. (2019). Halide perovskite photovoltaics: background, status, and future prospects. *Chemical Reviews* 119 (5): 3036–3103.

6 Kojima, A., Teshima, K., Shirai, Y., and Miyasaka, T. (2009). Organometal halide perovskites as visible-light sensitizers for photovoltaic cells. *Journal of the American Chemical Society* 131 (17): 6050–6051.

7 Gualdrón Reyes, A.F., Masi, S., and Mora-Seró, I. (2021). Progress in halide-perovskite nanocrystals with near-unity photoluminescence quantum yield. *Trends in Chemistry* 3 (6): 499–511.

8 Schmidt, L.C., Pertegás, A., González-Carrero, S. et al. (2014). Nontemplate synthesis of $CH_3NH_3PbBr_3$ perovskite nanoparticles. *Journal of the American Chemical Society* 136 (3): 850–853.

9 Protesescu, L., Yakunin, S., Bodnarchuk, M.I. et al. (2015). Nanocrystals of cesium lead halide perovskites ($CsPbX_3$, X = Cl, Br, and I): novel optoelectronic materials showing bright emission with wide color gamut. *Nano Letters* 15 (6): 3692–3696.

10 De Roo, J., Ibáñez, M., Geiregat, P. et al. (2016). Highly dynamic ligand binding and light absorption coefficient of cesium lead bromide perovskite nanocrystals. *ACS Nano* 10 (2): 2071–2081.

11 Liu, M., Zhang, H., Gedamu, D. et al. (2019). Halide perovskite nanocrystals for next-generation optoelectronics. *Small* 15 (28): 1900801.

12 Almeida, G., Goldoni, L., Akkerman, Q. et al. (2018). Role of acid–base equilibria in the size, shape, and phase control of cesium lead bromide nanocrystals. *ACS Nano* 12 (2): 1704–1711.

13 Polavarapu, L., Nickel, B., Feldmann, J., and Urban, A.S. (2017). Advances in quantum-confined perovskite nanocrystals for optoelectronics. *Advanced Energy Materials* 7 (16): 1700267.

14 Pan, A., He, B., Fan, X. et al. (2016). Insight into the ligand-mediated synthesis of colloidal $CsPbBr_3$ perovskite nanocrystals: the role of organic acid, base, and cesium precursors. *ACS Nano* 10 (8): 7943–7954.

15 Imran, M., Caligiuri, V., Wang, M. et al. (2018). Benzoyl halides as alternative precursors for the colloidal synthesis of lead-based halide perovskite nanocrystals. *Journal of the American Chemical Society* 140 (7): 2656–2664.

16 Protesescu, L., Yakunin, S., Kumar, S. et al. (2017). Dismantling the "Red Wall" of colloidal perovskites: highly luminescent formamidinium and formamidinium–cesium lead iodide nanocrystals. *ACS Nano* 11 (3): 3119–3134.

17 Locardi, F., Cirignano, M., Baranov, D. et al. (2018). Colloidal synthesis of double perovskite $Cs_2AgInCl_6$ and Mn-doped $Cs_2AgInCl_6$ nanocrystals. *Journal of the American Chemical Society* 140 (40): 12989–12995.

18 Hassanabadi, E., Latifi, M., Gualdrón-Reyes, A.F. et al. (2020). Ligand & band gap engineering: tailoring the protocol synthesis for achieving high-quality $CsPbI_3$ quantum dots. *Nanoscale* 12 (26): 14194–14203.

19 Lee, C., Shin, Y., Villanueva-Antolí, A. et al. (2021). Efficient and stable blue- and red-emitting perovskite nanocrystals through defect engineering: PbX_2 Purification. *Chemistry of Materials* 33 (22): 8745–8757.

20 Zhang, Y., Siegler, T.D., Thomas, C.J. et al. (2020). A "Tips and Tricks" practical guide to the synthesis of metal halide perovskite nanocrystals. *Chemistry of Materials* 32 (13): 5410–5423.

21 Wu, L., Hu, H., Xu, Y. et al. (2017). From nonluminescent Cs_4PbX_6 (X = Cl, Br, I) nanocrystals to highly luminescent $CsPbX_3$ nanocrystals: water-triggered transformation through a CsX-stripping mechanism. *Nano Letters* 17 (9): 5799–5804.

22 Grisorio, R., Fanizza, E., Allegretta, I. et al. (2020). Insights into the role of the lead/surfactant ratio in the formation and passivation of cesium lead bromide perovskite nanocrystals. *Nanoscale* 12 (2): 623–637.

23 Nedelcu, G., Protesescu, L., Yakunin, S. et al. (2015). Fast anion-exchange in highly luminescent nanocrystals of cesium lead halide perovskites ($CsPbX_3$, X = Cl, Br, I). *Nano Letters* 15 (8): 5635–5640.

24 Gualdrón-Reyes, A.F., Macias-Pinilla, D.F., Masi, S. et al. (2021). Engineering Sr-doping for enabling long-term stable $FAPb_{1-x}Sr_xI_3$ quantum dots with 100% photoluminescence quantum yield. *Journal of Materials Chemistry C* 9 (5): 1555–1566.

25 Das Adhikari, S., Behera, R.K., Bera, S., and Pradhan, N. (2019). Presence of metal chloride for minimizing the halide deficiency and maximizing the doping efficiency in Mn(II)-doped $CsPbCl_3$ nanocrystals. *The Journal of Physical Chemistry Letters* 10 (7): 1530–1536.

26 Pradhan, N. (2019). Tips and twists in making high photoluminescence quantum yield perovskite nanocrystals. *ACS Energy Letters* 4 (7): 1634–1638.

27 Fu, Y., Wu, T., Wang, J. et al. (2017). Stabilization of the metastable lead iodide perovskite phase via surface functionalization. *Nano Letters* 17 (7): 4405–4414.

28 Dutta, A., Dutta, S.K., Das Adhikari, S., and Pradhan, N. (2018). Tuning the size of $CsPbBr_3$ nanocrystals: all at one constant temperature. *ACS Energy Letters* 3 (2): 329–334.

29 Dutta, A., Behera, R.K., Dutta, S.K. et al. (2018). Annealing $CsPbX_3$ (X = Cl and Br) perovskite nanocrystals at high reaction temperatures: phase change and its prevention. *The Journal of Physical Chemistry Letters* 9 (22): 6599–6604.

30 Dutta, A., Behera, R.K., Pal, P. et al. (2019). Near-unity photoluminescence quantum efficiency for All $CsPbX_3$ (X=Cl, Br, and I) perovskite nanocrystals: a generic synthesis approach. *Angewandte Chemie International Edition* 58 (17): 5552–5556.

31 Chen, M., Yuan, Y., Wang, Z. et al. (2020). Optical property behaviors of $CsPbBr_3$ colloidal nanoparticles in a ligand-assisted reprecipitation process. *Crystal Growth & Design* 20 (8): 4855–4860.

32 Huang, H., Li, Y., Tong, Y. et al. (2019). Spontaneous crystallization of perovskite nanocrystals in nonpolar organic solvents: a versatile approach for their shape-controlled synthesis. *Angewandte Chemie International Edition* 58 (46): 16558–16562.

33 Jancik Prochazkova, A., Scharber, M.C., Yumusak, C. et al. (2020). Synthesis conditions influencing formation of $MAPbBr_3$ perovskite nanoparticles prepared by the ligand-assisted precipitation method. *Scientific Reports* 10 (1): 15720.

34 Thesika, K. and Vadivel Murugan, A. (2020). Microwave-enhanced chemistry at solid–liquid interfaces: synthesis of all-inorganic $CsPbX_3$ nanocrystals and unveiling the anion-induced evolution of structural and optical properties. *Inorganic Chemistry* 59 (9): 6161–6175.

35 Pan, Q., Hu, H., Zou, Y. et al. (2017). Microwave-assisted synthesis of high-quality "all-inorganic" $CsPbX_3$ (X = Cl, Br, I) perovskite nanocrystals and their application in light emitting diodes. *Journal of Materials Chemistry C* 5 (42): 10947–10954.

36 Kumar, S., Regue, M., Isaacs, M.A. et al. (2020). All-inorganic $CsPbBr_3$ nanocrystals: gram-scale mechanochemical synthesis and selective photocatalytic CO_2 reduction to methane. *ACS Applied Energy Materials* 3 (5): 4509–4522.

37 Zhang, T., Jiang, C., Luo, R. et al. (2021). Capping-ligand free grinding synthesis of luminescent lead halide perovskite nanocrystals. *Materials Today Communications* 26: 101926.

38 Wang, L., Ma, D., Guo, C. et al. (2021). $CsPbBr_3$ nanocrystals prepared by high energy ball milling in one-step and structural transformation from $CsPbBr_3$ to $CsPb_2Br_5$. *Applied Surface Science* 543: 148782.

39 Protesescu, L., Yakunin, S., Nazarenko, O. et al. (2018). Low-cost synthesis of highly luminescent colloidal lead halide perovskite nanocrystals by wet ball milling. *ACS Applied Nano Materials* 1 (3): 1300–1308.

40 Katan, C., Mercier, N., and Even, J. (2019). Quantum and dielectric confinement effects in lower-dimensional hybrid perovskite semiconductors. *Chemical Reviews* 119 (5): 3140–3192.

41 Butkus, J., Vashishtha, P., Chen, K. et al. (2017). The evolution of quantum confinement in $CsPbBr_3$ perovskite nanocrystals. *Chemistry of Materials* 29 (8): 3644–3652.

42 Shamsi, J., Urban, A.S., Imran, M. et al. (2019). Metal halide perovskite nanocrystals: synthesis, post-synthesis modifications, and their optical properties. *Chemical Reviews* 119 (5): 3296–3348.

43 Dong, Y., Qiao, T., Kim, D. et al. (2018). Precise control of quantum confinement in cesium lead halide perovskite quantum dots via thermodynamic equilibrium. *Nano Letters* 18 (6): 3716–3722.

44 Bekenstein, Y., Koscher, B.A., Eaton, S.W. et al. (2015). Highly luminescent colloidal nanoplates of perovskite cesium lead halide and their oriented assemblies. *Journal of the American Chemical Society* 137 (51): 16008–16011.

45 Imran, M., Di Stasio, F., Dang, Z. et al. (2016). Colloidal synthesis of strongly fluorescent $CsPbBr_3$ nanowires with width tunable down to the quantum confinement regime. *Chemistry of Materials* 28 (18): 6450–6454.

46 Zhao, Q., Hazarika, A., Schelhas, L.T. et al. (2020). Size-dependent lattice structure and confinement properties in $CsPbI_3$ perovskite nanocrystals: negative surface energy for stabilization. *ACS Energy Letters* 5 (1): 238–247.

47 Jiang, H., Cui, S., Chen, Y., and Zhong, H. (2021). Ion exchange for halide perovskite: from nanocrystal to bulk materials. *Nano Select* 2 (11): 2040–2060.

48 Akkerman, Q.A., D'Innocenzo, V., Accornero, S. et al. (2015). Tuning the optical properties of cesium lead halide perovskite nanocrystals by anion exchange reactions. *Journal of the American Chemical Society* 137 (32): 10276–10281.

49 Behera, R.K., Das Adhikari, S., Dutta, S.K. et al. (2018). Blue-emitting $CsPbCl_3$ nanocrystals: impact of surface passivation for unprecedented enhancement and loss of optical emission. *The Journal of Physical Chemistry Letters* 9 (23): 6884–6891.

50 Nenon, D.P., Pressler, K., Kang, J. et al. (2018). Design principles for trap-free $CsPbX_3$ nanocrystals: enumerating and eliminating surface halide vacancies with softer lewis bases. *Journal of the American Chemical Society* 140 (50): 17760–17772.

51 Bohn, B.J., Tong, Y., Gramlich, M. et al. (2018). Boosting tunable blue luminescence of halide perovskite nanoplatelets through postsynthetic surface trap repair. *Nano Letters* 18 (8): 5231–5238.

52 Sun, S., Yuan, D., Xu, Y. et al. (2016). Ligand-mediated synthesis of shape-controlled cesium lead halide perovskite nanocrystals via reprecipitation process at room temperature. *ACS Nano* 10 (3): 3648–3657.

53 Zhang, D., Eaton, S.W., Yu, Y. et al. (2015). Solution-phase synthesis of cesium lead halide perovskite nanowires. *Journal of the American Chemical Society* 137 (29): 9230–9233.

54 Hudait, B., Dutta, S.K., Bera, S., and Pradhan, N. (2021). Introducing B-site cations by ion exchange and shape anisotropy in $CsPbBr_3$ perovskite nanostructures. *Nano Letters* 21 (12): 5277–5284.

55 Tabassum, M., Zia, Q., Zhou, Y. et al. (2021). A review on advances in doping with alkali metals in halide perovskite materials. *SN Applied Sciences* 3 (12): 888.

56 Lou, Y., Niu, Y., Yang, D. et al. (2018). Rod-shaped thiocyanate-induced abnormal band gap broadening in SCN−doped $CsPbBr_3$ perovskite nanocrystals. *Nano Research* 11 (5): 2715–2723.

57 Liu, W., Lin, Q., Li, H. et al. (2016). Mn^{2+}-doped lead halide perovskite nanocrystals with dual-color emission controlled by halide content. *Journal of the American Chemical Society* 138 (45): 14954–14961.

58 He, M., Cheng, Y., Yuan, R. et al. (2018). Mn-Doped cesium lead halide perovskite nanocrystals with dual-color emission for WLED. *Dyes and Pigments* 152: 146–154.

59 Parobek, D., Dong, Y., Qiao, T., and Son, D.H. (2018). Direct hot-injection synthesis of Mn-doped $CsPbBr_3$ nanocrystals. *Chemistry of Materials* 30 (9): 2939–2944.

60 Ji, Y., Zhang, J.-B., Shen, H.-R. et al. (2021). Improving the stability of α-$CsPbI_3$ nanocrystals in extreme conditions facilitated by Mn^{2+} doping. *ACS Omega* 6 (21): 13831–13838.

61 Liu, H., Wu, Z., Shao, J. et al. (2017). $CsPb_xMn_{1-x}Cl_3$ perovskite quantum dots with high Mn substitution ratio. *ACS Nano* 11 (2): 2239–2247.

62 Guria, A.K., Dutta, S.K., Adhikari, S.D., and Pradhan, N. (2017). Doping Mn^{2+} in lead halide perovskite nanocrystals: successes and challenges. *ACS Energy Letters* 2 (5): 1014–1021.

63 Pan, G., Bai, X., Yang, D. et al. (2017). Doping lanthanide into perovskite nanocrystals: highly improved and expanded optical properties. *Nano Letters* 17 (12): 8005–8011.

64 Cohen, T.A., Milstein, T.J., Kroupa, D.M. et al. (2019). Quantum-cutting Yb^{3+}-doped perovskite nanocrystals for monolithic bilayer luminescent solar concentrators. *Journal of Materials Chemistry A* 7 (15): 9279–9288.

65 Bi, C., Wang, S., Li, Q. et al. (2019). Thermally stable copper(II)-doped cesium lead halide perovskite quantum dots with strong blue emission. *The Journal of Physical Chemistry Letters* 10 (5): 943–952.

66 Yong, Z.-J., Guo, S.-Q., Ma, J.-P. et al. (2018). Doping-enhanced short-range order of perovskite nanocrystals for near-unity violet luminescence quantum yield. *Journal of the American Chemical Society* 140 (31): 9942–9951.

67 Cai, T., Yang, H., Hills-Kimball, K. et al. (2018). Synthesis of all-inorganic Cd-doped $CsPbCl_3$ perovskite nanocrystals with dual-wavelength emission. *The Journal of Physical Chemistry Letters* 9 (24): 7079–7084.

68 Mondal, N., De, A., and Samanta, A. (2019). Achieving near-unity photoluminescence efficiency for blue-violet-emitting perovskite nanocrystals. *ACS Energy Letters* 4 (1): 32–39.

69 Yao, J.-S., Ge, J., Wang, K.-H. et al. (2019). Few-nanometer-sized α-$CsPbI_3$ quantum dots enabled by strontium substitution and iodide passivation for efficient red-light emitting diodes. *Journal of the American Chemical Society* 141 (5): 2069–2079.

70 Zou, S., Liu, Y., Li, J. et al. (2017). Stabilizing cesium lead halide perovskite lattice through Mn(II) substitution for air-stable light-emitting diodes. *Journal of the American Chemical Society* 139 (33): 11443–11450.

71 De, A., Mondal, N., and Samanta, A. (2017). Luminescence tuning and exciton dynamics of Mn-doped $CsPbCl_3$ nanocrystals. *Nanoscale* 9 (43): 16722–16727.

72 Akkerman, Q.A., Meggiolaro, D., Dang, Z. et al. (2017). Fluorescent alloy $CsPb_xMn_{1-x}I_3$ perovskite nanocrystals with high structural and optical stability. *ACS Energy Letters* 2 (9): 2183–2186.

73 Bera, S., Ghosh, D., Dutta, A. et al. (2019). Limiting heterovalent B-site doping in $CsPbI_3$ nanocrystals: phase and optical stability. *ACS Energy Letters* 4 (6): 1364–1369.

74 Xu, K. and Meijerink, A. (2018). Tuning exciton–Mn^{2+} energy transfer in mixed halide perovskite nanocrystals. *Chemistry of Materials* 30 (15): 5346–5352.

75 van der Stam, W., Geuchies, J.J., Altantzis, T. et al. (2017). Highly emissive divalent-ion-doped colloidal $CsPb_{1-x}M_xBr_3$ perovskite nanocrystals through cation exchange. *Journal of the American Chemical Society* 139 (11): 4087–4097.

76 Lu, M., Zhang, X., Bai, X. et al. (2018). Spontaneous silver doping and surface passivation of $CsPbI_3$ perovskite active layer enable light-emitting devices with an external quantum efficiency of 11.2%. *ACS Energy Letters* 3 (7): 1571–1577.

77 Yao, J.-S., Ge, J., Han, B.-N. et al. (2018). Ce^{3+}-doping to modulate photoluminescence kinetics for efficient $CsPbBr_3$ nanocrystals based light-emitting diodes. *Journal of the American Chemical Society* 140 (10): 3626–3634.

78 Mir, W.J., Mahor, Y., Lohar, A. et al. (2018). Postsynthesis doping of Mn and Yb into $CsPbX_3$ (X = Cl, Br, or I) perovskite nanocrystals for downconversion emission. *Chemistry of Materials* 30 (22): 8170–8178.

79 Huang, G., Wang, C., Xu, S. et al. (2017). Postsynthetic doping of $MnCl_2$ molecules into preformed $CsPbBr_3$ perovskite nanocrystals via a halide exchange-driven cation exchange. *Advanced Materials* 29 (29): 1700095.

80 Shapiro, A., Heindl, M.W., Horani, F. et al. (2019). Significance of Ni doping in $CsPbX_3$ nanocrystals via postsynthesis cation–anion coexchange. *The Journal of Physical Chemistry C* 123 (40): 24979–24987.

81 Fan, Q., Biesold-McGee, G.V., Ma, J. et al. (2020). Lead-free halide perovskite nanocrystals: crystal structures, synthesis, stabilities, and optical properties. *Angewandte Chemie International Edition* 59 (3): 1030–1046.

82 Sun, J., Yang, J., Lee, J.I. et al. (2018). Lead-free perovskite nanocrystals for light-emitting devices. *The Journal of Physical Chemistry Letters* 9 (7): 1573–1583.

83 Jellicoe, T.C., Richter, J.M., Glass, H.F.J. et al. (2016). Synthesis and optical properties of lead-free cesium tin halide perovskite nanocrystals. *Journal of the American Chemical Society* 138 (9): 2941–2944.

84 Men, L., Rosales, B.A., Gentry, N.E. et al. (2019). Lead-free semiconductors: soft chemistry, dimensionality control, and manganese-doping of germanium halide perovskites. *ChemNanoMat* 5 (3): 334–339.

85 Guo, Y., Lou, Y., Chen, J., and Zhao, Y. (2021). Lead-free $Cs_2AgSbCl_6$ double perovskite nanocrystals for effective visible-light photocatalytic C—C coupling reactions. *ChemSusChem* 15 (3): e202102334.

86 Yang, B., Chen, J., Hong, F. et al. (2017). Lead-free, air-stable all-inorganic cesium bismuth halide perovskite nanocrystals. *Angewandte Chemie International Edition* 56 (41): 12471–12475.

87 Yao, M.-M., Jiang, C.-H., Yao, J.-S. et al. (2019). General synthesis of lead-free metal halide perovskite colloidal nanocrystals in 1-dodecanol. *Inorganic Chemistry* 58 (17): 11807–11818.

88 Liu, S., Yang, B., Chen, J. et al. (2020). Efficient thermally activated delayed fluorescence from all-inorganic cesium zirconium halide perovskite nanocrystals. *Angewandte Chemie International Edition* 59 (49): 21925–21929.

89 Das Adhikari, S., Echeverría-Arrondo, C., Sánchez, R.S. et al. (2022). White light emission from lead-free mixed-cation doped Cs_2SnCl_6 nanocrystals. *Nanoscale* 14 (4): 1468–1479.

90 Wong, A.B., Bekenstein, Y., Kang, J. et al. (2018). Strongly quantum confined colloidal cesium tin iodide perovskite nanoplates: lessons for reducing defect density and improving stability. *Nano Letters* 18 (3): 2060–2066.

91 Wang, A., Guo, Y., Muhammad, F., and Deng, Z. (2017). Controlled synthesis of lead-free cesium tin halide perovskite cubic nanocages with high stability. *Chemistry of Materials* 29 (15): 6493–6501.

4

Dimensionality Modulation in Halide Perovskites

Akriti, Jee Yung Park, Shuchen Zhang, and Letian Dou

Purdue University, Davidson School of Chemical Engineering, 480 Stadium Mall Drive, West Lafayette, IN 47907, USA

Low-dimensional semiconductors offer a rich complexity of structural, physical, and electronic properties which has continued to interest the scientific community. Organic, inorganic, and hybrid low-dimensional semiconductors have been extensively researched for their intriguing structure-property correlations [1, 2]. In the field of halide perovskites, the focus on low dimensionality has been mainly motivated by the need to improve the performance reliability and environmental stability of their three-dimensional (3D) counterparts. Low-dimensional halide perovskites are promising materials for future optoelectronics owing to the improved crystallinity, better stability, superior charge transport properties and quantum confinement effects [3].

In this chapter, we will provide a review of the synthesis techniques, characterizations, applications, challenges and prospects of zero-dimensional (0D), one-dimensional (1D), and two-dimensional (2D) halide perovskites.

4.1 Classification of Low-Dimensional Perovskites

Before we dive into the details of synthesis and characterization of low-dimensional perovskites, it is important to spell out the nomenclature used in the chapter for the different classifications of low-dimensional perovskites since a standard protocol is not uniformly used throughout the literature [4–9].

4.1.1 Morphological Low-Dimensional Perovskites Through Size Reduction (ABX_3 Perovskites)

Morphologically low-dimensional perovskites are obtained by size reduction of the 3D perovskite. These low-dimensional perovskites have the same crystal structure, or ABX_3 (A = organic/inorganic cation, B = metal cation, X = Cl/Br/I) building block, as three-dimensional (3D) perovskites, but they are confined in one or

Halide Perovskite Semiconductors: Structures, Characterization, Properties, and Phenomena, First Edition. Edited by Yuanyuan Zhou and Iván Mora-Seró.

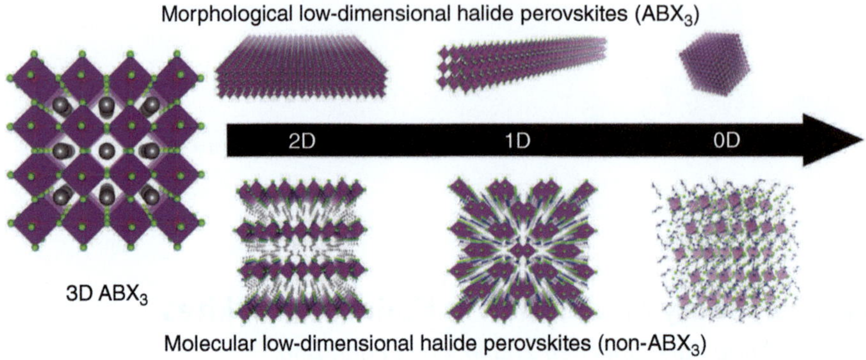

Figure 4.1 Structure of low-dimensional halide perovskites. Morphological low-dimensional perovskites with ABX_3 building blocks and molecular low-dimensional perovskites with $[PbX_6]^{4-}$ building blocks [4].

more directions. Figure 4.1 shows a schematic representation of the morphological low-dimensional perovskites.

4.1.2 Molecular Low-Dimensional Perovskites Through Structure Tuning (Non-ABX_3 Perovskites)

Molecular low-dimensional perovskites are obtained by altering the chemical structure of ABX_3 3D perovskites by introducing organic spacer cations. These low-dimensional perovskites have the same $[PbX_6]^{4-}$ inorganic octahedral building block as 3D perovskites, but the overall structure depends on the intercalating organic cations. Figure 4.1 shows a schematic representation of the molecular low-dimensional perovskites.

4.2 Synthesis and Characterization of Morphological Low-Dimensional (ABX_3) Halide Perovskites

Due to the extensive attention and interest from researchers, numerous synthesis and characterization techniques have been utilized to fabricate and analyze the properties of morphological low-dimensional halide perovskites. However, in this section, we will only highlight few of these innovative approaches. A detailed discussion on morphological low-dimensional perovskites can be found in Chapters 2 and 3.

4.2.1 0D Quantum Dots

0D quantum dots are obtained by size confinement of the ABX_3 perovskite in all three dimensions at the nanometer scale.

In 2016, Dirin et al. demonstrated a facile template-assisted non-colloidal synthesis method for 0D perovskite nanocrystals. The first step of synthesis involved impregnation of a mesoporous silica template with perovskite precursor solution. The excess solution is then removed from the template, and perovskite

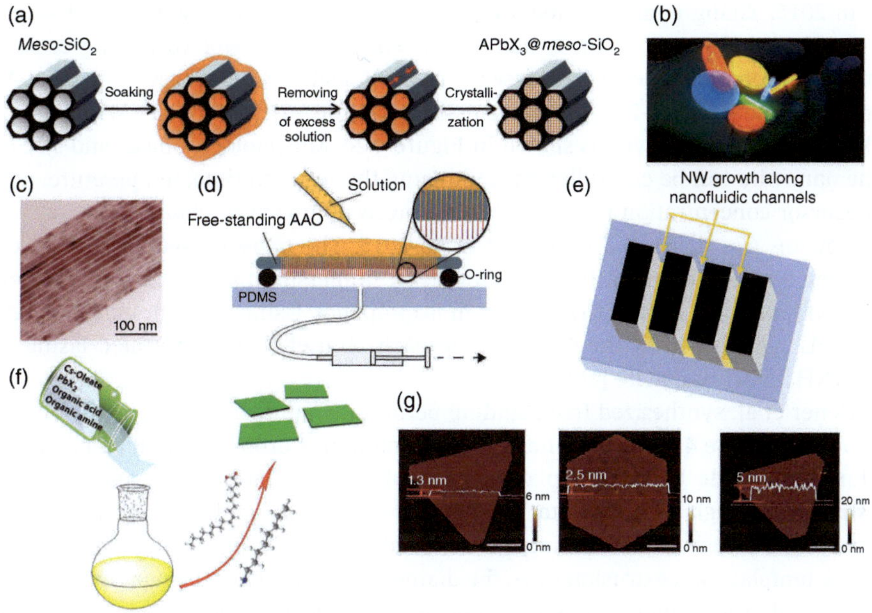

Figure 4.2 Morphologically low-dimensional halide perovskites. (a) Mesoporous template-assisted synthesis of $APbX_3$ nanodots [10]. (b) Hot injection synthesis of luminescent 0D nanocrystals embedded in the polymer matrix [11]. (c) TEM image of $CsPbBr_3$ nanowire synthesized using solution phase [12]. (d) Schematic of free-standing nanowire synthesis by extrusion from anodized aluminum oxide template [13]. (e) Perovskite nanowire synthesis using nanofluidic channels. (f) Ligand-assisted reprecipitation of perovskite nanoplatelets [14]. (g) AFM image of nanosheets synthesized using sequential hot casting and chemical vapor deposition approach [15].

nanocrystals are formed after the solution has dried. A schematic representation of this template-assisted synthesis is shown in Figure 4.2a. This synthesis method can also be used to control the photoluminescence (PL) emission, time-resolved PL, and PL quantum yield of the nanocrystals by varying the pore size of the template [10].

0D nanocrystals have also been synthesized using hot injection method. Protesescu et al. demonstrated the first synthesis of $CsPbX_3$ quantum dots by injecting lead halide (PbX_2) and Cs-oleate in organic solvent maintained at high temperature of 140–200 °C. Nucleation and growth kinetics of this reaction are quite fast, and size tunability of the nanocrystals can be attained by varying the reaction temperature. The robustness and versatility of the synthesis technique are evidenced by the incorporation of the nanocrystals into a poly(methylmethacrylate) (PMMA) matrix. The bright emissions from the perovskite-polymer composites are shown in Figure 4.2b [11].

4.2.2 1D Nanowires

1D halide perovskite nanowires and nanorods are obtained by nanometer-scale size confinement of the ABX_3 crystal across two dimensions.

In 2015, Zhang et al. reported a catalyst-free solution-phase synthesis of single crystalline $CsPbX_3$ (X = Br, I) nanowires. The reaction involved mixing cesium oleate with lead halide in the presence of oleic acid (OA) and oleylamine (OLA) in octadecene at 150–250 °C. Transmission electron microscopy (TEM) image of the synthesized nanowire is shown in Figure 4.2c. Morphology, phase, and size of the nanowire can be controlled by optimizing the reaction time, temperature, and precursor concentration [12]. Another synthesis approach to obtain single crystal nanowires was developed by Zhu et al. in 2015. The two-step synthesis involved coating a glass slide with $PbAc_2$, followed by immersing the coated glass slide in CH_3NH_3X (X = I, Br, and Cl) solution in isopropanol. Using the high-quality single crystals synthesized by the two-step approach, Zhu et al. demonstrated lasing in $CH_3NH_3PbBr_3$ nanowire [16].

Oener et al. synthesized free-standing perovskite nanowires using extrusion. As shown in Figure 4.2d, perovskite precursor solution is drop-casted on an anodized aluminum oxide (AAO) template with double-sided pores. The nanowires are extruded through the template by applying a pressure gradient. The perovskite phase is obtained after annealing the extruded nanowires. By varying the pore size of the template and extrusion time, the diameter and length of the nanowire can be controlled. Additionally, the AAO template can be cleaned and reused for multiple extrusion cycles [13].

A novel synthesis approach for 1D nanowire using open nanofluidic channels was shown by Spina et al. (Figure 4.2e). A supersaturated solution of $MAPbI_3$ in dimethyl formamide (DMF) is first dropped on an etched SiO_2 channeled substrate. After the formation of a nucleation site, a metastable precursor or solvatomorph phase of $MAPbI_3$-DMF is obtained. Finally, solvent-evaporation-induced guided growth leads to the formation of perovskite nanowires. A key merit of this synthesis technique is the ability to control size, cross-sectional shape, aspect ratio, and orientation of the nanowire, which can potentially be adapted for large-scale fabrication [17].

4.2.3 2D Nanoplatelets

Morphological 2D nanoplatelets, nanosheets, and nanodisks are obtained by limiting the size of the 3D halide perovskites in one dimension.

A room-temperature synthesis of perovskite nanoplatelets is shown in Figure 4.2f. The reaction mixture contains Cs-oleate, PbX_2, OA, and OLA. Large size and well-defined square nanoplatelets are grown using ligand-mediated reprecipitation. The choice of organic acid can help in further modulation of the shape and size of the synthesized nanoplatelets [14].

Liu et al. employed a two-step synthesis approach for fabrication of 2D nanosheets. The first step involves casting a saturated PbI_2 aqueous solution onto the substrate. Thermal annealing of the substrate at elevated temperature leads to formation of PbI_2 nanosheets. The second step involves intercalating the PbI_2 octahedron layers with methylammonium iodide (MAI) using chemical vapor deposition, which leads to the formation of 2D perovskite nanosheets. The shape and thickness control of the perovskite nanosheets can be obtained by modulating the annealing temperature, type of solvent, and choice of substrate, as shown in the atomic force microscopy (AFM) images of Figure 4.2g [15].

4.3 Synthesis and Characterization of Molecular Low-Dimensional (Non-ABX_3) Halide Perovskites

As 3D perovskites inherently suffer from phase structural instability due to their ionic nature, molecular low-dimensional perovskites offer a remedy to these limitations. By slicing along the (100) or (110) planes, 2D or corrugated-2D perovskites can be obtained, and 0D and 1D perovskites can be derived by further slicing these 2D structures. Table 4.1 summarizes and compares the different properties offered by these molecular low-dimensional perovskites, including large Stokes shifts and strong exciton binding energies. Thorough understanding of the differences and uniqueness of each dimensional phase is crucial in further expanding the horizons of their application in optoelectronic devices [18–20].

4.3.1 0D

Among molecular low-dimensional (non-ABX_3) halide perovskites, 0D perovskites possess the lowest dimension and highest quantum confinement. As 0D perovskites are basically isolated metal halide octahedra, they require novel synthesis methods different from the conventional methods employed to fabricate their 3D counterparts. Additionally, 0D perovskites exhibit unique photophysical properties derived from their structural peculiarities. Owing to their enhanced stability, the field has gathered more interest in 0D perovskites, especially due to the potential benefits they can offer in commercial applications. In this section, we dive into the details of new synthetic means devised to successfully realize 0D perovskites, analysis of their distinctive photophysical properties, and new prospects for their application in optoelectronic devices [21–23].

4.3.1.1 Synthesis and Properties of 0D Perovskites

Although the field of 0D perovskites is still considered to be in its infancy, various synthetic methods have already been demonstrated. 0D perovskite synthesis techniques can be classified into three main categories: single crystal growth method, colloidal synthesis method, and thin film deposition method. Table 4.2 summarizes the different synthesis methods utilized for 0D perovskites.

The cooling-down method is one of the most common approaches employed to obtain single crystals of 0D perovskites. By simply exploiting the solubility difference between the crystals at different temperatures, high-quality single crystals are obtained through slow cooling of a saturated precursor solution from a high temperature to a lower temperature. 0D $(CH_3NH_3)_3Bi_2I_9$ single crystals were the first of their kind to be synthesized through this cooling-down approach [24]. Following

Table 4.1 Property comparisons of 3D, quasi-2D, corrugated-2D, 1D, and 0D perovskites [18].

Parameter	3D	Quasi-2D	Corrugated-2D	1D	0D (except Cs_4PbBr_6)
Stokes shift (nm)	10–20	10–20	40–200	100–200	100–350
FWHM (nm)[a)]	12–40	10–30	90–250	100–200	100–200
PL QY (%)[b)]	≈100	≈80	≈20	≈30	≈100
E_b (meV)[c)]	20–60	100–400	100–400	100–500	150–600
PL mechanism	FEs[d)]	FEs	FEs and STEs[e)]	FEs and STEs	STEs
Stability	Low	Medium	Medium	Medium	High

a) FWHM – full width at half maximum.
b) PL QY – photoluminescence quantum yield.
c) E_b – exciton binding energy.
d) FE – free excitons.
e) STE – self trapped excitons.

the first attempt, many different 0D perovskites, including species such as Bi and Sb, were realized through the cooling-down method [24, 25].

Single crystals of 0D perovskites can also be realized through seed crystal-assisted constant-temperature evaporation method. In this technique, the precursors are first dissolved in a solution to obtain a seed crystal. The seed crystal is then added to another precursor solution maintained at a constant temperature for a long time to synthesize large high-quality crystals [23].

Hydrothermal method is another conventional method utilized to realize 0D perovskite crystals. For this method, the precursors are first dissolved in water and sealed in an autoclave to be heated to 220 °C for a day. The aqueous precursor solution is then cooled down to room temperature at a constant rate of $3\,°C\,h^{-1}$ [26].

While all the methods mentioned until now utilize temperature control to obtain crystals, precipitation of desired products can also be facilitated through solvent evaporation crystallization method [27]. Antisolvent vapor crystallization is another method devised to grow 0D perovskite crystals. The antisolvent for this method is added to a saturated precursor solution, as shown in Figure 4.3a [31]. The antisolvent can either be directly added to the solution or slowly added as vapor. Finally, a unique synthesis approach called the vertical Bridgman method can also be employed to obtain 0D perovskites. The vertical Bridgman method involves loading the precursors into quartz ampoules, which are then evacuated to 10^{-6} mbar pressure. The quartz ampoules are heated to 250 °C and put into the Bridgman crystal growth furnace to obtain the desired crystals. This method has been specifically reported for obtaining Cs_4EuX_6 crystals [32].

Colloidal synthesis methods provide a facile synthesis means to fabricate perovskite nanocrystals. Hot-injection is one of the conventionally utilized methods for 0D perovskite synthesis, as demonstrated in Figure 4.3b. At high temperature, Cs-oleate precursors are injected into a PbX_2 precursor solution and with fast

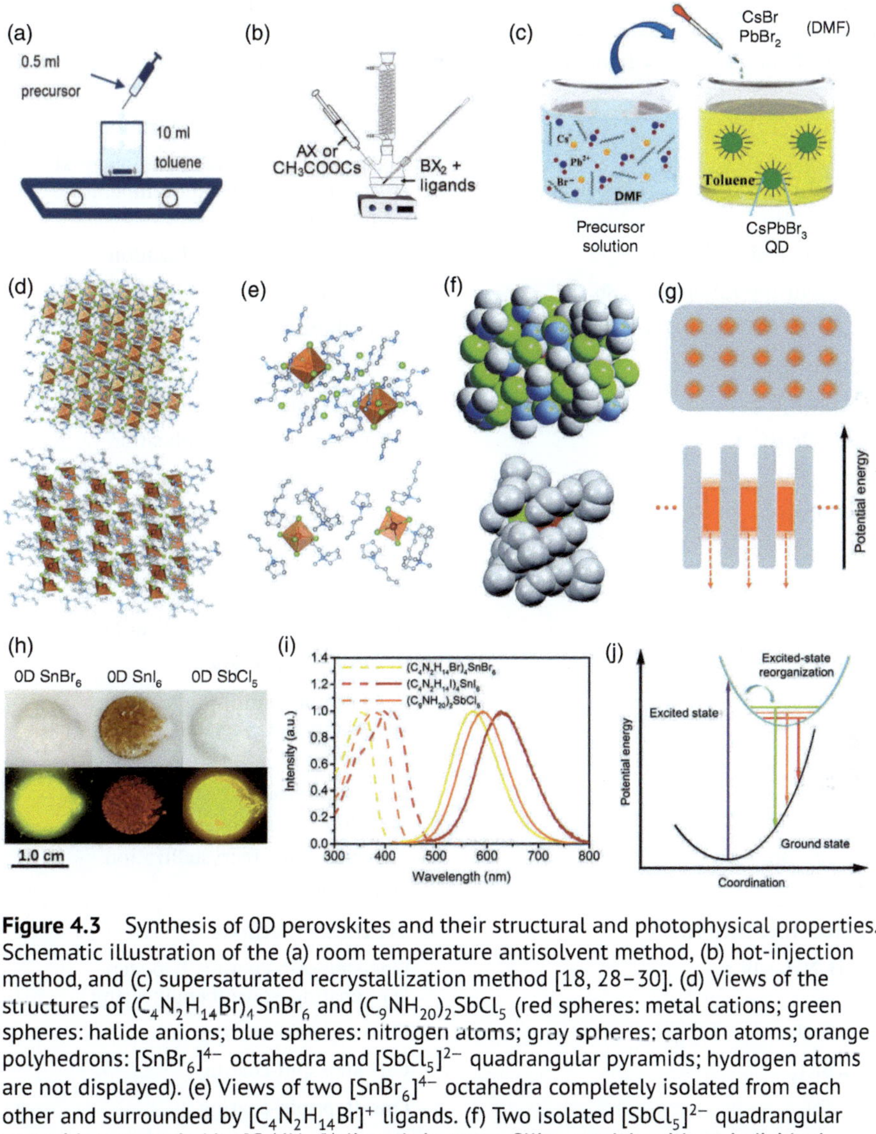

Figure 4.3 Synthesis of 0D perovskites and their structural and photophysical properties. Schematic illustration of the (a) room temperature antisolvent method, (b) hot-injection method, and (c) supersaturated recrystallization method [18, 28–30]. (d) Views of the structures of $(C_4N_2H_{14}Br)_4SnBr_6$ and $(C_9NH_{20})_2SbCl_5$ (red spheres: metal cations; green spheres: halide anions; blue spheres: nitrogen atoms; gray spheres: carbon atoms; orange polyhedrons: $[SnBr_6]^{4-}$ octahedra and $[SbCl_5]^{2-}$ quadrangular pyramids; hydrogen atoms are not displayed). (e) Views of two $[SnBr_6]^{4-}$ octahedra completely isolated from each other and surrounded by $[C_4N_2H_{14}Br]^+$ ligands. (f) Two isolated $[SbCl_5]^{2-}$ quadrangular pyramids surrounded by $[C_9NH_{20}]^+$ ligands in space-filling models with an individual $[SnBr_6]^{4-}$ completely covered by $[C_4N_2H_{14}Br]^+$ ligands and an individual $[SbCl_5]^{2-}$ covered by $[C_9NH_{20}]^+$ ligands. (g) Schematic drawing of a perfect host–guest system with the light-emitting species periodically embedded in an inert matrix and its potential energy diagram. (h) Images of 0D hybrid halide perovskites under ambient light and UV lamp irradiation (365 nm). (i) Excitation (dashed lines) and emission (solid lines) spectra at room temperature. (j) Mechanisms of excited state structural reorganization. The straight and curved arrows represent optical and relaxation transitions, respectively [21].

Table 4.2 Summary of the main synthesis methods for 0D perovskites [18].

Perovskite single crystals	Synthesis methods
$(C_4N_2H_{14}Br)_4SnX_6$ (X = Br or I)	Antisolvent vapor crystallization, solution growth
$(C_4N_2H_{14}Br)_4SnBr_xI_{6-x}$ ($x = 3$)	Antisolvent vapor crystallization
$(CH_3NH_3)_3Bi_2I_9$	Cooling-induced crystallization, seed crystal-assisted constant-temperature evaporation
$(C_8NH_{12})_4Bi_{0.57}Sb_{0.43}Br_7{\cdot}H_2O$	Cooling-induced crystallization
(N-methylpyrrolidinium)$_3Sb_2Cl_{9-9x}Br_{9x}$	Cooling-induced crystallization
$A_2InCl_5{\cdot}(H_2O)$ (A = Rb, Cs)	Hydrothermal
$Rb_7Bi_3Cl_{16}$	Hydrothermal
$Cs_3Sn_3F_2Cl_7$	Hydrothermal
$Cs_2InBr_5{\cdot}H_2O$	Cooling-induced crystallization
$Cs_3Bi_2I_9$	Cooling-induced crystallization
Cs_3BiBr_6	Cooling-induced crystallization
$Cs_3Cu_2I_5$	Solvent evaporation crystallization
Cs_4EuX_6 (X = Br, I)	Vertical Bridgman
Cs_4PbBr_6	Antisolvent vapor crystallization
Perovskite nanocrystals	**Synthesis methods**
Cs_4PbBr_6	Reverse microemulsion method, supersaturated recrystallization, hot injection
Cs_4SnBr_6	Supersaturated recrystallization, hot injection
Cs_4KSnBr_6	Supersaturated recrystallization
$Cs_3Bi_2I_9$	Hot injection
Perovskite thin films	**Synthesis methods**
$(CH_3NH_3)_3Bi_2I_9$	One-step spin-coating, two-step spin-coating, solvent engineering
$Cs_3Sb_2I_9$	One-step spin-coating
$Cs_3Bi_2I_9$	One-step spin-coating
Cs_2TeI_6	One-step spin-coating
$(CH_3NH_3)_3Sb_2I_9$	Two-step spin-coating, solvent engineering

cooling of the solution in an ice-water bath, 0D perovskites are formed [33]. Unlike the hot-injection method, room-temperature colloidal synthesis provides a simpler approach as it does not require an energy-intensive process. Reverse microemulsion method has been employed for synthesizing Cs_4PbBr_5 nanocrystals. Cs-oleate precursor, n-hexane, and OA are mixed in a three-neck flask with mild degassing and N_2 purging. $PbBr_2$ is then dissolved in DMF, HBr, OA, and OLA and injected into the flask, resulting in the swift formation of 0D perovskite nanocrystals [34]. The supersaturated recrystallization method is another room-temperature synthesis technique, as demonstrated in Figure 4.3c. All precursors are first dissolved in DMF solution and then the precursor solution is mixed with large amounts of toluene, which terminates the synthesis reaction [28, 35].

Finally, researchers have also devised methods to fabricate high-quality thin films of 0D perovskites for future device applications. The most common method used to obtain 0D perovskite films is spin-coating followed by annealing. The simplest approach to synthesizing 0D thin films is the one-step spin-coating method, in which the precursor solution is spun onto the substrate [36]. However, due to the chemical reactions and fast crystallization that might occur during the process, a more complex two-step spin-coating approach has also been devised. Two-step spin-coating involves depositing the precursor solution onto the substrate using different spin conditions with intermittent addition of toluene as an antisolvent to aid the crystallization process [37].

The various techniques developed to synthesize single crystals allow us to calculate the precise crystal structure and analyze the structural properties of 0D perovskites. Figure 4.3d–g shows the various views of the crystal structure of $(C_4N_2H_{14}Br)_4SnBr_6$ and $(C_9NH_{20})_2SbCl_5$ [21]. From the 0D perovskite crystal structure, it can be confirmed that complete isolation of the metal halide octahedra is achieved with insulating organic ligands surrounding the photoactive regions. Thus, structurally, in 0D perovskites, electronic bands among neighboring octahedra are not formed. 0D perovskites can also be considered as core-shell structures, which can be better visualized using a space-filling model.

As discussed earlier, the unique structure of 0D perovskites leads to distinctive photophysical properties. Continuing with the examples of $(C_4N_2H_{14}Br)_4SnBr_6$ and $(C_9NH_{20})_2SbCl_5$, it can be seen from Figure 4.3h–j that these 0D perovskites exhibit a very large Stokes shift and broad FWHMs. These broadband emissions are suggested to be from lower energy states, such as self-trapped excitons instead of direct excited states and recombination of free excitons. Generally, reduced dimensionality of materials results in larger bandgaps and higher exciton binding energies due to reduced orbital overlap and strong quantum confinement. The broad emission from different recombination pathways makes 0D perovskites more noteworthy for fundamental analysis and provides opportunities for future application in light-emitting devices and photodetectors.

4.3.1.2 0D Cesium Lead Halides

Although many 0D perovskites investigated early on were mostly lead-free, numerous 0D lead halide perovskites, such as cesium lead halides, have also gained extensive attention. 0D cesium lead halides are composed of isolated octahedra like other 0D perovskites, but instead of lengthy organic ligands, cesium ions stabilize the disconnected octahedral structure. Figure 4.4a–e shows the TEM images, elemental mapping, and crystal structure schematic for the 0D cesium lead halide nanocrystals. Due to the confined nature of 0D perovskites, it is worth contrasting the differences between 0D Cs_4PbX_6 and 3D $CsPbX_3$. It is to be noted that determining the true origins of the optical properties of 0D Cs_4PbX_6 is quite challenging due to contradictory research reports regarding their physical properties.

Contrasting the structural aspects, 3D Cs-based perovskites have a network of octahedra with corner sharing halide anions, but for 0D Cs-based perovskites, the octahedra are completely isolated, resulting in a rhombohedral phase with a greatly distorted and compressed system. The 3D and 0D structures are well represented by Figure 4.4f. Figure 4.4g shows the orbital overlap resulting from the crystal structure. It is notable that while significant orbital overlap exists within the 3D structure, the 0D phase shows no orbital overlap among different octahedra. Thus, as can be seen from Figure 4.4h,i, 0D cesium lead halides exhibit large bandgaps, high exciton binding energies, and distinct densities of states from their 3D counterparts [42].

Another question that has perplexed the research field is the green absorption observed for supposedly pure Cs_4PbBr_6. There have been different proposed explanations for this phenomenon. One explanation is that the formation of defect states leads to the existence of green PL in 0D Cs_4PbBr_6. This is a probable explanation, though it is somewhat contradictory to the description of halide perovskites having the general trait of being defect tolerant. Such defect-tolerant materials exhibiting dominant optical properties derived from defect states have made researchers question the validity of this claim. Another explanation is the formation of $CsPbBr_3$ nanocrystals within the 0D Cs_4PbBr_6 matrix. The small nanocrystals in the medium contribute to the dominant PL from these materials. However, this dispute remains unresolved as properties such as the PL lifetime do not exactly resemble those of a separate $CsPbBr_3$ nanocrystal. Therefore, the origin of the PL behavior could be a synergistic effect, which requires both the 0D and 3D phases [43–45].

Further developments can be expected from this sector by intentionally realizing a 0D/3D composite where the superior optical properties of 3D $CsPbX_3$ can be maintained and stabilized by the 0D Cs_4PbBr_6 components. Due to the insulating energy barriers and low conductivity, Cs_4PbBr_6 alone is considered undesirable for future device applications. $CsPbBr_3$ possesses high PLQY and lasing qualities; however, its stability is an issue. These two materials can also undergo a reversible reaction to be transformed from 0D to 3D and vice versa. Therefore, the potential effort to use them together so that they are complementary to each other seems like a promising approach. As it can be seen from Figure 4.4j–l, with great tunability of the cesium lead halide species, the field still possesses potentially promising qualities for fabrication of efficient optoelectronic devices [39, 46].

4.3.2 1D

1D perovskites, compared to 0D and 2D, remain relatively less investigated. Like 0D perovskites, strong quantum confinement and the easily formed self-trapped excitons lead to broad PL in 1D perovskites. This suggests that 1D perovskites can be potentially employed for devices that target white light emissions. However, the PL mechanism in 1D perovskites is not completely dominated by self-trapped excitons since, unlike in 0D perovskites, the octahedra are not completely isolated. This

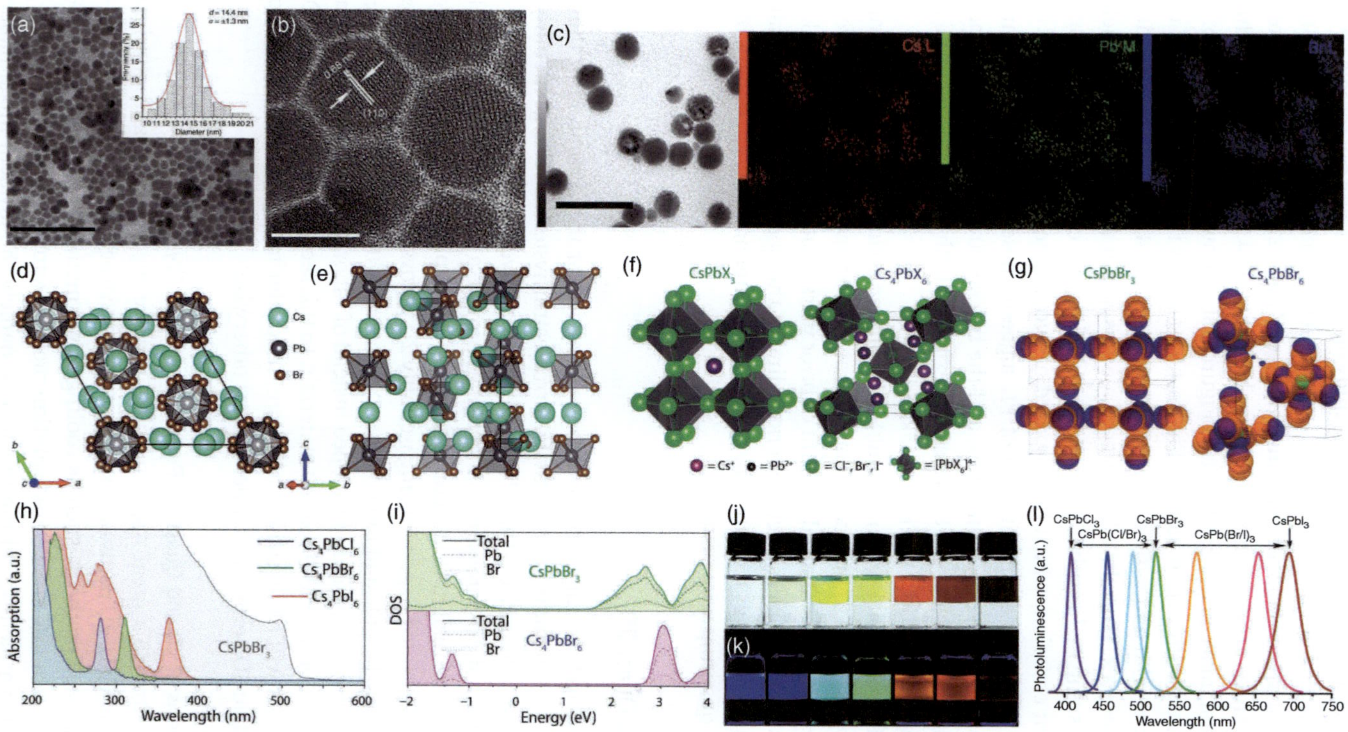

Figure 4.4 Unique properties of 0D cesium lead halides relative to their 3D counterpart. (a) TEM image and the size distribution of Cs_4PbBr_6 nanocrystals (NCs) with Gauss fitting. Scale bar is 100 nm. (b) High-resolution TEM (HRTEM) image of Cs_4PbBr_6 NCs. Scale bar is 10 nm. (c) High-angle annular dark field – STEM (HAADF-STEM) image and elemental mapping (Cs, Pb, and Br) of Cs_4PbBr_6 NCs. Scale bar is 50 nm. Schematic crystal structure of Cs_4PbBr_6 (d) along and (e) perpendicular to (001) [38]. (f) Crystal structures of cubic $CsPbX_3$ and rhombohedral Cs_4PbX_6 (shown in its primitive cubic cell). (g) Comparison of orbital overlap between the p orbitals of each Br^- anion and the s and p orbitals of Pb^{2+} cations, showing the strong orbital overlap of the $[PbBr_6]^{4-}$ clusters in $CsPbBr_3$ and the decoupling in Cs_4PbBr_6 [39]. (h) Comparison of the absorption spectra of Cs_4PbX_6 and $CsPbBr_3$ NCs, showing the large bandgaps and strong excitonic absorption for Cs_4PbX_6. (i) Density functional theory (DFT) and density of state (DOS) calculations for $CsPbBr_3$ and Cs_4PbBr_6 [40]. Photographs of $CsPbX_3$ NCs dispersed in hexane under (j) daylight and (k) UV light ($\lambda = 365$ nm) illumination. From left to right, the NCs are obtained using Cs_4PbCl_6, $Cs_4PbCl_{4.5}Br_{1.5}$, $Cs_4PbCl_3Br_3$, Cs_4PbBr_6, $Cs_4PbBr_{4.5}I_{1.5}$, $Cs_4PbBr_3I_3$, and Cs_4PbI_6 NCs as the precursors. (l) PL emission spectra ($\lambda_{exc} = 400$ nm for all but 350 nm for $CsPbCl_3$ sample) of the corresponding $CsPbX_3$ NCs dispersed in hexane. X is Cl, Cl/Br, Br, Br/I, or I [41].

section will be dedicated to looking at the 1D perovskite structure and its properties, which stands at a midpoint between the 2D phase and the 0D phase.

In 2017, Yuan et al. conducted a thorough study on both bulk and microscale crystalline 1D organic lead bromide perovskite. Looking at an example of $C_4N_2H_{14}PbBr_4$, where the edge-sharing octahedra are lined up linearly and surrounded by organic cations, needle-shape crystal morphology is realized. Figure 4.5a shows the photoluminescent optical image of the crystals, and Figure 4.5b,c shows the TEM image of a single crystal and its diffraction pattern. Figure 4.5d–f shows the crystal structure of the 1D perovskite with and without the ligands, and Figure 4.5g shows the powder X-ray diffraction (PXRD) results of the microscale crystal and the bulk crystal, which agree with the simulated results. Upon UV irradiation, the crystals exhibit strong blue–white light emission, which is composed of two distinct spectral features, as seen in Figure 4.5h–j. The long lifetime of PL suggests these broad emissions are likely from the self-trapped exciton states, and the two spectral features signal that free excitons and self-trapped excitons states coexist and both lead to distinct radiative recombination, as shown in Figure 4.5k,l [47].

Overall, the 1D perovskite system resembles the photophysical properties of corrugated 2D perovskites due to the coexistence of free exciton and self-trapped exciton states. However, due to the limited amount of prior work done regarding molecular 1D perovskites, the field lacks deep understanding of this specific phase. Studying 1D perovskites can serve as a ground for expansion of fundamental understanding regarding bulk quantum wire systems.

4.3.3 2D and Quasi-2D

High-quality perovskite single crystals are very helpful for the evaluation of their intrinsic properties, including conductivity, ferroelectric hysteresis, charge mobility, and carrier diffusion length. On the other hand, the interlayer van der Waals interactions in 2D and quasi-2D perovskite nanocrystals provide a platform to study novel physical phenomena related to structure, such as exciton physics. Therefore, many efforts have been devoted to growing high-quality 2D perovskite crystals. Due to the presence of organic ligands and ionic nature of 2D perovskites, solution-based synthesis methods have been widely developed instead of vapor-phase synthesis, which is typically used for other traditional 2D materials. In this section, the synthesis of 2D perovskite crystals will be discussed, and here, they refer to organic and inorganic hybrid lead halide 2D perovskites unless stated otherwise.

4.3.3.1 Synthesis of 2D and Quasi-2D Perovskites Single Crystal

Single crystals provide a great opportunity to analyze the structure and physical properties of materials. 2D perovskite single crystals have been realized using a variety of solution-based methods, such as cooling down method, co-solvent evaporation, solvent evaporation, and layered solution method, summarized in Table 4.3.

For cooling-down method, all the precursors, namely the divalent metal halide (MX_2) and organic amine halide ($RNH_2{\cdot}HX$), are mixed and dissolved in the water solution at high temperature. Once they are completely dissolved, a continuous

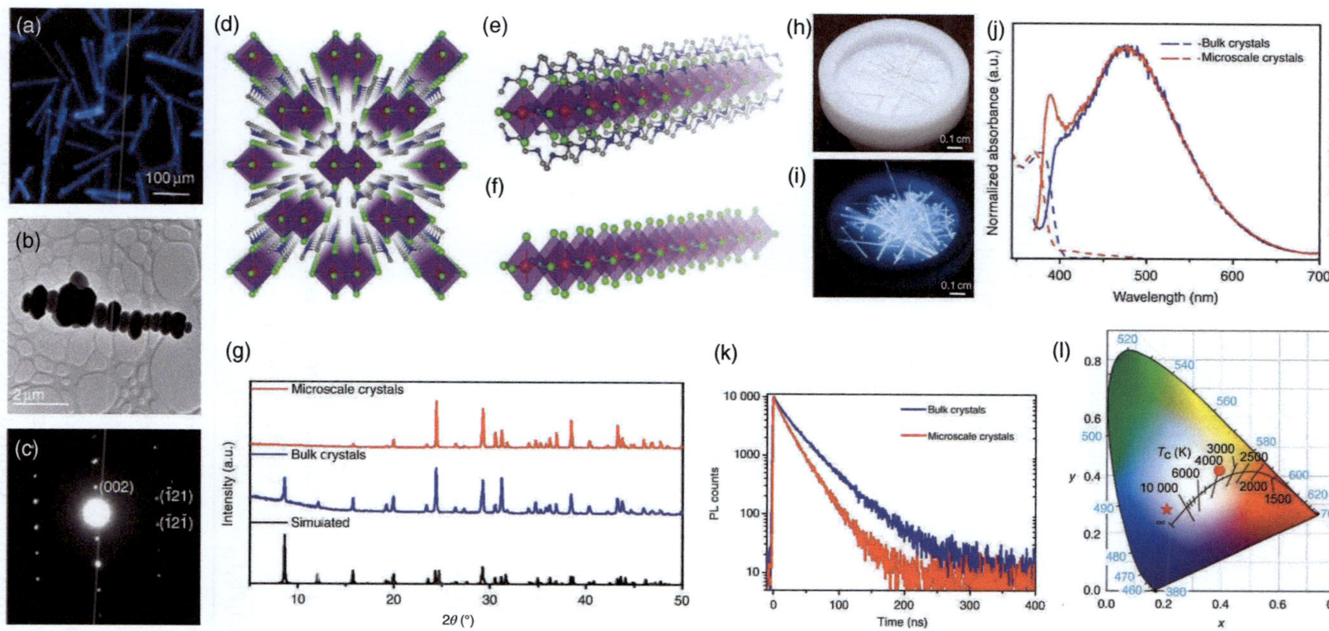

Figure 4.5 Structural and photophysical properties of 1D perovskites. (a) Optical image of needle-shaped 1D lead bromide perovskite single crystals. (b) TEM image of the microscale 1D perovskite crystals. (c) Electron diffraction pattern along [210] zone axis. (d) Structure of 1D perovskite $C_4N_2H_{14}PbBr_4$ (red spheres: lead atoms; green spheres: bromine atoms; blue spheres: nitrogen atoms; gray spheres: carbon atoms; purple polyhedrons: $[PbBr_6]^{4-}$ octahedrons; hydrogen atoms were hidden for clarity). (e) View of an individual lead-bromide quantum wire wrapped by the organic cations. (f) View of an individual lead-bromide quantum wire with edge sharing octahedra. (g) PXRD patterns of the bulk and microscale 1D perovskite crystals, as well as the simulated PXRD patterns based on the single crystal structure. Image of bulk perovskite crystals under (h) ambient light and (i) under UV light (365 nm). (j) Absorption (dashed lines) and emission (solid lines, excited at 360 nm) spectra of the bulk and microscale perovskite crystals at room temperature. (k) The photoluminescence decays in the bulk and microscale 1D perovskite crystals (measured at 475 nm) at room temperature [47]. (l) Comission Internationale de l'Eclairage (CIE) chromaticity coordinates of the 1D perovskites in this work (star), and the corrugated 2D perovskite $(EDBE)PbBr_4$ (circle) [48].

Table 4.3 Summary of synthesis methods for 2D and quasi-2D perovskites single crystals.

2D perovskites (*n* = 1)	Synthesis methods[a)]
$(C_4H_9NH_3)_2PbI_4$	CDM [49–54]; CSE [55]
$(C_6H_5CH_2NH_3)_2PbI_4$	SE [56]; LSM [57–59]
$(C_6H_5C_2H_4NH_3)_2PbI_4$	CDM [51, 60–62]; CSE [63]; LSM [57–59]
$(C_6H_5C_4H_8NH_3)_2PbI_4$	LSM [58]
$(C_6H_{13}NH_3)_2PbI_4$	SE (acetone) [64]
$(C_{10}H_{21}NH_3)_2PbI_4$	CSE (acetone/nitromethane) [65]
$[CH(NH_2)_2][C(NH_2)_3]PbI_4$	CDM [66]
$(C_4H_9NH_3)_2PbBr_4$	SE (DMF) [67]; CDM [51]; CSE [63, 68]
$(C_6H_5CH_2NH_3)_2PbBr_4$	SE (DMF) [67]; LSM [59]
$(C_{10}H_{21}NH_3)_2PbBr_4$	CSE (H_2O/nitromethane) [65]
(N-MEDA)[$PbBr_4$]	CSE (DMF/acetone) [69, 70]
$(C_6H_5CH_2NH_3)_2PbCl_4$	LSM [59]
$(C_6H_5C_2H_4NH_3)_2PbCl_4$	CSE (DMF/nitromethane) [71]; LSM [59]
$(C_{10}H_{21}NH_3)_2PbCl_4$	CDM [65]
Quasi-2D perovskites (*n* = 2)	**Synthesis methods[a)]**
$(C_4H_9NH_3)_2(CH_3NH_3)Pb_2I_7$	CDM [49–51]
$(C_6H_5C_2H_4NH_3)_2(CH_3NH_3)Pb_2I_7$	CDM [51, 60–62]
$(C_6H_{13}NH_3)_2(CH_3NH_3)Pb_2I_7$	SE (DMF) [64]
$(C_4H_9NH_3)_2(CH_3NH_3)Pb_2Br_7$	CDM [50, 51]
$(C_6H_5C_2H_4NH_3)_2(CH_3NH_3)Pb_2Br_7$	CSE (DMF/acetone) [62]; CDM (DMF/water) [72]
Quasi-2D perovskites (*n* = 3)	**Synthesis methods[a)]**
$(C_4H_9NH_3)_2(CH_3NH_3)_2Pb_3I1_0$	CDM [49, 50, 73, 74]
$(C_6H_5C_2H_4NH_3)_2(CH_3NH_3)_2Pb_3I_{10}$	CDM [60, 61]
$(CH_3CH_2NH_3)_4Pb_3Cl_{10}$	SE [75]
$(CH_3CH_2NH_3)_4Pb_3Br_{10}$	SE [75]
$(C_6H_{13}NH_3)_2(CH_3NH_3)Pb_3Br_{10}$	CSE (DMF/acetone) [72]
Quasi-2D perovskites (*n* = 4)	**Synthesis methods[a)]**
$(C_4H_9NH_3)_2(CH_3NH_3)_3Pb_4I_{13}$	CDM [49, 50, 76]
$(C_6H_5C_2H_4NH_3)_2(CH_3NH_3)_2Pb_3I_{13}$	CDM [60]

a) Cooling down method – CDM; Co-solvent evaporation – CSE; Solvent evaporation – SE; Layered solution method – LSM.

cooling rate is maintained. The crystal growth begins at a certain temperature, and the growth is finally quenched by bringing the solution to room temperature. For example, in 1996, Mitzi first reported the synthesis of BA_2PbI_4 using the cooling down crystallization method [77]. Sometimes, the divalent metal oxide (MO) is used in place of the divalent metal halide (MX_2) to grow the perovskite

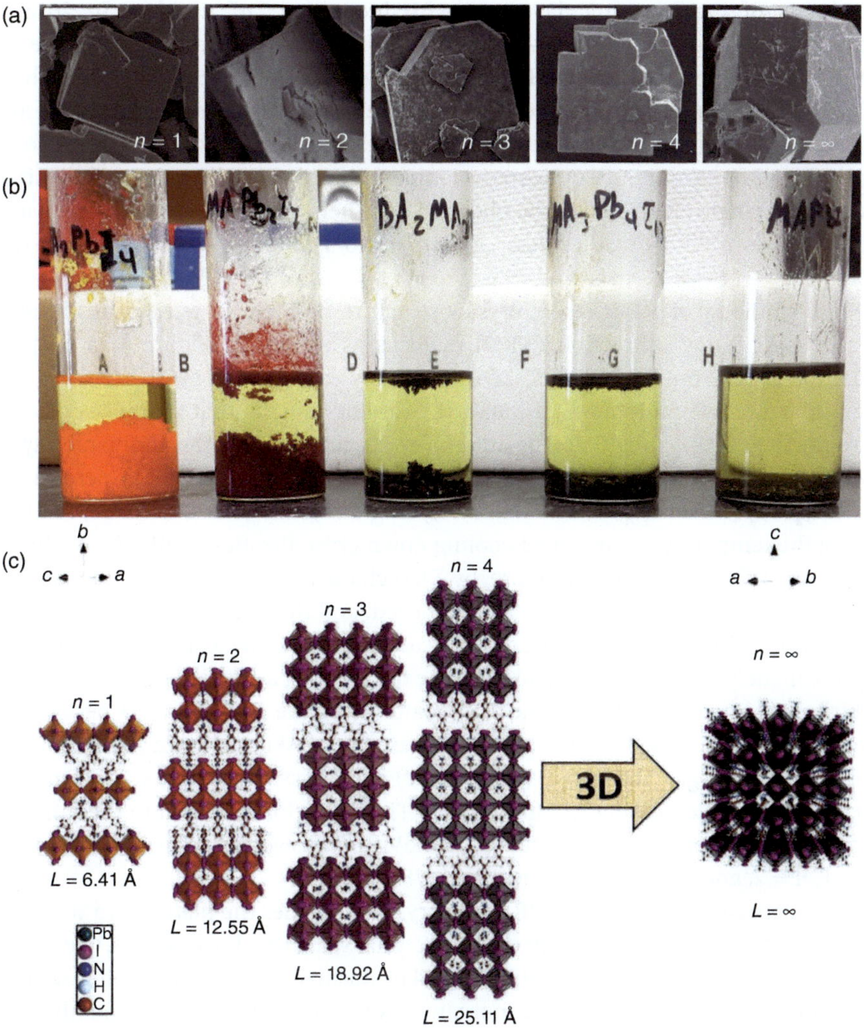

Figure 4.6 2D and quasi-2D perovskites single crystals. (a, b) SEM images (top panel) and photographs of the $(BA)_2(MA)_{n-1}Pb_nI_{3n+1}$ perovskite crystals (bottom panel). All scale bars are 200 μm. (c) Increasing inorganic layer thickness (*L*) of perovskite crystals [50].

crystal. By mixing neutralized base butylamine (BA = $C_4H_9NH_2$) with HI, H_3PO_2, and PbO in the hot aqueous water solution, Stoumpos et al. reported the synthesis of a series of Ruddlesden–Popper $(BA)_2(CH_3NH_3)_{n-1}Pb_nI_{3n+1}$ (n = 1, 2, 3, and 4) perovskites (RPPs) by cooling down crystallization method, as shown in Figure 4.6a [49]. SEM images indicate the square perovskite seeds, which complete crystallization in two hours. The synthesized perovskite crystals are the colorful rectangular-shape sheets, and their color changes from red to black as shown in Figure 4.6b. In this growth process, BA is essential to guide the reaction for accurate control of the stoichiometry. Figure 4.6c depicts the gradually increasing thickness of the inorganic layer (L) with increasing "n" number. In 2018, a temperature-programmed crystallization method was reported to realize a series of 2D perovskites $(BA)_2(CH_3NH_3)_{n-1}Pb_nI_{3n+1}$ (n = 1, 2, 3, 4) [50]. In this method, PbO, BAI, MAI, and HI (containing H_3PO_2) are mixed with varied mass ratios and heated with magnetic stirring to form uniform solution. Adopting a cooling down rate of 3 °C h^{-1} from 110 °C to room temperature, the final product, large-sized perovskite single crystal, could be obtained and then, separated out from the water solution. Perovskite single crystals of sizes up to 1 cm can be obtained from this method.

Post-synthesis of RPPs, their crystal structure and phase purity can be analyzed using X-ray diffraction (XRD) patterns as shown in Figure 4.7a [78]. In addition to BA organic cations, 2D perovskites can also be synthesized by utilizing large alkylammonium cations with varying lengths that mainly act as the structural guides to regulate the interlayer spacing among the inorganic layers [79]. Phenylethylamine (PEA = $C_8H_9NH_2$)-based 2D hybrid halide perovskites $[(PEA)_2PbI_4{\cdot}(MAPbI_3)_{n-1}$ (n = 1, 2, 3, 4)] single crystals of size 2–3 mm, realized using the temperature-controlled cooling down crystallization method, are shown in Figure 4.7b [60, 61]. Furthermore, by XRD characterizations, the diffraction patterns of $(PEA)_2PbI_4{\cdot}(MAPbI_3)_{n-1}$ were indexed, and the lattice distance of the first diffraction peak in all three patterns was calculated to show a stepwise increase of approximately 0.6 nm, corresponding to the thickness of a single $[PbI_6]^{4-}$ sheet [61]. This result is consistent with the XRD patterns of the BA-based 2D perovskites shown in Figure 4.7a. Both BA- and PEA-based perovskites show a monotonically increasing band-gap absorption (Figure 4.7c) and emission (Figure 4.7d) from 1.85 to 2.42 eV with decreasing "n" from 5 to 1. This behavior is related to the different exciton binding energies of 380 meV, 270 meV, and an average value of 220 meV for $n = 1$, $n = 2$, and $n > 2$ RPPs, respectively [78].

The successful synthesis of RPP single crystals provides the platform to study the correlation between optical properties and crystal structure. For example, Chen et al. found the spin coherence lifetime is affected by Rashba splitting and phonon scattering, both of which depend on the inorganic layer thickness [60]. Additionally, even though PEA-based perovskites present the similar emission trends as BA-based perovskites, Peng et al. believed that introducing larger organic cation PEA can lead to a decrease in the level of self-doping and crystal sizes through the defect inhibiting crystallization process compared to BA cations [61].

The cooling method based on water solutions has shown good results, but it is only limited to relatively small ligands, such as BA and PEA. Because of the solubility and structure rigidity of high "n" number quasi-2D perovskites, it is difficult to grow and obtain their high-quality single crystal. To solve this problem, some other polar organic solvents, such as DMF, can be used to grow perovskite crystals either using a DMF/water mixture or just DMF. For example, using only water as solvent, $(PEA)_2(MA)Pb_2Br_7$ perovskite is difficult to grow because of its rigid structure. L. Dou's research group added a little amount of DMF in water to modify the growth condition and successfully obtained the $(PEA)_2(MA)Pb_2Br_7$ perovskite crystal [62]. But sometimes, for the high solubility of precursors in

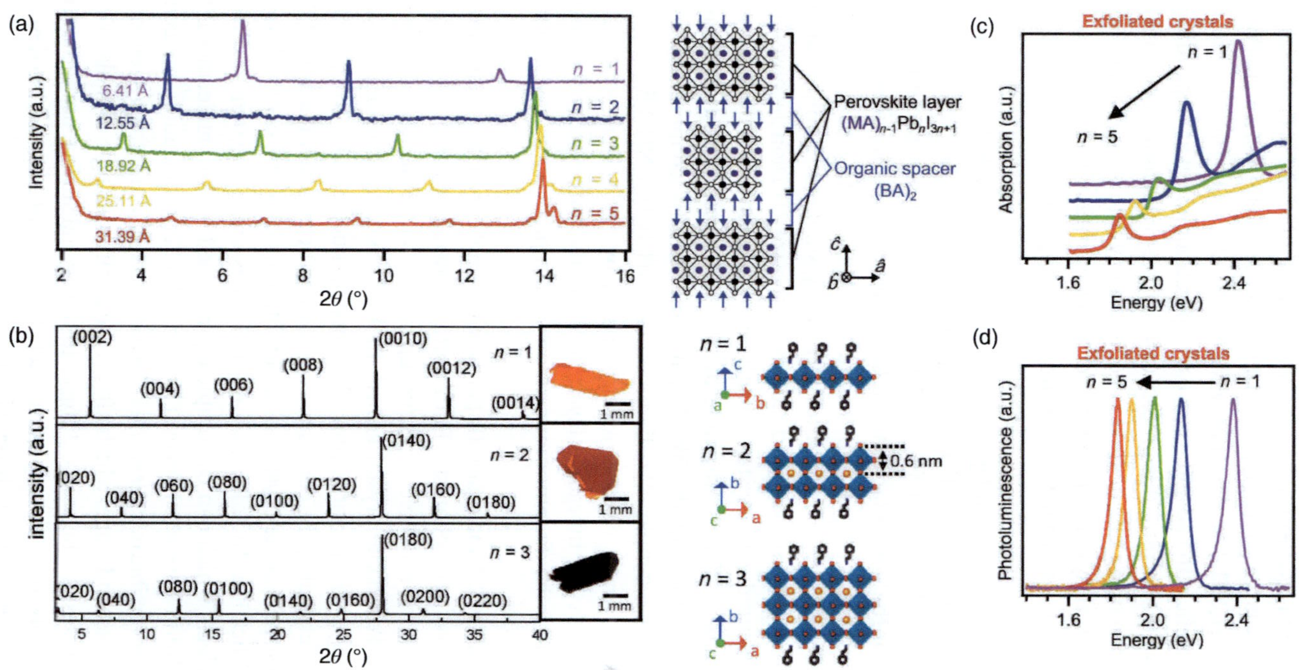

Figure 4.7 Structures and optical properties of 2D and quasi-2D perovskites. (a) XRD patterns for $(BA)_2(MA)_{n-1}Pb_nI_{3n+1}$ ($n = 1, 2, 3, 4, 5$). (b) XRD patterns of the freshly cleaved single crystals of $(PEA)_2PbI_4{\cdot}(MAPbI_3)_{n-1}$ ($n = 1$, 2, and 3) (left) and schematic illustrations of the layered structure and the corresponding orientation of 2D perovskite crystals showing that the thickness of a single perovskite sheet is ≈ 0.6 nm (right). (c, d) Absorption and PL of freshly exfoliated 2D and quasi-2D perovskites [61, 78].

organic polar solvents, 2D perovskites seed cannot be obtained in the solution even after cooling down to room temperature. Therefore, for these challenging systems, a solvent evaporation method based on organic solvents has been developed to realize perovskite precipitation. For example, by slowly evaporating the acetone solvent, we can obtain monolayer of $(C_6H_{13}NH_3)_2PbI_4$, $(C_6H_{13}NH_3)_2(CH_3NH_3)Pb_2Br_7$, and $(C_6H_{13}NH_3)_2(CH_3NH_3)Pb_3Br_{10}$ perovskites [64, 72]. The organic solvent evaporation method offers the advantage of easy operation; however, depending on the vapor pressure of the solvent, this process can be time-consuming. To accelerate the crystal growth rate using this method, a good strategy is to add another solvent to form a mixed solvent system and increase the evaporation rate. For instance, by adopting an acetone/nitromethane solution, a large-sized $4 \times 10 \times 0.1\,mm^3$ $(C_{10}H_{21}NH_3)_2PbI_4$ crystal was obtained [65]. The addition of nitromethane accelerates the evaporation of solvents and growth of single crystals, so that single crystals of large size can be grown in a relatively short time. Using the solvent evaporation method, $(C_{10}H_{21}NH_3)_2PbBr_4$ and $(C_{10}H_{21}NH_3)_2PbCl_4$ perovskites crystals were also prepared, but for the non-polar chain in ligand, acetone was replaced by water due to the poor solubility of ligand in the mixed solvent [65, 80].

In addition to the RP phase perovskites formed by single ammonium cations, diammonium cations can also be introduced to synthesize 2D Dion-Jacobson (DJ) phase perovskites using the above stated methods. Diammonium cations have the advantage that complex R′ is more liable to form 2D layers, and diammonium cations can eliminate the van der Waals gaps and directly connect the layers together [81, 82]. For example, 2D diammonium $NH_3(CH_2)_nNH_3PbI_4$ ($n = 4$, 6, and 8) perovskites single crystals were prepared by the solvent evaporation method [83]. Although the inorganic layers are slightly distorted by the spatial constraints imposed by the diammonium cations, these perovskites have a typical corner-sharing structure.

To obtain high quality 2D and quasi-2D perovskites single crystals, it is important to understand the growth mechanism. Both cooling down and solvent evaporation methods share the same principle of using supersaturated solution to precipitate perovskite crystals. Perovskite crystal growth in a supersaturated solution can be divided into two stages: (i) the formation of the crystal nucleus and (ii) the formation of a large single crystal by the continuous growth of the crystal nucleus [84]. The nucleation stage for perovskite crystals is quite different from the classical nucleation during solidification. In solidification, molecules come together to directly form a crystal nucleus, which can be completely reversed back to molecules if the experimental condition reverses. However, in perovskite crystallization, molecules assemble to form irreversible micelles, and then the micelles join to form a reversible crystal nucleus. The second stage, which affects the quality of single crystals, is controlled by two processes: (i) the solute diffuses to the crystal surface in the solution and (ii) the solute deposits on the crystal surface. Equation (4.1) is used to describe the stable nucleation rate on the substrate [85].

$$v = \omega \Gamma N_0 \exp\left(-\frac{\Delta G}{k_B T}\right) \tag{4.1}$$

where v is the steady nucleation rate, ω is the frequency at which the solute molecules diffuse from the solution to the critical nucleus, Γ is the nonequilibrium Zeldovich factor, N_0 is the density of adsorption sites on the crystal surface, ΔG is the Gibbs free energy change for the nucleus formation, also known as the nucleation barrier, k_B is the Boltzmann constant, and T is the solution temperature.

The driving force in the crystallization process of a supersaturated solution is related not only to the thermodynamic temperature, but also to the natural logarithmic ratio of the concentration of the supersaturated solution. Regardless of the concentration of the solution, there are small grains precipitated and dissolved in the solution. When the solution is in the supersaturation region, the precipitation is larger than the dissolution, which will immediately cause the formation of small grains with more defects. When the supersaturation is higher, the nucleation rate is faster.

The steady-state nucleation rate is used to predict the deposition of micelles on the crystal surface on the microscopic scale, and the crystal growth rate is used to control the macroscopic change in crystal quality. There is a general understanding that the lower the growth rate, the better the quality of the crystal [86, 87]. That is why cooling-down method and solvent evaporation method can grow perovskite crystals of very high quality. As stated above, the continuous growth of crystal includes two processes: solute diffusion from the solution to the crystal surface, and solute deposition on the crystal surface [87]. The solute diffusion rate varies exponentially with the temperature and therefore can be controlled by modulating the temperature [88]. The solute deposition process is more difficult to control than the diffusion rate. Crystallization kinetics shows that when solute concentration is controlled above the solubility curve (when slightly oversaturated), high-quality crystals can be grown on already-formed crystals or seeds without excess nucleation (nucleation point) [89].

Layered solution method (LSM) is another technique developed for growth of perovskite crystals. LSM involves dissolving two reactants in two solvents with certain mutual solubility and an obvious density difference between the two solvents. The different solubilities and densities lead to a clear interface formation between the two solvents. Under slow diffusion, large single crystal is precipitated at the interface. Specifically, Kamminga et al. used four phenyl alkylammonium cations with different alkyl chain lengths ranging from one to four carbons to prepare single crystal perovskites at room temperature by the layered solution technique [57, 90]. The obtained products have good stability and can be stored in low-humidity environment for several months without damage. Subsequently, Ye's group synthesized $(PMA)_2PbI_4$, $(PEA)_2PbI_4$, and $(PBA)_2PbI_4$ (PBA = $C_6H_5(CH_2)_4NH_3^+$) perovskites using the same method. The source of fluorescence and the behavior of excitons were confirmed by experiments, and the quantum confinement effect caused by the structural reorganization was demonstrated by calculation [58].

These solution methods have the advantages of stability, low cost, and easy operation, and are therefore widely used to grow various perovskites crystals. In addition to lead 2D and quasi-2D perovskites, many efforts have been made for the synthesis of nonlead 2D perovskites [91–94]. In 1994, Mitzi et al. prepared 2D Sn-based

Table 4.4 Summary of crystal parameters for 2D and quasi-2D perovskites single crystals.

2D Perovskites	Structure
$(BA)_2PbI_4$	Orthorhombic, *Pbca*; 8.8632(21), 8.6816(8), 27.570(2) [77]
$(PEA)_2PbI_4$	Triclinic, *P*1; 8.7389(2), 8.7403(2), 32.9952(6) [59]
$(PEA)_2PbCl_4$	Triclinic, *P*1; 11.1463(3), 11.2181(3), 17.6966(5) [90]
$(2T)_2PbI_4$	Triclinic, *P*1; 12.0933(7), 12.2631(7), 20.8347(12) [63]
$(2T)_2PbBr_4$	Triclinic, *P*1; 11.3032(13), 11.6338(13), 21.602 [63]
$(4Tm)_2PbI_4$	Monoclinic, *C*2*c*; 64.191(6), 6.0936(5), 12.2111 [63]
$(4Tm)_2SnI_4$	Triclinic, *P*1; 12.0929(5), 12.1407(6), 32.4960 [63]
$(PEA)_2PbBr_4$	Triclinic, *P*1; 11.6150(4), 11.6275(5), 17.5751(6) [96]
Quasi-2D perovskites	**Structure**
$(BA)_2(MA)Pb_2I_7$ ($n = 2$)	Orthorhombic, *Cc*2*m*; 8.9470(4), 39.347(2), 8.8589(6) [49]
$(BA)_2(MA)_2Pb_3I_{10}$ ($n = 3$)	Orthorhombic, *C*2*cb*; 8.9275(6), 51.959(4), 8.8777(6) [49]
$(BA)2(MA)_3Pb_4I_{13}$ ($n = 4$)	Orthorhombic, *Cc*2*m*; 8.9274(4), 64.383(3), 8.8816(4) [49]
$(PEA)_2(MA)Pb_2I_7$ ($n = 2$)	Orthorhombic, *Pbca*; 8.82, 8.99, 25.52 [97]

perovskite $(BA)_2(MA)_{n-1}Sn_nI_{3n+1}$ using cooling down crystallization method under argon atmosphere to prevent oxidation. When the precursor solution containing SnI_2, $C_4H_9NH_3I$, and CH_3NH_3I was cooled down at a rate of 2-5 °C h^{-1} from 90 to 10 °C, plate-like crystals were obtained. For $n = 3$, an orthorhombic structure was obtained [55]. Later, Mitzi reported the synthesis of layered Ge-based perovskite BA_2GeI_4 and studied its crystal structure and optical properties [77]. And recently, Han's group reported a 2D lead-free $(PEA)_2GeI_4$ perovskite, prepared by liquid phase crystallization, which was precipitated by cooling HI and H_3PO_2 mixed solutions containing stoichiometric GeO_2 and (PEA)I [95]. $(PEA)_2GeI_4$ perovskite has a direct bandgap of 2.12 eV and it was found that the introduction of PEA cation in the layered structure improved the perovskite stability.

In conclusion, 2D and quasi-2D perovskites are large material systems because of their diversity of ligands and crystal structures, as shown in Table 4.4. Currently, mainly three kinds of crystal structures have been reported in 2D and quasi-2D perovskites. Typically, for small ligands not affecting the Pb–I octahedron, the perovskite will be orthorhombic, such as $(BA)_2PbI_4$ [77]. But with the increase in ligand size and volume, a distortion happens in Pb–I octahedron, resulting in triclinic and even monoclinic structures, such as $(PEA)_2PbI_4$ and $(2T)_2PbI_4$ [59, 63]. For quasi-2D perovskites, increasing the "n" value of inorganic layers also enhances the rigidity of perovskite structure and weakens the lattice distortion, resulting in change of crystal structure from triclinic [$(PEA)_2PbI_4$, $n = 1$] to orthorhombic [$(PEA)_2(MA)Pb_2I_7$, $n = 2$] [59, 97]. So, the choice of growth methods is governed by the solubility of ligands, polarity of the ligand chain, and also the desired structure parameters of the target perovskites.

4.3.3.2 Synthesis of 2D and Quasi-2D Perovskites Nanocrystal

To prepare atomically thin 2D halide perovskite nanomaterials, various approaches have been investigated. Inspired by the methodology widely applied in 2D research, mechanical exfoliation was initially investigated. Niu et al. and Yaffe et al. reported the mechanical exfoliation of ultrathin $(BA)_2PbI_4$ 2D crystals and studied their optical properties via ultrafast spectroscopy [98, 99]. Large lateral-sized crystals can be obtained using this method, but the sheets are typically randomly shaped with a wide size distribution. Additionally, ultrathin sheets with thickness as small as 2.5 nm have been synthesized. However, majority of synthesized sheets are much thicker with a wide distribution in thickness. Although the exfoliation method is a simple and quick technique to obtain large single sheets, it is not suitable for high throughput production and high-density device integration. Recently, Dou et al. employed solution synthesis method to directly grow large-area atomically thin 2D halide perovskite sheets. In this method, a ternary solvent mixture is used, and the crystals grow during the solvent evaporation process [68]. Figure 4.8a shows the dark-field optical image of the as-synthesized 2D crystals with well-defined square shapes. TEM and AFM imaging was used to investigate the morphology of the individual 2D sheets. AFM image depicted that the thinnest sheet was a monolayer structure with a thickness of ~1.6 nm. The ternary solvent was found to be critical, because this solvent system can precisely control the solute solubility and solvent evaporation speed. Controllable crystallization was achieved during the evaporation process. It is to be noted that the precursors' solubility should be constantly low so that the solutes do not concentrate during the solvent evaporation. Compared with the exfoliation methods, better control over the size, shape, and composition of the 2D halide perovskite crystals can be achieved using the solvent evaporation method. Later, a more systematic study on solvent engineering during crystal growth was carried out by varying the solvent type and blending ratio [101]. It was found that the growth of the 2D halide perovskite crystals is governed by the synergetic effects of diffusion dominated branched growth and c-axis suppression. Using this method, larger lateral-size crystals of up to ~40 μm can be obtained. Very recently, 2D halide perovskites with different cations (such as PEA+) were prepared using similar methods [102, 103]. These studies suggest that solution processing might be more suitable to achieve high-quality perovskite 2D crystals than mechanical exfoliation or chemical vapor deposition (CVD) growth, in which the target nanocrystals form on substrate through vapor transport at a high temperature.

Most 2D halide perovskites employ simple organic cations, e.g. BA or PEA, which solely function as the insulating barriers in the 2D structure. Conversely, semiconducting organic cations (Figure 4.8b) could play a more active role in tuning the overall optical and electronic properties [100, 104]. Combining a host of semiconducting organic cations and previously reported nanocrystal growth methods, Gao et al. developed a series of perovskites with hybrid structures, named "organic semiconductor-incorporated perovskites," or simply "OSiPs" [100, 105]. Through judicious molecular design, a series of π-conjugated ligands with tunable energy levels and band gaps have been synthesized and successfully incorporated into the 2D perovskite lattice. A complete band alignment tuning from type-I

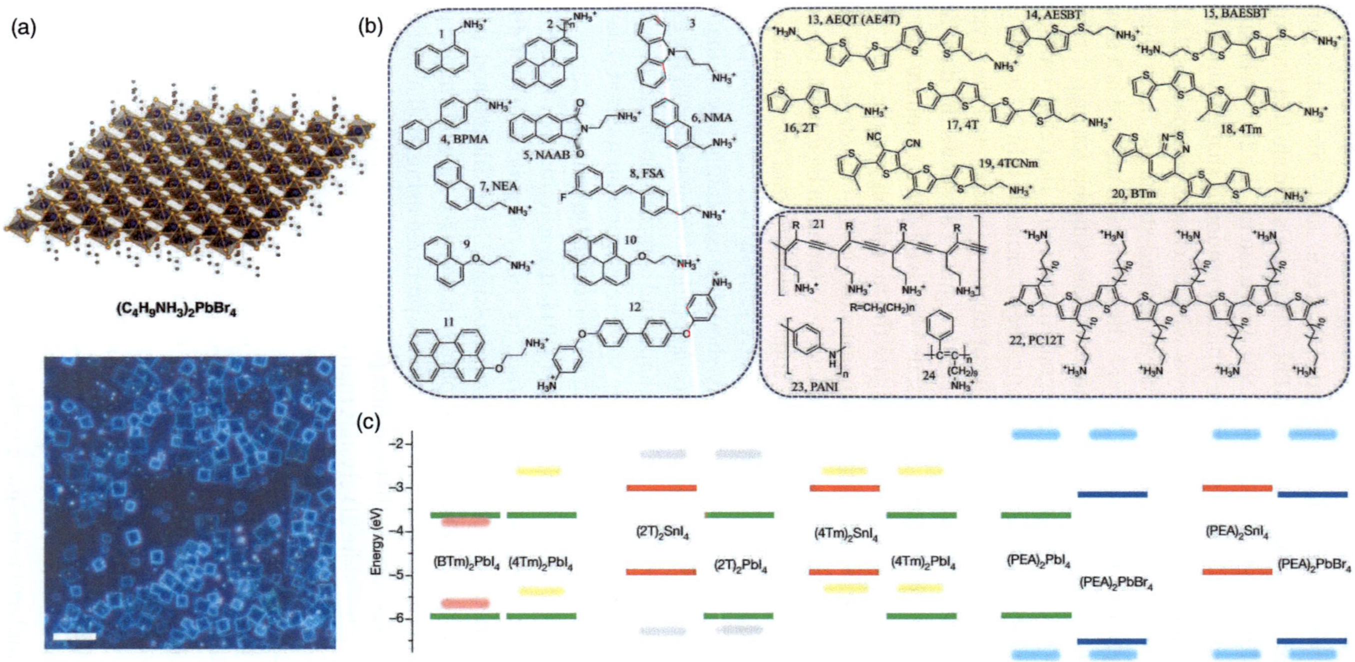

Figure 4.8 2D perovskites nanocrystals and ligands design. (a) Schematic illustration and dark-field optical image of 2D halide perovskite nanocrystal obtained by ternary solvent evaporation. Scale bar in optical image is 10 μm [68]. (b) Chemical structures of conjugated organic ligands for 2D halide perovskites [100]. (c) Proposed band alignments between the conjugated organic ligands and inorganic layers in 2D perovskites [63]. $(PEA)_2PbI_4$, $(PEA)_2SnI_4$, $(PEA)_2PbBr_4$, $(2T)_2PbI_4$, $(2T)_2SnI_4$ and $(4Tm)_2SnI_4$ belong to type I band alignment; $(4Tm)_2PbI_4$ belongs to type II band alignment; $(BTm)_2PbI_4$ belongs to reverse type I band alignment.

to type-II to reverse type-I and reverse type-II was demonstrated, as shown in Figure 4.8c. These organic–inorganic heterostructures provide a new platform to study the ultrafast charge and energy transfer across the heterogeneous interfaces using femtosecond spectroscopy tools [106]. Furthermore, the conjugated ligands not only encapsulate the perovskite layer but also provide a driving force for lattice stabilization, resulting in significantly improved chemical and thermal stability.

Different from the nanocrystal growth using organic solvents, Pan et al. developed a simple crystallization method at the air–water interface of the precursor solution droplets to synthesize large-area free-standing nanosheets of various RP perovskites [107]. In this method, a drop of warm saturated aqueous solution is taken from concentrated hydroiodic (or hydrobromic) acid with perovskite precursors and put on the glass slide. With the temperature decreased, the droplet becomes supersaturated, and nucleation and crystal growth begin. Because of the hydrophobic ligand chain, the perovskite self-assembles at the solution–air interface and serves as a soft template that lowers the nucleation barrier at the solution surface (to below that in the solution) and facilitates the highly anisotropic 2D growth in a "soft epitaxy" fashion. At the same time, rapid nucleation occurs on the water surface with crystallization along the basal plane, leading to the growth of floating nanosheets with lateral sizes of up to a few hundred micrometers within seconds. Systematic research shows the thickness of these sheets varies from ~1 μm down to a few nanometers depending on the growth time. The key to achieving monolayer or few-layer thicknesses while maximizing lateral growth is maintaining a proper supersaturation, which is carefully optimized by tuning the concentrations of each precursor and the starting cooling temperatures. Using this method, diverse RP perovskite systems [$L_2A_{n-1}Pb_nX_{3n+1}$] (Figure 4.9) were synthesized, using "L" site cations such as BA, *n*-pentylammonium (PA), *n*-hexylammonium (HA), phenylmethylammonium (PMA) or PEA; "A" site cations such as MA, formamidinium (FA) or guanidinium (GA); "X" site anion such as Br or I; and "*n*" inorganic layers from 1 to 5. This method is clearly versatile to grow many kinds of 2D and quasi-2D perovskites nanocrystals; however, its feasibility is limited to water–air interfaces and cannot be utilized for organic solvents. Therefore, this method remains to be further developed to grow perovskites nanocrystal with much larger organic cations, such as π-conjugated ligands.

4.4 Applications of Low-Dimensional Halide Perovskites

0D halide perovskites can be employed in the fabrication of a wide range of optoelectronic devices, including light-emitting diodes (LEDs), solar cells, photodetectors, X-ray detectors, scintillators, luminescent solar concentrators, and electrochemical capacitors [18]. In their review paper, Le et al. have discussed the progress, challenges and future prospects of perovskite quantum dot based LEDs. High color purity, narrow bandgap, tunable PL, and high PLQY (up to 100%) make 0D perovskites quite promising for future display technologies [108]. Zhao et al. utilized the compositional tunability of quantum dots to fabricate 0D heterojunction

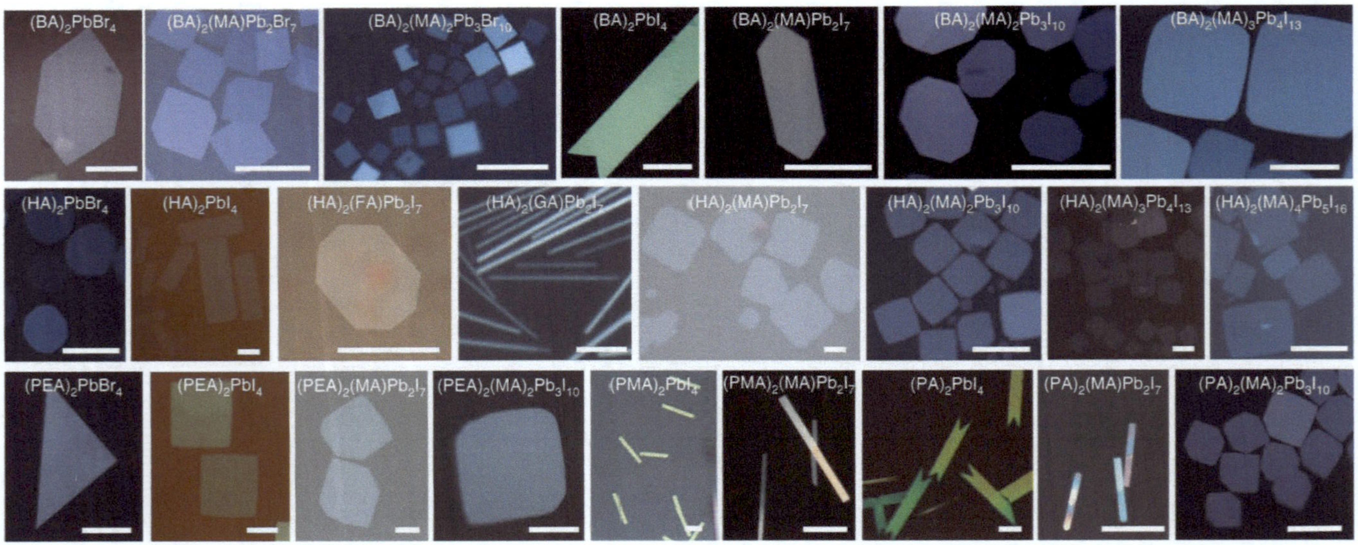

Figure 4.9 Interface method to synthesize 2D and quasi-2D perovskite nanocrystals. Optical images of various types of RP perovskite nanosheets (or occasionally nanoribbons) still floating on water's surface. All scale bars are 50 μm [107].

solar cells. By controlling the composition and position of heterojunction, solar cells with power conversion efficiency (PCE) of ~17.4% were fabricated [109]. In 2019, Liang et al. reviewed the unique properties of 0D perovskites and their applications in the field of photoelectrochemistry. The review paper highlights the advancements made in application of perovskite quantum dots in photocatalytic degradation, photocatalytic CO_2 reduction, photoelectrochemical sensing, and photovoltaic devices [110].

1D halide perovskites are quite promising for their potential in lighting applications owing to their large bandgap and high exciton binding energy [111]. Using perovskite nanowires, Zhu et al. demonstrated room-temperature and wavelength-tunable lasing with very low lasing thresholds (220 nJ cm^{-2}) and high quality factors (~3600) [16]. In 2018, Chen et al. fabricated hybrid phototransistor using organic semiconductor and $CsPbI_3$ nanorod bilayers. The perovskite layer provided efficient light absorption, and the organic layer provided efficient charge transport and photogating, resulting in a transistor with high photoresponsivity (4300 A W^{-1}), ultrahigh photosensitivity (2.2×10^6), and excellent stability (>1 month) [112]. The long carrier lifetimes and low nonradiative recombination rates in 1D perovskites, make them promising candidates for applications in nanophotonics, optical computing, and chemical/biological sensing [16].

Like their 1D and 0D counterparts, 2D and quasi-2D halide perovskites have been extensively used in fabrication of solar cells and LEDs. Typically, with an increase in "n" value or inorganic repeating layers, there is an increase in the short circuit current density and PCE of 2D ($n = 1$) and quasi-2D ($n > 1$) halide perovskite-based solar cells [113]. In 2018, Chen et al. demonstrated that further improvement in charge transport properties of $(BA)_2(MA)_3Pb_4I_{13}$ quasi-2D perovskite solar cells can be achieved by controlling the orientation of crystallization. Vertically orientated nucleation and growth of the quasi-2D films led to unhindered transport of carriers across the absorbing layer, resulting in higher short-circuit current density, lower series resistance, and higher PCE compared to non-orientated films [114]. In their review paper, Zhang et al. discuss the advancements in quasi-2D-based perovskite LEDs and their bright future in ultrahigh-definition displays, solid-state lighting, optical communications, etc. [115]. In addition to solar cells and LEDs, Sn-based 2D halide perovskites with conjugated organic cations have shown promise in thermoelectric device applications [116]. Xie et al. demonstrated the versatility of Li-doped $(PEA)_2PbBr_4$ scintillator as radiation detector for a wide range of energies from keV up to MeV [117]. Li et al. studied the light-enhanced ion migration in carbon nanotubes and 2D perovskite heterostructures. Using this heterostructure, the authors also realized the function of photomemory with a multivalue and erasable two-phase performance [118]. 2D halide perovskites have mixed electronic and ionic transport, resulting in low operating current requirements for resistive memory devices. This also makes 2D halide perovskites good candidates to be used as building blocks for neuromorphic circuits and systems [119]. The color tunability and solution processability of 2D halide perovskites also make them a promising gain media for lasing. In 2020, Qin et al. utilized quasi-2D perovskite to fabricate a stable green laser under continuous wave optical pumping in air at room temperature [120].

4.5 Current Challenges and Prospects of Low-Dimensional Halide Perovskites

The long list of synthesis techniques and wide range of applications detailed in this chapter are a testament to the tremendous research efforts in the field of low-dimensional halide perovskites. However, a comprehensive and systematic analysis of impact of structural changes on material properties and device performance is still lacking.

Low-dimensional perovskites offer higher compositional flexibility compared to their 3D counterparts. Figure 4.10a gives a few examples of the different organic capping ligands and smaller organic/inorganic cations that can be incorporated into the perovskite structure. The choice of an organic capping ligand can help in further improving the ambient stability of perovskite [100, 104]. Using Sn instead of Pb in halide perovskites or Ag/Bi in double perovskites can make perovskites environmentally friendly [124, 125]. Analyzing the impact of compositional alterations on the degradation mechanism can help resolve the stability and toxicity issues currently impeding the commercialization of low-dimensional perovskites. Additionally, compositional variations can have a significant impact on the overall crystal lattice. Figure 4.10b depicts three different connectivity modes in 2D halide perovskites – corner sharing, edge sharing, and face sharing. The most reported connectivity mode is corner sharing, and the edge/face sharing modes are obtained by using organic capping ligands that introduce strain into the lattice and distort the corner sharing connectivity mode [121]. A clear understanding of the impact of the various crystal lattices on photophysical properties like band gap, exciton binding energies, charge transport, etc. can help exploit the full potential of low-dimensional perovskites.

Understanding the crystallization kinetics of low-dimensional perovskites is another area that needs more detailed analysis. Zhou et al. investigated the crystal orientation of quasi-2D halide perovskites synthesized using different ratios of MA and FA organic cations. Figure 4.10c shows the grazing incidence wide-angle X-ray scattering (GIWAXS) patterns of $(BA)_2(MA)_3Pb_4I_{13}$ and $(BA)_2(MA_{0.4}FA_{0.6})_3Pb_4I_{13}$ perovskite films. The pure MA-based halide perovskite depicts much higher crystallinity compared to the MA/FA mixed cation perovskite. However, the incorporation of FA cation was found to increase the lifetime of charge carriers and PCE of the fabricated solar cells [122]. A systematic analysis of the impact of cations and anions on the crystallization kinetics and device performance will provide a great control lever for smart material design.

Another innovative technique to improve the mechanical robustness of halide perovskites is by looking into composite systems. By synthesizing a polymer-perovskite composite film, Finkenauer et al. were able to control the grain size of the $MAPbI_3$ absorbing layer, as shown in Figure 4.10d. The change in grain size of the polycrystalline film helped in modulating the PL, carrier lifetime, and efficiency of the fabricated solar cell [123]. A similar exploration of perovskite-polymer composites in low-dimensional domain can help in improving the reliability of low-dimensional based optoelectronic devices.

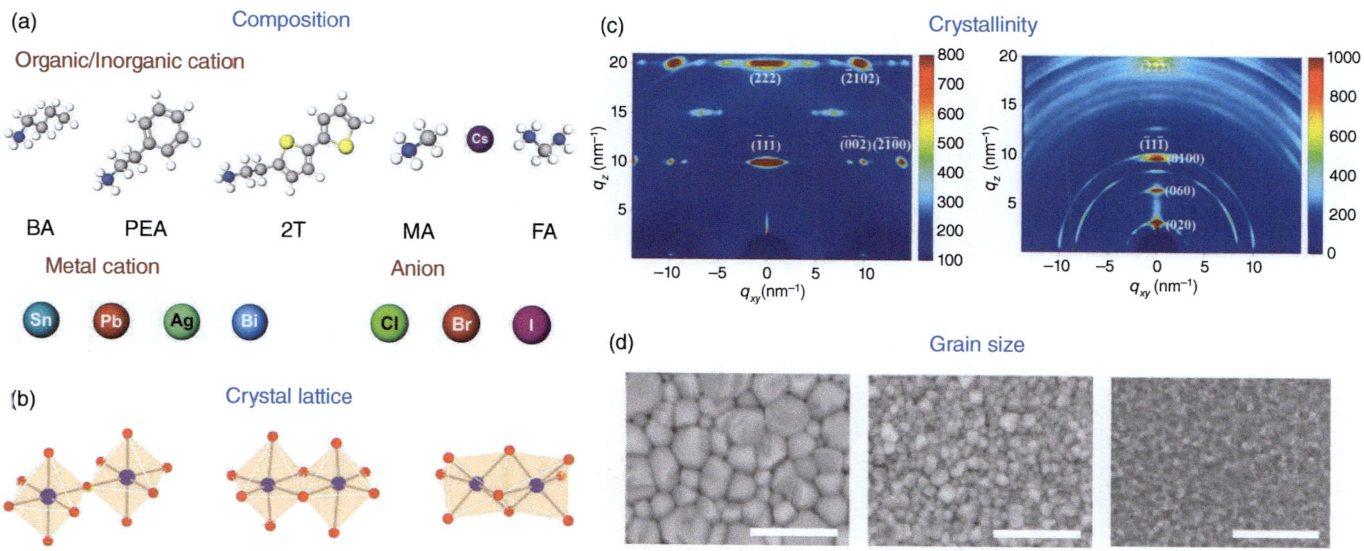

Figure 4.10 Structural tunability of halide perovskites. (a) Alternatives for cationic and anionic sites in halide perovskites. For the organic/inorganic cations, gray spheres represent carbon atoms, blue spheres represent nitrogen atoms, white spheres represent hydrogen atoms, and yellow spheres represent sulfur atoms. (b) Crystal lattice variations in halide perovskites depending on the connectivity mode of the octahedrons – corner-sharing (left), edge-sharing (center), and face-sharing (right) [121]. (c) GIWAXS patterns showing the difference in crystallinity of $(BA)_2(MA_{1-x}FA_x)_3Pb_4I_{13}$ perovskite film with $x = 0$ (left) and $x = 0.6$ (right) [122]. (d) Surface SEM images of polymer-perovskite composite films at device conditions using polymers with three different molecular weights – 15 000 g mol^{-1} (left), 24 000 g mol^{-1} (center), and 36 000 g mol^{-1} (right). All scale bars are 1 μm [123].

In general, further studies into the still unanswered fundamental material properties, underlying mechanisms, and predicting the material behavior under device operating conditions can launch the low-dimensional perovskites into wide-spread applications.

References

1 Panich, A.M. (2008). Electronic properties and phase transitions in low-dimensional semiconductors. *Journal of Physics: Condensed Matter* 20: 293202.

2 Papavassiliou, G.C. (1997). Three- and low-dimensional inorganic semiconductors. *Progress in Solid State Chemistry* 25: 125–270.

3 Hong, K., Le, Q.V., Kim, S.Y., and Jang, H.W. (2018). Low-dimensional halide perovskites: review and issues. *Journal of Materials Chemistry C* 6: 2189.

4 Zhou, C., Lin, H., He, Q. et al. (2019). Low dimensional metal halide perovskites and hybrids. *Materials Science and Engineering R* 137: 38–65.

5 Amerling, E., Lu, H., Larson, B.W. et al. (2021). A multi-dimensional perspective on electronic doping in metal halide perovskites. *ACS Energy Letters* 6: 1104–1123.

6 Zhu, P. and Zhu, J. (2020). Low-dimensional metal halide perovskites and related optoelectronic applications. *InfoMat* 2: 341–378.

7 Chu, Z., Chu, X., Zhao, Y. et al. (2021). Emerging low-dimensional crystal structure of metal halide perovskite optoelectronic materials and devices. *Small Structures* 2: 2000133.

8 Chen, P., Bai, Y., Lyu, M. et al. (2018). Progress and perspective in low-dimensional metal halide perovskites for optoelectronic applications. *Solar RRL* 2: 1700186.

9 González-Carrero, S., Galian, R.E., and Pérez-Prieto, J. (2015). Organometal halide perovskites: bulk low-dimension materials and nanoparticles. *Particle and Particle Systems Characterization* 32: 709–720.

10 Dirin, D.N., Protesescu, L., Trummer, D. et al. (2016). Harnessing defect-tolerance at the nanoscale: highly luminescent lead halide perovskite nanocrystals in mesoporous silica matrixes. *Nano Letters* 16 (9): 5866–5874.

11 Protesescu, L., Yakunin, S., Bodnarchuk, M.I. et al. (2015). Nanocrystals of cesium lead halide perovskites ($CsPbX_3$, X = Cl, Br, and I): novel optoelectronic materials showing bright emission with wide color gamut. *Nano Letters* 15 (6): 3692–3696.

12 Zhang, D., Eaton, S.W., Yu, Y. et al. (2015). Solution-phase synthesis of cesium lead halide perovskite nanowires. *Journal of the American Chemical Society* 137 (29): 9230–9233.

13 Oener, S.Z., Khoram, P., Brittman, S. et al. (2017). Perovskite nanowire extrusion. *Nano Letters* 17 (11): 6557–6563.

14 Sun, S., Yuan, D., Xu, Y. et al. (2016). Ligand-mediated synthesis of shape-controlled cesium lead halide perovskite nanocrystals *via* reprecipitation process at room temperature. *ACS Nano* 10 (3): 3648–3657.

15 Liu, J., Xue, Y., Wang, Z. et al. (2016). Two-dimensional $CH_3NH_3PbI_3$ perovskite: synthesis and optoelectronic application. *ACS Nano* 10 (3): 3536–3542.

16 Zhu, H., Fu, Y., Meng, F. et al. (2015). Lead halide perovskite nanowire lasers with low lasing thresholds and high quality factors. *Nature Materials* 14: 636–642.

17 Spina, M., Bonvin, E., Sienkiewicz, A. et al. (2016). Controlled growth of $CH_3NH_3PbI_3$ nanowires in arrays of open nanofluidic channels. *Scientific Reports* 6: 19834.

18 Sun, S., Lu, M., Gao, X. et al. (2021). 0D perovskites: unique properties, synthesis, and their applications. *Advanced Science* 8: 2102689.

19 Smith, M.D. and Karunadasa, H.I. (2018). White-light emission from layered halide perovskites. *Accounts of Chemical Research* 51: 619–627.

20 Zhou, C., Tian, Y., Khabou, O. et al. (2017). Manganese-doped one-dimensional organic lead bromide perovskites with bright white emissions. *ACS Applied Materials and Interfaces* 9: 40446–40451.

21 Zhou, C., Lin, H., Tian, Y. et al. (2018). Luminescent zero-dimensional organic metal halide hybrids with near-unity quantum efficiency. *Chemical Science* 9: 586–593.

22 McDonald, C., Ni, C., Švrček, V. et al. (2017). Zero-dimensional methylammonium iodo bismuthate solar cells and synergistic interactions with silicon nanocrystals. *Nanoscale* 9: 18759–18771.

23 Zheng, X., Zhao, W., Wang, P. et al. (2020). Ultrasensitive and stable X-ray detection using zero-dimensional lead-free perovskites. *Journal of Energy Chemistry* 49: 299–306.

24 Kawai, T., Ishii, A., Kitamura, T. et al. (1996). Optical absorption in band-edge region of $(CH_3NH_3)_3Bi_2I_9$ single crystals. *Journal of the Physical Society of Japan* 65 (5): 1464–1468.

25 Zhang, R., Mao, X., Yang, Y. et al. (2019). Air-stable, lead-free zero-dimensional mixed bismuth-antimony perovskite single crystals with ultra-broadband emission. *Angewandte Chemie, International Edition* 58: 2725.

26 Han, P., Luo, C., Yang, S. et al. (2020). All-inorganic lead-free 0D perovskites by a doping strategy to achieve a PLQY boost from <2% to 90%. *Angewandte Chemie, International Edition* 59: 12709.

27 Zhang, F., Zhao, Z., Chen, B. et al. (2020). Strongly emissive lead-free 0D $Cs_3Cu_2I_5$ perovskites synthesized by a room temperature solvent evaporation crystallization for down-conversion light-emitting devices and fluorescent inks. *Advanced Optical Materials* 8: 1901723.

28 Idosa, D.A., Abebe, M., Mani, D. et al. (2023). A blue-light-emitting 3 nm-sized $CsPbBr_3$ perovskite quantum dot with $ZnBr_2$ synthesized by room-temperature supersaturated recrystallization. *Photonics* 10: 802.

29 Chouhan, L., Ghimire, S., Subrahmanyam, C. et al. (2020). Synthesis, optoelectronic properties and applications of halide perovskites. *Chemical Society Reviews* 49: 2869–2885.

30 Yang, L., Li, D., Wang, C. et al. (2017). Room-temperature synthesis of pure perovskite-related Cs_4PbBr_6 nanocrystals and their ligand-mediated evolution into highly luminescent $CsPbBr_3$ nanosheets. *Journal of Nanoparticle Research* 19: 258.

31 Zhou, C., Tian, Y., Wang, M. et al. (2017). Low dimensional organic tin bromide perovskites and their photoinduced structural transformation. *Angewandte Chemie, International Edition* 56: 9018.

32 Wu, Y., Han, D., Chakoumakos, B.C. et al. (2018). Zero-dimensional Cs_4EuX_6 (X = Br, I) all-inorganic perovskite single crystals for gamma-ray spectroscopy. *Journal of Materials Chemistry C* 6: 6647–6655.

33 Tan, L., Wang, W., Li, Q. et al. (2020). Colloidal syntheses of zero-dimensional Cs_4SnX_6 (X = Br, I) nanocrystals with high emission efficiencies. *Chemical Communications* 56: 387–390.

34 Zhang, Y., Saidaminov, M.I., Dursun, I. et al. (2017). Zero-dimensional Cs_4PbBr_6 perovskite nanocrystals. *Journal of Physical Chemistry Letters* 8: 961–965.

35 Li, X., Gao, X., Zhang, X. et al. (2021). Lead-free halide perovskites for light emission: recent advances and perspectives. *Advanced Science* 8: 2003334.

36 Umar, F., Zhang, J., Jin, Z. et al. (2019). Dimensionality controlling of $Cs_3Sb_2I_9$ for efficient all-inorganic planar thin film solar cells by HCl-assisted solution method. *Advanced Optical Materials* 7: 1–9, article ID: 1801368.

37 Hebig, J.C., Kühn, I., Flohre, J., and Kirchartz, T. (2016). Optoelectronic properties of $(CH_3NII_3)_3Sb_2I_9$ thin films for photovoltaic applications. *ACS Energy Letters* 1: 309–314.

38 Ma, Z., Liu, Z., Lu, S. et al. (2018). Pressure-induced emission of cesium lead halide perovskite nanocrystals. *Nature Communications* 9: 4–11, article ID: 4506.

39 Akkerman, Q.A., Abdelhady, A.L., and Manna, L. (2018). Zero-dimensional cesium lead halides: history, properties, and challenges. *Journal of Physical Chemistry Letters* 9 (9): 2326–2337.

40 Akkerman, Q.A., Park, S., Radicchi, E. et al. (2017). Nearly monodisperse insulator Cs_4PbX_6 (X = Cl, Br, I) nanocrystals, their mixed halide compositions, and their transformation into $CsPbX_3$ nanocrystals. *Nano Letters* 17 (3): 1924–1930.

41 Wu, L., Hu, H., Xu, Y. et al. (2017). From nonluminescent Cs_4PbX_6 (X = Cl, Br, I) nanocrystals to highly luminescent $CsPbX_3$ nanocrystals: water-triggered transformation through a CsX-stripping mechanism. *Nano Letters* 17 (9): 5799–5804.

42 Hu, M., Ge, C., Yu, J., and Feng, J. (2017). Mechanical and optical properties of Cs_4BX_6 (B = Pb, Sn; X = Cl, Br, I) zero-dimension perovskites. *Journal of Physical Chemistry C* 121: 27053–27058.

43 Cha, J.H., Han, J.H., Yin, W. et al. (2017). Photoresponse of $CsPbBr_3$ and Cs_4PbBr_6 perovskite single crystals. *Journal of Physical Chemistry Letters* 8: 565–570.

44 Aceves, R., Babin, V., Flores, M.B. et al. (2001). Spectroscopy of $CsPbBr_3$ quantum dots in CsBr:Pb crystals. *Journal of Luminescence* 93 (1): 27–41.

45 Kondo, S., Nakagawa, H., Saito, T., and Asada, H. (2004). Extremely photo-luminescent microcrystalline $CsPbX_3$ (X = Cl, Br) films obtained by amorphous-to-crystalline transformation. *Current Applied Physics* 4: 439–444.

46 Tong, G., Li, H., Zhu, Z. et al. (2018). Enhancing hybrid perovskite detectability in the deep ultraviolet region with down-conversion dual-phase ($CsPbBr_3$-Cs_4PbBr_6) films. *Journal of Physical Chemistry Letters* 9 (7): 1592–1599.

47 Yuan, Z., Zhou, C., Tian, Y. et al. (2017). One-dimensional organic lead halide perovskites with efficient bluish white-light emission. *Nature Communications* 8: 14051.

48 Dohner, E.R., Jaffe, A., Bradshaw, L.R., and Karunadasa, H.I. (2014). Intrinsic white-light emission from layered hybrid perovskites. *Journal of the American Chemical Society* 136 (38): 13154–13157.

49 Stoumpos, C.C., Cao, D.H., Clark, D.J. et al. (2016). Ruddlesden–popper hybrid lead iodide perovskite 2D homologous semiconductors. *Chemistry of Materials* 28 (8): 2852–2867.

50 Leng, K., Abdelwahab, I., Verzhbitskiy, I. et al. (2018). Molecularly thin two-dimensional hybrid perovskites with tunable optoelectronic properties due to reversible surface relaxation. *Nature Materials* 17: 908–914.

51 Shi, E., Deng, S., Yuan, B. et al. (2019). Extrinsic and dynamic edge states of two-dimensional lead halide perovskites. *ACS Nano* 13 (2): 1635–1644.

52 Cortecchia, D., Neutzner, S., Kandada, A.R.S. et al. (2017). Broadband emission in two-dimensional hybrid perovskites: the role of structural deformation. *Journal of the American Chemical Society* 139 (1): 39–42.

53 Yin, T., Liu, B., Yan, J. et al. (2019). Pressure-engineered structural and optical properties of two-dimensional $(C_4H_9NH_3)_2PbI_4$ perovskite exfoliated nm-thin flakes. *Journal of the American Chemical Society* 141 (3): 1235–1241.

54 Cao, D.H., Stoumpos, C.C., Farha, O.K. et al. (2015). 2D homologous perovskites as light-absorbing materials for solar cell applications. *Journal of the American Chemical Society* 137 (24): 7843–7850.

55 Mitzi, D.B., Feild, C.A., Harrison, W.T.A., and Guloy, A.M. (1994). Conducting tin halides with a layered organic-based perovskite structure. *Nature* 369: 467–469.

56 Liu, S., Li, F., Han, X. et al. (2018). Preparation and two-photon photoluminescence properties of organic inorganic hybrid perovskites $(C_6H_5CH_2NH_3)_2PbBr_4$ and $(C_6H_5CH_2NH_3)_2PbI_4$. *Applied Sciences* 8 (11): 2286.

57 Kamminga, M.E., Fang, H.H., Filip, M.R. et al. (2016). Confinement effects in low-dimensional lead iodide perovskite hybrids. *Chemistry of Materials* 28 (13): 4554–4562.

58 Gan, L., Li, J., Fang, Z. et al. (2017). Effects of organic cation length on exciton recombination in two-dimensional layered lead iodide hybrid perovskite crystals. *Journal of Physical Chemistry Letters* 8 (20): 5177–5183.

59 Du, K.Z., Tu, Q., Zhang, X. et al. (2017). Two-dimensional lead(II) halide-based hybrid perovskites templated by acene alkylamines: crystal structures, optical properties, and piezoelectricity. *Inorganic Chemistry* 56 (15): 9291–9302.

60 Chen, X., Lu, H., Li, Z. et al. (2018). Impact of layer thickness on the charge carrier and spin coherence lifetime in two-dimensional layered perovskite single crystals. *ACS Energy Letters* 3 (9): 2273–2279.

61 Peng, W., Yin, J., Ho, K.T. et al. (2017). Ultralow self-doping in two-dimensional hybrid perovskite single crystals. *Nano Letters* 17 (8): 4759–4767.

62 Akriti, Zhang, S., Lin, Z.Y. et al. (2021). Quantifying anionic diffusion in 2D halide perovskite lateral heterostructures. *Advanced Materials* 33: 2105183.

63 Shi, E., Yuan, B., Shiring, S.B. et al. (2020). Two-dimensional halide perovskite lateral epitaxial heterostructures. *Nature* 580: 614–620.

64 Tanaka, K. and Kondo, T. (2003). Bandgap and exciton binding energies in lead-iodide-based natural quantum-well crystals. *Science and Technology of Advanced Materials* 4 (6): 599–604.

65 Gao, X., Zhang, X., Yin, W. et al. (2019). Ruddlesden–popper perovskites: synthesis and optical properties for optoelectronic applications. *Advanced Science* 6: 1900941.

66 Nazarenko, O., Kotyrba, M.R., Yakunin, S. et al. (2018). Guanidinium-formamidinium lead iodide: a layered perovskite-related compound with red luminescence at room temperature. *Journal of the American Chemical Society* 140 (11): 3850–3853.

67 Kawano, N., Koshimizu, M., Sun, Y. et al. (2014). Effects of organic moieties on luminescence properties of organic–inorganic layered perovskite-type compounds. *Journal of Physical Chemistry C* 118 (17): 9101–9106.

68 Dou, L., Wong, A.B., Yu, Y. et al. (2015). Atomically thin two-dimensional organic–inorganic hybrid perovskites. *Science* 349 (6255): 1518–1521.

69 Dohner, E.R., Hoke, E.T., and Karunadasa, H.I. (2014). Self-assembly of broadband white-light emitters. *Journal of the American Chemical Society* 136 (5): 1718–1721.

70 Hu, T., Smith, M.D., Dohner, E.R. et al. (2016). Mechanism for broadband white-light emission from two-dimensional (110) hybrid perovskites. *Journal of Physical Chemistry Letters* 7 (12): 2258–2263.

71 Thirumal, K., Chong, W.K., Xie, W. et al. (2017). Morphology-independent stable white-light emission from self-assembled two-dimensional perovskites driven by strong exciton–phonon coupling to the organic framework. *Chemistry of Materials* 29 (9): 3947–3953.

72 Tabuchi, Y., Asai, K., Rikukawa, M. et al. (2000). Preparation and characterization of natural lower dimensional layered perovskite-type compounds. *Journal of Physics and Chemistry of Solids* 61 (6): 837–845.

73 Tu, Q., Spanopoulos, I., Yasaei, P. et al. (2018). Stretching and breaking of ultrathin 2D hybrid organic–inorganic perovskites. *ACS Nano* 12 (10): 10347–10354.

74 Tsai, H., Nie, W., Blancon, J.C. et al. (2016). High-efficiency two-dimensional Ruddlesden–Popper perovskite solar cells. *Nature* 536: 312–316.

75 Mao, L., Wu, Y., Stoumpos, C.C. et al. (2017). White-light emission and structural distortion in new corrugated two-dimensional lead bromide perovskites. *Journal of the American Chemical Society* 139 (14): 5210–5215.

76 Li, H., Lu, J., Zhang, T. et al. (2018). Cation-assisted restraint of a wide quantum well and interfacial charge accumulation in two-dimensional perovskites. *ACS Energy Letters* 3 (8): 1815–1823.

77 Mitzi, D.B. (1996). Synthesis, crystal structure, and optical and thermal properties of $(C_4H_9NH_3)_2MI_4$ (M = Ge, Sn, pb). *Chemistry of Materials* 8 (3): 791–800.

78 Blancon, J.C., Tsai, H., Nie, W. et al. (2017). Extremely efficient internal exciton dissociation through edge states in layered 2D perovskites. *Science* 355 (6331): 1288–1292.

79 Lemmerer, A. and Billing, D.G. (2012). Synthesis, characterization and phase transitions of the inorganic–organic layered perovskite-type hybrids $[(C_nH_{2n+1}NH_3)_2PbI_4]$, $n = 7, 8, 9$ and 10. *Dalton Transactions* 41: 1146–1157.

80 Manser, J.S., Christians, J.A., and Kamat, P.V. (2016). Intriguing optoelectronic properties of metal halide perovskites. *Chemical Reviews* 116 (21): 12956–13008.

81 Lemmerer, A. and Billing, D.G. (2012). Lead halide inorganic–organic hybrids incorporating diammonium cations. *CrystEngComm* 14: 1954–1966.

82 Straus, D.B. and Kagan, C.R. (2018). Electrons, excitons, and phonons in two-dimensional hybrid perovskites: connecting structural, optical, and electronic properties. *Journal of Physical Chemistry Letters* 9 (6): 1434–1447.

83 Safdari, M., Svensson, P.H., Hoang, M.T. et al. (2016). Layered 2D alkyldiammonium lead iodide perovskites: synthesis, characterization, and use in solar cells. *Journal of Materials Chemistry A* 4: 15638–15646.

84 Di, J., Chang, J., and Liu, S. (2020). Recent progress of two-dimensional lead halide perovskite single crystals: crystal growth, physical properties, and device applications. *EcoMat* 2: e12036.

85 Frenkel, J. (1939). A general theory of heterophase fluctuations and pretransition phenomena. *The Journal of Chemical Physics* 7: 538–547.

86 Wang, K., Wu, C., Yang, D. et al. (2018). Quasi-two-dimensional halide perovskite single crystal photodetector. *ACS Nano* 12 (5): 4919–4929.

87 Liu, Y., Yang, Z., and Liu, S. (2018). Recent progress in single-crystalline perovskite research including crystal preparation, property evaluation, and applications. *Advanced Science* 5: 1700471.

88 Nernst, W. (1904). Theorie der Reaktionsgeschwindigkeit in heterogenen Systemen. *Zeitschrift für Physikalische Chemie* 47U (1): 52–55.

89 Dash, J.G. (1977). Clustering and percolation transitions in helium and other thin films. *Physical Review B* 15: 3136–3146.

90 Mitzi, D.B. (1999). A layered solution crystal growth technique and the crystal structure of $(C_6H_5C_2H_4NH_3)_2PbCl_4$. *Journal of Solid State Chemistry* 145 (2): 694–704.

91 Liao, Y., Liu, H., Zhou, W. et al. (2017). Highly oriented low-dimensional tin halide perovskites with enhanced stability and photovoltaic performance. *Journal of the American Chemical Society* 139 (19): 6693–6699.

92 Fu, P., Huang, M., Shang, Y. et al. (2018). Organic–inorganic layered and hollow tin bromide perovskite with tunable broadband emission. *ACS Applied Materials and Interfaces* 10 (40): 34363–34369.

93 Lanzetta, L., Marin-Beloqui, J.M., Sanchez-Molina, I. et al. (2017). Two-dimensional organic tin halide perovskites with tunable visible emission and their use in light-emitting devices. *ACS Energy Letters* 2 (7): 1662–1668.

94 Kagan, C.R., Mitzi, D.B., and Dimitrakopoulos, C.D. (1999). Organic–inorganic hybrid materials as semiconducting channels in thin-film field-effect transistors. *Science* 286 (5441): 945–947.

95 Cheng, P., Wu, T., Zhang, J. et al. (2017). $(C_6H_5C_2H_4NH_3)_2GeI_4$: a layered two-dimensional perovskite with potential for photovoltaic applications. *Journal of Physical Chemistry Letters* 8 (18): 4402–4406.

96 Shibuya, K., Koshimizu, M., Nishikido, F. et al. (2009). Poly[bis(phenethylammonium) [dibromidoplumbate(II)]-di-μ-bromido]. *Acta Crystallographica Section E: Structure Reports Online* E65: m1323–m1324.

97 Yin, J., Maity, P., Xu, L. et al. (2018). Layer-dependent Rashba band splitting in 2D hybrid perovskites. *Chemistry of Materials* 30 (23): 8538–8545.

98 Niu, W., Eiden, A., Prakash, G.V., and Baumberg, J.J. (2014). Exfoliation of self assembled 2D organic–inorganic perovskite semiconductors. *Applied Physics Letters* 104: 171111.

99 Yaffe, O., Chernikov, A., Norman, Z.M. et al. (2015). Excitons in ultrathin organic–inorganic perovskite crystals. *Physical Review B* 92: 045414.

100 Gao, Y., Shi, E., Deng, S. et al. (2019). Molecular engineering of organic–inorganic hybrid perovskites quantum wells. *Nature Chemistry* 11: 1151–1157.

101 Chen, J., Gan, L., Zhuge, F. et al. (2017). A ternary solvent method for large-sized two-dimensional perovskites. *Angewandte Chemie, International Edition* 56: 2390–2394.

102 Guo, R., Zhu, Z., Boulesbaa, A. et al. (2017). Synthesis and photoluminescence properties of 2D phenethylammonium lead bromide perovskite nanocrystals. *Small Methods* 1: 1700245.

103 Ma, D., Fu, Y., Dang, L. et al. (2017). Single-crystal microplates of two-dimensional organic–inorganic lead halide layered perovskites for optoelectronics. *Nano Research* 10: 2117–2129.

104 Gao, Y., Wei, Z., Hsu, S.N. et al. (2020). Two-dimensional halide perovskites featuring semiconducting organic building blocks. *Materials Chemistry Frontiers* 4: 3400–3418.

105 Gao, Y. and Dou, L. (2021). Organic semiconductor-incorporated two-dimensional halide perovskites. *National Science Review* 9 (5): nwab111.

106 Gélvez-Rueda, M.C., Fridriksson, M.B., Dubey, R.K. et al. (2020). Overcoming the exciton binding energy in two-dimensional perovskite nanoplatelets by attachment of conjugated organic chromophores. *Nature Communications* 11: 1901.

107 Pan, D., Fu, Y., Spitha, N. et al. (2021). Deterministic fabrication of arbitrary vertical heterostructures of two-dimensional Ruddlesden–Popper halide perovskites. *Nature Nanotechnology* 16: 159–165.

108 Le, Q.V., Hong, K., Jang, H.W., and Kim, S.Y. (2018). Halide perovskite quantum dots for light-emitting diodes: properties, synthesis, applications, and outlooks. *Advanced Electronics Materials* 4: 1800335.

109 Zhao, Q., Hazarika, A., Chen, X. et al. (2019). High efficiency perovskite quantum dot solar cells with charge separating heterostructure. *Nature Communications* 10: 2842.

110 Liang, J., Chen, D., Yao, X. et al. (2020). Recent progress and development in inorganic halide perovskite quantum dots for photoelectrochemical applications. *Small* 16: 1903398.

111 Rahaman, M.Z., Ge, S., Lin, C. et al. (2021). One-dimensional molecular metal halide materials: structures, properties, and applications. *Small Struct.* 2: 2000062.

112 Chen, Y., Wu, X., Chu, Y. et al. (2018). Hybrid field-effect transistors and photodetectors based on organic semiconductor and $CsPbI_3$ perovskite nanorods bilayer structure. *Nano-Micro Letters* 10: 57.

113 Gao, P., Yusoff, A.R.B.M., and Nazeeruddin, M.K. (2018). Dimensionality engineering of hybrid halide perovskite light absorbers. *Nature Communications* 9: 5028.

114 Chen, A.Z., Shiu, M., Ma, J.H. et al. (2018). Origin of vertical orientation in two-dimensional metal halide perovskites and its effect on photovoltaic performance. *Nature Communications* 9: 1336.

115 Zhang, L., Sun, C., He, T. et al. (2021). High-performance quasi-2D perovskite light-emitting diodes: from materials to devices. *Light: Science and Applications* 10: 61.

116 Hsu, S.N., Zhao, W., Gao, Y. et al. (2021). Thermoelectric performance of lead-free two-dimensional halide perovskites featuring conjugated ligands. *Nano Letters* 21 (18): 7839–7844.

117 Xie, A., Hettiarachchi, C., Maddalena, F. et al. (2020). Lithium-doped two-dimensional perovskite scintillator for wide-range radiation detection. *Communications Materials* 1: 37.

118 Li, Y.T., Ding, L., Li, J.Z. et al. (2019). Light-enhanced ion migration in two-dimensional perovskite single crystals revealed in carbon nanotubes/two-dimensional perovskite heterostructure and its photomemory application. *ACS Central Science* 5 (11): 1857–1865.

119 Tian, H., Zhao, L., Wang, X. et al. (2017). Extremely low operating current resistive memory based on exfoliated 2D perovskite single crystals for neuromorphic computing. *ACS Nano* 11 (12): 12247–12256.

120 Qin, C., Sandanayaka, A.S.D., Zhao, C. et al. (2020). Stable room-temperature continuous-wave lasing in quasi-2D perovskite films. *Nature* 585: 53–57.

121 Mao, L., Stoumpos, C.C., and Kanatzidis, M.G. (2019). Two-dimensional hybrid halide perovskites: principles and promises. *Journal of the American Chemical Society* 141 (3): 1171–1190.

122 Zhou, N., Shen, Y., Li, L. et al. (2018). Exploration of crystallization kinetics in quasi two-dimensional perovskite and high performance solar cells. *Journal of the American Chemical Society* 140 (1): 459–465.

123 Finkenauer, B.P., Gao, Y., Wang, X. et al. (2021). Mechanically robust and self-healable perovskite solar cells. *Cell Reports Physical Science* 2 (2): 100320.

124 Pitaro, M., Tekelenburg, E.K., Shao, S., and Loi, M.A. (2022). Tin halide perovskites: from fundamental properties to solar cells. *Advanced Materials* 34: 2105844.

125 Connor, B.A., Leppert, L., Smith, M.D. et al. (2018). Layered halide double perovskites: dimensional reduction of $Cs_2AgBiBr_6$. *Journal of the American Chemical Society* 140 (15): 5235–5240.

126 Zhang, X., Wang, H., Wang, S. et al. (2020). Room temperature synthesis of all inorganic lead-free zero-dimensional Cs_4SnBr_6 and Cs_3KSnBr_6 perovskites. *Inorganic Chemistry* 59 (1): 533–538.

5

Halide Double Perovskites

Carina Pareja-Rivera[1], Dulce Zugasti-Fernández[1], Paul Olalde-Velasco[2], and Diego Solis-Ibarra[1]

[1] *Universidad Nacional Autónoma de México, Instituto de Investigaciones en Materiales, Laboratorio de Fisicoquímica y Reactividad de Superficies (LaFReS), Circuito Interior s/n, Cd. Universitaria, Coyoacán, Ciudad de México 04510, México*
[2] *Universidad Nacional Autónoma de México, Departamento de Ciencias Químicas, Facultad de Estudios Superiores Cuautitlán, Av. Primero de Mayos s/n, Cuautitlán Izcalli, Estado de México 54740, México*

Halide perovskites have tremendous potential for their use in optoelectronic applications and, perhaps, most notably, as absorbers in solar cells [1]. The superb properties of halide perovskites, particularly lead-based perovskites, have motivated the search for alternative materials that share similar properties while reducing the drawbacks of these perovskites [2]. Perhaps some of the most notable drawbacks are the intolerance to light and moisture and the known toxicity of lead that is ubiquitous in most relevant materials [3]. While several different metals have been explored as a potential replacement for lead, only a handful have shown some promise [4]. Among those examples are tin and germanium perovskites, such as $[MA]SnI_3$ and $[MA]GeI_3$ (where MA is methylammonium). Indeed, devices with tin and germanium perovskites have reached outstanding power conversion efficiencies (PCE) of up to 10.37% and 0.57%, respectively [5, 6]. However, tin and germanium perovskites have limited stability since both metals are easily oxidized to their respective +4 oxidation state [6].

Beyond tin and germanium halide perovskites, other single-metal perovskites have shown little to no potential for optoelectronic applications, and thus, the search for alternatives has intensified in the past few years [7]. In halide double perovskites, or HDPs, a combination of two metals –hence double– substitute the single B-site metal. The implications of substituting one metal for two are huge. First, it opens the door to many other metals, for which a 2+ oxidation state is not viable, being incorporated in perovskite structures. Second, it vastly expands the pool of possible halide perovskites, as now it is not only a matter of a single metal in a single oxidation state. Instead, it consists of a pool of permutations between different metals in different oxidation states, thus exponentially increasing the number of possible materials and potential applications [8].

Halide Perovskite Semiconductors: Structures, Characterization, Properties, and Phenomena, First Edition. Edited by Yuanyuan Zhou and Iván Mora-Seró.

Akin to other halide perovskites, HDPs have been known for long time, with the first report of an HDP dating back to the late nineteenth century (K_2NaAlF_6). Also, similarly to other halide perovskites, there has been a recent renaissance of HDPs, at first driven by the search for new materials for optoelectronic applications and more recently propelled by their diversity and potential for many other applications. This chapter will begin by defining and examining HDPs, their structure and properties, current and prospective applications, and conclude with an outlook for this interesting family of materials.

5.1 Definition and Structure

HDPs, also known as elpasolites, are materials with the general chemical formula $A_2BB'X_6$. They are obtained from the substitution of the divalent metal B by a monovalent cation B(I) and another trivalent B′(III) within the crystallographic sites of a single perovskite, ABX_3 (X = Cl, Br, I) [9].

In HDPs, metal cations B and B′ alternate within lattice sites, presenting a rock salt-type arrangement forming FCC sublattices that are intertwined and displaced half a unit cell along the three dimensions [10]. Just as in any perovskite structure, the octahedrons $[BX_6]^{n-}$ and $[BX_6]^{n-}$ are corner-sharing in a three-dimensional (3D) network (see Figure 5.1).

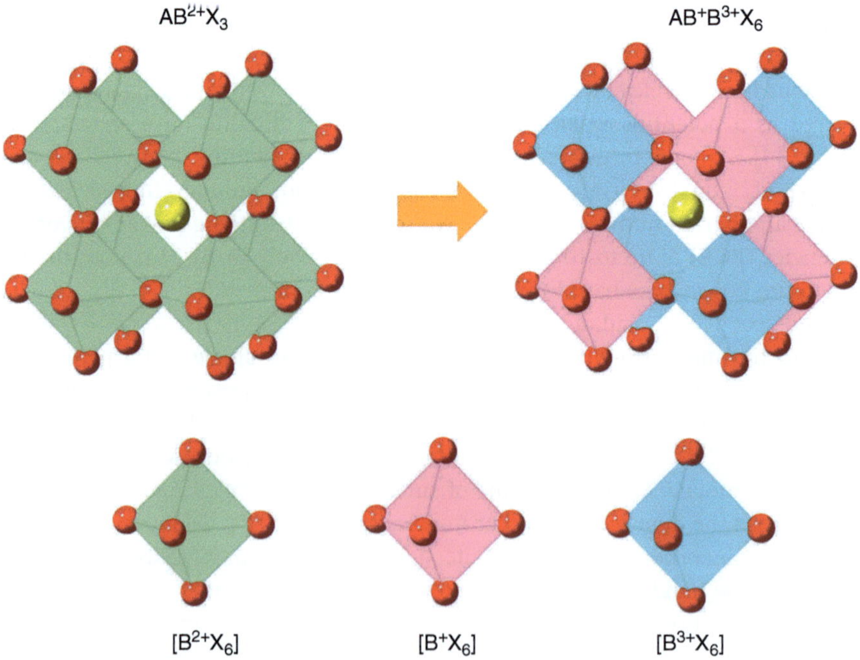

Figure 5.1 Structure of single metal, $AB^{2+}X_3$ (left) and its substitution to form double perovskites, $A_2B^+B^{3+}X_6$. The A^+, B^+, B^{3+}, and X^- are represented by the yellow ball, pink and blue octahedron, and red balls, respectively. Source: Zhang et al. [11]/Royal Society of Chemistry.

As discussed before, introducing different metals within the halide perovskite structure and the possibility of making combinations of metals makes it possible to obtain a wide range of materials with a broader set of electronic properties [9, 12]. However, another option for substitution is using the same type of atom B but changing its oxidation state; such is the case with the double perovskite $Cs_2Au^{1+}Au^{3+}Cl_6$ [13]. The criteria for metal selection will be detailed at the end of this section.

It is worth mentioning that due to distortions in the octahedrons, some perovskites can crystallize either in the tetragonal or the rhombohedral systems, such as $Cs_2Au^{1+}Au^{3+}X_6$ ($X = Cl^-$, Br^-, and I^-) and $(MA)_2KBiCl_6$ (see Figure 5.2) [15].

Changing the composition of B and B′ makes it possible to obtain subnetworks that will depend on the charge difference (ΔQ) between B and B′ , which in turn allows achieving either random, rock-salt, or layered orderings of the B sites in the 3D network. Double perovskites present a rock-salt type order, and this happens when $\Delta Q \geq 2$. In the random ordering, as its name suggests, B and B′ randomly occupy some place B, while in the layered double perovskites, B and B′ form alternating planes of octahedrons (Figure 5.3). Although, in general, the term layered double perovskite is used to refer to other types of perovskites, like Ruddlesden–Popper,

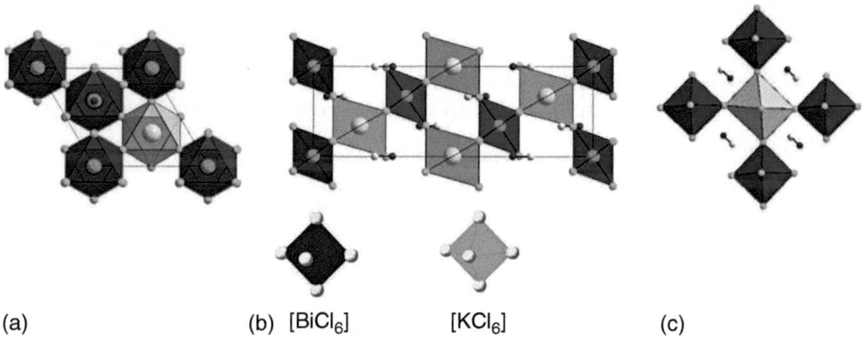

Figure 5.2 Crystal structure of rhombohedral $(MA)_2KBiCl_6$ (MA = methylammonium), obtained from single-crystal X-ray diffraction, is viewed along (a) the *c* axis, (b) the *a* axis, and (c) tilted to show the MA location. Source: Wei et al. [14]/Royal Society of Chemistry.

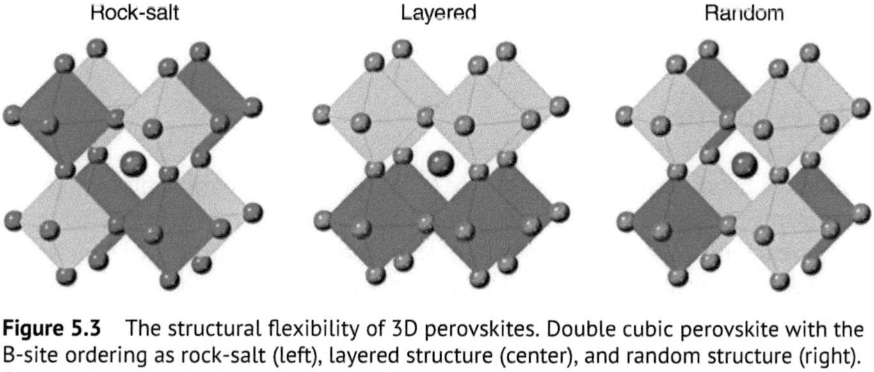

Figure 5.3 The structural flexibility of 3D perovskites. Double cubic perovskite with the B-site ordering as rock-salt (left), layered structure (center), and random structure (right). Source: Yin et al. [16]/Royal Society of Chemistry.

Dion–Jacobson, and (111) layered perovskites, whose characteristics will be detailed in Section 5.5.

It is essential to emphasize that when we talk about double perovskites, we refer to double B-site perovskites since there is a variation in the composition of double perovskites when the A cation is replaced by two different types of cations, i.e. $AA'BB'X_6$ [17]. When A and A′ are different sizes, another configuration could be generated, as will be discussed later in the chapter.

It is worth mentioning that other heterovalent substitutions to the B-site exist, where it is possible to maintain the charge balance presented by the conventional ABX_3 perovskites, such as in the case of the structures $A_3B_2{}^{3+}\square X_6$ [3] or $A_2B^{4+}\square X_6$, where □ represents the vacancy of a B-site [18]. However, vacancies result in a lack of connectivity between the octahedrons. These materials will be considered in a particular case and discussed in Section 5.5.

When there is a discrepancy between the sizes of cations and anions, distortions will be generated in the cubic structure. These distortions can be of such magnitude that the perovskite structure is no longer retained, thus it is necessary to define a parameter that allows us to determine whether certain compounds are likely to form perovskites phases or not. For simple perovskites, Goldschmidt's tolerance factor is an excellent approximation to predict whether perovskites will be structurally stable or not. However, for HDPs, it has been shown that the Goldschmidt factor is insufficient to predict the structural stability of these compounds, so a new probabilistic factor that predicts the formation of both simple perovskites and double perovskites has been proposed. This new factor τ is given by

$$\tau = \frac{r_X}{r_B} - n_A\left(n_A - \frac{r_A/r_B}{\ln\left(\frac{r_A}{r_B}\right)}\right)$$

where n_A is the oxidation state of A, and r_i corresponds to the ionic radius, by definition $r_A > r_B$. Within the studies that have been done to validate this new factor, it has been observed that there is a monotonic dependence between the stability of the perovskite and τ: the probability of obtaining a structurally stable perovskite increases as τ decreases. So, for $\tau < 4.18$ the perovskite formation is most probable. This new factor τ presents an experimentally proven accuracy of 92% [19].

5.2 Properties

Lead-based halide perovskites such as $MAPbI_3$ have optoelectronic properties suitable for applications in photovoltaic and optoelectronic devices. Therefore, materials that can emulate some of the properties of lead perovskites with reduced toxicity and improved stability are heavily sought-after materials.

The properties of $MAPbI_3$ are due to electronic transitions between the maximum of the valence band (VBM) and the minimum of the conduction band (CBM). The VBM is mainly composed by Pb^{2+} 5s and I^- 5p orbitals, while the CBM is primarily formed by Pb^{2+} 6p orbitals. Thus, the first natural attempts to replace Pb^{2+} involved

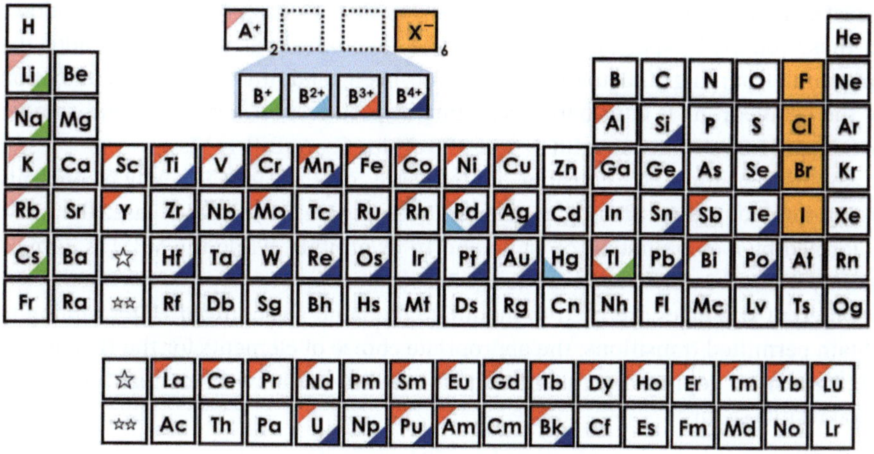

Figure 5.4 The elements that can be used in the composition of HDPs. Pink, green, cyan, red, blue, and orange denote the elements incorporated in the sites A^+, B^+, B^{2+}, B^{3+}, B^{4+} and X^-, respectively. Source: Wolf et al. [15]/John Wiley & Sons.

the same group IV elements, such as Sn^{2+} and Ge^{2+}. However, both Sn^{2+} and Ge^{2+} tend to oxidize in the presence of molecular oxygen (O_2) into Sn^{4+} and Ge^{4+}.

In the case of HDPs, by implementing both trivalent and monovalent B sites, the possible substitutions for less toxic elements are augmented, which allows the modulation of the bandgap and, improves stability. For this reason, HDPs are candidates to produce materials with similar characteristics to those of $MAPbI_{3.}$

Given the variety of elements that can be used for the substitution of sites A, B^+, B^{3+}, and X, we present in Figure 5.4 the elements, which can occupy each site according to their oxidation state when incorporated into HDPs [8].

Depending on their composition, mainly B-site metals, HDPs can present either direct or indirect bandgaps with parity-forbidden transitions [15]. For example, the two most studied cases of HDPs are $Cs_2AgBiBr_6$ and $Cs_2AgBiCl_6$, which both present indirect bandgaps of 1.95 and 2.77 eV, respectively. Although these values are considered inappropriate for solar cells with a single absorbent material, these materials may be useful in tandem solar cells [20]. In both cases, the CBM is derived from the 6p orbitals of Bi, while the VBM of $Cs_2AgBiCl_6$ is formed by hybridization between the Ag 5d and Cl 3p orbitals [11], and the VBM of $Cs_2AgBiBr_6$ mainly derives from the 6s orbitals Bi, which is a metal-to-metal charge transfer (MMCT) bandgap [21]. This type of transfer is the most common among HDPs.

Some variations have been observed in the reported values of the bandgap both experimentally and in theoretical calculations for HDPs. In particular, for $Cs_2AgBiCl_6$, the dispersion is up to 0.57 eV in the experimental values, while it is up to 0.8 eV in the calculations [14, 22]. The discrepancy in the experimental values reported is attributed to the synthesis, the role of disorder, the techniques, and the measurement processes used to determine the bandgap. In the case of calculations, different bandgap values are obtained when using different density functionals; however, this does not change its nature, namely whether the bandgap is either direct or indirect. Moreover, it has been proven that the calculated values are closer to those reported experimentally when the theoretical calculations consider the spin–orbit coupling. Calculations have also been made in which the inclination of the octahedrons is considered, resulting in pseudocubic structures and smaller bandgaps (1.26 eV for $Cs_2AgBiCl_6$ and 0.88 eV for $Cs_2AgBiBr_6$) [23].

Another example to consider is the case of $Cs_2AgInCl_6$, which presents parity-forbidden direct bandgap of 2.1 eV (see Figure 5.5). In this case, the CBM is formed by a combination of Cl 3p and In 5s orbitals, while the VBM is derived mainly from Cl 3p and Ag 4d orbitals [17, 25].

Different strategies are used to modulate the band structure in HDPs aiming to obtain permitted transitions: the appropriate choice of elements for the B^+ and B^{3+} sites, chemical doping, and the introduction of disorder between cations have given good results so far.

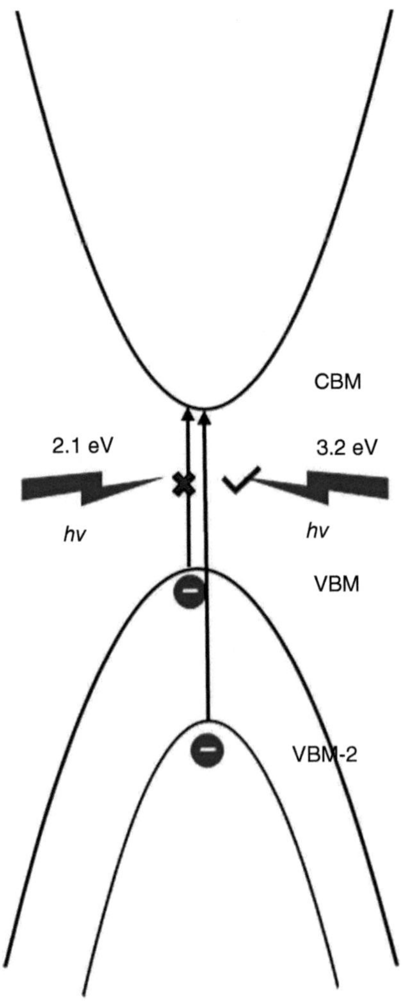

Figure 5.5 Parity-forbidden transition between VBM and CBM and parity-allowed transition between VBM-2 and CBM in $Cs_2AgInCl_6$. Source: Luo et al [24]/American Chemical Society.

5.2.1 Chemical Doping

Chemical doping with monovalent or trivalent metals can change parity-forbidden transitions. In isovalent alloys, stable solid solutions are formed, while in alliovalent alloys, the dopants produce cation vacancies, which achieve nonradiative recombination of those carriers derived from interstitial defects due to the low lattice mismatch [15, 26].

An example of this is the bandgap modulation of $Cs_2AgBiBr_6$ by doping Tl at low concentrations (Tl <1 atom %) to obtain $Cs_2(Ag_{1-a}Bi_{1-b})Tl_xBr_6$, where it was possible to reduce the bandgap from 1.96 eV for $Cs_2AgBiBr_6$ to 1.5 eV for $Cs_2(AgBi_{1-x})Tl_xBr_6$ ($x = 0.075$) by substituting Tl^{3+} in the Bi^{3+} sites while maintaining an indirect bandgap. On the other hand, substituting the Ag^+ sites with Tl^+ results in a bandgap reduction of 0.1 eV and turns it direct [27]. Before doping, the bands in this perovskite are formed by Bi s states at the VBM that interact with the Ag d states along the [100] direction. This interaction leads to an indirect bandgap in $CsAgBiCl_6$ when doping is carried out with Tl^+, as it allows the Tl 6s orbitals to change the direction of the interaction, resulting in a direct bandgap and lowering the band energy (Figure 5.6).

$Cs_2AgTlCl_6$ does not present detectable emission in PL because the carriers thermalize in the forbidden region and remain there; they do not recombine radiatively. In contrast, doping HDPs can also improve emission efficiency, as occurs with perovskites Sb^{3+}:$Cs_2NaInCl_6$, where doping with Sb^{3+} breaks the parity-forbidden transition and optimizes the photoluminescence quantum yield (PLQY) up to 75.89%, which is high considering that the highest blue PLQY recorded is 78.9% for Cs_2SnCl_6 [26].

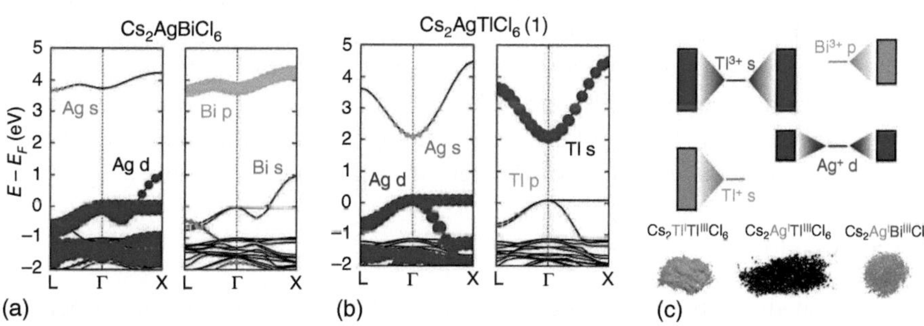

Figure 5.6 Band structures of (a) $Cs_2AgBiCl_6$ and (b) $Cs_2AgTlCl_6$, calculated with DFT-HSE06+SOC in their primitive unit cells using measured lattice parameters. Band structures are shown in duplicate with only metal orbital contributions shown; chloride orbitals are present but omitted for clarity. The conduction bands are rigidly shifted to match the experimental bandgaps, and all bands are shifted to align the Cs 5s bands. (c) Top: Schematic representation of the relative energies of the atomic orbitals (lines) and bands (rectangles) for closely related double perovskites, demonstrating the dominant metal-to-metal charge-transfer character of the bandgaps and the origin of the anomalously low bandgaps in $Cs_2AgTlCl_6$ and $Cs_2AgTlBr_6$. Bottom: Photographs of $Cs_2Tl^+Tl^{3+}Cl_6$ (left), $Cs_2AgTlCl_6$ (center), and $Cs_2AgBiCl_6$ (right). Source: Slavney et al. [27]/John Wiley & Sons.

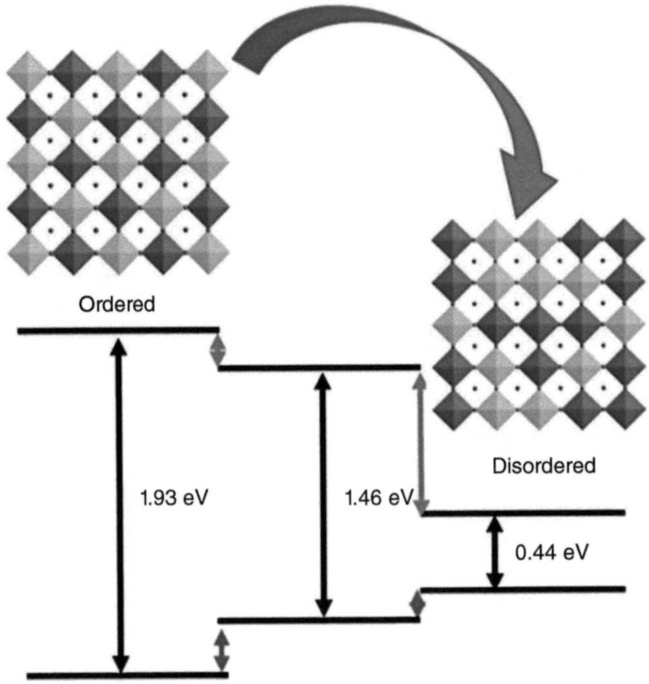

Figure 5.7 Bandgap evolution of the ordered, partially disordered, and fully disordered $Cs_2AgBiBr_6$. Source: Yang et al. [28]/American Chemical Society.

5.2.2 Random Ordering

Modifying the ordering of the cations in a semiconductor alloy significantly changes its band structure. A case study in this regard is the ordered compound $Cs_2AgBiBr_6$, where simulating the change in the random B sites of the Bi^{3+} and Ag^+ ions results in a bandgap reduction from 1.93 to 0.44 eV. However, it would be difficult to carry out this experiment experimentally since this would only be possible when the temperature is increased up to 800 K, where a phase change begins [28]. Recent experiments have been carried out with a maximum temperature of 150 °C (Figure 5.7) [29].

5.2.3 Stability

As mentioned before, one of the purposes of HDPs is to achieve better stability than their lead counterparts. However, before reaching that point, it is necessary to analyze a vast amount of experimental and theoretical results. Theoretical stability studies are another valuable tool for the selection of HDPs whose synthesis is thermodynamically favorable. For this purpose, the energy above the convex hull E_{hull} is used, which measures the decomposition energy of HDPs in a linear combination of stable phases present in the phase diagram. When a compound is stable, E_{hull} is zero; that is, it is in the convex hull and in equilibrium phases. While more positive values of E_{hull} indicate decreasing stability [30].

From these kinds of studies, it is important to signal those compounds with a small positive value of E_{hull}, although theoretically unstable, can be synthesized

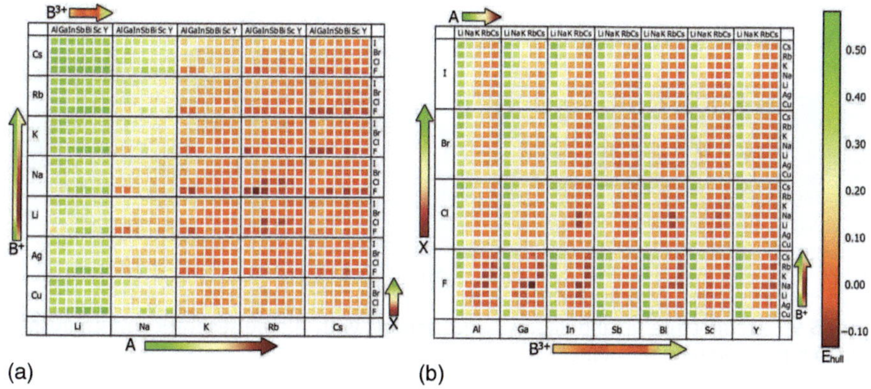

Figure 5.8 Color map of the calculated energy above the convex hull E_{hull} (in eV per atom) of 980 $A_2B^+B^{3+}X_6$ double perovskites. The order of the A^+, B^+, B^{3+}, and X^- elements is different in (a) and (b), as shown by the arrows. Since the color in (a) is mainly green on the left side and mainly red on the right side, the B^+ arrows have two parts: green–yellow and red–yellow, respectively. Source: Zhang et al. [31]/American Chemical Society.

and are stable for a certain period of time (perhaps small compared to other perovskites).

Recent calculations show that the elements that most influence the thermodynamic stability of HDPs are those used in A, X and B^+ sites, while the trivalent metal, B^{3+} negligibly effects on the stability. It has also been found that stability increases gradually as the mass of A increases (from Li to Cs) while that of X decreases (from I to F); on the other hand, stability decreases when B^+ increases (from Li to Cs) and increases when B^+ changes from Cu to Ag. This can be seen in more detail in Figure 5.8 [31].

$Cs_2AgBiBr_6$ has shown stability in light and humidity for up to 15 days, while in darkness for up to 30 days [25]. Another case that we must consider is $Cs_2AgInCl_6$, which as powders is stable for up to three months, while as single crystals for up to five months [25].

5.3 Applications in Solar Cells and LEDs

Because many HDPs possess promising optoelectronic properties, such as high defect tolerance, good environmental stability, tunable bandgap, and others, natural applications explored for HDPs are in solar cells and light-emitting diodes (LEDs). Currently, the most studied HDPs are $Cs_2AgBiBr_6$ and $Cs_2AgInCl_6$. Both have been implemented in solar cells and LEDs, and many other HDPs will likely be incorporated shortly in such devices.

5.3.1 Photovoltaic Solar Cells

One of the most important applications of HDPs is perovskite-based solar cells (PSCs). In a photovoltaic process, light is absorbed by a perovskite, and the

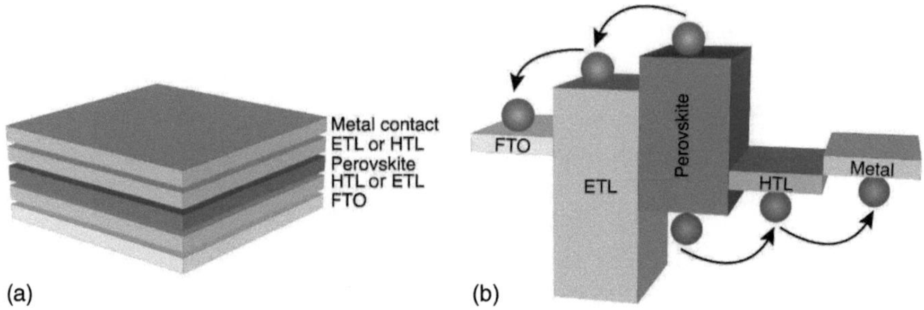

Figure 5.9 Architecture (a) and energy band diagram with charge separation mechanism (b) for a typical perovskite solar cell.

photogenerated electrons are selectively injected into the conduction band of an electron transporter layer (ETL) and extracted through a transparent conducting oxide (TCO). At the same time, holes are blocked due to the band alignment. In the same way, holes are injected into a hole-transporting material. The process mentioned above is depicted in Figure 5.9. However, the development of $Cs_2AgBiBr_6$–based PSCs is hampered by their relatively large and indirect bandgap and the difficulty in preparing homogeneous thin films. In 2017, Bein et al. reported $Cs_2AgBiBr_6$ films synthesized by spin-coating as an absorber layer for PSCs. These films showed excellent stability with a power conversion efficiency (PCE) close to 2.5% [32]. Subsequently, different methods for deposition [33] and interface engineering were adopted. The case of planar heterojunction synthesized by a low-pressure assisted method showed a PCE of ~1.44% [34], while an inverted planar heterojunction PSC fabricated by the antisolvent drip method showed a PCE of 2.23% [35].

From the compositional point of view, Zhu et al. observed that replacing Bi with Sb reduces the bandgap of $Cs_2AgSb_xBi_{1-x}Br_6$ from 2.22 eV ($x = 0$) to 1.97 eV ($x = 0.75$), achieving PCE of 0.19% and 0.25%, respectively. The poor PCE values are attributed to the lower quality of the thin films and their large pin-hole defects [36]. On the other hand, Tang et al. have studied the incorporation of alkali metals (Li^+, Na^+, K^+, and Rb^+), which have improved the quality of $Cs_2AgBiBr_6$ thin films by reducing defects such as pin-holes, favoring greater tolerance to high temperatures and air conditions without encapsulation, increasing PCEs up to 2.57% [37].

Generally, the PCE of $Cs_2AgBiBr_6$ is limited by inherent and extrinsic defects in thin films. Thus, additional buffering, transporting, or intermediate layers will certainly be key for developing efficient HDP-based solar cells. For example, Shao et al. introduced an intermediate layer of the dye N719 (di-tetrabutylammonium cis-bis(isothiocyanate)bis(2,2′-bipyridyl-4,4′ dicarboxylate)ruthenium(II)) on the surface. The dye broadens the spectrum for light absorption, reduces thin film defects, and promotes hole extraction, resulting in a PCE of 2.84% [38]. Independently, Wang et al. combined dye sensitization with double perovskite $Cs_2AgBiBr_6$ to prepare organic dye/perovskite heterojunction solar cells. With this scheme, using chlorophyll derivatives as a hole transporting layer [39] and indoline dyes, solar cells with maximum PCEs of 3.11% and 4.23% were prepared successively [40, 41].

In addition to $Cs_2AgBiBr_6$ and its derivatives, a Cs_2NaBiI_6 solar cell was also fabricated. Because it presents a small bandgap (1.66 eV), it was used as the absorbent material, which can be deposited by evaporation [42] or by spin coating [43]. This last method presented PCE of 0.42% and high stability to humidity and oxygen in the environment. These last reports do not rule out the possibility of using other types of HDPs in PSC.

5.3.2 Light-Emitting Diodes (LEDs)

Generally speaking, there are three possible uses for HDP in LEDs. In the first approach, HDP LEDs are the only emitting material. The second approach uses materials whose composition allows for emission in different wavelengths, particularly red, green, and blue (RGB), since combining these allows for white light. The third approach combines a single LED device (UV or blue) coated with a phosphor. In this structure, the light from the LED excites the phosphor, and the two emissions combine to generate white light, see Figure 5.10. However, this approach suffers from a low color rendering index (CRI) due to red deficiency and high correlated color temperature (CCT). To overcome this problem, a combination of additional phosphors can be integrated [44].

The luminescence types found in HDP-based LED devices' are both the narrow-band emission derived from free exciton (FE) and the broad-band white-light emission from self-trapped exciton (STE). [45] These are favorable for high external quantum efficiency (EQE) and CRI [46, 47]. Most HDPs reported to date only contain chloride as the halide component for high quantum efficiency and stability [48, 49].

5.3.3 White-LEDs

In 2019, Yella et al. [50] demonstrated that by synthesizing 30% Bi^{3+} doped $Cs_2AgInCl_6$ nanocrystals, their bandgap changed to a direct transition and decreased compared to $Cs_2AgInCl_6$. Dispersing nanocrystals in a polymethylmethacrylate (PMMA) polymer matrix showed white light emission with an excellent CRI (CRI ≈ 91). This report determined that HDPs could be implemented as an emitting material in white light-emitting diodes (WLEDs) due to their broad

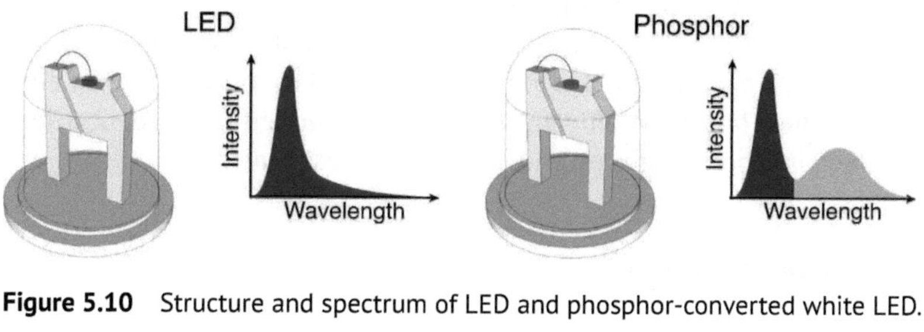

Figure 5.10 Structure and spectrum of LED and phosphor-converted white LED.

emission. So far, there is no reported case of HDPs as an emitting material in LEDs, so we believe this report may be promising.

5.3.4 Phosphorus

An alternative to implementing HPDs in WLEDs is to use them as phosphors by combining them with UV LED chips to generate high efficiency and CRI to produce white light. Such is the case for Bi^{3+}–Er^{3+} co-doped $Cs_2AgInCl_6$, where co-doping modifies the host's optical absorption and emission spectra. Bi^{3+} co-doping gives rise to a new absorption channel at 372 nm, suitable for excitation with commercial UV LEDs [51]. Tang et al. reported that Ho^{3+} doped $Cs_2(Ag,Na)InCl_6$ HDPs form energy transfer channels from STEs to Ho^{3+}, which favors the emission of warm white light. The crystals were ball milled into a fine powder, and the powder was painted onto commercial UV LED chips without encapsulation [52].

An ideal phosphor for LED applications would retain the high quantum efficiency and broad STE emission of $Cs_2Ag_{1-x}Na_xInCl_6$ while shifting the excitation spectrum to longer wavelengths where it can be excited by $Ga_{1-x}In_xN$ LEDs. It would also be desirable to red-shift the emission band to obtain warmer white light. For this purpose, phosphors such as $Cs_2Ag_{0.6}Na_{0.4}InCl_6$:$Bi^{3+}$ are used. Here, the Na alloy reduced the electronic dimensionality and broke the parity-forbidden transition while doping with traces of Bi^{3+} to suppress recombination centers. Hence simultaneously enhancing the radiative recombination and suppressing the nonradiative recombination, leading to the record PLQY for white phosphors [53]. Unlike Na^+, Bi^{3+} can act as a sensitizer, which extends the excitation spectrum into the visible region, making it very attractive for use in LEDs [54].

5.3.5 Two or More Phosphorus

An alternative approach for developing WLEDs is to use different near-UV (NUV) or UV-emitting phosphors. Recently, Liu et al. used the combination of the HDP $Cs_2AgInCl_6$:Cr^{3+}, which exhibits broadband near-infrared (NIR) emission, with green and red phosphors in a UV LED to obtain a WLED [55]. On the other hand, Woodward et al. employed the HDP $Cs_2NaInCl_6$:Sb^{3+} as a blue phosphor with the combination of green and red phosphors in a UV LED, generating a device with CIE color coordinates of (0.890, 0.4009). This was shown to be a potential alternative to blue phosphors for applications involving near-UV [56].

From this, it has been demonstrated that doping HDPs is an efficient way to modulate and enhance their photoluminescence [56]. Although more materials have been reported, their broad and indirect bandgap is not suitable for implementation in devices. The challenge for HDPs is to be implemented as a sole source of emission, increasing their luminous efficiency in LEDs. Although their application is very recent, there is scope to develop and implement these materials.

5.4 Other Applications

Given the favorable and versatile properties of HDPs, such as solar cells and LEDs, HDPs continue to demonstrate considerable potential for their use in a broader range

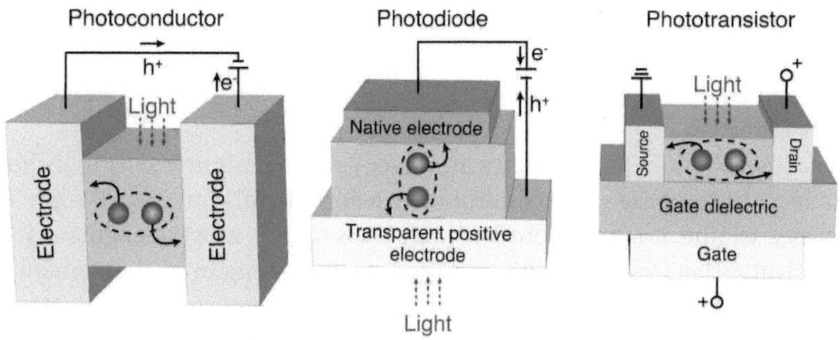

Figure 5.11 Commonly used device configurations for photodetectors: photoconductor, photodiode, and phototransistors. Source: Liu et al. [57]/American Chemical Society.

of other opto/electronics applications and beyond. Those applications include photodetectors for UV and X-ray radiation, memristors, photocatalysts, humidity sensors, and thermochromic sensors. Now, we will discuss some of the lesser-known applications of HDPs.

5.4.1 Photodetectors

A photodetector is a device that converts light signals into electrical pulses, acting like a human eye. Photodetectors can be divided into three main types: photodiodes, photoconductors, and phototransistors see Figure 5.11 [57]. Photodiodes feature a p–n, p–i–n, or a Schottky junction, where low dark current can be obtained due to the junction barrier. The photo-generated carriers can be separated by a shallow bias and even with zero applied voltage. Photodiodes exhibit fast response speed and high sensing capability; however, they have low EQE and low responsivity [58]. On the other hand, a typical photoconductor is composed of a semiconductor and two metal contacts, where an applied voltage is required to aid charge separation. These devices present a high dark current that limits their detection sensibility. Lastly, phototransistors have three terminal electrodes: source, drain, and gate. In addition, a thin dielectric film is arranged between the semiconductor channel and the gate electrode, which allows adjusting the charge transport of the phototransistor depending on the gate polarization [59]. The efficiency and performance of photodetectors are evaluated through several parameters: responsivity (R, $\mathrm{A\,W^{-1}}$), noise-equivalent power ($W\,[\mathrm{Hz}^{1/2}]^{-1}$), specific detectivity (D^*, Jones), gain (G), linear dynamic range (dB), and response speed [60].

The detection range of halide perovskites was extended from ultraviolet–visible–near-infrared (UV–vis–NIR) to X/γ photons. The former has been in high demand for various applications, such as wearable electronics, computer vision, machine vision, and biosensing [59]. Not surprisingly, as can probably be inferred from their good photovoltaic behavior, many halide perovskite materials can absorb photons within a specific wavelength range and efficiently convert them into photocurrent [61]. Also, akin to the field of perovskite photovoltaics, HDPs are a promising alternative to traditional lead-based perovskites as photodetectors toward the high-energy side of the electromagnetic spectrum. The latter is because HDPs can have good light absorption, long charge carrier lifetimes, and large charge carrier mobilities, albeit with reduced toxicity and improved stability when compared to lead perovskites.

5.4.1.1 UV Detectors

For the detection of UV light, trap states introduce gain and improve sensitivity but, in turn, reduce the response speed of the photodetectors. In this context, $Cs_2AgInCl_6$ single crystals exhibit low trap density, limited absorption at wavelengths below 400 nm, and ultra-long carrier lifetime. Based on this, Tang et al. implemented high-quality $Cs_2AgInCl_6$ single crystals into a photoconductive planar structure with gold contacts [24]. They showed the highest calculated ratio between detection and response time when compared to metal oxide-based devices based on ZnO, SnO_2, and TiO_2. Their performance was superior to that reported for $CH_3NH_3PbCl_3$ thin films and single crystals [62, 63]. It was shown that oxygen contamination could influence the stability of the devices. However, such degradation could be remedied mainly by storing the devices in a vacuum to remove surface contaminants.

Subsequently, Shi et al. [64] prepared $Cs_2AgBiBr_6$ photodetectors as thin films with the Au/$Cs_2AgBiBiBr_6$/Au structure. The photocurrent gradually increases with light illumination and electric field strength. Although it presents a low Schottky barrier at the Au/$Cs_2AgBBiBr_6$ interface, the external polarization compensates for the energy barrier, so the transport of photogenerated carriers from the $Cs_2AgBiBr_6$ to the electrodes increases the photocurrent drastically [64]. The device's performance displayed remarkably high responsiveness and fast response speed. The proposed $Cs_2AgBiBr_6$ thin-film photodetectors without encapsulation demonstrate remarkable operational stability over the aging test in ambient air. Moreover, after storage for two weeks in ambient air, the performance of the photodetectors was efficiently sustained.

On the other hand, Gao et al. assembled a photodiode device on $Cs_2AgBiBr_6$ thin films with enlarged grain sizes and reduced trap densities, which were obtained by optimizing the solution–processing method. They achieved high-sensitivity and fast response for a lead-free perovskite photodetector under 450 nm [65]. They optimized the electron and hole transport interlayers; when spiro-OMeTAD is used as HTL, high dark currents are obtained for devices with ETL of TiO_2 and SnO_2. In contrast, devices with TFB as HTL exhibit significantly lower dark current under the same bias conditions [66].

In contrast, Xiao et al. fabricated a planar heterojunction photodetector, ITO/SnO_2/$Cs_2AgBiBr_6$/Au, to detect efficiently UV (320–400 nm) and deep blue (435 nm) photons [67]. At the SnO_2/$Cs_2AgBiBr_6$ interface, the separation of electron–hole pairs from photogenerated carriers is favored. It exhibits a high responsivity of 0.11 $A\,W^{-1}$ at 350 nm and a quick response time of less than 3 ms. This device, stored at ambient conditions without encapsulation, showed no apparent degradation after six months. Due to their low toxicity and high stability, HDPs as active materials in photodetectors are promising for this application.

5.4.1.2 X-Ray Detectors

Currently, attention is focused on detecting X-rays for medical diagnostics, safety, and industrial quality control. This radiation can be measured directly or indirectly;

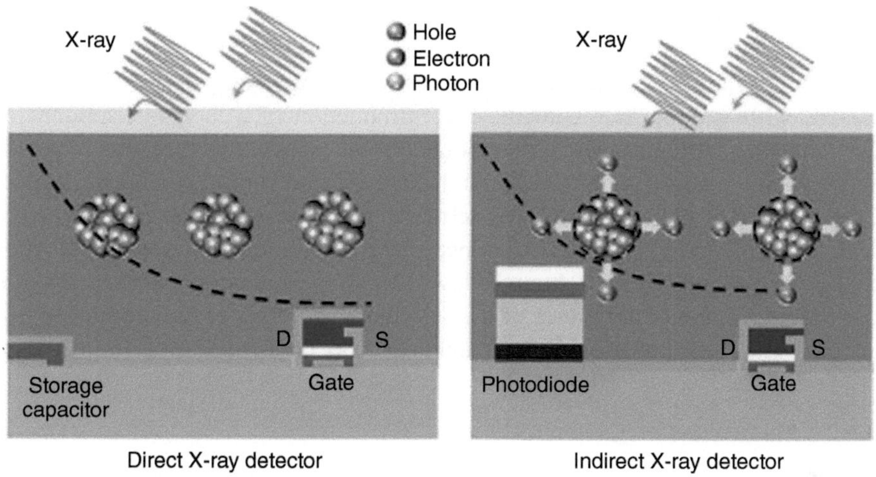

Figure 5.12 Devices for direct and indirect detection of X-rays. Source: Wu et al. [68]/with permission of Elsevier.

see Figure 5.12. The latter is done using scintillators that convert X-rays into visible photons which are then absorbed by a common photodetector [69]. HDPs are suitable candidates for X-ray detection due to their desirable properties, such as high X-ray attenuation coefficient, defect tolerance, large mobility lifetime product ($\mu\tau$), and high resistivity [70]. Two challenges with X-ray photodetectors are the large noise and baseline lift from ionic migration, limiting detection, and imaging performance.

5.4.1.2.1 Direct X-Ray Detector In 2017, Tang et al. published the first report of an HDP-based X-ray detector using a $Cs_2AgBiBr_6$ single crystal. For X-ray detection purposes, $Cs_2AgBiBr_6$ has many desirable properties needed for this application, such as high average atomic number, high resistivity, and low ionization energy. Indeed, the devices based on $Cs_2AgBiBr_6$ showed remarkable sensitivity and good operational stability, and more importantly, they demonstrated the potential that HDPs have for their use in X-ray detectors [71]. To eliminate ion migration on polycrystalline $Cs_2AgBiBr_6$ wafers, bismuth oxybromide (BiOBr) was added as a heteroepitaxial passivator. Which suppressed ion migration, achieving signal drift an order of magnitude lower than any previously reported study [72]. Subsequently, Yu et al. reported films of a $Cs_2AgBiBr_6$ composite embedded in polyvinyl alcohol (PVA). Favoring the photodetector's homogeneity, flexibility, and sensitivity, compared to the reported sensitivity using $Cs_2AgBiBr_6$ single crystals (40 versus 105 $\mu C\,Gy_{air}^{-1}\,cm^{-2}$, respectively) [73].

5.4.1.2.2 Indirect X-Ray Detector In indirect X-ray detectors, a material first absorbs an X-ray photon to later generate visible light. HPDs have proven to be promising X-ray scintillators and have received much attention recently. Tang et al. showed that incorporating lanthanide ions (Ln = Tb^{3+} or Eu^{3+}) in $Cs_2NaLnCl_6$, improves the light yield after X-ray absorption. The f–f transitions of lanthanide cations benefit the PL properties of these double perovskites. The light yield of $Cs_2NaTbCl_6$ powder crystals under X-ray irradiation reaches 46 600 photons per MeV, much higher than $(Lu,Y)_2SiO_5:Ce^{3+}$ crystals and other lead-based perovskites [74].

Recently, Yang et al. prepared a series of $Cs_2Ag_{0.6}Na_{0.4}In_{1-y}Bi_y\ Cl_6$ single crystals with varying Bi^{3+} content. The introduction of a moderate amount of Bi^{3+} enhances radioluminescence production and accelerates radiative recombination, resulting in a high scintillator light yield and rapid light decay. With a $Cs_2Ag_{0.6}Na_{0.4}In_{0.85}Bi_{0.15}Cl_6$ wafer as scintillator, high-quality static and dynamic images of different objects were obtained under low-dose X-ray irradiation. After heat treatment at 85 °C for 50 hours followed by X-ray irradiation for other 50 hours in ambient air, the performance of the scintillator in terms of radioluminescence intensity and X-ray image quality remained virtually unchanged [69].

5.4.2 Memristors

Memristors (memory + resistor) are two-terminal passive devices whose resistance states can be precisely modulated upon electrical stimulation. They were proposed by Leon Chua in 1971 as the fourth basic circuit element, given the relationship between flux and charge [75]. The memristor has a vast potential for applications in information storage, neuromorphic computing, and memory computing applications [76]. Traditionally, they have been implemented in information storage devices, where they must have a high ON/OFF ratio, low operating voltages, and high resistance [77]. These devices can generally be found in one of the following three structures: the sandwich, the panel, or the cross-array, as shown in Figure 5.13 [78].

The first reported case of a perovskite-based memristor belongs to $Cs_2AgBiBr_6$ which is highly stable and has excellent memory performance, including 1000 "writing" and "erasing" cycles, 105 seconds of reading, and 104 times of mechanical bending. The fabricated $Au/Cs_2AgBiBr_6/ITO$ memristor retained resistive switching performance even after exposure to very harsh environments, including high humidity (from 10% to 80%), high temperatures (up to 453 K), burning for 10 seconds with an alcohol flame, and γ-ray irradiation from ^{60}Co. In addition, HDP memristors have a relatively small ON/OFF ratio because of the hard charge migration. The remarkable stability was attributed to strong Ag—Br bonding and good crystallinity of $Cs_2AgBiBr_6$ [79]. Reports on double perovskite lead-free memristors are scarce, and how the structure influences the memory characteristics remains to be uncovered.

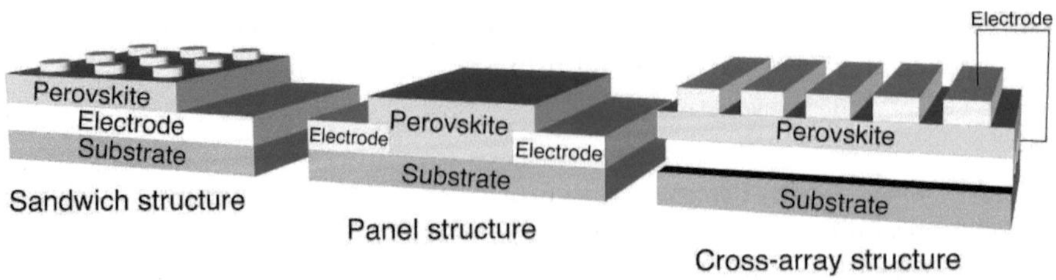

Figure 5.13 Typical structures in memristors: Sandwich, planar, cross-array. Source: Fang et al. [78]/American Chemical Society.

5.4.3 Photocatalysis

Few HDPs have shown the ability to be used as photocatalysts. Interestingly, the double perovskite $Cs_2AgBiBr_6$ has demonstrated significant potential in this area. This is due to its broad absorption spectrum, long carrier lifetimes, and long operational stability. For instance, $Cs_2AgBiBr_6$ has been implemented in several photoelectrochemical reactions such as hydrogen evolution reaction (HER), CO_2 reduction reaction (CRR), and degradation of organic pollutants [80].

The first HDP application as photocatalyst was that of $Cs_2AgBiBr_6$ for HER in a saturated aqueous solution of HBr and H_3PO_2. Since $Cs_2AgBiBr_6$ has a low photocatalytic activity (6.1 μmol g^{-1} per 10 hours), a composite with 2.5% reduced graphene oxide (RGO) as cocatalyst was created. The addition of RGO favors the separation of the photogenerated electron–hole pairs; then, electrons can be rapidly transferred to the conductive RGO by suppressing charge recombination, which enhances HER by about 50 times, reaching 489 μmol g^{-1}, with good stability [81]. On the other hand, theoretical studies have shown that $Cs_2BiAgCl_6$ is potentially promising for photocatalytic water splitting, whereas making $Cs_2InAgCl_6$ and $Cs_2SbAgCl_6$ suitable for water splitting would require modification of their energy levels [82].

Additionally, Kuang et al. used $Cs_2AgBiBr_6$ nanocrystals (NCs) synthesized by the hot injection method for CRR. Achieving a total electron consumption of 105 μmol g^{-1} under an air mass (AM) 1.5 G illumination for six hours, which was 18.5 times larger than the total electron consumption of the bulk material, displaying the dramatic increase by nanostructuring [83]. $Cs_2AgBiBr_6$ NCs showed remarkable stability with moisture and low-polarity solvents. The organic layer on the surface of the NCs decreases the transport of photogenerated charges to the reactants, limiting the catalyst yield and causing poisoning [84]. In the case of $Cs_2AgBiBr_6$ with alcohol-based photocatalytic system, almost complete degradation of the organic dye Rhodamine B (~98%) was achieved in 120 minutes, showing a linear dependence on irradiation time. It was found that $Cs_2AgBiBr_6$ preserves its high chemical stability in ethanol during photocatalytic processes. At the same time, no apparent photocatalytic oxidation of ethanol was observed. Its present high chemical stability and high density of surface catalytic sites (Ag or Bi) can facilitate the activation of either dye or O_2 molecules and accelerate the reaction rate between dye molecules and active species [85].

5.4.4 Sensors

In 2019, Zhan et al. developed a humidity sensor using $Cs_2BiAgBr_6$ thin films as the active material. When the sensor is exposed to an environment with higher relative humidity (RH), water molecules in the thin film can be physisorbed. In this process, there is a proton transfer on the surface, and this charge movement causes an increase in the electrical conductivity. This sensor showed a superfast response (1.78 seconds) and recovery times (0.45 seconds) in the RH range of 5–75% and stability up to 110 hours [86]. On the other hand, Gao et al. reported fully reversible thermochromism in $Cs_2BiAgBr_6$ thin films and single crystals. It was determined from theoretical calculations that the temperature change generates length changes in the Ag—Br and Bi—Br bonds, inducing a direct modification of the bandgap. Moreover, the sensors showed excellent thermal and environmental stability after several heating/cooling cycles. These findings paved the way for a new class of intelligent windows, temperature sensors, and visual thermometers based on lead-free halide double perovskites [87].

5.4.5 Future Applications

This chapter shows that HDPs have developed a wide range of applications, well beyond solar cells and LEDs. While these applications are clearly in their infancy, HDPs have shown great promise and thus offer multiple areas of opportunity which need further exploration. For instance, there are still many applications where halide perovskites have been employed while HDPs have not. HDPs will likely venture into many of these applications and find new fields of opportunity. These applications include piezo- and ferroelectricity, batteries, spintronics, and lasing.

5.5 Related Materials: Layered Double Perovskites and Vacancy Ordered Double Perovskites

The emergence of halide double perovskites and layered perovskites, and the lessons learned from their development, has motivated the search for lower-dimensional double perovskites. The dimensional reduction of HDPs has the potential to leverage the advantages of lower-dimensional materials while at the same time vastly expanding the chemical diversity and properties of the resulting materials. Many different subclasses of HDPs have been discussed in different works. This section will consider two subclasses of double perovskites, those formed from dimensionality reduction (Section 5.5.1) and those with ordered vacancies (Section 5.5.1) [88, 89].

5.5.1 Dimensional Reduction

Akin to the development of halide perovskites with lower dimensionality, the corresponding HDPs have also been explored. Like halide perovskites, low dimensional HDPs come from the dimensional reduction of 3D perovskites. The most established and studied low-dimensional HDPs are 2D structures. They have the general formula, $A'_4A_{2n-2}B_nB'_mX_{6n+2}$ (A = cation in the inorganic layers, A′ = cation in the organic layers), where n is the number of octahedral layers forming the inorganic sheets [90]. These 2D perovskite structures consist of flat sheets of corner-sharing, alternating $[BX_6]^{n-}$ and $[B'X_6]^{m-}$ octahedra (Figure 5.14). To date, all 2D HDPs reported belong to the (001) family with $n = 1$ and 2; besides presenting local structural distortions, such as tilting and elongation of some octahedra, these generate more flexible and stable structures. In this type of 2D HDPs, Ruddlesden–Popper [91] and Dion–Jacobson [92] type structures can be obtained. For example, in HDPs, Cu(I)-based perovskites present significantly distorted sixfold coordination spheres responsible for maintaining the perovskite-like structure [93].

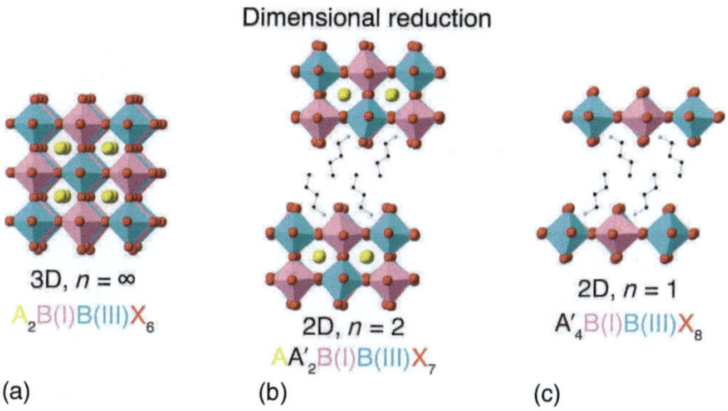

Figure 5.14 Structures of (a) $A_2B(I)B(III)X_6$ and two lower-dimensional derivatives: (b) $AA'_2B(I)B(III)X_7$ ($n = 2$) and (c) $A'_4B(I)B(III)X_8$ ($n = 1$).

Karunadasa et al. studied for the first time the effect of 2D confinement in the HDPs, $(BA)_4AgBiBr_8$ and $(BA)_2CsAgBiBr_7$ (BA = butylammonium). From 3D to 2D, these HDPs exhibit an indirect to direct band transition driven by dimensional confinement, although structural distortions further separate the direct and indirect gaps. In addition, they show strong characteristic absorption at similar energies, assigned to an Ag to Bi charge transfer, responsible for the similar yellow color in these solids [94]. Subsequently, they synthesized from the 3D HDPs $Cs_2AgTlBr_6$, the 2D derivatives with inorganic bilayer $AA'_2B(I)B(III)X_7$ ($n = 2$, Figure 5.14b) and monolayer $A'_4B(I)B(III)X_8$ ($n = 1$, Figure 5.14c) as films. It is observed that features in sharp absorption and broad luminescence with higher n-dependence are dominant in the Ag–Tl analogs. Suggesting a delocalized electronic character that is more sensitive to quantum confinement. For $n = 2$, one has an indirect transition, while $n = 1$ presents a direct transition [90].

Subsequently, Mitzi et al. incorporated a bifunctionalized organic molecule, AE2T (5,5′-diylbis(aminoethyl)-[2,2′-bithiophene), where aromatic edge-to-face interactions helped to stabilize $(AE2T)_2AgBiI_8$ [95]. In this case, the AE2T boundary orbitals are very close to the inorganic band edges, resulting in type I/II boundary quantum alignment. In contrast, the inorganic-derived bands are much less dispersed than those in lead-based 2D HPs [96]. 2D HDPs are promising because, unlike their 3D analogs, a wide range of organic cations can be used to stabilize these structures [97]. Furthermore, it is also possible to modulate their properties by varying the thickness of the inorganic layers [98], changing the combination of metals and using multiple halides [99]. The combination of all these possibilities generates an almost infinite variety of possible 2D HDPs. With it, an endless chemical space that should allow to produce materials for many applications, many of them not even yet envisioned.

5.5.2 Vacancy Ordered Perovskites

The so-called ordered-vacancy perovskite is a type of perovskite that has recently received much attention [100]. Their similarity to double perovskites is that

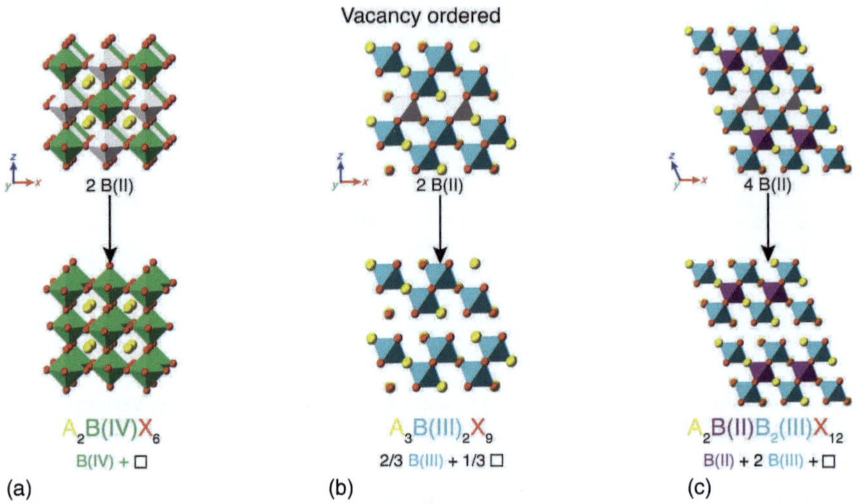

Figure 5.15 Schematic of the relationship between the archetypal double perovskite ($A_2BB'X_6$) and vacancy-ordered double perovskite structure, (a) $A_2B(IV)X_6$. (b) $A_3B(III)X_9$ (c) Family of (111)-oriented perovskites with the general formula: $A_3B(II)B_2(III)X_{12}$.

the B-site is partially replaced with a vacancy □ [9]. In this group, we have the $A_2B(IV)□X_6$, where half of the sites are occupied by M(IV) and the other half by vacancies. Some of the known members of this family include Cs_2SnI_6, Cs_2PdBr_6, Cs_2TeI_6, and Cs_2TiBr_6 [88]. In another subgroup, we have those formed from trivalent metals (such as In^{3+}, Sb^{3+}, and Bi^{3+}), where we have an occupancy of two M(III) by one □, giving the stoichiometry of $A_3B(III)□X_9$, such as $Cs_3Sb_2I_9$ [101]. An extension of this family is the one in which an additional metal is included to yield the general formula $A_4B(II)B(III)□X_{12}$, which now combines metals in a 2+ and 3+ oxidation state along with vacancies to form a layered double perovskite (Figure 5.15c) [102, 103].

5.5.2.1 $A_2B(IV)X_6$: B(IV) Substitution + Vacancies

Perovskites with tetravalent cations, an extension of ABX_3 perovskites, have the general formula $A_2B(IV)□X_6$ and are known as K_2PtCl_6-type perovskites. They start from the relation 2B(II) = B(IV) + □. Due to the absence of connectivity between $[BX_6]$ octahedra, they are classified as 0D perovskites; see Figure 5.15a. In this type of classification are A_2SnI_6 (A = Cs, MA) [104–106], Cs_2PdBr_6 [107], and Cs_2TiBr_6 [108, 109]. In general, the VBMs of this family of materials are mostly composed of the unbonded p orbitals of the halides. On the other hand, the CBMs are predominantly formed by both metals, the lowest-energy net-antibonding state is one where one metal is bonding, and the other is antibonding with the halides [10]. Due to the isolated octahedra $[BX_6]$, the CBMs are dispersed by the charge distribution of the B cations. As a result, the mobility of the electron in these materials tends to be significantly higher than that of the holes, and thus, they are intrinsic n-type semiconductors [110]. There is one report on this type of perovskites as absorbers in solar cells, that of Cs_2TiBr_6 with a maximum PCE of 3.3% [111].

5.5.2.2 $A_3B(III)X_9$: B(III) Substitution + Vacancies

Heterovalent substitutions with trivalent B(III) cations (e.g. Bi^{3+} and Sb^{3+}) have a general formula $A_2B(III)\square X_9$, derived by removing 1/3 of the cations from the B-site. An exciting feature of this type of compound is that it presents two polymorphs [112]. The first is in 0D dimer form, where isolated face-shared $[B_2X_9]$ dioctahedral, and the second is layered 2D perovskite structure, which can be seen as a perovskite of ordered vacancies. The preferential formation of the polymorph can be favored by the judicious choice of A, B, and X [113, 114].

As mentioned before, the 2D structures $A_3B_2X_9$ are formed from vacancies along the (111) plane [115]. To maintain their charge balance, they are formed from the ratio $2B(II) = 2/3B(III) + 1/3\square$, see Figure 5.15b. They exhibit higher electron dimensionality compared to dimers [116]. Sarapov et al. studied $Cs_3Sb_2I_9$, in its dimer form; it displays an indirect bandgap of 2.40 eV, while the layered polymorph presents a direct bandgap of 2.04 eV [101]. The difference is because the polymorph is more dispersive in the VBM and CBM, favoring the charge transport properties. Unfortunately, these types of devices present PCE up to 3.34% [117, 118]. As mentioned above, smaller A-site cations, such as Rb^+ can be used where the layered polymorph is favored. Due to its charge transport anisotropy, PCEs of 0.66% were obtained [113].

5.5.2.3 $A_2B(II)B_2(III)X_{12}$: B(II), B(III) Substitution + Vacancies $\square$

Materials with formula $A_2B(II)B_2(III)\square X_{12}$ are known as quadruple perovskites [102]. These structures can be visualized as the extended structure of (111)-oriented perovskites with formula $A_3B(II)B_2(III)X_{12}$. An additional layer of $B(II)X_6$ octahedra has been inserted between the two $B(III)X_6$ slabs to form a three-octahedra thick inorganic slab [119]. The first report of such halide perovskites was that of $Cs_4CuSb_2Cl_{12}$. The latter showed a remarkably small bandgap of 1.0 eV, which is particularly remarkable compared to the single-metal, parent perovskite: α-$Cs_3Sb_2Cl_9$ (3.0 eV), see Figure 5.15c [120]. The low toxicity, high abundance of copper and antimony, direct bandgap nature, and superior structural stability render this material extremely promising for a wide range of optoelectronic applications [121–123]. Since the first report, several other members of this family of materials have been reported. To date, the reported members contain metals such as Cu(II), Mn(II), Cd(II), Sb(III), and Bi(III). Up to now, this type of material has only been reported with chloride as the halide. However, recently, it has been suggested that incorporating larger B(III) ions, such as lanthanides, allows for the incorporation of bromide or iodide [124]. Due to their recent discovery and exciting properties, they could be further explored for applications such as LEDs, phosphors, and spintronics.

5.6 Conclusions

As discussed in this chapter, halide double perovskites have recently re-emerged as a very large and diverse family of materials that hold great promise for a wide set of applications. Originally, double perovskites were studied as possible replacements for solar cell absorbers, but recent investigations have shown that they have a lot of potential in many different areas and applications well beyond photovoltaics. While the development of most such applications is still in its infancy, there are significant efforts and advances in quite a few areas, such as photodetection, chemosensory, catalysis, spintronics, and many others. Simultaneously, and while all these applications are explored, the renaissance of halide double perovskites has also inspired the development of a tremendous amount of new double perovskites and perovskite-inspired materials, which will surely continue to be a trend in the foreseeable future.

As outlined in this chapter, the future of HDPs is bright and diverse. In the short term, the expansion of these materials will allow for a better understanding of the structure–properties relationship, subsequently facilitating the targeted design of these materials. In the medium term, the combination of application and material development will synergistically push double perovskites into new frontiers, whether that be pushing the performance of current devices or expanding into entirely new or unexplored applications. Finally, in the long term, we envision HDPs and related materials to follow the path of halide perovskites and be able to incursion into successful commercial devices.

References

1 Wang, K., Yang, D., Wu, C. et al. (2019). Recent progress in fundamental understanding of halide perovskite semiconductors. *Progress in Materials Science* 106. https://www.sciencedirect.com/science/article/pii/S0079642519300623?dgcid=rss_sd_all.

2 Zhumekenov, A.A., Saidaminov, M.I., Mohammed, O.F., and Bakr, O.M. (2021). Stimuli-responsive switchable halide perovskites: taking advantage of instability. *Joule* 1–20. https://doi.org/10.1016/j.joule.2021.07.008.

3 Lanzetta, L., Aristidou, N., and Haque, S.A. (2020). Stability of lead and tin halide perovskites: the link between defects and degradation. *The Journal of Physical Chemistry Letters* 11 (2): 574–585. http://feedproxy.google.com/~r/acs/jpclcd/~3/vENVHzptZr0/acs.jpclett.9b02191?utm_source=researcher_app&utm_medium=referral&utm_campaign=RESR_MRKT_Researcher_inbound.

4 Glück, N. and Bein, T. (2020). Prospects of lead-free perovskite-inspired materials for photovoltaic applications. *Energy and Environmental Science*. http://feeds.rsc.org/~r/rss/EE/~3/JADKS8g3HSU/D0EE01651A?utm_source=researcher_app&utm_medium=referral&utm_campaign=RESR_MRKT_Researcher_inbound.

5 Meng, X., Wu, T., Liu, X. et al. (2020). Highly reproducible and efficient $FASnI_3$ perovskite solar cells fabricated with volatilizable reducing solvent. *Journal of Physical Chemistry Letters* 11 (8): 2965–2971.

6 Kopacic, I., Friesenbichler, B., Hoefler, S.F. et al. (2018). Enhanced performance of germanium halide perovskite solar cells through compositional engineering. *ACS Applied Energy Materials* 1 (2): 343–347.

7 Fu, H. (2019). Review of lead-free halide perovskites as light-absorbers for photovoltaic applications: from materials to solar cells. *Solar Energy Materials and Solar Cells* 193 (December 2018): 107–132. https://doi.org/10.1016/j.solmat.2018.12.038.

8 Jin, Z., Zhang, Z., Xiu, J. et al. (2020). A critical review on bismuth and antimony halide based perovskites and their derivatives for photovoltaic applications: recent advances and challenges. *Journal of Materials Chemistry A* 8 (32): 16166–16188.

9 Akkerman, Q.A. and Manna, L. (2020). What defines a halide perovskite? *ACS Energy Letters* 5 (2): 604–610. http://dx.doi.org/10.1021/acsenergylett.0c00039?utm_source=researcher_app&utm_medium=referral&utm_campaign=RESR_MRKT_Researcher_inbound.

10 Slavney, A.H., Connor, B.A., Leppert, L., and Karunadasa, H.I. (2019). A pencil-and-paper method for elucidating halide double perovskite band structures. *Chemical Science* 10 (48): 11041–11053. http://feeds.rsc.org/~r/rss/SC/~3/TXA6AiFogoo/C9SC03219C?utm_source=researcher_app&utm_medium=referral&ut_campaign=RESR_MRKT_Researcher_inbound.

11 Zhang, P., Yang, J., and Wei, S.H. (2018). Manipulation of cation combinations and configurations of halide double perovskites for solar cell absorbers. *Journal of Materials Chemistry A* 6 (4): 1809–1815.

12 Slavney, A.H., Hu, T., Lindenberg, A.M., and Karunadasa, H.I. (2016). A bismuth-halide double perovskite with long carrier recombination lifetime for photovoltaic applications. *Journal of the American Chemical Society* 138 (7): 2138–2141.

13 Elliott, N. and Pauling, L. (1938). The crystal structure of cesium aurous auric chloride, $Cs_2AuAuCl_6$, and cesium argentous auric chloride, $Cs_2AgAuCl_6$. *Journal of the American Chemical Society* 60 (8): 1846–1851.

14 Wei, F., Deng, Z., Sun, S. et al. (2016). The synthesis, structure and electronic properties of a lead-free hybrid inorganic-organic double perovskite $(MA)_2KBiCl_6$ (MA = methylammonium). *Materials Horizons* 3 (4): 328–332.

15 Wolf, N.R., Connor, B.A., Slavney, A.H., and Karunadasa, H.I. (2021). Doubling the stakes: the promise of halide double perovskites. *Angewandte Chemie, International Edition* 60 (30): 16264–16278.

16 Yin, W.J., Weng, B., Ge, J. et al. (2019). Oxide perovskites, double perovskites and derivatives for electrocatalysis, photocatalysis, and photovoltaics. *Energy and Environmental Science* 12 (2): 442–462.

17 Ji, K., Alharbi, K.N., Solana-Madruga, E. et al. (2021). Double double to double perovskite transformations in quaternary manganese oxides. *Angewandte Chemie, International Edition* 60 (41): 22248–22252.

18 Faizan, M., Bhamu, K.C., Murtaza, G. et al. (2021). Electronic and optical properties of vacancy ordered double perovskites A_2BX_6 (A = Rb, Cs; B = Sn, Pd, Pt; and X = Cl, Br, I): a first principles study. *Scientific Reports* 11 (1): 1–9. https://doi.org/10.1038/s41598-021-86145-x.

19 Bartel, C.J., Sutton, C., Goldsmith, B.R. et al. (2019). New tolerance factor to predict the stability of perovskite oxides and halides. *Science Advances* 5 (2): 1–10.

20 Chakraborty, S., Xie, W., Mathews, N. et al. (2017). Rational design: a high-throughput computational screening and experimental validation methodology for lead-free and emergent hybrid perovskites. *ACS Energy Letters* 2 (4): 837–845.

21 Du, K.Z., Meng, W., Wang, X. et al. (2017). Bandgap engineering of lead-free double perovskite $Cs_2AgBiBr_6$ through trivalent metal alloying. *Angewandte Chemie, International Edition* 56 (28): 8158–8162.

22 Filip, M.R., Hillman, S., Haghighirad, A.A. et al. (2016). Band gaps of the lead-free halide double perovskites $Cs_2BiAgCl_6$ and $Cs_2BiAgBr_6$ from theory and experiment. *Journal of Physical Chemistry Letters* 7 (13): 2579–2585.

23 Faghihnasiri, M., Beheshtian, J., Shayeganfar, F., and Shahsavari, R. (2020). Phase transition and mechanical properties of cesium bismuth silver halide double perovskites (Cs_2AgBiX_6, X = Cl, Br, I): a DFT approach. *Physical Chemistry Chemical Physics* 22 (10): 5959–5968.

24 Luo, J., Li, S., Wu, H. et al. (2018). $Cs_2AgInCl_6$ double perovskite single crystals: parity forbidden transitions and their application for sensitive and fast UV photodetectors. *ACS Photonics* 5 (2): 398–405.

25 Volonakis, G., Haghighirad, A.A., Milot, R.L. et al. (2017). $Cs_2InAgCl_6$: a new lead-free halide double perovskite with direct band gap. *Journal of Physical Chemistry Letters* 8 (4): 772–778.

26 Zeng, R., Zhang, L., Xue, Y. et al. (2020). Highly efficient blue emission from self-trapped excitons in stable Sb^{3+}-doped $Cs_2NaInCl_6$ double perovskites. *Journal of Physical Chemistry Letters* 11 (6): 2053–2061.

27 Slavney, A.H., Leppert, L., Saldivar Valdes, A. et al. (2018). Small-band-gap halide double perovskites. *Angewandte Chemie, International Edition* 57 (39): 12765–12770.

28 Yang, J., Zhang, P., and Wei, S.H. (2018). Band structure engineering of $Cs_2AgBiBr_6$ perovskite through order-disordered transition: a first-principle study. *Journal of Physical Chemistry Letters* 9 (1): 31–35.

29 Ji, F., Klarbring, J., Wang, F. et al. (2020). Lead-free halide double perovskite $Cs_2AgBiBr_6$ with decreased band gap. *Angewandte Chemie, International Edition* 59 (35): 15191–15194.

30 Li, W., Jacobs, R., and Morgan, D. (2018). Predicting the thermodynamic stability of perovskite oxides using machine learning models. *Computational Materials Science* 150 (January): 454–463. https://doi.org/10.1016/j.commatsci.2018.04.033.

31 Zhang, T., Cai, Z., and Chen, S. (2020). Chemical trends in the thermodynamic stability and band gaps of 980 halide double perovskites: a high-throughput first-principles study. *ACS Applied Materials and Interfaces* 12 (18): 20680–20690.

32 Greul, E., Petrus, M.L., Binek, A. et al. (2017). Highly stable, phase pure $Cs_2AgBiBr_6$ double perovskite thin films for optoelectronic applications. *Journal of Materials Chemistry A* 5 (37): 19972–19981. https://doi.org/10.1039/C7TA06816F.

33 Wang, M., Zeng, P., Bai, S. et al. (2018). High-quality sequential-vapor-deposited $Cs_2AgBiBr_6$ thin films for lead-free perovskite solar cells. *Solar RRL* 2 (12): 1–6.

34 Wu, C., Zhang, Q., Liu, Y. et al. (2018). The dawn of lead-free perovskite solar cell: highly stable double perovskite $Cs_2AgBiBr_6$ film. *Advanced Science* 5 (3): 2–9.

35 Gao, W., Ran, C., Xi, J. et al. (2018). High-quality $Cs_2AgBiBr_6$ double perovskite film for lead-free inverted planar heterojunction solar cells with 2.2% efficiency. *ChemPhysChem* 19 (14): 1696–1700.

36 Liu, Y., Zhang, L., Wang, M. et al. (2019). Bandgap-tunable double-perovskite thin films by solution processing. *Materials Today* 28 (September): 25–30. https://doi.org/10.1016/j.mattod.2019.04.023.

37 Li, J., Duan, J., Du, J. et al. (2020). Alkali metal ion-regulated lead-free, all-inorganic double perovskites for HTM-free, carbon-based solar cells. *ACS Applied Materials and Interfaces* 12 (42): 47408–47415.

38 Yang, X., Chen, Y., Liu, P. et al. (2020). Simultaneous power conversion efficiency and stability enhancement of $Cs_2AgBiBr_6$ lead-free inorganic perovskite solar cell through adopting a multifunctional dye interlayer. *Advanced Functional Materials* 30: 1–8, Article ID: 2001557.

39 Wang, B., Yang, L., Agnese, C.D. et al. (2020). Photoactive Zn-chlorophyll hole transporter-sensitized lead-free $Cs_2AgBiBr_6$ perovskite. *Solar Cells* 4: 2–8, Article ID: 2000166.

40 Wang, B., Li, N., Yang, L. et al. (2021). Chlorophyll derivative-sensitized TiO_2 electron transport layer for record efficiency of $Cs_2AgBiBr_6$ double perovskite solar cells. *Journal of the American Chemical Society* 143 (5): 2207–2211.

41 Wang, B., Li, N., Yang, L. et al. (2021). Organic dye/$Cs_2AgBiBr_6$ double perovskite heterojunction solar cells. *Journal of the American Chemical Society* 143 (36): 14877–14883.

42 Li, P., Gao, W., Ran, C. et al. (2019). Post-treatment engineering of vacuum-deposited Cs_2NaBiI_6 double perovskite film for enhanced photovoltaic performance. *Physica Status Solidi (A) Applications and Materials Science* 216 (23): 1–8.

43 Zhang, C., Gao, L., Teo, S. et al. (2018). Design of a novel and highly stable lead-free Cs_2NaBiI_6 double perovskite for photovoltaic application. *Sustainable Energy and Fuels* 2 (11): 2419–2428.

44 Guner, T. and Demir, M.M. (2018). A review on halide perovskites as color conversion layers in white light emitting diode applications. *Physica Status Solidi (A) Applications and Materials Science* 215 (13): 1–11.

45 Luo, J., Hu, M., Niu, G., and Tang, J. (2019). Lead-free halide perovskites and perovskite variants as phosphors toward light-emitting applications. *ACS Applied Materials and Interfaces* 11 (35): 31575–31584.

46 Tang, H., Xu, Y., Hu, X. et al. (2021). Lead-free halide double perovskite nanocrystals for light-emitting applications: strategies for boosting efficiency and stability. *Advanced Science* 8 (7): 1–23.

47 Zhao, X.G., Yang, D., Ren, J.C. et al. (2018). Rational design of halide double perovskites for optoelectronic applications. *Joule* 2 (9): 1662–1673. https://doi.org/10.1016/j.joule.2018.06.017.

48 Locardi, F., Sartori, E., Buha, J. et al. (2019). Emissive bi-doped double perovskite $Cs_2Ag_{1-x}Na_xInCl_6$ nanocrystals. *ACS Energy Letters* 4 (8): 1976–1982.

49 Liu, Y., Jing, Y., Zhao, J. et al. (2019). Design optimization of lead-free perovskite $Cs_2AgInCl_6$:Bi nanocrystals with 11.4% photoluminescence quantum yield. *Chemistry of Materials* 31 (9): 3333–3339.

50 Manna, D., Das, T.K., and Yella, A. (2019). Tunable and stable white light emission in Bi^{3+}-alloyed $Cs_2AgInCl_6$ double perovskite nanocrystals. *Chemistry of Materials* 31 (24): 10063–10070.

51 Arfin, H., Kaur, J., Sheikh, T. et al. (2020). Bi^{3+}-Er^{3+} and Bi^{3+}-Yb^{3+} codoped $Cs_2AgInCl_6$ double perovskite near-infrared emitters. *Angewandte Chemie, International Edition* 59 (28): 11307–11311.

52 Li, S., Hu, Q., Luo, J. et al. (2019). Self-trapped exciton to dopant energy transfer in rare earth doped lead-free double perovskite. *Advanced Optical Materials* 7 (23). https://onlinelibrary.wiley.com/doi/abs/10.1002/adom.201901098?af=R&utm_source=researcher_app&utm_medium=referral&utm_campaign=RESR_MRKT_Researcher_inbound.

53 Luo, J., Wang, X., Li, S. et al. (2018). Efficient and stable emission of warm-white light from lead-free halide double perovskites. *Nature* 563 (7732): 541–545. https://doi.org/10.1038/s41586-018-0691-0.

54 Gray, M.B., Majher, J.D., Strom, T.A., and Woodward, P.M. (2019). Broadband white emission in $Cs_2AgIn_{1-x}Bi_xCl_6$ phosphors. *Inorganic Chemistry* 58 (19): 13403–13410.

55 Zhao, F., Song, Z., Zhao, J., and Liu, Q. (2019). Double perovskite $Cs_2AgInCl_6$:Cr^{3+}: broadband and near-infrared luminescent materials. *Inorganic Chemistry Frontiers* 6 (12): 3621–3628.

56 Gray, M.B., Hariyani, S., Strom, T.A. et al. (2020). High-efficiency blue photoluminescence in the $Cs_2NaInCl_6$:Sb^{3+}double perovskite phosphor. *Journal of Materials Chemistry C* 8 (20): 6797–6803.

57 Liu, K., Jiang, Y., Jiang, Y. et al. (2019). Chemical formation and multiple applications of organic-inorganic hybrid perovskite materials. *Journal of the American Chemical Society* 141 (4): 1406–1414.

58 Ahmadi, M., Wu, T., and Hu, B. (2017). A review on organic–inorganic halide perovskite photodetectors: device engineering and fundamental physics. *Advanced Materials* 29 (41): 1–24.

59 Wangyang, P., Gong, C., Rao, G. et al. (2018). Recent advances in halide perovskite photodetectors based on different dimensional materials. *Advanced Optical Materials* 6 (11): 1–30.

60 Pareja-Rivera, C., Morett, D., Barreiro-Argüelles, D. et al. (2021). Lead-free halide perovskites, beyond solar cells and LEDs. *Journal of Physics: Energy* 3 (3): 032014. https://iopscience.iop.org/article/10.1088/2515-7655/ac01bf.

61 Deng, J., Li, J., Yang, Z., and Wang, M. (2019). All-inorganic lead halide perovskites: a promising choice for photovoltaics and detectors. *Journal of Materials Chemistry C* 7 (40): 12415–12440. http://pubs.rsc.org/en/Content/ArticleLanding/2019/TC/C9TC04164H?utm_source=researcher_app&utm_medium=referral&utm_campaign=RESR_MRKT_Researcher_inbound.

62 Adinolfi, V., Ouellette, O., Saidaminov, M.I. et al. (2016). Fast and sensitive solution-processed visible-blind perovskite UV photodetectors. *Advanced Materials* 28 (33): 7264–7268.

63 Babayigit, A., Ethirajan, A., Muller, M., and Conings, B. (2016). Toxicity of organometal halide perovskite solar cells. *Nature Materials* 15 (3): 247–251. https://doi.org/10.1038/nmat4572.

64 Lei, L.Z., Shi, Z.F., Li, Y. et al. (2018). High-efficiency and air-stable photodetectors based on lead-free double perovskite $Cs_2AgBiBr_6$ thin films. *Journal of Materials Chemistry C* 6 (30): 7982–7988.

65 Yang, J., Bao, C., Ning, W. et al. (2019). Stable, high-sensitivity and fast-response photodetectors based on lead-free $Cs_2AgBiBr_6$ double perovskite films. *Advanced Optical Materials* 7 (13): 1–8.

66 Bao, C., Yang, J., Bai, S. et al. (2018). High performance and stable all-inorganic metal halide perovskite-based photodetectors for optical communication applications. *Advanced Materials* 30 (38): 1–8.

67 Wu, C., Du, B., Luo, W. et al. (2018). Highly efficient and stable self-powered ultraviolet and deep-blue photodetector based on $Cs_2AgBiBr_6/SnO_2$ heterojunction. *Advanced Optical Materials* 6 (22): 1–7.

68 Wu, H., Ge, Y., Niu, G., and Tang, J. (2021). Metal halide perovskites for X-ray detection and imaging. *Matter* 4 (1): 144–163. https://doi.org/10.1016/j.matt.2020.11.015.

69 Zhu, W., Ma, W., Su, Y. et al. (2020). Low-dose real-time X-ray imaging with nontoxic double perovskite scintillators. *Light: Science and Applications.* http://feeds.nature.com/~r/lsa/rss/current/~3/sw-xgQcuwiI/s41377-020-00353-0?utm_source=researcher_app&utm_medium=referral&utm_campaign=RESR_MRKT_Researcher_inbound.

70 Bibi, A., Lee, I., Nah, Y. et al. (2021). Lead-free halide double perovskites: toward stable and sustainable optoelectronic devices. *Materials Today* 49 (October): 123–144. https://doi.org/10.1016/j.mattod.2020.11.026.

71 Pan, W., Wu, H., Luo, J. et al. (2017). $Cs_2AgBiBr_6$ single-crystal X-ray detectors with a low detection limit. *Nature Photonics* 11 (11): 726–732. https://doi.org/10.1038/s41566-017-0012-4.

72 Yang, B., Pan, W., Wu, H. et al. (2019). Heteroepitaxial passivation of $Cs_2AgBiBr_6$ wafers with suppressed ionic migration for X-ray imaging. *Nature Communications* 10 (1): https://doi.org/10.1038/s41467-019-09968-3.

73 Li, H., Shan, X., Neu, J.N. et al. (2018). Lead-free halide double perovskite-polymer composites for flexible X-ray imaging. *Journal of Materials Chemistry C* 6 (44): 11961–11967.

74 Hu, Q., Deng, Z., Hu, M. et al. (2018). X-ray scintillation in lead-free double perovskite crystals. *Science China: Chemistry* 61 (12): 1581–1586.

75 Chua, L.O. (1971). Memristor—the missing circuit element. *IEEE Transactions on Circuit Theory* 18 (5): 507–519.

76 Zhao, X., Xu, H., Wang, Z. et al. (2019). Memristors with organic-inorganic halide perovskites. *InfoMat* 1 (2): 183–210.

77 Liu, Y., Li, F., Chen, Z. et al. (2016). Resistive switching memory based on organic/inorganic hybrid perovskite materials. *Vacuum* 130: 109–112. https://doi.org/10.1016/j.vacuum.2016.05.010.

78 Fang, Y., Zhai, S., Chu, L., and Zhong, J. (2021). Advances in halide perovskite memristor from lead-based to lead-free materials. *ACS Applied Materials and Interfaces* 13 (15): 17141–17157.

79 Cheng, X.F., Qian, W.H., Wang, J. et al. (2019). Environmentally robust memristor enabled by lead-free double perovskite for high-performance information storage. *Small* 15 (49): 1–8.

80 Xu, X., Zhong, Y., and Shao, Z. (2019). Double perovskites in catalysis, electrocatalysis, and photo(electro)catalysis. *Trends in Chemistry* 1 (4): 410–424. https://doi.org/10.1016/j.trechm.2019.05.006.

81 Wang, T., Yue, D., Li, X., and Zhao, Y. (2020). Lead-free double perovskite $Cs_2AgBiBr_6$/RGO composite for efficient visible light photocatalytic H_2 evolution. *Applied Catalysis B: Environmental* 268 (November 2019): 118399. https://doi.org/10.1016/j.apcatb.2019.118399.

82 Volonakis, G. and Giustino, F. (2018). Surface properties of lead-free halide double perovskites: possible visible-light photo-catalysts for water splitting. *Applied Physics Letters* 112 (24): 243901.

83 Zhou, L., Xu, Y.F., Chen, B.X. et al. (2018). Synthesis and photocatalytic application of stable lead-free $Cs_2AgBiBr_6$ perovskite nanocrystals. *Small* 14 (11): 1–7.

84 Yui, T., Kan, A., Saitoh, C. et al. (2011). Photochemical reduction of CO_2 using TiO_2: effects of organic adsorbates on TiO_2 and deposition of Pd onto TiO_2. *ACS Applied Materials and Interfaces* 3 (7): 2594–2600.

85 Zhang, Z., Liang, Y., Huang, H. et al. (2019). Stable and highly efficient photocatalysis with lead-free double-perovskite of $Cs_2AgBiBr_6$. *Angewandte Chemie, International Edition* 58 (22): 7263–7267.

86 Weng, Z., Qin, J., Umar, A.A. et al. (2019). Lead-free $Cs_2BiAgBr_6$ double perovskite-based humidity sensor with superfast recovery time. *Advanced Functional Materials* 29 (24): 1–9.

87 Ning, W., Zhao, X.G., Klarbring, J. et al. (2019). Thermochromic lead-free halide double perovskites. *Advanced Functional Materials* 29 (10): 1–8.

88 Xiao, Z., Song, Z., and Yan, Y. (2019). From lead halide perovskites to lead-free metal halide perovskites and perovskite derivatives. *Advanced Materials* 31 (47): 1–22.

89 Igbari, F., Wang, Z.K., and Liao, L.S. (2019). Progress of lead-free halide double perovskites. *Advanced Energy Materials* 9 (12): 1–32.

90 Connor, B.A., Biega, R.I., Leppert, L., and Karunadasa, H.I. (2020). Dimensional reduction of the small-bandgap double perovskite $Cs_2AgTlBr_6$. *Chemical Science* 11 (29): 7708–7715.

91 McClure, E.T., McCormick, A.P., and Woodward, P.M. (2020). Four lead-free layered double perovskites with the $n = 1$ Ruddlesden-Popper structure. *Inorganic Chemistry* 59 (9): 6010–6017.

92 Mao, L., Teicher, S.M.L., Stoumpos, C.C. et al. (2019). Chemical and structural diversity of hybrid layered double perovskite halides. *Journal of the American Chemical Society* 141 (48): 19099–19109.

93 Castro-Castro, L. and Guloy, A. (2003). Organic-based layered perovskites of mixed-valent Gold(I)/Gold(III) iodides. *Angewandte Chemie* 115 (24): 2877–2880.

94 Connor, B.A., Leppert, L., Smith, M.D. et al. (2018). Layered halide double perovskites: dimensional reduction of $Cs_2AgBiBr_6$. *Journal of the American Chemical Society* 140 (15): 5235–5240.

95 Jana, M.K., Janke, S.M., Dirkes, D.J. et al. (2019). Direct-bandgap 2D silver-bismuth iodide double perovskite: the structure-directing influence of an oligothiophene spacer cation. *Journal of the American Chemical Society* 141 (19): 7955–7964.

96 Liu, C., Huhn, W., Du, K.-Z. et al. (2018). Tunable semiconductors: control over carrier states and excitations in layered hybrid organic-inorganic perovskites. *Physical Review Letters* 121 (14): 146401. https://link.aps.org/doi/10.1103/PhysRevLett.121.146401.

97 Ruan, H., Guo, Z., Lin, J. et al. (2021). Structure and optical properties of hybrid-layered-double perovskites $(C_8H_{20}N_2)_2AgMBr_8$ (M = In, Sb, and Bi). *Inorganic Chemistry* 60 (19): 14629–14635.

98 Martín-García, B., Spirito, D., Biffi, G. et al. (2021). Phase transitions in low-dimensional layered double perovskites: the role of the organic moieties. *Journal of Physical Chemistry Letters* 12 (1): 280–286.

99 Connor, B.A., Smaha, R.W., Li, J. et al. (2021). Alloying a single and a double perovskite: a $Cu^{+/2+}$ mixed-valence layered halide perovskite with strong optical absorption. *Chemical Science* 12 (25): 8689–8697.

100 Maughan, A.E., Ganose, A.M., Bordelon, M.M. et al. (2016). Defect tolerance to intolerance in the vacancy-ordered double perovskite semiconductors Cs_2SnI_6 and Cs_2TeI_6. *Journal of the American Chemical Society* 138 (27): 8453–8464.

101 Saparov, B., Hong, F., Sun, J.P. et al. (2015). Thin-film preparation and characterization of Cs_3Sb2I_9: a lead-free layered perovskite semiconductor. *Chemistry of Materials* 27 (16): 5622–5632.

102 Lin, Y.P., Hu, S., Xia, B. et al. (2019). Material design and optoelectronic properties of three-dimensional quadruple perovskite halides. *Journal of Physical Chemistry Letters* 10 (17): 5219–5225.

103 Gray, M.B., Majher, J.D., Holzapfel, N.P., and Woodward, P.M. (2021). Exploring the stability of mixed-halide vacancy-ordered quadruple perovskites. *Chemistry of Materials* 33 (6): 2165–2172.

104 Stoumpos, C.C., Malliakas, C.D., and Kanatzidis, M.G. (2013). Semiconducting tin and lead iodide perovskites with organic cations: phase transitions, high mobilities, and near-infrared photoluminescent properties. *Inorganic Chemistry* 52 (15): 9019–9038. https://doi.org/10.1021/ic401215x.

105 Clark, S.J., Donaldson, J.D., and Harvey, J.A. (1995). Evidence for the direct population of solid-state bands by non-bonding electron pairs in compounds of the type $CsM^{II}X_3$ (M^{II} = Ge, Sn, Pb; X = Cl, Br, l). *Journal of Materials Chemistry* 5 (11): 1813–1818.

106 Pitaro, M., Tekelenburg, E.K., Shao, S., and Loi, M.A. (2022). Tin halide perovskites: from fundamental properties to solar cells. *Advanced Materials* 34 (1): 2105844.

107 Sakai, N., Haghighirad, A.A., Filip, M.R. et al. (2017). Solution-processed cesium hexabromopalladate(IV), Cs_2PdBr_6, for optoelectronic applications. *Journal of the American Chemical Society* 139 (17): 6030–6033.

108 Ju, M.G., Chen, M., Zhou, Y. et al. (2018). Earth-abundant nontoxic titanium(IV)-based vacancy-ordered double perovskite halides with tunable 1.0 to 1.8 eV bandgaps for photovoltaic applications. *ACS Energy Letters* 3 (2): 297–304.

109 Maughan, A.E., Ganose, A.M., Scanlon, D.O., and Neilson, J.R. (2019). Perspectives and design principles of vacancy-ordered double perovskite halide semiconductors. *Chemistry of Materials* 31 (4): 1184–1195.

110 Funabiki, F., Toda, Y., and Hosono, H. (2018). Optical and electrical properties of perovskite variant $(CH_3NH_3)_2SnI_6$. *Journal of Physical Chemistry C* 122 (20): 10749–10754.

111 Chen, M., Ju, M.G., Carl, A.D. et al. (2018). Cesium titanium(IV) bromide thin films based stable lead-free perovskite solar cells. *Joule* 2 (3): 558–570. https://doi.org/10.1016/j.joule.2018.01.009.

112 Chang, J.H., Doert, T., and Ruck, M. (2016). Structural variety of defect perovskite variants $M_3E_2X_9$ (M = Rb, Tl, E = Bi, Sb, X = Br, I). *Zeitschrift für Anorganische und Allgemeine Chemie* 642 (13): 736–748.

113 Harikesh, P.C., Mulmudi, H.K., Ghosh, B. et al. (2016). Rb as an alternative cation for templating inorganic lead-free perovskites for solution processed photovoltaics. *Chemistry of Materials* 28 (20): 7496–7504.

114 Correa-Baena, J.P., Nienhaus, L., Kurchin, R.C. et al. (2018). A-site cation in inorganic $A_3Sb_2I_9$ perovskite influences structural dimensionality, exciton binding energy, and solar cell performance. *Chemistry of Materials* 30 (11): 3734–3742.

115 Sun, S., Tominaka, S., Lee, J.H. et al. (2016). Synthesis, crystal structure, and properties of a perovskite-related bismuth phase, $(NH_4)_3Bi_2I_9$. *APL Materials* 4 (3): https://doi.org/10.1063/1.4943680.

116 Vishnoi, P., Zuo, J.L., Strom, T.A. et al. (2020). Structural diversity and magnetic properties of hybrid ruthenium halide perovskites and related compounds. *Angewandte Chemie* 132 (23): 9059–9066.

117 Yang, Y., Liu, C., Cai, M. et al. (2020). Dimension-controlled growth of antimony-based perovskite-like halides for lead-free and semitransparent photovoltaics. *ACS Applied Materials and Interfaces* 12 (14): 17062–17069.

118 Ahmad, K. and Mobin, S.M. (2020). Recent progress and challenges in $A_3Sb_2X_9$-based perovskite solar cells. *ACS Omega* 5 (44): 28404–28412.

119 Vargas, B., Rodrĺguez-López, G., and Solis-Ibarra, D. (2020). The emergence of halide layered double perovskites. *ACS Energy Letters* 5 (11): 3591–3608.

120 Vargas, B., Ramos, E., Pérez-Gutiérrez, E. et al. (2017). A direct bandgap copper-antimony halide perovskite. *Journal of the American Chemical Society* 139 (27): 9116–9119.

121 Cai, T., Shi, W., Hwang, S. et al. (2020). Lead-free $Cs_4CuSb_2Cl_{12}$ layered double perovskite nanocrystals. *Journal of American Chemical Society* 142 (27): 11927–11936. http://feedproxy.google.com/~r/acs/jacsat/~3/x-xPwQMO-bs/jacs.0c04919?utm_source=researcher_app&utm_medium=referral&utm_campaign=RESR_MRKT_Researcher_inbound.

122 Wang, X.D., Miao, N.H., Liao, J.F. et al. (2019). The top-down synthesis of single-layered $Cs_4CuSb_2Cl_{12}$ halide perovskite nanocrystals for photoelectrochemical application. *Nanoscale* 11 (12): 5180–5187.

123 Singhal, N., Chakraborty, R., Ghosh, P., and Nag, A. (2018). Low-bandgap $Cs_4CuSb_2Cl_{12}$ layered double perovskite: synthesis, reversible thermal changes, and magnetic interaction. *Chemistry – An Asian Journal* 13 (16): 2085–2092.

124 Fedorovskiy, A.E., Drigo, N.A., and Nazeeruddin, M.K. (2020). The role of Goldschmidt's tolerance factor in the formation of A_2BX_6 double halide perovskites and its optimal range. *Small Methods* 4 (5). https://onlinelibrary.wiley.com/doi/abs/10.1002/smtd.201900426?af=R&utm_source=researcher_app&utm_medium=referral&utm_campaign=RESR_MRKT_Researcher_inbound.

6

Tin Halide Perovskite Solar Cells

Xianyuan Jiang, Zihao Zang, and Zhijun Ning

ShanghaiTech University, School of Physical Science and Technology, 393 Middle Huaxia Road, Pudong, Shanghai 201210, China

6.1 Introduction

Inorganic–organic metal halide perovskite solar cell is generally regarded as the most promising next-generation photovoltaic technology due to its various advantages such as high efficiency, tuneability, and low cost. The power conversion efficiency (PCE) of lead perovskite solar cells (LPSCs) has risen from 3.8% to over 25% in the last decade (Figure 6.1) [1], approaching the Shockley–Queisser limit [5]. The fast growth of lead perovskite aroused great interest in the search for novel perovskite materials. To surpass the efficiency limitation, the development of new perovskites with a longer absorption region is needed. On the other hand, the use of poisonous metal lead brings environmental concern, which could influence its commercial application.

Tin perovskite material is an ideal candidate to address the above challenges due to its small bandgap and heavy metal-free character. Hybrid tin perovskite solar cell was firstly reported by Kanatzidis's and Snaith's groups independently. However, the efficiency of tin perovskite, in the beginning, is low and the device is extremely unstable. In the early stage before 2016, the efficiencies of tin perovskite solar cells (TPSCs) are all below 6% and the reproducibility is poor. By employing an inverted device architecture (p–i–n), substituting organic cation methylammonium (MA) with formamidinium (FA), and using SnF_2 to manipulate the crystal growth and prevent oxidation, the efficiency and stability of TPSCs are much improved [6]. In 2017, Liao et al. introduced large molecules to build low-dimensional TPSCs and manipulated the crystallinity of tin perovskite, which significantly improved the stability and reproducibility of the device [7]. Based on low- dimensional structures and interface engineering to reduce surface defect and carrier recombination, Ning and Loi's groups reported efficiency of over 9%, respectively [8, 9]. By introducing large amine molecules to manipulate the crystal growth and reduce carrier recombination, Diau group reported efficiency above 9% as well [10].

Halide Perovskite Semiconductors: Structures, Characterization, Properties, and Phenomena, First Edition. Edited by Yuanyuan Zhou and Iván Mora-Seró.

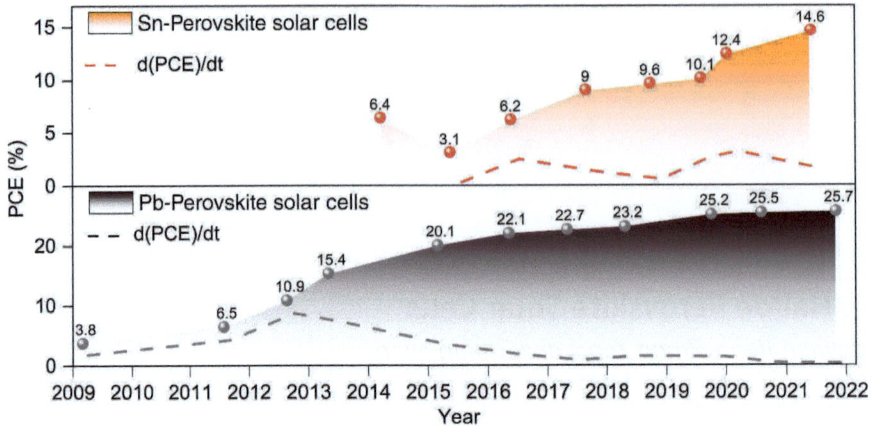

Figure 6.1 The development PCE of LPSC and TPSC solar cells. The dashed line means the derivative of PCE to time. Source: Adapted from Refs. [1–4].

Recently, a jump in device performance was made based on device structural engineering. Jiang et al. explored indene-C_{60} bisadduct (ICBA) as an electron transporting layer (ETL) to improve band alignment between tin perovskite [11]. This work leads to a jump of the open-circuit voltage (V_{OC}) up to 0.94 V, indicating that high voltage is achievable for tin perovskite solar cells. This work pushes the efficiency of TPSCs to over 12%. In combination with the improvement of the orientation and carrier diffusion length, the efficiency was further improved to 14.6% that is certified by an independent agency [12]. Along with efficiency improvement, the stability of TPSCs is much improved as well based on passivating layer design. For example, Zhou et al. introduced a native-oxide layer into inorganic tin perovskite film leading to a stable operation time up to 500 hours under one-sun illumination [13]. These progresses indicate tin perovskite solar cell is a promising candidate for Pb-free, low-cost, and high-efficiency thin-film solar cells.

In this chapter, we firstly give an introduction of the property of tin perovskite and the pros and cons of it for solar cells application. Then, the main strategies to improve the performance of tin perovskite are discussed including composition engineering, additive manipulation, and device structural modification. Finally, we summarize device performance and suggest potential methods for further enhancing its performance.

6.2 Tin Perovskite Properties

6.2.1 Crystal Structure

Perovskite generally has ABX_3 structure, in which B and X are coordinated and form $[BX_6]^{4-}$ octahedra, where A occupies the cavity created by the corner-sharing $[BX_6]^{4-}$ octahedral framework (Figure 6.2a). To realize a stable three-dimensional (3D) perovskite structure, the geometrical tolerance factor

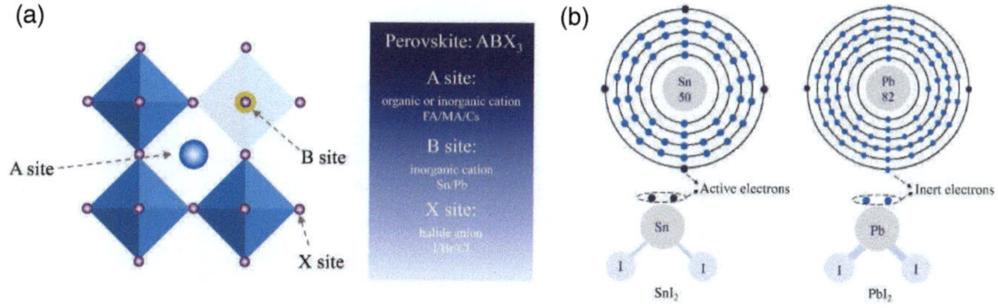

Figure 6.2 (a) Diagram of ABX_3 perovskite structure. (b) The orbital structure of Sn and Pb, and the schematic of SnI_2 and PbI_2. Source: Jiang et al. [2]/Reproduced with permission of American Chemistry Society.

($t = (R_A + R_B)/[\sqrt{2}(R_X + R_B)]$) should range from 0.8 to 1.0, where R_A, R_B, and R_X are the ionic radii of A, B, and X sites, respectively [14]. The radius of Sn^{2+} (118 pm) is similar to that of Pb^{2+} (119 pm) [15], allowing it to fulfill the ion size requirement (Figure 6.2b). When $0.8 < t < 0.9$, a distorted structure with rhombohedral, orthorhombic, or tetragonal symmetry could be formed; if $0.9 < t < 1.0$, a perfect cubic structure can be formed. The partial substitution of I^- by Br^- in $CsSnI_{3-x}Br_x$ perovskites enables the crystal structure to evolve from orthorhombic to cubic [16]. $FASnI_3$ has an orthorhombic crystal structure with a space group of *Amm*2. No phase transition was observed for $FASnI_3$, which is different from $FAPbI_3$ possessing a nonperovskite yellow phase and a black phase [17]. Since the radius size of MA (217 pm) is smaller than FA (253 pm), $MASnI_3$ has a tetragonal structure with a space group of *P*4*mm* [18].

6.2.2 Band Structure and Oxidation

The band structure of tin perovskite is similar to lead perovskite since it is in the same group and has similar valence electron structures. The bandgap of tin perovskite is between 1.2 and 1.4 eV. The simulated band structure of tin perovskite is shown in Figure 6.3a. It shows sharp conduction band and valence band edges, giving rise to small effective masses of electrons and holes. Since the conduction band minimum (CBM) and valence band maximum (VBM) of tin perovskite has same *k* value, it is a direct bandgap semiconductor. Due to the large density of state in edges, it shows large absorption coefficients of 10^4 to 10^5 cm^{-1}.

From band structure simulation (Figure 6.3b), the CBM is mainly constituted by Sn 5*p* orbit, while the VBM is mainly by I 5*p* and Sn 5*s* orbits. Since both 5*s* orbit and 5*p* orbit of Sn are higher than Pb, its energy levels are lifted in comparison to Pb perovskite. The modification of molecules in the A-site changes the band structure. $MASnI_3$ has a bandgap of 1.2 eV, while $FASnI_3$ has a bandgap about 1.4 eV. The smaller bandgap of $MASnI_3$ can be ascribed to the smaller lattice constant and better orbit coupling. The use of Cs to replace FA molecule reduces the bandgap as well, e.g. $CsSnI_3$ has a bandgap around 1.3 eV. The variation of strain in perovskite structure can induce the change of lattice structure [23]. Due to its relatively narrow bandgap, the absorption range of tin perovskite is extended to over 900 nm [24]. The theoretic

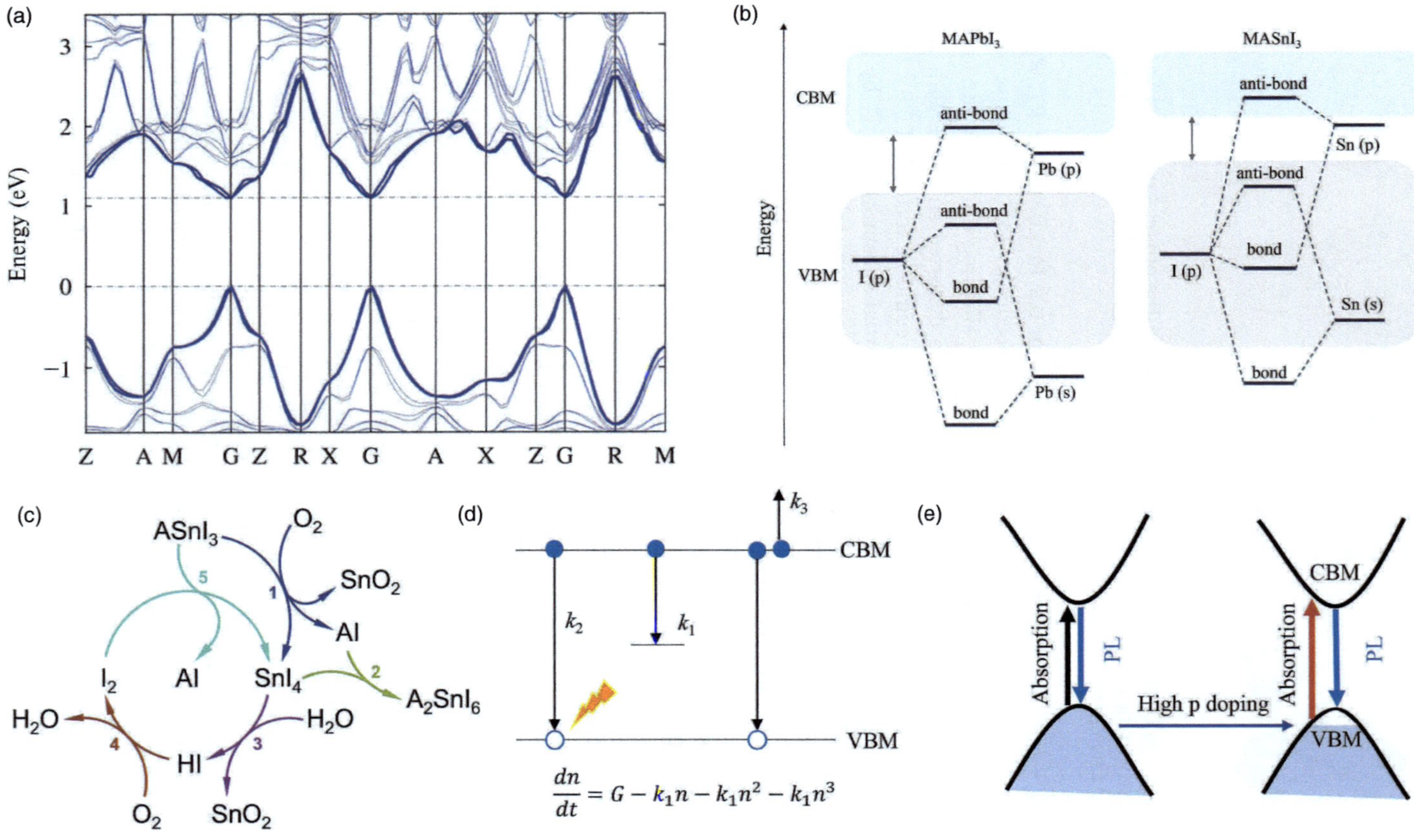

Figure 6.3 (a) Calculated band structure of $MASnI_3$ [19]. Source: Umari et al. [19]/Reproduced with permission of Springer-Nature. (b) The schematic depicts the band structures of $MAPbI_3$ and $MASnI_3$ [20]. Source: Goyal et al. [20]/Reproduced with permission of American Chemistry. (c) Proposed cyclic degradation mechanism of a tin iodide perovskite under ambient air exposure [21]. Source: Lanzetta et al. [21]/Reproduced with permission of Springer Nature. (d) Recombination of monomolecular (k_1), bimolecular (k_2), and Auger (k_3). (e) Schematic representation of the Burstein–Moss effect. Source: Adapted from Handa et al. [22].

efficiency of TPSCs based on Shockley–Queisser equation is up to 33%, higher than that of lead perovskite.

Due to the absence of lanthanide contraction, the energy level of 5*s* orbit is lifted and 5*s* electrons are active and easy to lose. Thus, the tin perovskite is easy to be oxidized and forms Sn^{4+}, and the possible reaction pathway is shown in Figure 6.3c. The reaction between oxygen and tin perovskite forms SnO_2 and A_2SnI_6. The simulation indicates that oxidation is difficult to occur inside the lattice, while it is easy to occur on surface defects [21]. Oxidation can cause undesired self-doping effects and nonradiative recombination. On the other hand, active 5*s* electrons of Sn^{2+} result in strong Lewis acidity.

6.2.3 Electrical Properties and Defects

Fast carrier transport is critical for improving the efficiency of solar cells. The exciton binding energy of tin perovskite is generally smaller than lead perovskite. The simulated data of $CsSnI_3$ are 12.2 meV, and the experimentally tested data are 18 meV [25], smaller than those of lead perovskite [26]. $MASnI_3$ and $CsSnI_3$ exhibit hole mobilities of 322 $cm^2\ V^{-1}\ s^{-1}$ [17] and 585 $cm^2\ V^{-1}\ s^{-1}$ [27], respectively. This is a very high value even in comparison to lead perovskite.

Carrier lifetime is another important factor for carrier transport. Up to now, the reported carrier lifetimes of tin perovskite are all below 50 ns [28], which is much smaller than lead perovskite. As a result, TPSC has a much shorter diffusion length than lead perovskite, generally lower than 1.2 μm [29]. One important factor for the small carrier lifetime is the high defect density of the film deriving from oxidation and Sn vacancy. Tin perovskite shows monomolecular recombination constants (k_1) of 10^9 to 10^{10} s^{-1} due to its high defect density (Figure 6.3d) [30]. The bimolecular (k_2) recombination constant of tin perovskite (1.4×10^{-9} $cm^3\ s^{-1}$) is comparable with lead perovskite [31], and it owns a relatively high photoluminescence quantum yield (PLQY) [32]. In comparison to LPSC, the Auger (k_3) recombination for TPSC is more than an order of magnitude lower [18].

The easy oxidation of tin perovskite generates a p-type character film. When a large amount of holes occupies the trap states near the VBM, the film shows the Burstein–Moss effect, which causes a blue shift of the absorption onset (Figure 6.3e) and a low absorption coefficient [22]. By reducing oxidation of the film, e.g. introducing 2D perovskites, this effect can be prevented [33]. Recent simulation indicates, at the material surface, an iodine-rich environment could facilitate the oxidation of Sn^{2+} and the formation of tin vacancies, whereas Sn-rich condition is more difficult to form defects due to their higher formation energies [34, 35].

6.3 Perovskite Composition Engineering

6.3.1 Three-Dimensional TPSC

In the early stage of the development of TPSCs, most works focused on the composition engineering of tin perovskite (Figure 6.4). In 2014, Snaith et al. reported on

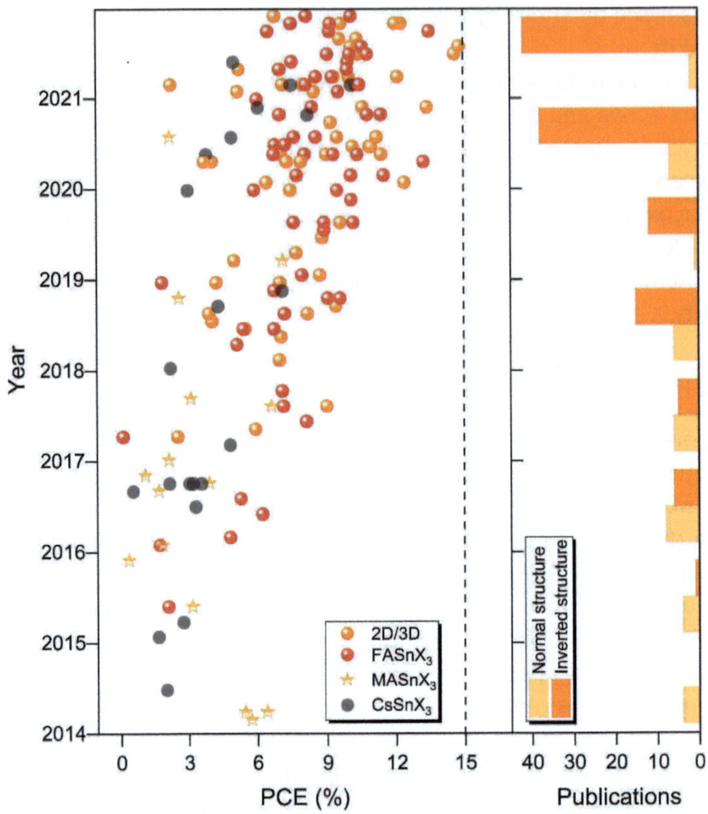

Figure 6.4 The efficiency statistics of TPSC. Source: Adapted from Refs. [2, 18, 36].

$MASnI_3$ solar cells [31]; however, the low crystal quality and oxidation of TPSCs led to an efficiency of only 6.4% with poor reproducibility. New film growth methods were then developed to improve the quality of $MASnX_3$ perovskite films. Kanatzidis et al. prepared $MASnI_{3-x}Br_x$ solid crystals and then dissolved products in a solvent to fabricate perovskite film [24]. When switching from the pure $MASnI_3$ to pure $MASnBr_3$ perovskite, the PCE is improved from 4.27% to 5.73%. Based on a modified approach using a SnI_2-dimethyl sulfoxide (DMSO) as an intermediate phase to reduce the crystallization speed, a PCE of 5.79% is achieved [37]. Wang et al. explored the low-temperature vapor-assisted solution process (LT-VASP) for film growth via ion exchange/insertion reactions, generating a PCE of 7.78%, which is one of the best values for $MASnX_3$-based perovskite solar cells [38].

Many efforts were made to develop $FASnX_3$ TPSC in recent years since it shows better stability against oxidation. Koh et al. firstly fabricated $FASnI_3$ solar cells; however, due to the rough morphology and serious charge recombination, the PCE is only 2.1% [39]. To improve device efficiency, Seok et al. introduced pyrazine into the precursor to obtain a SnF_2–pyrazine complex, which improved film morphology and reduced oxidation, giving rise to a PCE up to 4.8% [40]. Based on inverted structure, SnF_2 additive, and antisolvent method, Liao et al. achieved a milestone

PCE of 6.22% [41]. The use of mixed cations (MA/FA) generates a PCE of 8.12% [42]. Diau et al. incorporated ethylenediammonium diiodide ($EDAI_2$) additive into $FASnI_3$ to passivate grain boundaries, boosting the PCE to 9.6% [10]. Han et al. used poly(vinyl alcohol) (PVA) to form hydrogen bond with $FASnI_3$, which prevented the iodide migration. The efficiency of TPSC was kept well after 400 hours of operation at the maximum power point (MPP) [43]. The device containing a tin triple-halide amorphous covering layer kept 95% of its initial PCE after 1000 hours operation at MPP [44], demonstrating comparable stability to state-of-the-art LPSC. Hayase et al. introduced ethylammonium cation into $FASnI_3$ perovskite structure, which brings much lower defect density and better stability of the film, generating an efficiency over 13% [45].

Inorganic tin perovskite has received much attention in recent years due to its high formation energy, which has potential to improve the stability of TPSCs against oxidation [46]. In 2012, Chen and coworkers fabricated a Schottky solar cell based on $CsSnX_3$ achieving a PCE of 0.9% [47]. Moghe et al. grew a $CsSnBr_3$ layer by the sequential deposition of $SnBr_2$, CsBr, and fluoride passivating dopants. The film retained nearly 100% of its absorbance even after exposing to ambient air for 24 hours [48]. Hao et al. compared several SnX_2 compounds and found that $SnBr_2$ is the best additive to improve the stability of $CsSnI_3$. The corresponding device shows a prolonged storage lifetime of 100 hours with a high PCE of 4.3% [49]. Song used excess SnI_2 as a compensator to suppress Sn vacancies, achieving a maximum PCE of 4.81% [50]. Zhou et al. established a simple one-step vapor-deposition approach for fabricating $CsSn_{0.5}Ge_{0.5}I_3$ perovskite thin film. The device shows an efficiency of 7.11% and better stability thanks to the stable native-oxide GeO_2 protecting layer [13]. Li et al. introduced a liquid-like molecule with large polaron into $CsSnI_3$ to prevent the approach of carriers to defect centers. It reduced carrier recombination and increased efficiency to 8.20% [51]. Ye et al. suppressed the undesirable oxidation of $CsSnI_3$ by engineering the localized electron density using phthalimide (PTM) additive, achieving a champion efficiency of 10.1% [52]. Overall, the efficiency of inorganic tin perovskite is still lower than hybrid structure, especially the voltage, which can be partly ascribed to the high defect density of the film.

6.3.2 Low-Dimensional TPSC

Low-dimensional structures are generally applied for perovskite solar cells to improve device stability. Density function simulation indicates that low-dimensional tin perovskite shows smaller oxidation enthalpy compared with 3D structure. In addition, bulky organic cations like phenylethylamine (PEA) could isolate perovskite slabs from the ambient oxygen and water molecules (Figure 6.5a). Furthermore, the quantum well structures will shield the Sn 5*s*-electrons in the valence shell effectively by reducing the antibonding coupling between Sn 5*s* and I 5*p* [4]. Hence, the inclusion of low-dimensional structures is an effective strategy to suppress oxidation.

Low-dimensional perovskites are generally fabricated by including large amine molecules in A site. As Figure 6.5b shows, the 2D structure can be categorized into

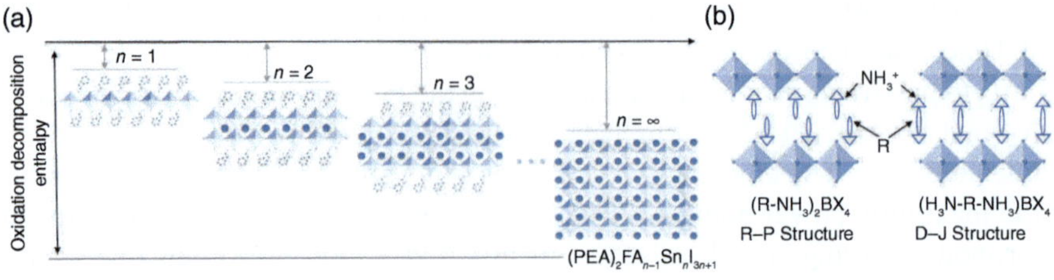

Figure 6.5 (a) Oxidation decomposition enthalpy of tin perovskites. Large enthalpy of tin perovskites indicates that they are easy to oxidize [2]. Source: Jiang et al. [2]/Reproduced with permission of American Chemistry Society. (b) Schematic of R–P and D–J perovskite structures.

two types as Ruddlesden–Popper (R–P) phase and Dion–Jacobson (D–J) phase. Generally, molecules with a single amine unit form R–P structure, while molecules with two amine units enable the formation of D–J structure. PEA is the most frequently applied molecule for preparing R–P phase structure, with a formula of $(PEA)_2FA_{n-1}Sn_nI_{3n+1}$, and n represents the number of layers of the inorganic slabs. Based on PEA and butylamine, Ning group and Kanatzidis group explored low-dimensional structure for TPSCs almost at the same time, respectively. The inclusion of PEA brings the formation of R–P phase low-dimensional perovskite, and the prepared film exhibits reduced oxidation compared to their 3D counterparts, giving rise to improved efficiency of 5.94% [7]. In the end, it needs to be noted that the introduction of large amine molecules can modify the crystal growth kinetics, which is helpful to improve the crystallinity of the film.

Based on low-dimensional structure, in combination with perovskite interface passivation, device engineering, and crystal quality improvement, the efficiencies of TPSCs were improved to 9.4%, 12.4%, and 14.6%, respectively [8, 11, 12]. The use of low-dimensional structures generates a V_{OC} of 0.94 V, close to the theoretic maximum value. Theoretic simulation indicates 2D structure on the surface increases the formation energy of Sn vacancy defects, which could be another important factor for the high voltage [33]. Other molecules are introduced for the construction of R–P phase perovskite, including 2-thiopheneethylammonium iodide [53] and 5-ammonium valeric acid iodide [54]. Low-dimensional structure effectively improves the performance of TPSCs.

D–J structure using molecules with double amine units to connect the layered structure is potential to further improve structural stability. Padture et al. first employed 4-(aminomethyl)piperidinium to fabricate TPSC [55]. The optimization of the ratio of 4AMP/FAI yielded a maximum PCE of 10.9% [56]. 1,4-butanediamine (BEA) is used to fabricate $(BEA)FA_2Sn_3I_{10}$ D-J structure. The solar cell based on it delivers a PCE of 6.43% with negligible hysteresis [57].

Recently, low-dimensional structure is included in inorganic TPSCs [58]. The 2D structure grows firstly and it acts as a template to induce the growth of 3D structure. It improves crystal orientation and crystallinity, bringing a much-improved device efficiency and stability in comparison to the control film.

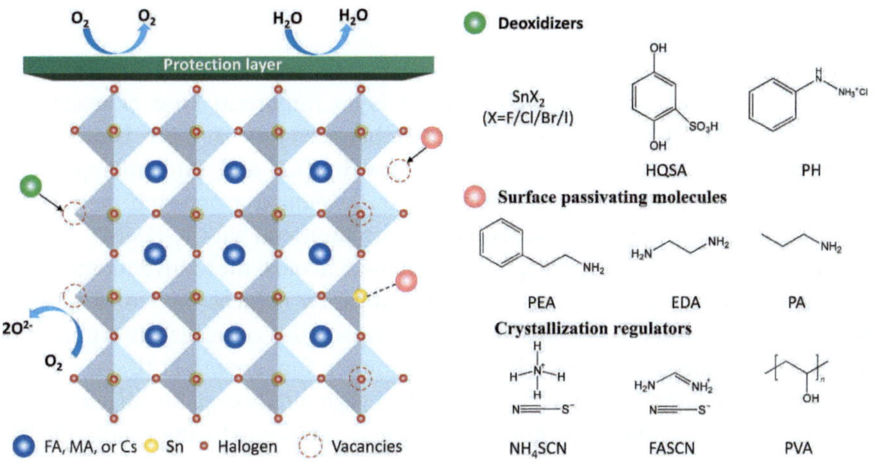

Figure 6.6 Schematic of the defects and functional molecules. Source: Jiang et al. [2]/Reproduced with permission of American Chemistry Society.

6.4 Additives Manipulation

To improve the quality of tin perovskite film, many functional additives were explored to manipulate crystal growth and reduce defect density. They can be briefly divided into three categories: crystallization regulators, deoxidizers, and interfaces passivating molecules (Figure 6.6).

6.4.1 Crystallization Regulators

High crystallinity with large grain size and less boundary is critical for enhancing carrier transport and reducing defect density. Oriented crystal growth perpendicular to the substrate can promote carrier transport and the performance of low-dimensional TPSC. To improve crystallinity and orientation, additives are added to the precursor to manipulate crystal growth kinetics. Wang et al. fabricated a 2D-quasi-2D–3D hierarchy structure using NH_4SCN as a coordinative additive [8]. This additive is also helpful to improve the crystallinity of the film. Kim et al. introduced FASCN into PEAI-based 2D/3D tin perovskite, leading to improved crystallinity and out-of-plane orientation [59]. Using NH_4Cl as an additive improved the orientation and morphology of tin perovskite as well [60].

The rapid crystal growth accelerates the formation of defects. The use of large-size Lewis base molecules to coordinate with Sn^{2+} can slow down the crystallization speed. A compact $FASnI_3$ perovskite layer with less defects was fabricated using π-conjugated Lewis base molecules with high electron density to regulate the crystallization. The fabricated device shows impressive stability under operation [61]. Poly(vinyl alcohol) (PVA) was added in the precursor to form hydrogen bond between $FASnI_3$ that slows down the crystal growth. It reduces trap states and suppresses the migration of the iodide ions [43].

6.4.2 Deoxidizers

Since tin perovskite is extremely easy to be oxidized, reducing additives are explored to suppress oxidation. Element Sn is an important kind of reagent to reduce oxidation. Disproportionation reaction between Sn^{4+} and Sn^0 is utilized to reduce oxidized species in solution, e.g. Sn powder [62] and Sn(0) nanoparticles [63] were added in the precursor. Jiang et al. developed a one-step method for synthesizing SnI_2-solution by a reaction between Sn metal with I_2-DMSO solution ($Sn + I_2 + x\text{DMSO} \rightarrow SnI_2 \cdot \text{DMSO}_x$). This method avoids the conventional complicated SnI_2 purification process, which significantly reduces the cost for TPSCs and oxidation. Furthermore, this method increases the coordination number of Sn-DMSO, which prohibits SnI_2 phase segregation in film growth [12].

Because of the large electronegativity, fluorine is generally used for tin perovskite, although its working mechanism is not clear yet. Volatile reducing agents such as formic acid are explored to reduce the oxidation of $FASnI_3$, which can be easily removed from the perovskite layer [64]. Strong reducing reagent N_2H_4 is utilized to inhibit oxidation as well; however, it is explosive and requires cautious handling [65]. Yan et al. added hydroxybenzene sulfonic acid or salt into the perovskite precursor solution along with excess $SnCl_2$ to generate an *in situ* encapsulation complex, rendering a significantly improved stability of the $FASnI_3$ film [66]. Wang et al. introduce phenylhydrazine hydrochloride into TPSC, which exhibits excellent storage stability and the device efficiency is even improved after air exposure [67].

6.4.3 Interfaces Passivating Materials

Surface defects and oxidation can significantly affect the efficiency and stability of TPSCs. To passivate the interface between perovskite and substrate, Chen et al. introduced an ultrathin PEABr interlayer on the substrate, which reduces background carrier density [68]. An encapsulating layer can passivate the surface dangling bond and form a barrier preventing perovskite from oxygen. For example, 2D structure is used to passivate surface defects of TPSCs. Amorphous-polycrystalline structure [44], a thin layer of poly(methyl methacrylate) [69], and a self-sealing polymer poly(ethylene-*co*-vinyl acetate) are explored for surface passivation of TPSCs as well [70]. Surface passivation has become a general approach to improve device performance.

6.5 Device Architecture Engineering

Device architecture plays an important role in carrier transport in TPSCs. The development of TPSCs in the early stage generally employed the normal structure. However, the normal structure and carriers transporting layers used for lead perovskite cannot match well with TPSCs; hence, device structural engineering was explored.

6.5.1 Normal and Inverted Structures

The normal structure generally uses organic molecules as hole transporting layer (HTL) that require an oxidation process in air to enhance their conductivity.

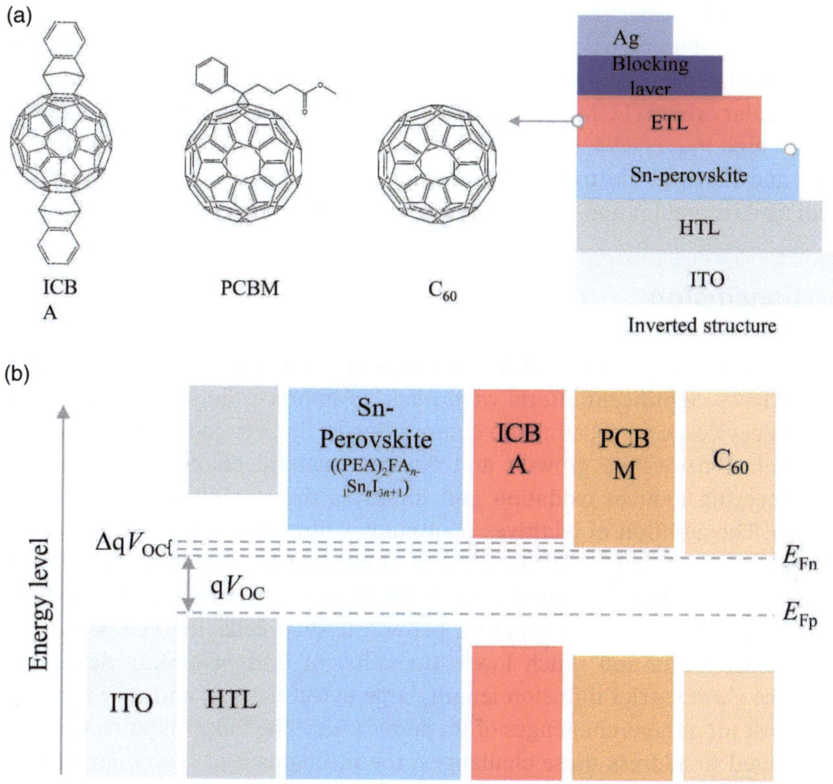

Figure 6.7 (a) Schematic of inverted structure, ICBA, PCBM, and C_{60}. (b) Energy levels of tin perovskite and transporting layers. Source: Jiang et al. [2]/Reproduced with permission of American Chemistry Society.

However, this process induces oxidation of tin perovskite. Furthermore, HTLs such as 2,2′,7,7′-tetrakis-(*N*,*N*-di-4-methoxyphenylamino)-9,9′-spiro-bifluorene (Spiro-OMeTAD) and poly[bis(4-phenyl)(2,4,6-trimethylphenyl)amine] (PTAA) typically need to be doped with metal salts that would disrupt the Sn–I bonding result in surface defects [71]. As a result, TPSCs based on normal structure generally show low efficiency and poor reproducibility. Inverted structure using PEDOT:PSS or NiO_x as HTL allows effective carrier transfer and does not require extra processes and additives, facilitating the fabrication of high-quality tin perovskites (Figure 6.7a). Nowadays, almost all high-efficiency TPSCs are based on the inverted device structure.

6.5.2 Band Alignment

Interface band alignment is critical for carrier transfer in solar cells. As seen in Figure 6.3b, due to the higher energy level of Sn 5*s* lone-pair, both the CBM and VBM of tin perovskite are shallower than lead perovskite. Nishikubo et al. measured the CBM of $FASnI_3$ (−3.73 eV) by employing a gas-tight capsule to reduce oxidation during sample transport [72]. The CBM is about 0.2 eV higher than that of $FAPbI_3$, consistent with the density functional theory calculation [19]. The deep energy level of electron transporting layers brings voltage loss. The V_{OC} of TPSCs based on [6,6]-phenyl-C_{61}-butyric acid methyl ester (PCBM) or C_{60} is generally lower than 0.7 V. To alleviate the energy level offset, ICBA with the shallower lowest unoccupied molecular orbital (LUMO) level of −3.74 eV is utilized for TPSCs (Figure 6.7b), bringing a high V_{OC} of 0.94 V [11]. The high LUMO level enlarges quasi-Fermi level splitting and alleviates electron transfer from iodide anions, contributing to reduced nonradiative recombination as evidenced by electroluminescence measurement.

6.6 Conclusion

This chapter summarized the basic properties of tin perovskite and the development of TPSCs. Significant efforts were made to improve the performance of tin perovskite devices, e.g. adjusting the composition of tin perovskite, using additives to manipulate perovskite growth, and device structural engineering. Composition engineering reduces oxidation and improves the stability of tin perovskite solar cells. The addition of additives manipulates film growth and reduces defect density and film oxidation. Device structural engineering including architecture modification and band alignment leads to enhanced carrier transport and device voltage. Despite the fast progress of tin perovskite solar cells, the PCE is far from the theoretical value and much lower than that of lead perovskite devices. At present, the short carrier diffusion length, large defect density, and easy oxidation are the most prominent challenges of tin perovskites. The following directions can be considered to address these challenges: the mechanism of crystal growth and degradation needs to be studied in-depth; the structure of tin perovskite film needs to be precisely controlled, especially for low-dimensional structural films; creation of a perovskite protective layer on the surface without compromise carrier transfer; design and fabricate new carriers transporting layers.

References

1 Ma, C.Q. and Park, N.G. (2020). A realistic methodology for 30% efficient perovskite solar cells. *Chem* 6: 1254–1264.

2 Jiang, X., Zang, Z., Zhou, Y. et al. (2021). Tin halide perovskite solar cells: an emerging thin-film photovoltaic technology. *Account Materials Research* 2: 210–219.

3 Zang, Z.H., Li, H.S., Jiang, X.Y., and Ning, Z.J. (2021). Progress and perspective of tin perovskite solar cells. *Acta Physico-Chimica Sinica* 37: 2007090.

4 Zhang, T., Sun, Q., Zhang, X.L. et al. (2021). Minimizing energy loss in two-dimensional tin halide perovskite solar cells-A perspective. *APL Materials* 9: 020906.

5 Shockley, W. and Queisser, H.J. (1961). Detailed balance limit of efficiency of p-n junction solar cells. *Applied Physics Letters* 32: 510–519.

6 Diau, E.W.-G., Jokar, E., and Rameez, M. (2019). Strategies to improve performance and stability for tin-based perovskite solar cells. *ACS Energy Letters* 4: 1930–1937.

7 Liao, Y., Liu, H., Zhou, W. et al. (2017). Highly oriented low-dimensional tin halide perovskites with enhanced stability and photovoltaic performance. *Journal of the American Chemical Society* 139: 6693–6699.

8 Wang, F., Jiang, X.Y., Chen, H. et al. (2018). 2D-quasi-2D-3D hierarchy structure for tin perovskite solar cells with enhanced efficiency and stability. *Joule* 2: 2732–2743.

9 Shao, S.Y., Liu, J., Portale, G. et al. (2018). Highly reproducible Sn-based hybrid perovskite solar cells with 9% efficiency. *Advanced Energy Materials* 8: 1702019.

10 Jokar, E., Chien, C.H., Tsai, C.M. et al. (2019). Robust tin-based perovskite solar cells with hybrid organic cations to attain efficiency approaching 10. *Advanced Materials* 31: e1804835.

11 Jiang, X., Wang, F., Wei, Q. et al. (2020). Ultra-high open-circuit voltage of tin perovskite solar cells via an electron transporting layer design. *Nature Communications* 11: 1–7.

12 Jiang, X., Li, H., Zhou, Q. et al. (2021). One-step synthesis of SnI_2.$(DMSO)_x$ adducts for high-performance tin perovskite solar cells. *Journal of the American Chemical Society* 143: 10970–10976.

13 Chen, M., Ju, M.G., Garces, H.F. et al. (2019). Highly stable and efficient all-inorganic lead-free perovskite solar cells with native-oxide passivation. *Nature Communications* 10: 1–8.

14 Goldschmidt, V.M. (1926). Die Gesetze der Krystallochemie. *Die Naturwissenschaften* 14: 477–485.

15 Zhou, Y.Y. and Zhao, Y.X. (2019). Chemical stability and instability of inorganic halide perovskites. *Energy & Environmental Science* 12: 1495–1511.

16 Sabba, D., Mulmudi, H.K., Prabhakar, R.R. et al. (2015). Impact of anionic Br-substitution on open circuit voltage in lead free perovskite ($CsSnI_{3-x}Br_x$) solar cells. *Journal of Physical Chemistry C* 119: 1763–1767.

17 Stoumpos, C.C., Malliakas, C.D., and Kanatzidis, M.G. (2013). Semiconducting tin and lead iodide perovskites with organic cations: phase transitions, high mobilities, and near-infrared photoluminescent properties. *Inorganic Chemistry* 52: 9019–9038.

18 Pitaro, M., Tekelenburg, E.K., Shao, S., and Loi, M.A. (2021). Tin halide perovskites: from fundamental properties to solar cells. *Advanced Materials* e2105844.

19 Umari, P., Mosconi, E., and De Angelis, F. (2014). Relativistic GW calculations on $CH_3NH_3PbI_3$ and $CH_3NH_3SnI_3$ perovskites for solar cell applications. *Scientific Reports* 4: 4467.

20 Goyal, A., McKechnie, S., Pashov, D. et al. (2018). Origin of pronounced nonlinear band gap behavior in lead-tin hybrid perovskite alloys. *Chemistry of Materials* 30: 3920–3928.

21 Lanzetta, L., Webb, T., Zibouche, N. et al. (2021). Degradation mechanism of hybrid tin-based perovskite solar cells and the critical role of tin (IV) iodide. *Nature Communications* 12: 1–11.

22 Handa, T., Yamada, T., Kubota, H. et al. (2017). Photocarrier recombination and injection dynamics in long-term stable lead-free $CH_3NH_3SnI_3$ perovskite thin films and solar cells. *Journal of Physical Chemistry C* 121: 16158–16165.

23 Nishimura, K., Hirotani, D., Kamarudin, M.A. et al. (2019). Relationship between lattice strain and efficiency for Sn-perovskite solar cells. *ACS Applied Materials & Interfaces* 11: 31105–31110.

24 Hao, F., Stoumpos, C.C., Cao, D.H. et al. (2014). Lead-free solid-state organic-inorganic halide perovskite solar cells. *Nature Photonics* 8: 489–494.

25 Huang, L.-y. and Lambrecht, W.R.L. (2013). Electronic band structure, phonons, and exciton binding energies of halide perovskites $CsSnCl_3$, $CsSnBr_3$, and $CsSnI_3$. *Physical Review B* 88: 165203.

26 Li, B., Long, R., Xia, Y., and Mi, Q. (2018). All-inorganic perovskite $CsSnBr_3$ as a thermally stable, free-carrier semiconductor. *Angewandte Chemie, International Edition* 57: 13154–13158.

27 Chung, I., Lee, B., He, J. et al. (2012). All-solid-state dye-sensitized solar cells with high efficiency. *Nature* 485: 486–489.

28 Zhu, Z. and Mi, Q. (2021). Substituted thiourea as versatile ligands for crystallization control and surface passivation of tin-based perovskite. *Cell Reports Physical Science* 3: 100690.

29 Ng, C.H., Nishimura, K., Ito, N. et al. (2019). Role of GeI_2 and SnF_2 additives for SnGe perovskite solar cells. *Nano Energy* 58: 130–137.

30 Milot, R.L., Klug, M.T., Davies, C.L. et al. (2018). The effects of doping density and temperature on the optoelectronic properties of formamidinium tin triiodide thin films. *Advanced Materials* 30: e1804506.

31 Noel, N.K., Stranks, S.D., Abate, A. et al. (2014). Lead-free organic-inorganic tin halide perovskites for photovoltaic applications. *Energy & Environmental Science* 7: 3061–3068.

32 Poli, I., Kim, G.W., Wong, E.L. et al. (2021). High external photoluminescence quantum yield in tin halide perovskite thin films. *ACS Energy Letters* 6: 609–611.

33 Mahata, A., Meggiolaro, D., Gregori, L., and De Angelis, F. (2021). Suppression of tin oxidation by 3D/2D perovskite interfacing. *Journal of Physical Chemistry C* 125: 10901–10908.

34 Ambrosio, F., Meggiolaro, D., Almutairi, T.M., and De Angelis, F. (2021). Composition-dependent struggle between iodine and tin chemistry at the surface of mixed tin/lead perovskites. *ACS Energy Letters* 6: 969–976.

35 Meggiolaro, D., Ricciarelli, D., Alasmari, A.A. et al. (2020). Tin versus lead redox chemistry modulates charge trapping and self-doping in tin/lead iodide perovskites. *Journal of Physical Chemistry Letters* 11: 3546–3556.

36 Yan, Y., Pullerits, T., Zheng, K., and Liang, Z. (2020). Advancing tin halide perovskites: strategies toward the $ASnX_3$ paradigm for efficient and durable optoelectronics. *ACS Energy Letters* 5: 2052–2086.

37 Hao, F., Stoumpos, C.C., Guo, P. et al. (2015). Solvent-mediated crystallization of $CH_3NH_3SnI_3$ films for heterojunction depleted perovskite solar cells. *Journal of the American Chemical Society* 137: 11445–11452.

38 Wang, P., Li, F., Jiang, K.J. et al. (2020). Ion exchange/insertion reactions for fabrication of efficient methylammonium tin iodide perovskite solar cells. *Advancement of Science* 7: 1903047.

39 Koh, T.M., Krishnamoorthy, T., Yantara, N. et al. (2015). Formamidinium tin-based perovskite with low Eg for photovoltaic applications. *Journal of Materials Chemistry A* 3: 14996–15000.

40 Lee, S.J., Shin, S.S., Kim, Y.C. et al. (2016). Fabrication of efficient formamidinium tin iodide perovskite solar cells through SnF(2)-pyrazine complex. *Journal of the American Chemical Society* 138: 3974–3977.

41 Liao, W., Zhao, D., Yu, Y. et al. (2016). Lead-free inverted planar formamidinium tin triiodide perovskite solar cells achieving power conversion efficiencies up to 6.22%. *Advanced Materials* 28: 9333–9340.

42 Zhao, Z., Gu, F., Li, Y. et al. (2017). Mixed-organic-cation tin iodide for lead-free perovskite solar cells with an efficiency of 8.12. *Advancement of Science* 4: 1700204.

43 Meng, X., Lin, J., Liu, X. et al. (2019). Highly stable and efficient $FASnI_3$-based perovskite solar cells by introducing hydrogen bonding. *Advanced Materials* 31: e1903721.

44 Liu, X., Wang, Y., Wu, T. et al. (2020). Efficient and stable tin perovskite solar cells enabled by amorphous-polycrystalline structure. *Nature Communications* 11: 1–7.

45 Nishimura, K., Kamarudin, M.A., Hirotani, D. et al. (2020). Lead-free tin-halide perovskite solar cells with 13% efficiency. *Nano Energy* 74: 104858.

46 Wei, Q., Ke, Y.Q., and Ning, Z.J. (2020). Theoretical study of using kinetics strategy to enhance the stability of tin perovskite. *Energy & Environmental Science* 3: 541–547.

47 Chen, Z., Wang, J.J., Ren, Y. et al. (2012). Schottky solar cells based on $CsSnI_3$ thin-films. *Applied Physics Letters* 101: 093901.

48 Moghe, D., Wang, L.L., Traverse, C.J. et al. (2016). All vapor-deposited lead-free doped $CsSnBr_3$ planar solar cells. *Nano Energy* 28: 469–474.

49 Heo, J.H., Kim, J., Kim, H. et al. (2018). Roles of SnX_2 (X = F, Cl, Br) additives in tin-based halide perovskites toward highly efficient and stable lead-free perovskite solar cells. *Journal of Physical Chemistry Letters* 9: 6024–6031.

50 Song, T.B., Yokoyama, T., Aramaki, S., and Kanatzidis, M.G. (2017). Performance enhancement of lead-free tin based perovskite solar cells with reducing atmosphere-assisted dispersible additive. *ACS Energy Letters* 2: 897–903.

51 Li, B., Di, H.X., Chang, B.H. et al. (2021). Efficient passivation strategy on sn related defects for high performance all-inorganic $CsSnI_3$ perovskite solar cells. *Advanced Energy Materials* 31: 2007447.

52 Ye, T., Wang, X.Z., Wang, K. et al. (2021). Localized electron density engineering for stabilized B-gamma $CsSnI_3$-based perovskite solar cells with efficiencies > 10%. *ACS Energy Letters* 6: 1480–1489.

53 Lin, J.T., Chu, T.C., Chen, D.G. et al. (2021). Vertical 2D/3D heterojunction of tin perovskites for highly efficient HTM-free perovskite solar cell. *ACS Applied Energy Materials* 4: 2041–2048.

54 Kayesh, M.E., Matsuishi, K., Kaneko, R. et al. (2018). Coadditive engineering with 5-ammonium valeric acid iodide for efficient and stable Sn perovskite solar cells. *ACS Energy Letters* 4: 278–284.

55 Chen, M., Ju, M.-G., Hu, M. et al. (2018). Lead-free Dion–Jacobson tin halide perovskites for photovoltaics. *ACS Energy Letters* 4: 276–277.

56 Chen, M., Dong, Q., Eickemeyer, F.T. et al. (2020). High-performance lead-free solar cells based on tin-halide perovskite thin films functionalized by a divalent organic cation. *ACS Energy Letters* 5: 2223–2230.

57 Li, P., Liu, X., Zhang, Y. et al. (2020). Low-dimensional Dion-Jacobson-phase lead-free perovskites for high-performance photovoltaics with improved stability. *Angewandte Chemie, International Edition* 59: 6909–6914.

58 Li, H.S., Jiang, X.Y., Wei, Q. et al. (2021). Low-dimensional inorganic tin perovskite solar cells prepared by templated growth. *Angewandte Chemie, International Edition* 60: 16330–16336.

59 Kim, H., Lee, Y.H., Lyu, T. et al. (2018). Boosting the performance and stability of quasi-two-dimensional tin-based perovskite solar cells using the formamidinium thiocyanate additive. *Journal of Materials Chemistry A* 6: 18173–18182.

60 Xu, H.Y., Jiang, Y.Z., He, T.W. et al. (2019). Orientation regulation of tin-based reduced-dimensional perovskites for highly efficient and stable photovoltaics. *Advanced Functional Materials* 29: 1807696.

61 Wu, T., Liu, X., He, X. et al. (2019). Efficient and stable tin-based perovskite solar cells by introducing π-conjugated Lewis base. *Science China. Chemistry* 63: 107–115.

62 Gu, F.D., Ye, S.Y., Zhao, Z.R. et al. (2018). Improving performance of lead-free formamidinium tin triiodide perovskite solar cells by tin source purification. *Sol. RRL* 2: 1800136.

63 Nakamura, T., Yakumaru, S., Truong, M.A. et al. (2020). Sn(IV)-free tin perovskite films realized by in situ Sn(0) nanoparticle treatment of the precursor solution. *Nature Communications* 11: 1–8.

64 Meng, X., Wu, T., Liu, X. et al. (2020). Highly reproducible and efficient $FASnI_3$ perovskite solar cells fabricated with volatilizable reducing solvent. *Journal of Physical Chemistry Letters* 11: 2965–2971.

65 Song, T.B., Yokoyama, T., Stoumpos, C.C. et al. (2017). Importance of reducing vapor atmosphere in the fabrication of tin-based perovskite solar cells. *Journal of the American Chemical Society* 139: 836–842.

66 Tai, Q., Guo, X., Tang, G. et al. (2019). Antioxidant grain passivation for air-stable tin-based perovskite solar cells. *Angewandte Chemie, International Edition* 58: 806–810.

67 Wang, C., Zhang, Y., Gu, F. et al. (2021). Illumination durability and high-efficiency Sn-based perovskite solar cell under coordinated control of phenylhydrazine and halogen ions. *Matter* 4: 709–721.

68 Chen, K., Wu, P., Yang, W. et al. (2018). Low-dimensional perovskite interlayer for highly efficient lead-free formamidinium tin iodide perovskite solar cells. *Nano Energy* 49: 411–418.

69 Yin, Y., Wang, M., Malgras, V., and Yamauchi, Y. (2020). Stable and efficient tin-based perovskite solar cell via semiconducting–insulating structure. *ACS Applied Energy Materials* 3: 10447–10452.

70 Liu, G., Liu, C., Lin, Z. et al. (2020). Regulated crystallization of efficient and stable tin-based perovskite solar cells via a self-sealing polymer. *ACS Applied Energy Materials* 12: 14049–14056.

71 Konstantakou, M. and Stergiopoulos, T. (2017). A critical review on tin halide perovskite solar cells. *Journal of Materials Chemistry A* 5: 11518–11549.

72 Nishikubo, R., Ishida, N., Katsuki, Y. et al. (2017). Minute-scale degradation and shift of valence-band maxima of $(CH_3NH_3)SnI_3$ and $HC(NH_2)_2SnI_3$ perovskites upon air exposure. *Journal of Physical Chemistry C* 121: 19650–19656.

7

Fundamentals and Synthesis Methods of Metal Halide Perovskite Thin Films

Mingwei Hao[1], *Tanghao Liu*[1], *Yalan Zhang*[2], *Tianwei Duan*[1], *and Yuanyuan Zhou*[1,2]

[1] *Hong Kong Baptist University, Department of Physics, Kowloon Tang, Hong Kong, SAR 999077, China*
[2] *The Hong Kong University of Science and Technology, Department of Chemical and Biological Engineering, Clear Water Bay, Hong Kong, SAR 999077, China*

7.1 Introduction

Perovskite is a family of materials with the formula of ABX_3, where A and B are cations and X is an anion. The first perovskite, $CaTiO_3$, was discovered as an oxide form by Gustav Rose in 1839 [1]. Later, this general crystal structure was named after the Russian mineralogist Lev Perovski. The majority of the earliest perovskites discovered are oxides with ferroelectric, piezoelectric, and pyroelectric properties, but they are not good semiconductor materials [2–4]. In contrast to oxide perovskites, metal halide perovskites (MHPs) feature high absorption coefficients, direct and tunable bandgaps, high carrier mobility, and near-ambient fabrication process, which opens up new application possibilities for the perovskite family [5–8]. These qualities promote the advancement of MHP thin-film fabrication for optoelectronic devices via solution processing or other low-cost processing methods.

MHP's thin-film pioneering work started in the fields of transistors and light-emitting diodes (LEDs) [9, 10]. Since the use of MHPs in solar cells in 2009 [11, 12], the power conversion efficiency (PCE) of perovskite solar cells (PSCs) has been rapidly increased, which demonstrates unprecedented success in the field of photovoltaics (PVs) [13–15]. In the past decade, halide perovskite thin films have always been active in various fields, such as photodetectors [16–19], lasers [20–24], and spintronics [25–27], owing to their ease of processing, inexpensive fabrication process, and excellent optoelectronic properties. MHPs' inherent optical and electrical characteristics are determined by their crystal structures, chemical compositions, and microstructures, all of which can be controlled during the film growth process. Although the low formation energy of halide perovskites facilitates large-scale thin-film synthesis, this "soft" materials nature also has an impact on the stability of perovskites in device applications.

This chapter provides a comprehensive review of MHP processing from basic knowledge to synthetic approaches. Here, Section 7.2 introduces the fundamentals

Halide Perovskite Semiconductors: Structures, Characterization, Properties, and Phenomena, First Edition. Edited by Yuanyuan Zhou and Iván Mora-Seró.

of perovskite structure, compositions, and microstructures; Section 7.3 provides the current understandings on the thin-film growth theories of MHPs; Sections 7.4–7.6 highlight the updated fabrication methods from solution or vapor precursors to achieve high-quality thin films in the lab-to-fab scales; Section 7.7 illustrates the effects of post-deposition treatments, including solvent annealing and vacuum exposure. Finally, Section 7.8 offers an overview of the field of halide perovskite thin-film deposition.

7.2 Fundamentals of MHPs Thin Films

7.2.1 Crystal Structures and Compositions

Currently, various MHP structural derivates have been discovered based on this three-dimensional (3D) cubic ABX_3 perovskite structure (Figure 7.1a), including distorted perovskite phase (Figure 7.1b), lead-free perovskites (e.g. double perovskites) (Figure 7.1c), and low-dimensional perovskite (Figure 7.1d). The perovskite crystal structure and composition have a direct effect on the electronic bandgap structure of a material, and even small variations in the lattice can result

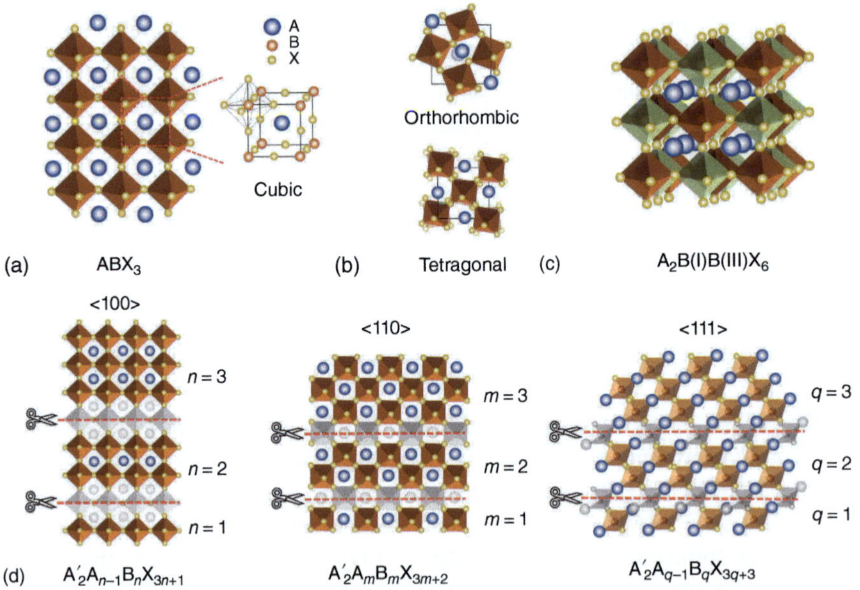

Figure 7.1 The structures for typical 3D perovskite and their variants. (a) Standard cubic ABX_3 perovskite. (b) Orthorhombic and tetragonal disordered perovskites resulting from the octahedral tilting. (c) Double perovskite $A_2B(II)B(III)X_6$, as an example of lead-free perovskite. (d) Low-dimensional organic–inorganic perovskites derived from 3D perovskite structural sections. From left to right, three models are <100>, <110>, and <111>-oriented perovskites, where n, m, and q represent different oriented families of 2D perovskites, with chemical formulas $A'_2A_{n-1}B_nX_{3n+1}$, $A'_2A_mB_mX_{3m+2}$, and $A'_2A_{q-1}B_qX_{3q+3}$, respectively. Source: Reproduced from Saparov and Mitzi [28] with permission of American Chemical Society.

in substantial changes in physical properties. Thus, the abundant variants in MHP libraries play an important role in optoelectronics.

7.2.1.1 3D MHPs

3D MHPs are based on the stoichiometric ABX_3, generally in which A is a monovalent cation (e.g. Cs^+, Rb^+, MA^+/methylammonium, and FA^+/formamidinium), B is a divalent metal cation (e.g. Pb^{2+} or Sn^{2+}), and X is a halide anion (I^-, Br^-, or Cl^-). A-site cation is coordinated to 12-fold X anions forming a cuboctahedron, and B-site cation is coordinated to sixfold X anions forming an octahedron (Figure 7.1a). Such unit cells are made up of A-site cations in cube corner positions, B-site cations in body-centered positions, and X anions in face-centered positions.

The cation and anion candidates of ABX_3 perovskite empirically complied with the Goldschmidt tolerance factor (t) [29]:

$$t = \frac{R_A + R_X}{\sqrt{2}(R_B + R_X)} \tag{7.1}$$

where R_A, R_B, and R_X are the ionic radii for the A, B, and X ions, respectively. The tolerance factor was empirically found satisfied between 0.8 and 1.0 for the majority of 3D perovskite [28].

The distortion of perovskite was caused by the size disparity between the R_A, R_B, and R_X. As a result, the orthorhombic or tetragonal phase of perovskite is formed due to the tilting and distortion of the octahedral networks as well as rotations of the molecular cations (Figure 7.1b) [30–32]. Notably, the method's credibility for predicting unique perovskite structures is less reliable due to assumptions made regarding the precise ionic radii of the cations and anions.

The octahedral factor (μ) is also important for the formability of stable perovskite, as follows:

$$\mu = \frac{R_B}{R_X} \tag{7.2}$$

where R_B and R_X are the ionic radii of B and X, respectively. An optimum value of μ is between 0.4 and 0.9 [33].

The A, B, and X ion sizes in perovskites have a significant impact on the structure and stability of perovskites, as shown by the Goldschmidt tolerance factor (t) and octahedral factor (μ). The size of the body-centered A cation has a substantial impact on the optical properties of the material since its size can cause the entire lattice to expand or contract, and changes in the B–X bonds lead to variations in the bandgap energy. Figure 7.2a presents a graphical representation of the formation constraints in the form of a 2D map of ionic radii A and X. The region within the contour lines represents the MHP formability conditions for the Pb and Sn families of compounds. Marked along the ordinate and abscissa, respectively, are the ionic radii of F^-, Cl^-, Br^-, and I^-, also involving the commonly A-site cations, such as Cs^+, $CH_3NH_3^+$(MA^+), $HC(NH_2)_2^+$(FA^+), and $CH_3CH_2NH_3^+$(EA^+) [34]. Given the components that have been specified, their intersection can be used to make a prediction about whether or not a perovskite will likely result. The perovskite

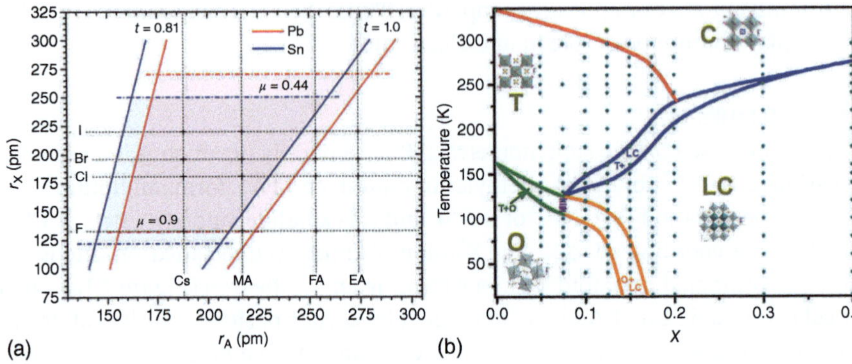

Figure 7.2 The ion radius and compositions influence on 3D MHPs. (a) Formability of Pb- (red) and Sn- (blue) halide perovskites is determined by A-site cation and X-site halide anion radius. The limits of the tolerance and octahedral factors are shown by solid and dashed lines, respectively. Six-coordinate Pb^{2+} and Sn^{2+} have ionic radii of 119 and 110 pm, respectively. Source: Adapted from Manser et al. [34] with permission. Copyright 2016, American Chemical Society. (b) Phase diagram for $MA_{1-x}FA_xPbI_3$ perovskites at different temperatures and x values, where C stands for cubic, T for tetragonal, LC for large-cell cubic, and O for orthorhombic. Source: Reproduced from Mohanty et al. [35] with permission of American Chemical Society.

structure is preferred if the intersection is located within the limits of formation, which are represented by the shaded regions. If it lies significantly outside of these boundaries, other motifs will emerge. Altering A-, B-, and/or X-site ions in multi-component perovskites can also be used to modify the phase structure. A comprehensive temperature-composition phase diagram for $MA_{1-x}FA_xPbI_3$ was established by Mohanty et al., and four phases were identified, including cubic (C: *Pm-3m*), tetragonal (T: *I4/mcm*), orthorhombic (O: *Pnma*), and large-cell cubic (LC: *Im-3*), by varying the composition (x) and temperatures (Figure 7.2b) [35].

7.2.1.2 Lead-free MHPs

Pb toxicity is a health and environmental concern that must be addressed in sustainable technology. Lead-free, such as halide double perovskite (Figure 7.1c), has been extensively researched to address this problem. Typically, apart from standard perovsktie containing Pb^{2+} replaced with Sn^{2+}, the formation of other lead-free MHPs can occur via one of three lead substitution methods [36]: (i) substitution with heterovalent cations, usually in the form of an equal number of monovalent and trivalent metal cations to balance the total charge and valence; (ii) substitution of lead with higher-valent metal cations and vacancies to neutralize the total charge. (iii) substitution with isoelectronic trivalent cations (Sb^{3+}, Bi^{3+}) forming only low-dimensional perovskite. These approaches give rise to four chemical formulas, including standard perovskite $AB^{III}X_3$, double perovskite $A_2B^IB^{III}X_6$, vacancy-ordered double perovskite $A_2B^{VI}X_6$, and layered/dimer perovskite $A_3B^{III}X_9$ (Figure 7.3).

The standard MHP structure $AB^{III}X_3$ commonly replaces Pb^{2+} with Sn^{2+}. The first Sn-based perovskite, $CsSnI_3$, was served as the inorganic hole transporter

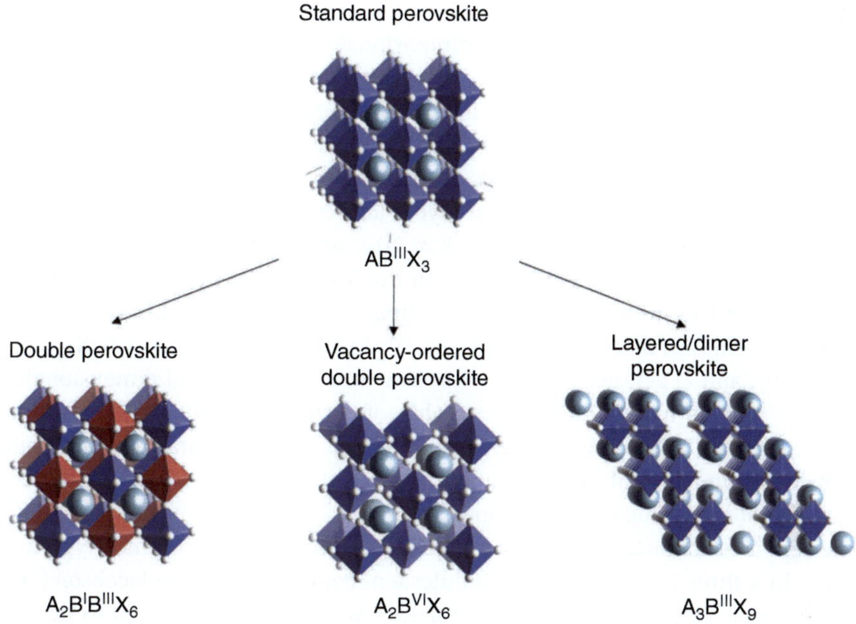

Figure 7.3 Illustration for four lead-free MHPs crystal structures. Derived from standard perovskite ABX_3, the other three lead-free perovskites are double perovskite, vacancy-ordered double perovskite, and layered/dimer perovskite. Source: Adapted from Giustino and Snaith [36] with permission. Copyright 2016, American Chemical Society.

in solid-state dye-sensitized solar cells. In 2014, $MASnI_3$ and $MASn(I_{1-x}Br_x)_3$ were demonstrated as organic–inorganic lead-free MHPs with energy conversion efficiencies of 6.4% and 5.73%, respectively [37]. However, Sn^{2+} is prone to oxidize into Sn^{4+} in ambient air, leading to Sn-based perovskites exhibiting self-doping effects and structural instabilities [37, 38]. Compared to the standard lead-free MHP, double perovskite $A_2B^IB^{III}X_6$ was termed due to the double occupancy of the B-site cation. The new $A_2B^IB^{III}X_6$ structure is composed of two split B-cations that form a network structure that includes the B^IX_6 and $B^{III}X_6$ octahedrons, with the A-site cation located in the center of the four octahedrons. $Cs_2AgBiBr_6$ is one of the pioneer compounds of lead-free halide double perovskites, enabling its synthesis under ambient conditions [39]. Other members of double perovskites, such as $Cs_2InSbCl_6$, $Cs_2AgInBr_6$, $Rb_2AgInBr_6$, and $Rb_2CuInCl_6$, have been predicted to exhibit high stability and good optoelectronic capabilities [40–42]. In comparison to double perovskites, tetravalent-metal vacancy-ordered double perovskites $A_2B^{VI}X_6$ have half of the tetravalent atoms replaced by a vacancy. Compounds with an $A_2B^{VI}X_6$ anionic lattice, such as Cs_2SnI_6 and Cs_2TiBr_6, have strong stability and moisture resistance [43]. Last but not least, the layered/dimer perovskite is formed by substituting Pb^{2+} with Bi^{3+} or Sb^{3+}, resulting in a 0D or 2D structure. However, these layered/dimer perovskites exhibit poor charge carrier mobility, which is detrimental to device performance [44, 45].

7.2.1.3 2D MHPs

A broader extended family of low-dimensional perovskites breaks the rigid structural constraints of typical ABX_3 perovskites, which enables them amazing structural tunability. Similarly, low-dimensional perovskites are also structurally composed of corner-shared BX_6 octahedra, like their parent 3D structure. The 2D MHPs were created by inserting a large organic cation A′ at the A-site, which opens the chemical link between octahedral layers and connects adjacent octahedrons by van der Waals force. As shown in Figure 7.1d, organic cation cuts 3D MHPs in a specified direction. The shear direction in the 3D MHP crystal is closely related to the structural formula of 2D MHPs. Cutting along three crystal directions will yield three formulas, that is $A'_2A_{n-1}BX_{3n+1}$, $A'_2A_mBX_{3m+2}$, and $A'_2A_{q-1}B_qX_{3q+3}$ for ⟨100⟩, ⟨110⟩, and ⟨111⟩, respectively. This structural flexibility and dimensionality tunability provide a rich and fertile archive for the preparation of interesting crystal structures with diverse physical properties. The structural flexibility and dimensional tunability of 2D MHP allow for the fabrication of crystal formations with different intriguing physical properties.

The derivates obtained along the <100> direction are the most common type, and they produce three 2D MHP phases: Ruddlesden–Popper (RP), Dion–Jacobson (DJ), and alternating cations in the interlayer space (ACI), which correspond to the formulas $A'_2A_{n-1}B_nX_{3n+1}$, $A'A_{n-1}B_nX_{3n+1}$, and $A'A_nB_nX_{3n+1}$, respectively (Figure 7.4) [46, 47]. The phase structure of the 2D MHPs is controlled by the spacer molecules A′, which have an organic tail and an amino head with monovalent or divalent cation. RP phase is made by putting in monovalent cations with the amino heads

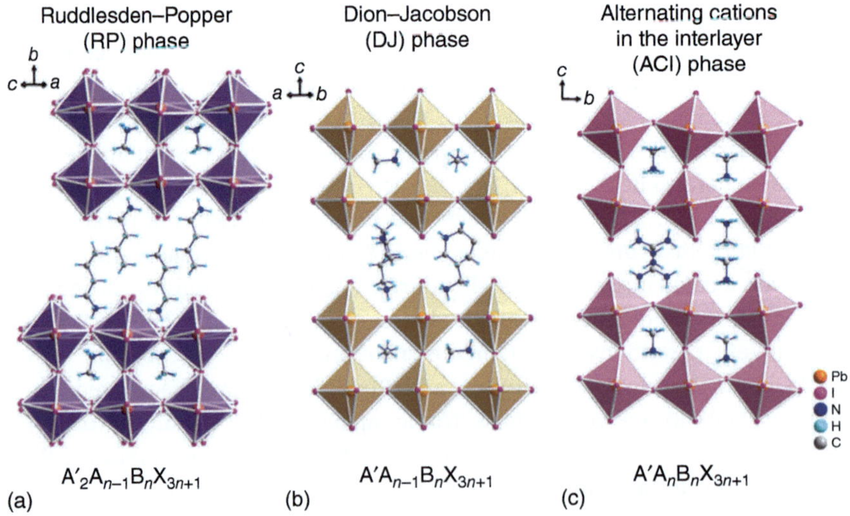

Figure 7.4 Structural comparison of three 2D MHPs with frequently used organic spacer. (a) Ruddlesden–Popper phase formed by mono-valent cations with amino head. (b) Dion–Jacobson phase formed by di-valent cations with double amino heads. (c) Alternating cations in the interlayer phase formed by tri-valent cations with three amino heads. Source: Reproduced from Li et al. [46] with permission of American Chemical Society.

facing the inorganic layer and tying them together with hydrogen bonds. Since many monovalent organic spacers can be used in halide perovskites, the RP phase is the most studied 2D MHP. The shape of the DJ phase is set by divalent spacer cations that connect the vicinal organic sheets with hydrogen bonds. Compared to the RP phase, the organic spacers that you can choose to use in DJ phase are much less. The ACI phase is the least common of these three types because only the guanidinium (GA^{3+}) cation can be used.

Beyond chemistry variations of organic molecules, the structures of layered perovskite can also be tuned by mixing large (A′) and small (A) cations in such a way that the structures are made up of layers of different thicknesses of the 3D perovskite that are segregated by the large organic cations. This allows for the properties of layered perovskites to be modified in a number of different ways. By varying the ratio of large to small organic cations, it is possible to achieve a smooth transition from $n = 1$ to $n = \infty$ perovskites in terms of their physical properties, providing a useful method for tuning the optoelectronic properties of thin films. Larger-n ($n > 3$) RP perovskites unfavored to lie flat during film deposition, which improves electrical connection across the film thickness. Due to the low energy barrier between RP family members with neighboring n values, single-phase thin films are difficult to produce [47].

7.2.2 Microstructures

MHP thin films, either prepared through the solution or vapor precursor, were frequently in the form of polycrystalline. Consequently, MHP thin films have grains and grain boundaries (GBs) where grains contact each other. A grain is the smallest unit in a microstructure, usually a single crystal, that is surrounded by GBs, interfaces, or surfaces. This definition applies regardless of whether the MHPs have a 3D crystal structure or other low-dimensional crystal structures. MHP thin film microstructure is determined by GB type, grain size and distribution, and grain crystallographic orientations (texture). The above-mentioned factors of microstructures have a great influence on device performance, attributes, and stability.

7.2.2.1 Types of the GBs

The GBs geometry between two grains 1 and 2 can be characterized by the tilt (θ) and twist (φ) misorientation angles (Figure 7.5a) [49]. As a result, GBs are able to be explained using thermodynamic quantities. GB creates disorder and high free volume, which can be charged in complicated multi-cations/anionic materials, such as MHPs, due to ion or impurity segregation (Figure 7.5b) [50].

2D defects in solid-state GBs are referred to as atomically bound single-crystal grains that have arbitrary crystallographic orientations (Figure 7.5c). Such GBs have 5 degrees of freedom, with three angular coordinates indicating grain misorientation and two indicating GB orientation. Generally, the GBs are formed by randomly oriented grains. In some cases, the GBs are the sub-boundaries inside single-crystal grains, where the crystal phase on each side of the sub-boundary is coupled by a certain symmetry operation (Figure 7.5d). Such an example is the existence of the twin boundary in some MHP thin films. There are also GBs with distinct second phases, which might be amorphous, crystalline, molecular, or polymer based (Figure 7.5e)

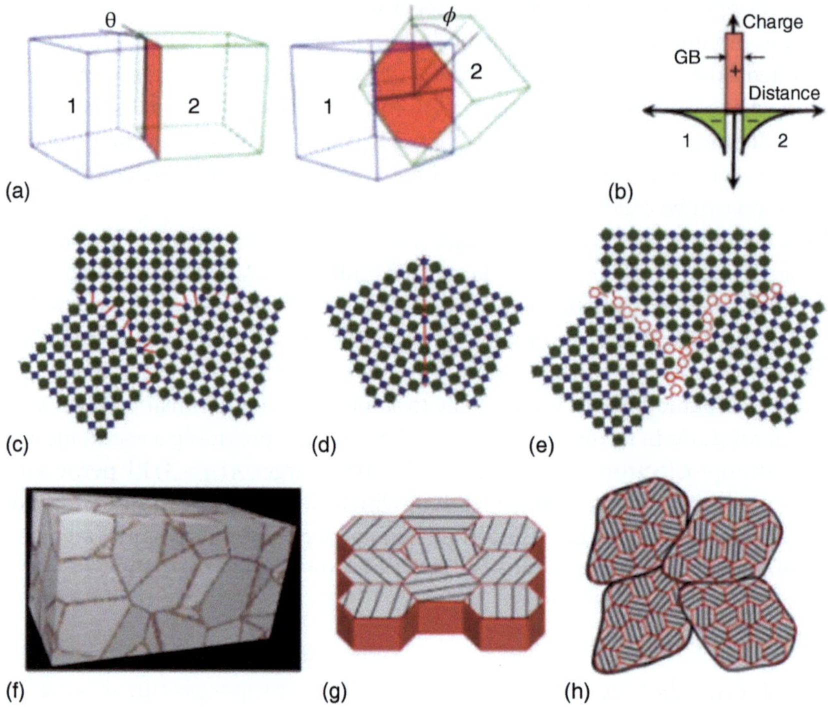

Figure 7.5 Illustration for grain boundaries. (a) The characterization for grains 1 and 2 with tilt (θ) and twist (φ) angles. (b) The relationship between charged distribution and distance related to GBs. (c) Random GBs between three different orientations, with bonding shown by red lines. (d) Twin boundary between two grains. (e) Functionalized GBs with molecules. (f) Traditional 3D random grains in bulk materials. (g) The arrangement of grains in MHP thin films. (h) Microstructures within the aggregates. Source: Reproduced with permission from Zhou and Padture [48], © 2022, John Wiley & Sons.

[48]. The scenario for GBs in the solid state is usually a 3D network of interconnected GBs, similar to a soap film in foam (Figure 7.5f) [51]. MHP thin film grain size is similar to film thickness, suggesting the GB network is not totally 3D (Figure 7.5g) [52]. Besides, MHP thin films can also include hierarchical microstructures consisting of aggregates, which are a collection of smaller single-crystal grains and related GBs (Figure 7.5h). However, aggregate borders are not GBs and cannot be determined by thermodynamic quantities.

7.2.2.2 Grain Size and Distribution

The grain size is a usual illustrative measure of a microstructure. The method for measuring the grain size of MHP thin films is to analyze scanning electron microscope (SEM) or atomic force microscopy (AFM) pictures of the top surface. This implies that the GB grooves observed on the top surface represent the intersection of every GB with that surface. The GBs are not formed intentionally by thermal or chemical etching rather, they are the result of natural processes [53, 54]. Nevertheless, post-treatment of MHP thin film surfaces will leave GB grooves on grain surfaces, which are not shown in as-formed perovskite films during processing. Besides, there also exist some wrinkles on the grain surface, which are sometimes confused with GB grooves, leading to an overestimation of the grain size.

Grain size distribution is also highly essential, but it is generally overlooked in MHP literature, where it is often shown as the mean grain size. Normal grain growth typically results in a reasonably narrow grain size distribution and satisfies the "Hillert" criterion [55], which suggests abnormal or excessive grain growth. The grain's form can also be significant; for example, anisotropic grain development is indicated by microstructures with plate-like or needle-like grains.

7.2.2.3 Crystallographic Orientations

The crystallographic orientations of grains in a microstructure are often chaotic. Nonetheless, the microstructure is said to have a texture if there is a preferred orientation between neighboring grains, among a group of grains, or among all grains. For devices that are built on MHP thin films, it is still unclear to what degree performance and film characteristics are determined by these aspects that are inherent to the orientation of the grains and how these orientations interact with other layers [56–59].

As for layered MHP thin films, the impact of crystallographic orientation may be much more significant. Because of the evident anisotropy of the crystal structure, electron transport can be performed easily along the planes of corner-sharing metal halide octahedra, but it is challenging across these planes. As a result, it is of the utmost importance to make certain that the orientations of stacked perovskite films in relation to the substrate are adjusted in accordance with the specific needs of the various applications. For example, in-plane charge transport is required for parallel transistors, and out-of-plane charge transport is needed for perpendicularly structured solar cells and LEDs.

7.3 Thin Film Growth Mechanism

7.3.1 Crystal Nucleation Mechanism

The crystallization process is involved in the production of high-quality MHP films. According to the classical theory, crystallization process can be separated into nucleation and growth processes. Because these two processes are sensitive to diverse growing conditions, such as supersaturation and air/liquid/solid interfaces, the final films have significantly varied morphologies. As a basis, it is first necessary to summarize the fundamental elements of nucleation and growth theory. In actuality, nucleation and growth can overlap in time, but here they are considered separately for simplicity. Furthermore, the intermediates which show great influence on the formation of high-quality films will also be introduced in this section.

7.3.1.1 Nucleation Theory

Nucleation determines the nature of future growth processes and the final product's properties. Understanding nucleation behavior is crucial for controlling the quality of MHP films. In classical nucleation theory, the saturation of the precursor often leads to the creation of nucleating phase molecules, which may condense into small single crystal nuclei. Heterogeneous and homogeneous nucleation are two traditional nucleation approaches, during which these nuclei are created at the substrate/precursor solution contact and/or inside the precursor solution, respectively.

Homogeneous nucleation means that the newly formed nuclei are spread out evenly in the parent phase. Heterogeneous nucleation, on the other hand, usually happens at the surface or interface of the container (Figure 7.6a) [60]. When the concentration (C) of a perovskite solution surpasses its solubility (C_s), nuclei are formed. If the supersaturation is not high enough for nucleation; however, these nuclei generated by thermal fluctuations are redissolved. The nucleation is a thermodynamic process that is driven by the change in total Gibbs free energy (ΔG).

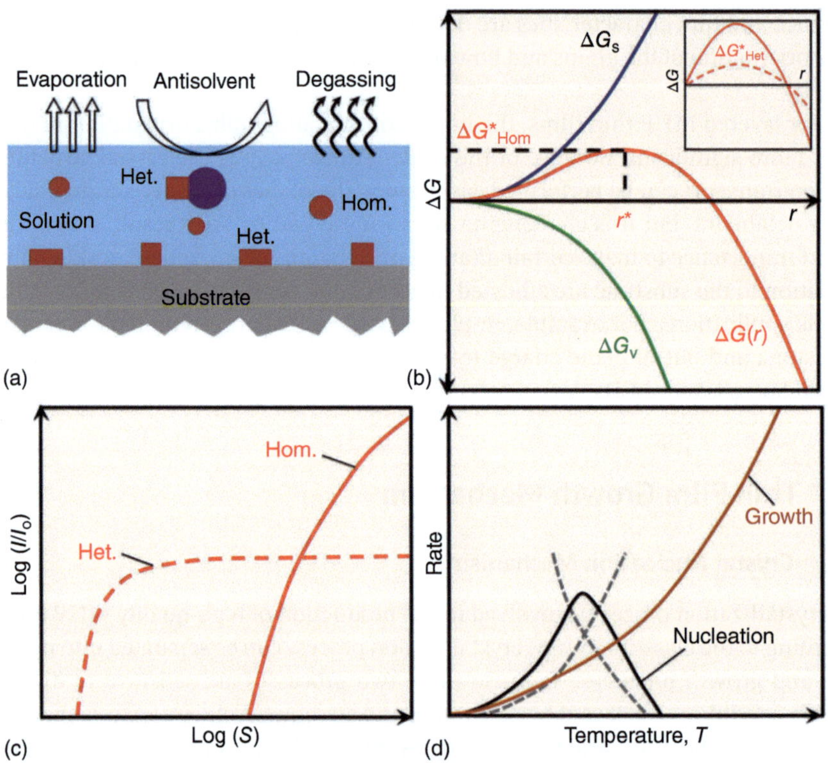

Figure 7.6 Illustration for classical nucleation process. (a) Homogeneous and heterogeneous nucleation approaches of a thin film in supersaturation solution. (b) The schematic illustration for the relationship among free energy change (ΔG), a sum of surface (ΔG_S), and volume (ΔG_V) free energy changes with increasing nucleus radius (r). (c) The schematic illustration for the relationship between normalized nucleation rate (I/I_0) as a function of supersaturation ratio (S) for homogenous and heterogeneous nucleation. (d) The schematic illustration for nucleation and growth rates as a function of temperature (T). Source: Reproduced from Zhou et al. [54] with permission of American Chemical Society.

ΔG equals the sum of the changes in volume free energy (ΔG_V) and surface free energy (ΔG_S). ΔG_V is related to the transformation of a unit volume of precursor solution into the crystalline nucleus. ΔG_S is related to the formation of a new nucleus/precursor solution interface per unit region. The total free-energy change ΔG, as a function of nucleus (spherical) radius, r, is as follows [61]:

$$\Delta G(r) = \Delta G_V + \Delta G_S = \left(-\frac{4\pi r^3}{3V_M}\right) RT\ln(S) + 4\pi r^2 \gamma_{CL} \tag{7.3}$$

where V is the nucleus' volume, A is the nucleus' area, and γ_{CL} is the energy at the crystalline-liquid nucleus interface. In more specific expression, V_M is the molar volume of the molecules, R is the gas constant, and T is the absolute temperature. $S = C/C_s$ is the saturation ratio, where C is the solute concentration in the precursor solution and C_s is the equilibrium solubility limit.

In a homogeneous nucleation process, as shown in Figure 7.6b, the maximum ΔG (ΔG^*_{Hom}) appears at the point of critical nucleus radius (r^*). The nuclei smaller than r^* dissolve back into the precursor solution, whereas those larger than r^* are thermodynamically stable and continue to grow. At constant temperature, $r^* \propto 1/\ln(S)$. As nuclei grow on foreign surfaces, the nucleation process is heterogeneous, and the energy barrier ΔG^*_{Het} is greatly lowered by the effective lowering of the interface energy, $\gamma \times f(\theta)$, $0 < f(\theta) < 1$ (Figure 7.6b, inset). The parameter $f(\theta)$ is represented by [54, 62]:

$$f(\theta) = \frac{1}{4}(2 + \cos\,\theta)(1 - \cos\theta)^2 \tag{7.4}$$

$$\cos\,\theta = \frac{\gamma_{SL} - \gamma_{SC}}{\gamma_{CL}} \tag{7.5}$$

where θ is the contact angle, and γ_{SL} and γ_{SC} are the surface energies of the substrate–liquid and substrate–crystalline interfaces, respectively. In heterogeneous nucleation, the substrate's accessible surface area and colloidal particles in the precursor solution or liquid–gas interface limit nucleation sites. Classical nucleation theory predicts the nucleation rate as follows [54, 63]:

$$I \propto \exp\left(\frac{-Q_D}{RT}\right) \exp\left(\frac{-\Delta G^*}{RT}\right) \tag{7.6}$$

where $I = I_{Hom}$ for $\Delta G^* = \Delta G^*_{Hom}$ ($f(\theta) = 1$), and $I = I_{Het}$ for $\Delta G^* = \Delta G^*_{Het}$ ($0 < f(\theta) < 1$). Q_D is the activation energy for precursor diffusion to the nucleus/solution interface.

Figure 7.6c demonstrates that it is easy to achieve heterogeneous nucleation at low supersaturation (S); however, it calls for high S for homogeneous nucleation at a constant temperature. In the scenario of heterogeneous nucleation, the nucleation rate is high at low S because of the lowered energy barrier, but it saturates at high S because of the scarcity of nucleation sites. The formation of MHP thin films commonly involves both types of nucleation events, with heterogeneous nucleation predominating at low S and homogeneous nucleation predominating at high S. As depicted schematically in Figure 7.6d, the maximum rate of nucleation occurs at an intermediate temperature [64]. At constant S, the first exponential component increases with rising temperature, whereas the second exponential term decreases. As a consequence, the overall nucleation rate as a function of temperature reaches a maximum. Despite a significant thermodynamic driving force for nucleation at low temperatures, diffusion slows nucleation. High temperatures weaken nucleation while speeding up diffusion [64].

7.3.1.2 Influences on Nucleation

The basic nucleation theory can explain the process in simple systems, as water droplets form from water vapor. Nucleation of crystalline MHPs from a precursor solution, on the other hand, is a much more involved process that belongs in the category of "non-classical" nucleation. The existence of numerous ions in perovskite and the complexity of the polar aprotic solvents utilized cause a wide range of interactions in the precursor solution during saturation, leading to the nucleation behavior [65, 66]. In addition, different chemicals are frequently added to the precursor solution in order to generate perovskite thin films with high-quality crystalline. Precursor saturation leads to the creation of intermediate entities, existing in the form of an amorphous or crystalline solid phase. All of these intermediates' processes are system- and condition-dependent, precluding generalization and necessitating additional research.

The colloidal nature and presence of a solvated phase are significant because they influence the formation of perovskite by regulating the nucleation, delaying the crystallization of perovskites, and affecting the defect density of the resulting film [67, 68]. For example, a Lewis-acidic Pb^{2+} center can be coordinated by a Lewis-base from six directions; hence, there is competition between I^- ions and the organic solvent, depending on their binding strength. As depicted in Figure 7.7a, the coordinating ability of a solvent can be defined by its Gutmann donor number (D_N). According to the findings of Hamill et al. the coordination of Pb and solvent becomes substantial when D_N surpasses 18.0 kcal mol^{-1} [69]. Solvated phases arise in the solution before the perovskite phase in these instances. As depicted in Figure 7.7b, the crystal structures of PbI_2-solvent and MAI-PbI_2-solvent were determined for routinely used solvents, like dimethylformamide (DMF) and dimethyl sulfoxide (DMSO). These solvated phases feature PbI_6 octahedral chains with 1D edge sharing that are separated by solvent molecules via Pb–O interactions. This crystal forms a 1D fibrous structure because of its rapid growth in the [100] direction [70–73].

7.3.2 Crystal Growth Mechanism

7.3.2.1 Basic Growth Theory

The crystal growth procedure contains two phases: (i) molecules diffuse from the solution to the surface of the crystal; (ii) molecules deposit on the crystal, leading to further growth. In principle, the nucleation and homogeneous formation process of perovskites can be loosely divided into three stages, as illustrated by the LaMer diagram in Figure 7.8a [74]. In the initial phase for homogeneous nucleation, the solution collects small molecules. In the second stage, solvent evaporation and temperature rise can create a supersaturated environment in which nucleation might begin. During saturation, concentration (C) rises linearly with time until

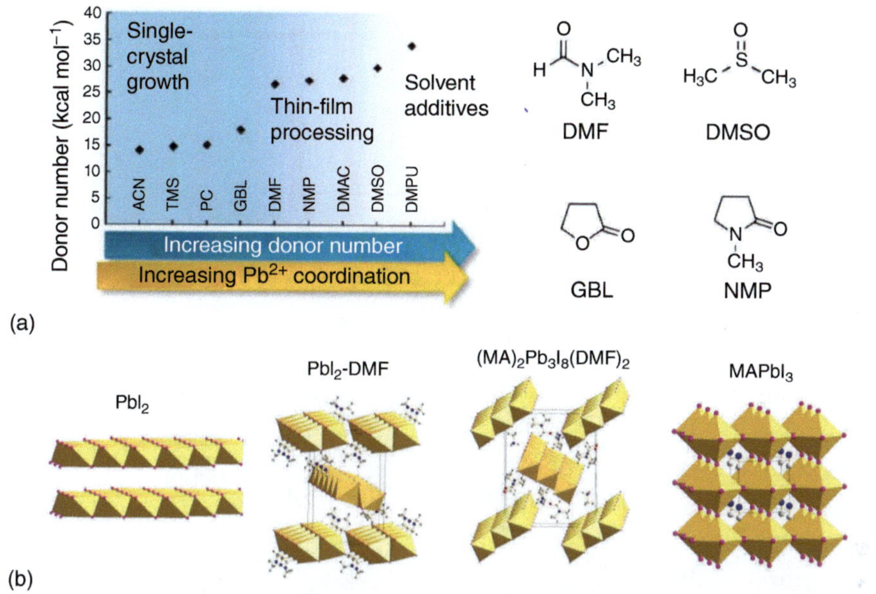

Figure 7.7 The solvent influences on perovskite intermediates. (a) Gutmann donor number (D_N) of different solvents-Pb^{2+} coordination and corresponding perovskite crystal growth preference. Source: Adapted from Hamill et al. [69] with permission. Copyright 2018, American Chemical Society. (b) Crystal structures of PbI_2, PbI_2–DMF solvated phase, $(MA)_2Pb_3I_8(DMF)_2$ solvated phase, and $MAPbI_3$ perovskite. Source: Reproduced from Ke et al. [70] with permission of IOP Publishing.

a supersaturation limit is reached, surpassing minimum concentration for nucleation (C_{ss}). The third step is crystal growth and ripening. Due to the depletion of "monomers" caused by nucleus formation, concentration decreases, causing it to fall below C_{ss} and inhibiting further nucleation. When the solute consumption rate exceeds the solvent evaporation rate, the solution concentration falls below the critical radius value. The growth of nucleation has superseded nucleation stage. Theoretically, the growth of MHP films might be controlled by regulating nucleation or growth. The occurrence of heterogeneous nucleation at the criterion of $C \ll C_{ss}$ (Figure 7.8b). In the case of thin film nucleation, secondary heterogeneous nucleation on colloidal particles is also feasible, with these nuclei settling on the substrate and being absorbed into the thin film during future growth [75].

As long as the solute concentration exceeds the solubility limit, stable nuclei will grow, decreasing the total free energy. As existing nuclei continue to expand, new stable nuclei will continue to nucleate. As shown in Figure 7.8c, there are three principal classical mechanisms for nuclei growth, that is island growth (Volmer–Weber model), layer growth (Frank–van der Merwe model), and layer-island growth (Stranski–Krastanov model) [77]. Through the attachment of monomers, nuclei expand both vertically and laterally during island formation. In this case, the bonding between the "monomers" of the growth phase is substantially greater than the bonding with the substrate surface. If the precursor film has enough nutrients, the islands eventually combine to create a dense polycrystalline thin film.

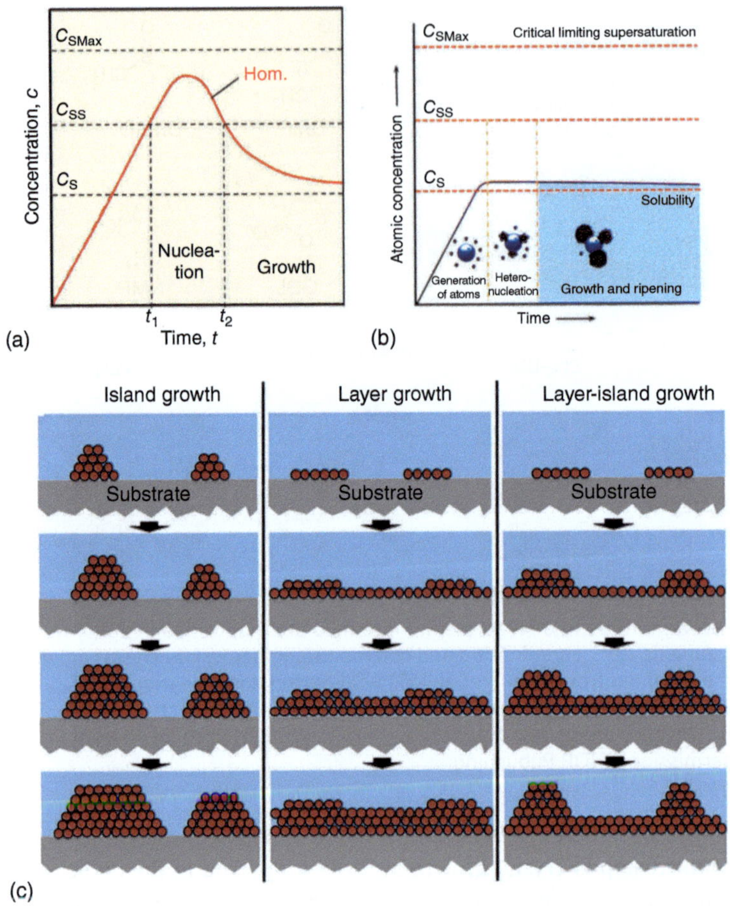

Figure 7.8 Schematic illustration for crystal growth mechanism. (a) LaMer diagram for homogeneous nucleation, where C_s is the solubility, C_{ss} is the minimum concentration for nucleation, and C_{sMax} is the maximum concentration for nucleation. Source: Reproduced from LaMer and Dinegar [74] with permission of American Chemical Society. (b) LaMer diagram for the heterogeneous nucleation. Source: Reproduced from Sun [75] with permission of Royal Society of Chemistry. (c) Schematic illustration of island growth (Volmer–Weber), layer growth (Frank–van der Merwe), and layer-island growth (Stranski–Krastanov) growth mechanisms. Source: Reproduced from Dunlap-Shohl et al. [76] with Permission of American Chemical Society.

Layer growth is more common in epitaxial systems, where the monomer-substrate connection is much stronger. Layer-island growth may occur in epitaxial systems as well, but only when the monomer-substrate link is weak. The island development method is most likely applicable to perovskite thin films since it does not normally need or involve epitaxy.

7.3.2.2 Grain-coarsening Theory

Solid-state grain growth and matrix-phase-mediated Ostwald ripening are the two most significant classical grain-coarsening mechanisms pertinent to MHP

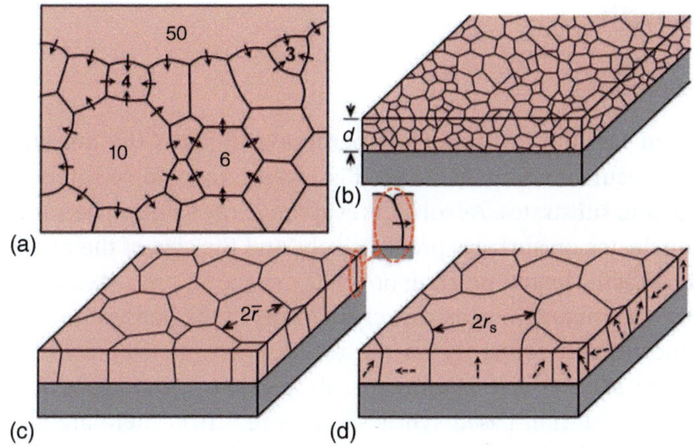

Figure 7.9 Illustrations for grain coarsening. (a) 2D microstructure of GBs motion. (b) The fine-grained thin film. (c) The coarse-grained stagnant thin film, the inset showing cross-section GB groove dragging a GB to the right. (d) The secondary grain growth, with dashed arrows indicating the same domain crystallographic orientation. Source: Reproduced from Dunlap-Shohl et al. [76] with permission of American Chemical Society.

films. In solid-state grain growth, the topology of individual grains is a particularly important factor to consider. Ostwald ripening is a subcategory of coarsening and relates in particular to coarsening in the presence of a continuous second phase. Both mechanisms derive their thermodynamic impetus from the disparities in grain curvature between small and large grains.

Figure 7.9a depicts a 2D diagram of a polycrystalline material with varying grain sizes and coordination numbers [78]. Grain stability is achieved by having exactly six sides, while grains with fewer than six sides experience shrinkage due to their concave GBs at the expense of nearby grains with more than six sides. Because monomer molecules in the smaller grain will diffuse across the solid-state GB that separates the two grains, the GB will move in the opposite direction in this particular scenario. As a consequence, the convex curvature becomes more prominent, and an increasing number of species move to the bigger grain. Also, the smaller grain eventually disappears, which results in the expansion of the larger grain. In contrast to bulk 3D materials, coarsening in thin films occurs slowly and generally comes to a standstill when the average grain size ($2\bar{r}$) reaches the film thickness (d), which results in all GBs crossing the top surface and bottom contact with the substrate (Figure 7.9b,c) [52]. This coarsening stagnation results from presumed isotropic GB energy, which is from drag forces exerted by the surface and interface on the moving GBs. GB grooves at the surface are expected to exert significant drag forces. Then, secondary coarsening occurs later in anisotropic thin films. In this instance, a few favorably oriented grains rapidly develop into large grained, textured thin films, and their low-energy crystallographic planes form the thin film surface and the contact with the substrate (Figure 7.9d) [52]. Through the growth of favorably oriented grains, the system has a tendency to maximize the area of those surfaces and interfaces by sacrificing less favorably oriented grains in the process.

7.4 One-step Growth

7.4.1 Growth From Solutions

The most straightforward method for producing perovskite films is the one-step solution technique. All precursor components are dissolved into a single solution, which is then deposited onto substrates. As solvents evaporate, the solution becomes saturated, perovskite nucleates on surfaces progressively, and the size of the nuclei increases to form a film. Lastly, heat annealing or other extraction processes could eliminate the remaining solvent and increase the crystallinity of the perovskite film. In this section, we concentrated primarily on the well-known one-step solution methods in the laboratory, such as spin-coating and drop-casting, and introduced the key factors influencing solution-based synthesis (Figure 7.10). There are also some solution-based methods for large-scale preparation of MHP film, which will be introduced in the subsequent Section 7.6.

7.4.1.1 Spin-coating

In the laboratory, spin-coating is the most preferred method for MHP thin film synthesis. In this method, after the precursor solution has been deposited onto the substrate, the substrate is rapidly rotated in order to distribute the solution evenly. Both the speed of the spinning and the amount of time involved can have an effect on the characteristics of the film. However, the films created using this process often have incomplete surface coverage, due to fast crystallization induced by the strong ionic reactions between PbI_2 and ammonium iodide, which further influences the performance of devices. Due to problems in film quality control, spin-coating seldom achieves mass production criteria, such as perovskite composition and phase purity, complete substrate coverage, pinhole removal, interface optimization, controlled grain size, and tailored GBs. To increase the quality of the spin-coated MHP

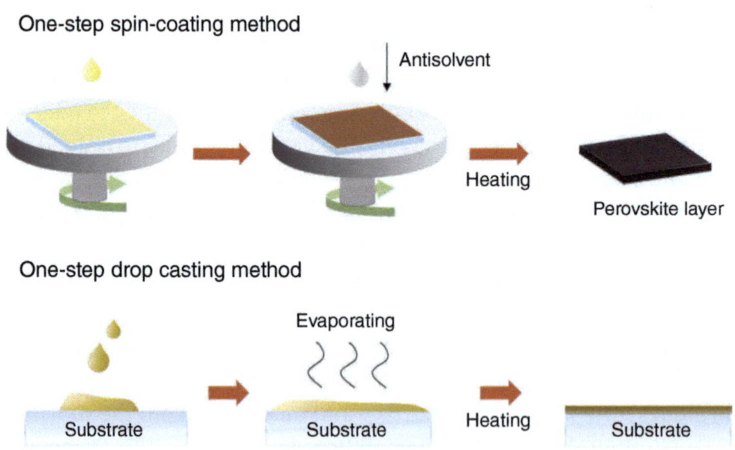

Figure 7.10 Schematic illustration for one-step solution-based growth methods in the lab, including spin-coating and drop-casting methods.

films, researchers investigated a variety of elements that influence the crystallization process, such as solvents, antisolvents, additives, and processing environment conditions (e.g. temperature and solvent vapor pressure).

In the deposition of perovskite films, DMF and DMSO are both extensively utilized solvents. Compared to DMSO, DMF evaporates more quickly. Consequently, the deposition of a DMF-based solution may result in a considerably thicker film. When the antisolvent is dropped from the top, DMF at the top can be eliminated swiftly, resulting in an immediate compact surface. In the meantime, residual DMF may exist at the bottom of the perovskite film. The subsequent thermal annealing procedure would result in the formation of voids at the bottom of the perovskite film due to the evaporation of residual DMF. DMSO, on the other hand, could form stronger bonds with perovskite precursors. The significantly delayed formation of the compact surface permits the complete removal of DMSO from the entire film and the avoidance of cavities. In order to improve the formation of perovskite films, other solvents with slow evaporation rates, such as gamma-butyrolactone (GBL) and *N*-Methylpyrrolidone (NMP), were also utilized. Currently, the most popular solvent is a combination of DMF and DMSO.

To address the problems of the crystallization process, some solvent extraction approaches were used to rapidly remove the solvent from the precursor solution, hence enhancing crystallization. The antisolvent washing method was firstly reported by Seok et al. Through this method, the production of a dense, crystalline MAI-PbI_2-DMSO intermediate-phase film was obtained as a result of the toluene antisolvent being washed onto the liquid precursor film (Figure 7.11a) [79]. This solvent washing method overcame the barrier of solubility restrictions

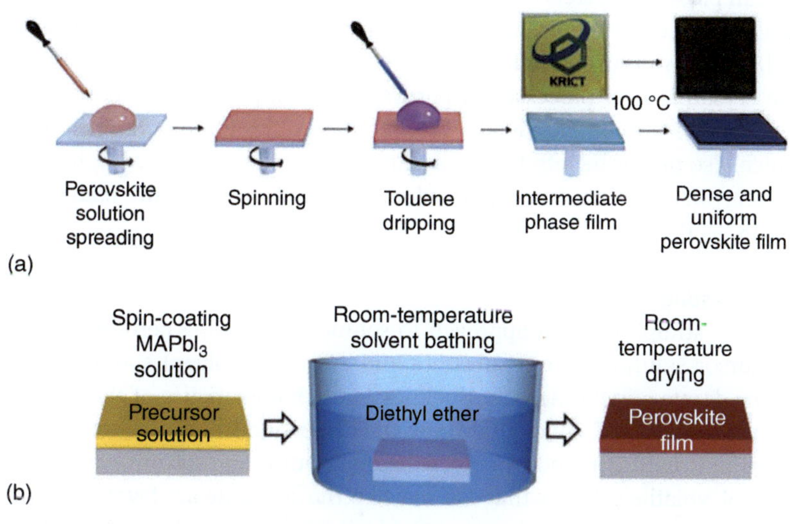

Figure 7.11 MHP crystallization process control methods in one-step synthesis. (a) Antisolvent washing process. Source: Reproduced with permission from Jeon et al. [79], Springer Nature. (b) Antisolvent bathing process. Source: Reproduced with permission from Zhou et al. [80], © 2015, Royal Society of Chemistry.

by accelerating the precursor supersaturation when a nonpolar antisolvent was dynamically dripped onto the precursor film. When antisolvent can be applied to the spinning film or used to wash the wet perovskite film after deposition, the host solvent can be rapidly eliminated, resulting in a high nucleus density. Consequently, perovskite grains could grow equally and pack tightly. In recent years, a variety of antisolvents, such as chlorobenzene, toluene, and dither ether, have been adopted.

When compared to dropping antisolvent during spin coating, antisolvent bathing process reduces the danger of harming the film and eliminates the need for exact antisolvent dripping timing. The antisolvent bathing method was firstly reported by Padture et al. (Figure 7.11b) [80]. This bathing method used NMP and diethyl ether as solvents and antisolvents, respectively. This method eliminates the complexity of precise timing, evenly applies the antisolvent, and reduces film damage, which is caused by the high-speed washing process between the solvent and the spinning film.

The crystallization dynamics of perovskite are also regulated by temperature. Heating the substrate prior to precursor deposition can substantially affect the film morphology. A "hot-casting" technique was reported by Mohite et al., in which the substrate is heated to 180 °C before the spin-coating of a 1 : 1 PbI_2: MACl solution in DMF, with the solution kept at 70 °C, and then annealed at 100 °C. This process creates a distinctive morphology in which leaf-like structures with large grain sizes cover the substrate [81].

Furthermore, various additives have been proposed during the past decade to modify the crystalization of perovskite films. For example, the polymers benefit the morphology control of perovskite films. In perovskite films, fullerene and its derivatives can passivate defects and inhibit ion migration. Metal cations (Li^+, Na^+, K^+) could also reduce the defect density in perovskites without altering their lattice structures [82, 83]. Anions of halide (Br^-, Cl^-) or pseudohalide (Ac^-, SCN^-) could modify the crystallization processes of perovskites to improve their morphology [84–86]. Ion liquids are ion-based salts that remain liquid at temperatures below 100 °C, which also proves benefit for perovskite film. By enhancing hydrophobicity and restricting MA^+ cation, ion liquid additives could enhance the stability of perovskites [87–89].

7.4.1.2 Drop-casting

One alternative, straightforward approach to solution processing perovskites is to deposit precursor solution onto a substrate and evaporate solvent spontaneously or under a moderate annealling. The thickness and characteristics of the film are affected by the volume and concentration of the dispersion solution. Other factors influencing film structure include droplet substrate wetting and solvent evaporation rate. In general, volatile solvents that can moisten the substrate are favored for this procedure. However, this approach can not typically provide well-formed thin films compared to spin-coating method.

This cast-coating technique has been successfully implemented in devices with an all-mesoporous architecture, in which the mesoporous electron transport layer

(ETL), the porous graphitic carbon electrode, and the non-conductive layer that separates the ETL from the electrode are all included within a mesoporous scaffold on an fluorine-doped tin oxide (FTO)/glass substrate. The early drop-casting method for fabricated efficient MHP film-based solar cell was reported by Han and coworkers [90]. The mixed-cation perovskite $(5\text{-AVA})_x(MA)_{1-x}PbI_3$ (5-AVA is 5-ammoniumvaleric acid) was synthesized and deposited on TiO_2 and ZrO_2 scaffolds by drop-coating a precursor solution. It is remarkable that the solar cell based on $(5\text{-AVA})_x(MA)_{1-x}PbI_3$-$TiO_2$ is more efficient compared to $MAPbI_3$-TiO_2, resulting from a longer exciton lifetime and a higher quantum yield for photoinduced charge separation [90].

Recently, this cast-coating method has also achieved success in $MAPbI_3$ PSCs manufactured under 88% humidity, delivering a PCE of 18.17%. In this casting method, the $MAPbI_3$ solution is put onto the center of a substrate on a 60 °C hot plate, where it spreads spontaneously. The wet coating then forms in a spherical, uniform shape. After being transferred to a hot plate with a higher temperature, the film quickly turns black. $MAPbI_3$ crystals will have varied orientations when prepared at different temperatures (Figure 7.12a). The crystals are oriented <110> and <200> at low preparation temperatures (60 °C) and high preparation temperatures (>120 °C), respectively. The $MAPbI_3$ PSCs can be fabricated below 90 °C to avoid pinhole formation (Figure 7.12b) [91].

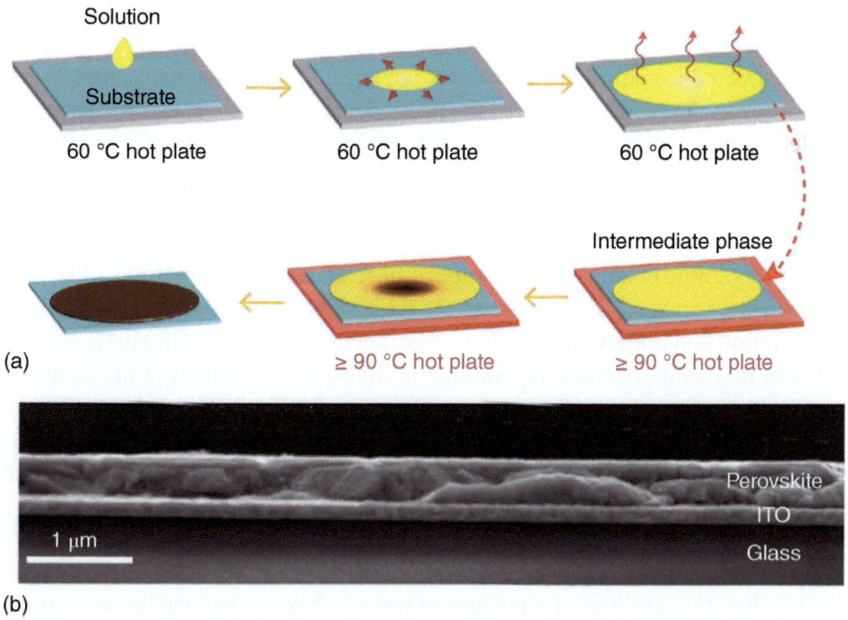

Figure 7.12 Drop casting methods for solar cell fabrication. (a) Illustration of the drop-casting method for preparing $MAPbI_3$ film. (b) Cross-section SEM image for the $MAPbI_3$ film made at 150 °C. Source: Reproduced with permission from Zuo and Ding [91], © 2021, John Wiley & Sons.

7.4.2 Growth from Vapor Phase

Vapor-based perovskite thin-film deposition techniques typically include dual-/single-source vapor deposition (SSVD) and pulsed laser deposition. The one-step vapor-based deposition method is mostly based on dual-source co-evaporation in a vacuum condition, where the vapor mixture condenses on the substrate, typically held at room temperature [92, 93]. Single-source evaporation is another facile vapor synthesis method to obtain MHP films; besides, dual-source co-evaporation provides better control over deposition than single source.

7.4.2.1 Thermal Evaporation

The dual-source co-evaporation approach for fabricating perovskite films was proposed in 2013. In this process, two precursor components were evaporated from two separate sources and then reacted with the substrate. Typically, a vacuum chamber is used for the evaporation process. Throughout this procedure, each source was monitored by a quartz sensor that revealed the evaporation rate, which was temperature dependent. The ratio of two evaporation rates and precursor materials dictated the composition of the perovskite film. The evaporation time determined the thickness of the film. After evaporation, the film was removed from the chamber and heated to facilitate the reaction of two precursor components, resulting in the final perovskite film [93]. The heating temperature and time, which have a significant impact on film quality, must be carefully optimized. Compared to solution processing, dual-source vapor deposition of MAI and $PbCl_2$ sources results in excellent uniformity of MHP films over a range of length scales, which leads to a substantial improvement in solar cell performance [93].

Dual-source co-evaporation has been a powerful strategy for halide perovskite thin-film deposition and has enabled the production of solar cells, LEDs, lasers, and photodetectors. Furthermore, its adaptability led to the use of multiple sources for depositing more complex perovskite compositions, such as bandgap tunable compositions for tandem devices, such as $MAPb(Br_{0.2}I_{0.8})_3$ and $FA_{0.7}Cs_{0.3}Pb(I_{0.9}\ Br_{0.1})$ [95, 96].

An early example of single-source vapor-phase deposition is synthesis of layered perovskite structures on a tantalum sheet. This method is a simplified version of co-evaporation and is known as single-source thermal ablation (SSTA) [97]. Because precursors may evaporate and recombine on the substrate without a major breakdown, rapid heating is crucial to the effectiveness of SSTA. When using SSTA, the film thickness is controlled by the quantity of precursor material deposited onto the tantalum sheet at the outset and source-sample distance during deposition. This method was used to achieve MHP films of $(BA)_2SnI_4$, $(PEA)_2PbI_4$, and $(PEA)_2PbBr_4$, which show photoluminescent or electroluminescent properties [97]. Unlike SSTA, the evaporation process of SSVD takes place within minutes instead of seconds. SSVD involves the addition of a powder of pre-synthesized perovskite or a combination of its precursors to a thermal crucible covered in a heating coil. SSVD was used to create fully inorganic $CsPbX_3$ perovskites with X = Br and Cl, and $CsSnI_3$ perovskites [98, 99]. Moreover, the lead-free $CsSn_{0.5}Ge_{0.5}I_3$ perovskite can also be achieved via the SSVD process, in which precursor was created through a solid-state reaction in evacuated Pyrex tubes between mixed powder precursors $CsI:SnI_2:GeI_2$ (2 : 1 : 1) (Figure 7.13) [94].

Figure 7.13 Illustration for single-source thermal evaporation methods. (a) Precursor and synthesis process of single-source evaporation deposition for lead-free perovskite $CsSn_{0.5}Ge_{0.5}I_3$. (b) As-made $CsSn_{0.5}Ge_{0.5}I_3$ perovskite film. Source: Reproduced from Chen et al. [94]/Springer Nature/CC BY 4.0.

7.4.2.2 Pulsed Laser Deposition

The problem of precisely regulating the pace at which the organic component is deposited may be mitigated by using pulsed laser deposition (PLD). In the PLD process, a UV laser impacting the material is used to provide the energy necessary to vaporize the precursors (Figure 7.14a) [100]. However, using a UV laser adds a layer of complexity since energetic lablated particles might spatter from the substrate's surface. By mounting substrates in either an on-axis or an off-axis orientation, this issue can be reduced (Figure 7.14a) [100]. $MAPbI_3$ films with pure phases can be developed at an $MAI:PbI_2$ stoichiometry of 4 : 1 due to the off-axis configuration's reduction of the impacts of high-energy particles on the film. As achieved by the PLD method, these films' morphology reveals a pillared structure.

The resonant infrared matrix-assisted pulsed laser evaporation (RIR-MAPLE) technique is predicated on a more nuanced design of the target–laser interaction (Figure 7.14b) [101]. Ablation of the target solution should be gentler if the target is a precursor solution rather than a pellet of precursors, and if the laser energy is resonant with one of the solvent bonds, generally needing an IR wavelength. The solvent absorbs most of the heat and evaporates to carry the precursors to the substrate. This method gives more control over the film's composition since the target solution's precursor stoichiometry is transmitted to the substrate. Films made in this way have a morphology similar to that of

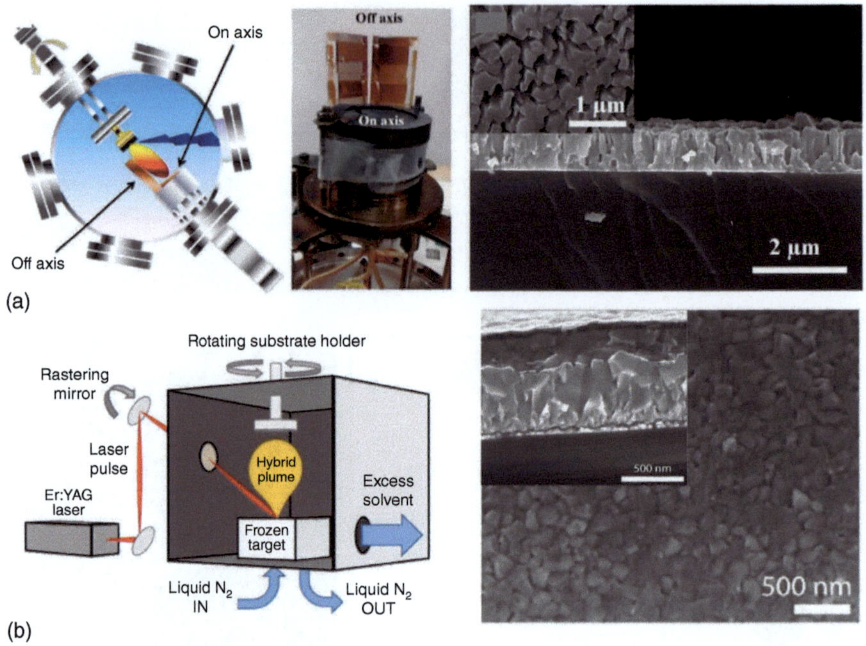

Figure 7.14 Illustration for pulse laser deposition methods. (a) The schematic illustration of the pulsed laser deposition method settings and as-fabricated $MAPbI_3$ SEM photos. Source: Reproduced with permission from Bansode et al. [100], © 2015, American Chemical Society. (b) The schematic illustration of the resonant infrared matrix-assisted pulsed laser evaporation settings and as-fabricated perovskite film SEM photo. Source: Reproduced with permission from Barraza et al. [101], © 2018, Springer Nature.

PLD and thermal co-evaporation, with seemingly tiny but densely packed grains (Figure 7.14b) [102].

7.5 Two-step Growth

The quality of the MHP thin-film layer is the key factor in the fabrication of high-performance perovskite solar cells. Even though one-step spin-coating method has been developed to be effective, the crystallization rate is hard to control, which leads to the inhomogeneity of MHP layer due to the different solubilities inorganic and organic components in DMF and DMSO. Meanwhile, in dual-source vapor deposition process, it is difficult to control stoichiometric ratio of organic salts and inorganic components for perovskite transformation. Compared to one-step method, the two-step method has the advantage of good repeatability and uniformity of the film.

Generally, two-step method relies on the following steps: firstly, the inorganic component (usually the metal precursor, e.g. Pb halide, Pb oxides, and chalcogenides) is deposited on the substrate via methods including but not limited to

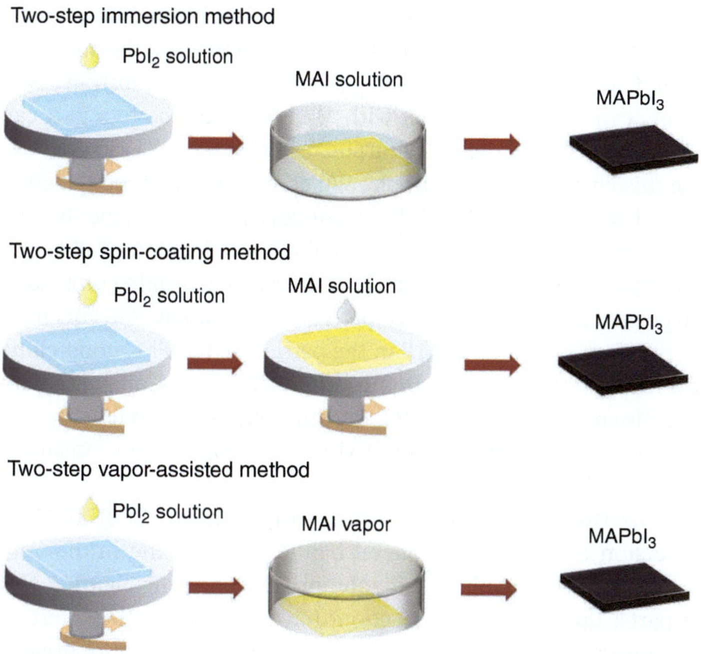

Figure 7.15 Schematic illustrations for $MAPbI_3$ perovskite film prepared by two-step methods, including immersion, spin-coating, and vapor-assisted methods.

spin-coating and evaporation; after that, the substrate is exposed to vapor, liquid, or solid organic salts to drive the reaction that produces the MHPs (Figure 7.15).

7.5.1 Growth from Solutions

7.5.1.1 Immersion Method

Mitzi and coworkers [103] firstly reported the two-step method for fabricating MHP films in 1998. Firstly, the metal iodides were deposited on a substrate using vacuum evaporation or spin-coating, then the substrate was immersed in the organic ammonium iodide. In 2003, Grätzel and coworkers [104, 105] firstly applied the two-step immersion method in the fabrication of MHP solar cells device with a PCE of 15%. In this method, the metal halide is coated on the substrate, followed by the immersion process in organic ammonium salts. Usually, the organic ammonium salt is dissolved in isopropanol due to its low solubility in the ready-prepared metal halide layer, which could avoid its degeneration. To remove the organic ammonium salt left over from the reaction after immersion and drying, isopropanol can be used to wash the surface of the MHP layer. The MHP is formed at the second step in the reaction of metal halides and organic ammonium salts. The key factors that determine the quality of the final MHP film are the metal halide film morphology, the concentration of the organic ammonium salt solution, and the reaction time in the second step.

Tailoring metal halide film morphology: When PbI_2 is deposited on a mesoporous substrate, MAI could easily penetrate into PbI_2 film through the pores and finish the reaction. However, when PbI_2 is deposited on a planar substrate, only the top part of PbI_2 can contact and react with MAI while the bottom part can hardly contact MAI. Modulating the PbI_2 morphology to be porous is an effective way to overcome this challenge. For example, Wu et al. [106] replaced the commonly used solvent DMF with DMSO. DMSO possesses a relatively higher boiling point and lower evaporation rate, thus retarding the crystallization of PbI_2 and leading to the amorphous PbI_2 film. Liu et al. [107] developed an aging process by placing the as-deposited PbI_2 film in a closed petri dish to modulate the PbI_2 morphology. Due to the trace amount of DMF in the petri dish, the PbI_2 grain cloud grows larger and leaves voids, resulting in the porous PbI_2 film. Compared with the compact PbI_2 film on a planar substrate, the amorphous or porous PbI_2 film could provide channels for the diffusion of MAI into the inner of PbI_2, resulting in complete reaction and pure MHP film.

The concentration of the organic ammonium salt solution: Bi et al. [108] studied the conversion mechanisms of metal halide to perovskite in different concentrations of organic ammonium salt solution, as shown in Figure 7.16a. It includes two parts: the solid–liquid reaction that happened at the interface and the dissolution-recrystallization growth. When the isopropanol-organic ammonium salt immerses on the surface of the metal halide layer, a solid–liquid

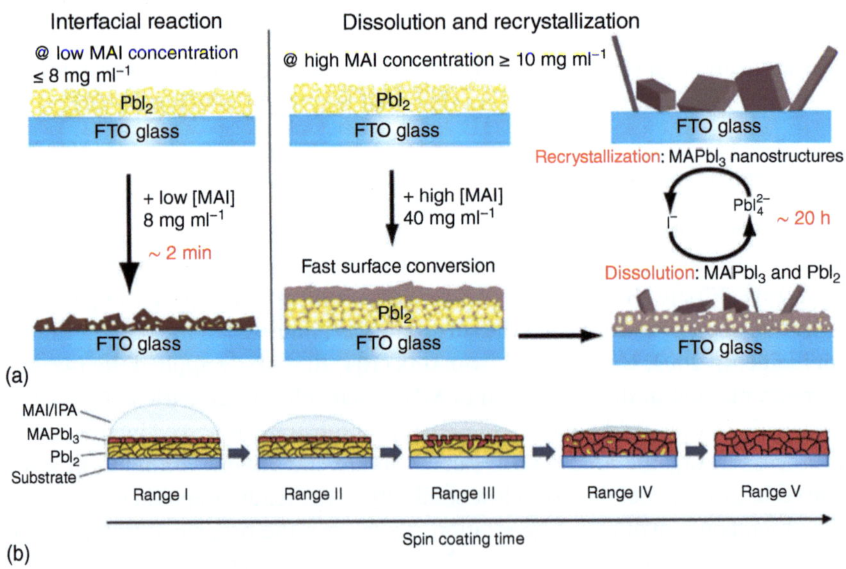

Figure 7.16 Mechanisms illustration for two-step immersion growth and spin-coating method. (a) The interfacial reaction mechanism at lower MAI concentrations and dissolution-recrystallization mechanism at higher MAI concentrations. Source: Reproduced from Fu et al. [108] with permission of American Chemical Society. (b) A model for the successive film formation. Source: Reproduced from Chauhan et al. [109] with permission of Royal Society of Chemistry.

interface conversion reaction occurs. At low concentrations of organic ammonium salt (e.g. MAI in isopropanol; here, take $MAPbI_3$ formation as an example), the conversion of MHP is triggered by the solid–liquid reaction where MAI diffuses to the structured PbI_2 layer and forms MHP directly. The conversion equation is shown in Eq. (7.7)

$$PbI_2(s) + CH_3NH_3^+(sol) + I^- (sol) \rightarrow CH_3NH_3PbI_3 (s) \tag{7.7}$$

When the concentration of organic ammonium salt is higher (>10 mg ml^{-1}), $MAPbI_3$ is immediately formed on the top surface of PbI_2, which prevents further diffusion of MAI and results in incomplete MHP conversion. Unreacted PbI_2 tends to form the lead–iodine complex PbI_4^{2-} under the high MAI concentration. This excess I^- then drives to dissolve the initially formed $MAPbI_3$ and reacts with the unreacted PbI_2 with Eqs. (7.8) and (7.9).

$$CH_3NH_3PbI_3 (s) + I^- (sol) \rightarrow CH_3NH_3^+ (sol) + PbI_4^{2-} (sol) \tag{7.8}$$

$$PbI_2 (s) + 2I^- (sol) \rightarrow PbI_4^{2-} (sol) \tag{7.9}$$

When this reaction is saturated, PbI_4^{2-} complexes will react with $CH_3NH_3^+$ ions to form $MAPbI_3$ (Eq. 7.10)

$$PbI_4^{2-} (sol) + CH_3NH_3^+ (sol) \rightarrow CH_3NH_3PbI_3 (s) + I^- (sol) \tag{7.10}$$

Reaction time in the second step: The relationship was found between MHP structure and immersion time in the second step by Wei and coworkers [110] PbI_2 film reacted with MAI solution and formed $MAPbI_3$ crystals, followed by a dissolution-recrystallization reaction with the increasing soaking time in the second step. The film surface gradually developed defects as the immersion time increased, followed by the dissolving and shedding of MHPs.

7.5.1.2 Spin-coating Method

Similarly, in the spin-costing method, the metal iodides were deposited on the substrate first. Then the organic ammonium salt solution (dissolved in isopropanol) was spin-coated on the metal iodide layer. This process is a solid–liquid heterogeneous reaction at the interface where solid metal iodides and organic ammonium salt react directly. In contrast, faster crystallization occurs when the organic ammonium salt solution is applied, which results in more uniform, compact, and smooth films compared to the immersion method, resulting in better solar cells and making it a more efficient processing strategy [111]. Based on this method, Zhao et al. [13] achieved a thermally stable device with a high PCE of 25.6% by doping RbCl and then converting the excess PbI_2 to the inactive $(PbI_2)_2RbCl$.

The possible model for successive film formation was derived from Panzer and coworkers [109] by analyzing the in-situ spectral features during the two-step spin coating process (Figure 7.16b). They identified five consecutive steps as follows (take $MAPbI_3$ as an example): (i) the formation of $MAPbI_3$ capping; (ii) as the MAI solution is vaporized over several seconds, further crystallization of $MAPbI_3$ is prevented by the capping layer; (iii) as the solvent evaporates, the iodine

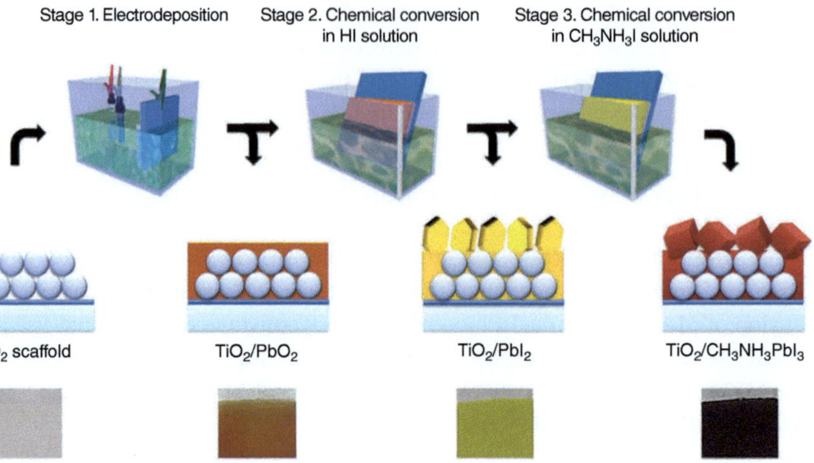

Figure 7.17 Illustration for electrochemical bath deposition process. Stage 1: PbO_2 is deposited electrochemically or chemically onto the substrate; Stage 2: PbO_2 is converted into the PbI_2; Stage 3: TiO_2/PbI_2 is immersed in MAI solution to form the MHP. Source: Reproduced with permission from Chen et al. [112], © 2015, Elsevier.

concentration increases, leading to the dissolution of the $MAPbI_3$ capping layer; (iv) in this process, most of the $MAPbI_3$ forms during dissolution and recrystallization; and (v) it is achieved a stable state once PbI_2 is fully transferred into $MAPbI_3$.

7.5.1.3 Electro/Chemical Bath Deposition

Chemical or electrochemical bath deposition is an effective method of large-scale MHP device fabrication. Yang and coworkers [112] describe a simple, versatile, scalable, and roll-to-roll compatible electrodeposition technique for forming MHP layers. Figure 7.17 illustrates the fabrication processes of $CH_3NH_3PbI_3$ ($MAPbI_3$) on TiO_2 porous scaffold based on electrodeposition, which includes three stages: Stage 1: At room temperature, the PbO_2 layer was electrodeposited in a standard three-electrode system; Stage 2: PbO_2 layer was transformed into PbI_2 layer at room temperature using HI ethanol solution; Stage 3: IPA solution of MAI was used to convert PbI_2 layer into $MAPbI_3$ layer at room temperature, followed by 15 minutes of heating at 100 °C. Using this method, a carbon electrode-based solar cell device with PCE>10% can be obtained. The other researchers have improved this method to directly convert the PbO_2 layer to $MAPbI_3$ by immersion in MAI isopropanol solution [113]. This improved method's trap density of $MAPbI_3$ film was much lower than that of spin-coating.

7.5.2 Growth From Vapor Phase

7.5.2.1 Vapor-assisted Solution Processing

As noted above, in the second step of the two-step method, the vapor can also be applied to convert the metal halide film to MHPs. Take $MAPbI_3$ as an example, spin-coated PbI_2 films were exposed to MAI vapor by sprinkling MAI powder

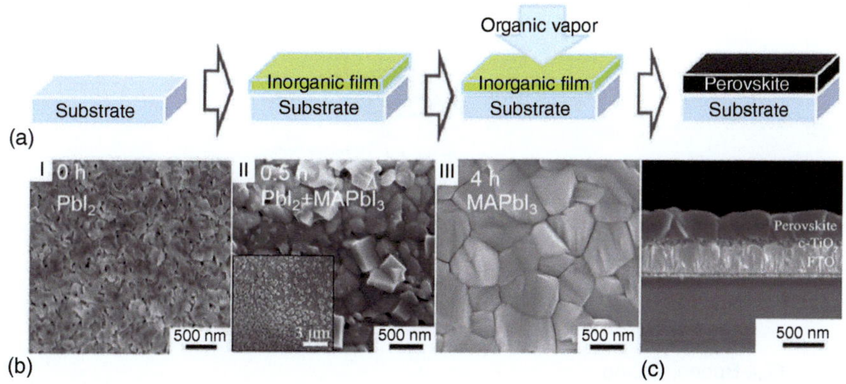

Figure 7.18 $MAPbI_3$ fabricated from a vapor-assisted solution process. (a) Schematic illustration of vapor-assisted solution process. (b) The top-view SEM images of the initial stage at 0 hour (I), (II) the intermediate stage at 0.5 hour, and (III) the post-stage at four hours. (c) The cross-sectional SEM image of MHP film made via VASP. Source: Reproduced from Chen et al. [114], © 2014, American Chemical Society.

around them in a capped Petri dish and heating it to 150 °C (Figure 7.18a) [114]. This vapor-assisted solution process (VASP) is a typical solid-vapor reaction. The MHP grains grow from heterogeneous nucleation sites while maintaining the smooth morphology of the original PbI_2 film. With the reaction time increasing, the PbI_2 and MHP phases coexisted in the film, and finally all the PbI_2 converted to MHP in the complete stage (Figure 7.18b). As shown in Figure 7.18c, the cross-section of the MHP film shows the rearrangement of PbI_2 and MAI via intensive diffusion during film growth. Even though this method is proven to be effective, however, the whole process takes around four hours, which limits its further application. Based on the VASP method, Li et al. [115] applied a low-pressure MAI vapor simultaneously during the second step to produce the desired halide MHPs. The reaction time has been shortened to two hours compared to the traditional VASP method.

7.5.2.2 Sequential Vapor Deposition

In most cases, high-efficiency PSCs have been prepared using spin-coating, which is only suitable for laboratory-scale devices. However, solvents used in solution processing can damage underlying functional layers and pollute environment. A vacuum evaporation process controls the film thickness as the precursors are deposited evenly and uniformly on substrates, enabling scalable fabrication. Through layer-by-layer sequential vapor deposition, lead halide and organic ammonium salt are evaporated separately in different vacuum chambers, thus avoiding cross-contamination risks associated with co-evaporation.

The first sequential vapor deposition work was published by Lin et al. [116] The schematic procedure for sequential vapor deposition is shown in Figure 7.19a. First, $PbCl_2$ films were thermally sublimed onto PEDOT:PSS-coated indium tin oxide (ITO) glass, followed by CH_3NH_3I evaporation. As CH_3NH_3I evaporated, the film's color changed from transparent to dark, indicating the formation of an MHP layer. Recently, Yi and coworkers [117] reported a sequential vacuum deposition

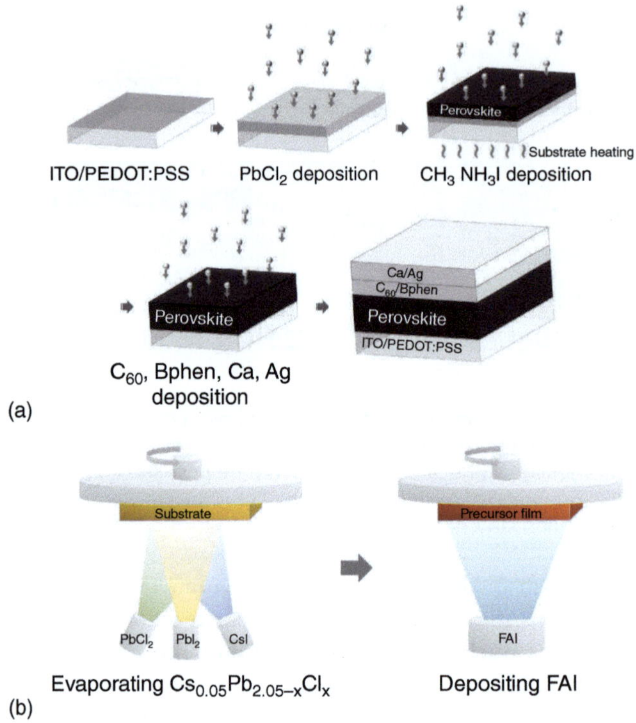

Figure 7.19 Illustration for sequential vapor deposition methods. (a) Schematic illustration of MHP solar cells fabricated by sequential layer-by-layer vacuum deposition. Source: Reproduced from Chen et al. [116] with permission of John Wiley & Sons. (b) The scheme presents the Cl-containing alloy–mediated sequential vacuum deposition approach.

method based on Cl-alloys for high-efficiency PSCs with 24.42% (Figure 7.19b), which is the champion efficiency of the thermally evaporated MHP solar cells. In this study, it is believed that the chlorine combines with lead halide to cause the MHP to align face-on with the chemical element. Additionally, a Cl embedded into the MHP promotes solid-state diffusion of organic ammonium halide salt and facilitates phase transition from the δ phase to the α phase, thereby increasing the crystallinity of MHP nanocrystals and resulting in a film with fewer defects.

7.6 Scalable Growth Methods

In general, the highest PCEs for PSCs have been reported in cells with small areas, from 0.04 to 0.2 cm^2, as well as for cells manufactured by spin coating methods with areas of 0.1 cm^2. With the increase in device area, the film quality and uniformity could decrease, which would then lead to PCE decreasing dramatically [118]. Therefore, it is crucial that the PCEs of large-area cells (>1 cm^2) should be comparable to those of small-area cells for practical commercial applications. In the past few years, many researchers have devoted themselves to improving large-scale MHP fabrication methods. Recent advances in process compatibility with large-area deposition have proven that achieving reasonable PCE for commercial devices is possible.

7.6.1 Blade Coating

Blade coating, also known as knife coating or bar coating, is a typical large-scale fabrication method with advantages like low cost, suitable for rigid or flexible substrates, and environment friendly (Figure 7.20a). The typical procedures are as follows: Firstly, dropping the precursor on the substrate and in front of the blade; after that, the substrate was swept forward by the blade, and then the homogenous wet thin film was coated on the substrate.

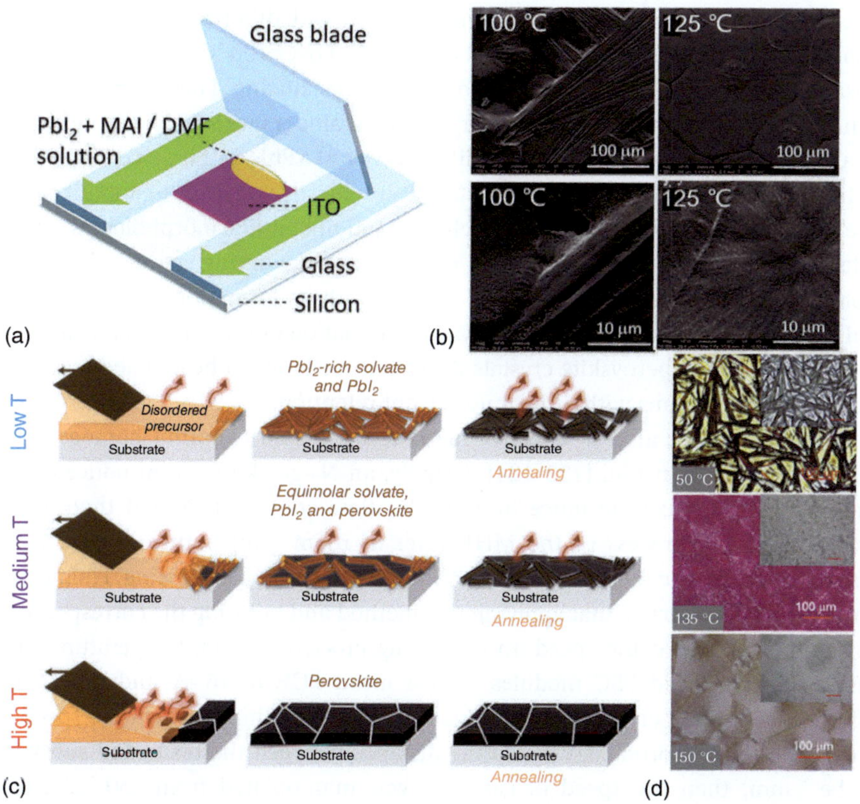

Figure 7.20 Illustration for blade-coating methods. (a) The experimental setup for doctor-blading to deposit perovskite films. (b) SEM images of $MAPbI_3$ films formed at low temperatures of 100 and 125 °C. Source: Reproduced with permission from Deng et al. [119], © 2015, Royal Society of Chemistry. (c) Schematic representation of the $MAPbI_3$ perovskite film formation mechanism under different temperatures. (d) Polarized and unpolarized (inset) optical micrographs of the $MAPbI_3$ perovskite representing at 50 °C, 135 °C, and 150 °C. Source: Reproduced with permission from Zhong et al. [120], © 2018, American Chemical Society.

Despite the concentration of MHP precursor, the meniscus that the solution forms between blade and substrate is also a critical factor influencing the thickness of the film. The meniscus is determined by the moving speed of the blade, the viscosity of the precursor, the geometry of the blade, and the substrate wettability [121].

In 2015, blade coating was first introduced by Huang et al. [119] for PSC fabrication with a 15.0% device efficiency. Till now, various studies have explored process methods to improve film quality. Preheating the substrate has been proven to be an effective method. When the temperature of substrate is much lower than the boiling point of the precursor solvent, the crystallization time is too long, which makes the film very rough and discontinuous (Figure 7.20b). Huang et al. used a high temperature deposition to guide the nucleation and grain growth and explore the relationship between temperature and film morphology. They found that the low intermediate temperatures (25–80 °C) yield solvates with differing compositions and poor PCEs; when the temperature was increased to over 100 °C, the solvated compositions cannot be observed, leading to directly crystallizing. They then obtained a continuous MHP film with large-size crystalline grains (Figure 7.20c,d) [119]. Zhao et al. [120] have investigated the solidification of $MAPbI_3$ in blade coating. They found that at low temperature (below 80 °C), the slow drying leads to not full coverage of the substrate, accompanied by rich PbI_2 and ribbon-like morphologies. At the intermediate temperature (80–100 °C), even though the drying speed is faster, the ribbon-like equimolar solvate was still present with its formation competing with PbI_2. An annealed $MAPbI_3$ film displays a combination of ribbon-like and compact morphologies. The perovskite crystals develop directly when heated above 100 °C, forming compact films with little or no solvent retention.

Gas quenching is also an effective method in the fabrication of homogeneous large-area MHP thin film [122–124]. Usually, an N_2 gas knife is introduced after the blade with a fixed distance to apply N_2 flow continuously and then accelerate the drying process of the MHP layer at room temperature. Figure 7.21a shows the schematic illustration of the N_2-knife assistant blade coating. Huang and coworkers [122] showed that applying this method and tailoring the corresponding solvent component could speed up the drying process at room temperature. This innovation produced PSC modules with a mean PCE of 16.4% and a 63.7 cm^2 effective area. Yang et al. [125] investigated the relationship between the speed of airflow and film morphology. Fixed the distance between the blade and substrate to be 1 mm, then the speed of gas flow was manipulated from 150, 225, and 300 l min^{-1}. As shown in Figure 7.21b, by increasing the airflow from 150 to 225 l min^{-1}, the film exhibited smaller dendrite grains and a reduced number of pin holes. When the speed of gas was up to 300 l min^{-1}, the morphology of the MHP was the most compact and free of pinholes, which were due to the faster solvent evaporation leading to higher supersaturation levels and burst nucleation. The PCEs of the corresponding devices were up to 11.7% (1.0 cm^2) and 17.71% (0.1 cm^2).

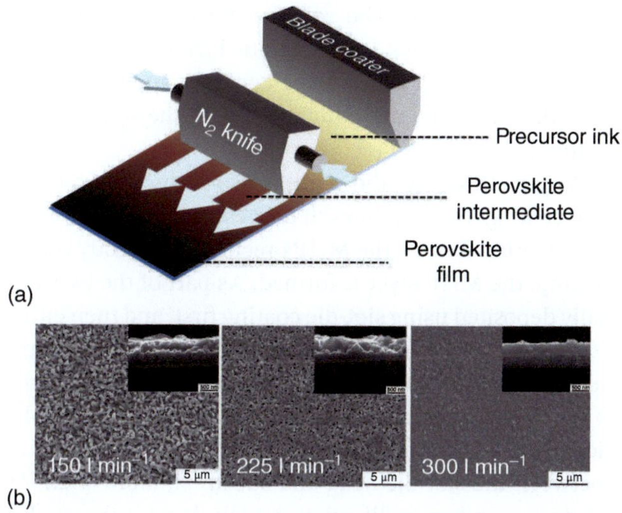

Figure 7.21 Illustration for gas quenching in blade coating. (a) Schematic illustration for N_2-knife–assisted blade coating of MHP films. Source: Adapted from Deng et al. [122] with permission. Copyright 2019, the American Association for the Advancement of Science (AAAS). (b) MHP films fabricated by N_2-knife assistant blade coating under the speed of air: 150, 225, and 300 l min^{-1}, inset figures are the cross-sectional SEM images of the corresponding MHP films. Source: Reproduced with permission from Gao et al. [125], © 2017, Royal Society of Chemistry.

7.6.2 Slot-die Coating

Slot-die coating is especially suitable for roll-to-roll processes because it provides a continuous ink supply. Slot-die heads are composed of two independently movable metal blades that form slits on the substrate's bottom side. A solution pumping system is connected to a reservoir integrated into the slot-die head (Figure 7.22).

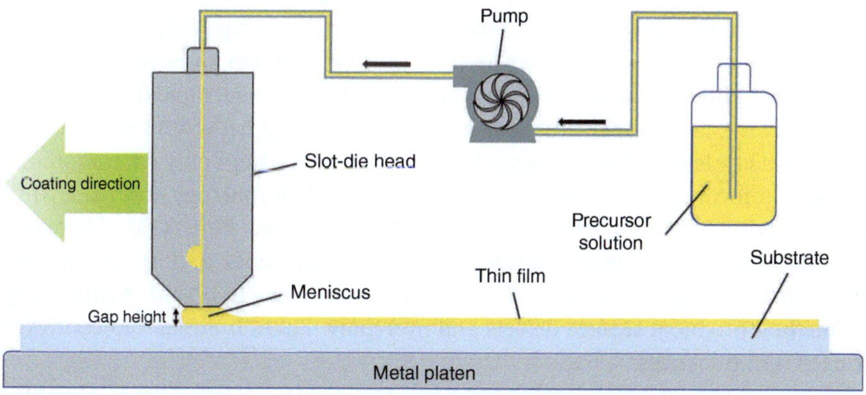

Figure 7.22 Schematic illustration for slot-die coating. Source: Reproduced from Whitaker et al. [126] with permission of Royal Society of Chemistry.

This system supplies ink at a constant rate. The solution forms upstream- and downstream-meniscuses between the head lips and substrate during deposition. As well as the parameters discussed in the part about blade coatings, such as the gap between the slot-die head and the substrate, the pumping rate of the solution also impacts the thickness of the film [126].

Similarly, like the spin-coating method, slot-die coating also includes two types: one-step coating and two-step coating; the step in each of these two modes is made by the blade of slot-die coating. For the former, the MHP precursor is directly coated on the substrate. After annealing, the MHP layer is formed. As part of the two-step process, a PbI_2 layer is typically deposited using slot-die coating first, and then either an organic ammonium salt is applied to the PbI_2 layer or the substrate is simply soaked in such an organic ammonium salt solution, which induces MHPs to form.

There are many works on the slot-die fabrication of high-performance MHP solar cells. Despite the optimization of coating parameters, such as ink supply rate, coating speed, and the gap height between the slot-die head and substrate, more attention is paid to the controlling of drying and crystallization, which decides the quality of MHP thin film. The parameters like solvent exchange [79], elevated temperature [127], solvent quenching [128], gas quenching [129], and vacuum-assisted crystallization [130] have been studied. Deng et al. [131] used small amounts of surfactants to alter the fluid drying dynamics, which increased the adhesion of the MHP precursor to the underlying non-wetting charge transport layer. Using methylammonium lead iodide containing chlorine as the precursor material, Zhu and coworkers [128] extended the precursor processing window up to eight minutes, allowing a high-quality MHP film with a large uniformity to be produced.

7.6.3 Spray Coating

Spray-coating is a typical technique for the fabrication of MHP solar cells. Compared to other large-scale fabrication methods, spray-coating can be applied to nonplanar surfaces, which are suitable for a variety of scenarios. Besides, spray-coating speed is much faster than other methods because the spray head moves at a speed of up to 5 m min^{-1}, which is much faster than slot-die coating [132]. Furthermore, spray coating offers several advantages over the conventional antisolvent dropping method, including a reduced amount of antisolvent being used, rapid distribution of the antisolvent, faster supersaturation and nucleation, and better uniformity.

There is a low level of MHP film quality across all single-step spray-coating studies with nonuniform crystallization and mostly dendritic morphology observed, while the underlying transport layer does not cover the entire film [133]. Therefore, PSCs have lower open-circuit voltages, short-circuit currents, and fill factors than spin-coated PSCs, possibly due to lower shunt resistance [134, 135]. Additionally, this method produces surface defects and overspray makes it difficult to control in terms of film thickness.

The procedures for spray-coating can be divided into four steps. Firstly, along the movement of the spray head, the precursor sheared into a mist of micrometer-sized droplets. Traditionally, this is achieved by forcing a solution through a narrow nozzle or aperture. Different methods can be used to generate atomization, such as high-flow gases, ultrasonic stimulation, or cavitation [135]. Now ultrasonic coaters

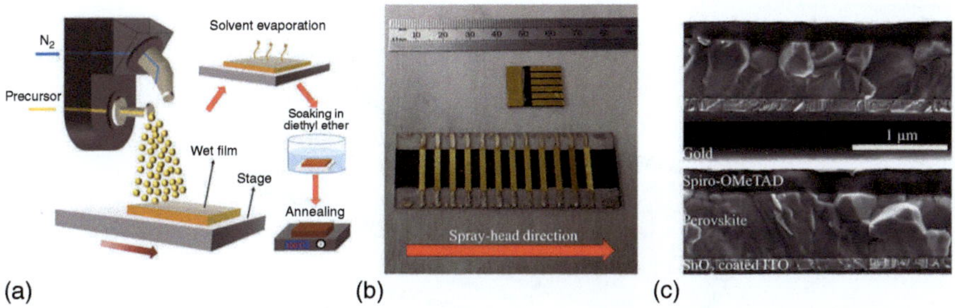

(a) (b) (c)

Figure 7.23 Illustration for spray coating methods. (a) Schematic representation of the spray-coating setup and vacuum-assisted solution processing. Source: Adapted from Bishop et al. [137]. Copyright 2020, American Chemical Society. (b) The photographs of fully spray-coated small area and large-area MHP device. (c) The cross-sectional SEM images of the spray-coated MHP device, whereas the transport layers of the upper one are made by spin-coating, while the bottom one is fully spray-coated. Source: (b) and (c) Reproduced from Bishop et al. [136], © 2020, Springer Nature / CC BY 4.0.

with piezoelectric transducers are more common on the spray system to make a uniform mist. Secondly, a mist of precursor droplets is directed to the substrate with the N_2 gas jet. Thirdly, a uniform and continuous wet film is formed when the droplets reach the substrate surface. Finally, the wet film dries on the substrate as the solvent evaporating.

The first study of spray-coating made MHP solar cells was published in 2014 with a PCE of 11% by Lidzey et al. [134] The highest PCE of the spray-coated MHP solar cell till now is 19.4% for small-area substrates, 16.3% for 15.4 mm^2, and 12.7% for 1.08 cm^2 [136]. The devices were made by antisolvent exposure process and vacuum-assisted solution processing, which demonstrates the possibility for spray-coating to fabricate high-efficiency and low-cost MHP solar cells at speed. The typical spray-coating setup and vacuum-assisted solution processing were illustrated in Figure 7.23a [137]. Firstly, the MHP precursor was sprayed from an ultrasonic coater onto the substrate (heated to 40 °C). Moderate temperatures could enhance wetting without causing crystallization of the film. Then the substrate with the wet film was transferred to a vacuum box (80 Pa) for five minutes. During this process, the color of the wet film changed to black, which means the crystallization was started. Following this, the film was transferred to a hot plate and annealed under N_2 in order to remove all remaining solvents to form the black MHP phase. Figure 7.23b shows the photograph of the small and large-area fully spray-coated MHP solar cells. Figure 7.23c shows the cross-sectional SEM images where the transport layers (SnO_2 and spiro-OMeTAD) of upper device are made by spin coating, while the bottom device is fully spray-coated.

7.6.4 Meniscus-assisted Solution Printing

Meniscus-assisted solution printing (MASP) is a modification of the blade-coating process that combines elements of both blade coating and slot-die coating. Figure 7.24a illustrates the schematic of the MASP. Specifically, a meniscus forms by restricting perovskite ink between a lower substrate and an upper plate due to capillary action [127]. As the solvent evaporates quickly at the edge of the meniscus,

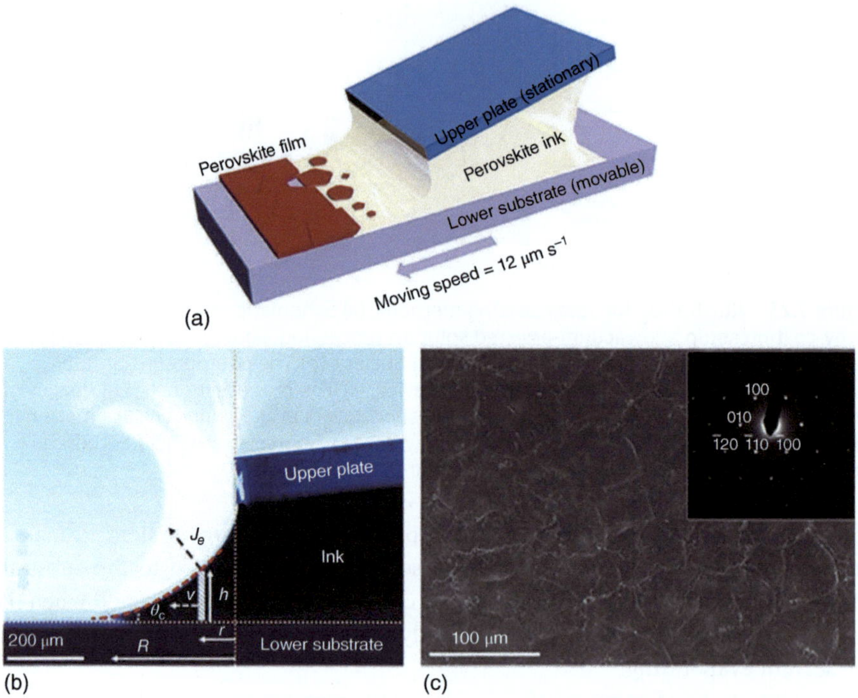

Figure 7.24 Illustration for meniscus-assisted solution printing. (a) The schematic illustration of the meniscus-assisted solution printing. (b) Image of a side-view meniscus ink constrained between a movable lower substrate and a stationary upper substrate. (c) The top-view SEM image of the MHP film crafted by MASP. Source: (a)–(c) Reproduced from He et al. [127], © 2017, Springer Nature/CC BY 4.0.

the coffee ring effect transports the perovskite solutes toward the contact line to replenish the evaporative loss of solvent (Figure 7.24b) [127]. Meniscus shape is determined by the geometry of the die/print head and the relative motion between it and the substrate. The shape of the meniscus between the die/print head and the substrate is determined by the fine control of its geometry and its relative motion. As the solvent evaporates, a convective flow of solution forms at this meniscus, transporting solute to the contact line between a substrate, solution, and air, where it precipitates. By placing the lower substrate on a programmed translation stage, it is possible to sweep the receding meniscus slowly and progressively across the entire substratum, and then a continuous MHP film can be fabricated (Figure 7.24c). Lin et al. [127] applied the MASP method in the fabrication of the compatible $FA_{0.85}MA_{0.15}PbI_{2.55}Br_{0.45}$ component solar cells with a PCE of nearly 20%. The uniform morphology of the MHP is controlled by the moving speed of the substrate. The moving speed should be close to the receding speed of the meniscus edge due to solvent evaporation. When the coating speed is faster than the meniscus receding speed, the meniscus edge quickly reaches the edge of the upper plate, leading to the perovskite crystals bridging the upper plate and the lower substrate. However, a liquid layer rather than a solid film is coated on the substrate when the coating speed exceeds the receding speed of the meniscus.

7.6.5 Inkjet Printing

The inkjet printing method is a noncontact printing method in which the solution is ejected from a nozzle, but the resulting jet can be guided carefully (in some cases, drop by drop) for precise patterning. Inkjet printing is a method through which a microfluidic cavity is subjected to pressure change, thereby causing the solution to jet out of a microfluidic nozzle.

For inkjet printing, there are two common methods of generating ink droplets: (i) continuous inkjet printing (CIP) and (ii) drop-on-demand inkjet printing (DOD) [138]. The CIP process (Figure 7.25a) consists of ejecting nonstop flows of fluid

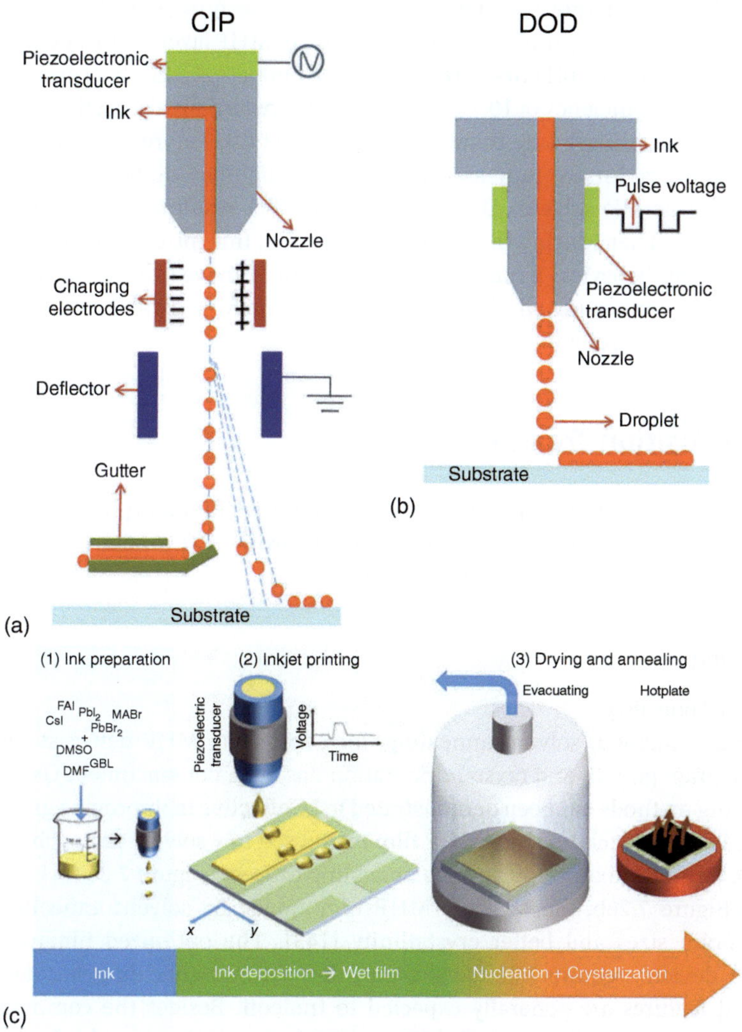

Figure 7.25 Schematic diagrams for the two main inkjet-printing methods. (a) Continuous inkjet printing and (b) drop-on-demand inkjet printing. Source: Reproduced from Karunakaran et al. [138] with permission of Royal Society of Chemistry. (c) Schematic illustration of the key steps involved in inkjet printing of triple-cation multi-crystalline MHP thin films. Source: Reproduced from Eggers et al. [139] with permission of John Wiley & Sons.

through a nozzle, which then separates under the influence of surface tension. In this method, droplets are continuously steered to position themselves on a target substrate as a result of Rayleigh instability. When a fluid is ejected from the nozzle, it breaks down into small droplets that fall toward the substrate. Each droplet receives a small charge as it descends through the nozzle. A deflection plate steers the direction of the droplets as they pass between the charged droplets. The DOD process is the most common type (Figure 7.25b). DOD systems move or accurately settle the print head on the desired surface by means of computer program or substrate motion. As the printer nozzle generates pressure pulses, a series of events accumulate to cause liquid to be ejected. The pressure pulse is produced by a sudden contraction in the chamber volume, causing the ink to escape the nozzle. In 2014, Yang et al. [140] first applied inkjet printing to the fabrication of MHP layers with a PCE of 11.6%. In 2018, Gheno et al. [141] firstly reported a fully inkjet-printed PSC (excluding electrodes) with an efficiency of 10.7%. Till now, a new record-breaking efficiency of over 21% has been achieved by inkjet-printed PSCs [139]. Figure 7.25c shows the fabrication process. Firstly, they used a mixture of high-boiling point solvent GBL and the polar-aprotic solvents DMF as well as DMSO, resulting in improved homogenous drying. Then, during the ink deposition part, this mixed solvent realized optimal wetting behavior. Finally, the wetting film is moved to the vacuum chamber, followed by annealing on a hotplate. The MHP thin films exhibit columnar perovskite crystal structures with few defects.

7.7 Postdeposition Treatments

Postdeposition treatment is an important method to improve film morphology and crystallizing process. Here, we considered the postdeposition treatment methods from the perspective of annealing and organic gas dosing.

7.7.1 Annealing

7.7.1.1 Solvent Annealing

A highly effective method, solvent annealing, for controlling MHP defect states and promoting grain growth and crystal orientation has been demonstrated. Using solvent annealing methods has been demonstrated to be effective in improving grain quality. Typically, in this method, the MHP film is exposed to a solvent atmosphere (DMF, DMSO, or their mixture) during the annealing process (Figure 7.26a) [142]. As shown in Figure 7.26b, the $MAPbI_3$ MHP films made by solvent annealing show higher grain sizes and better crystallinity [143]. The optimized films for optoelectronic devices require tailored solvents for solvent annealing because microstructural features are generally expected to tradeoff. Besides the common single-component solvents like DMF, DMSO, and GBL, the alcohol-based solvents (methanol, ethanol, and isopropanol) and their mixtures. Solvent annealing of PSCs and related optoelectronic devices requires optimizing the solvents for solvent annealing since there is a trade-off in microstructural characteristics in the final films.

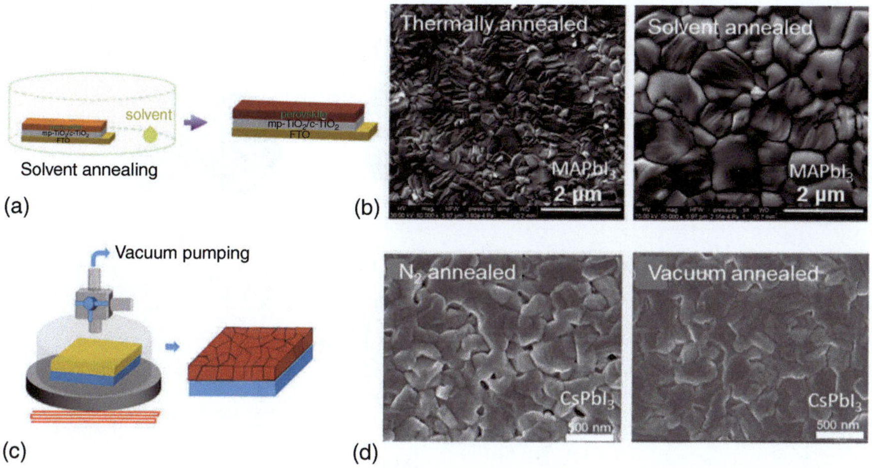

Figure 7.26 Illustration for solvent annealing methods. (a) The schematic diagram of solvent annealing process. Source: Adapted from Masawa et al. [142]/MDPI/CC BY 4.0. (b) The SEM images of the thermally annealed $MAPbI_3$ films and solvent annealed $MAPbI_3$ films. Source: (a) and (b) Reproduced with permission from Xiao et al. [143], © 2014, John Wiley & Sons. (c) The schematic diagram of the vacuum-assisted annealing method. (d) The SEM images of the $CsPbI_3$ films were annealed in N_2 and vacuum. Source: (c) and (d) Reproduced with permission from Yu et al. [144], © 2022, John Wiley & Sons.

7.7.1.2 Vacuum-assisted Annealing

During the MHP layer annealing, vacuum can be used as additional processing to crystalize by impacting the chemical conversion of the precursor and then impacting the crystallization process. Besides, this method can also be used to remove excess components, which are difficult to be removed via the thermal annealing approach [144, 145]. Generally, the wet MHP film was annealed in a vacuum chamber. Song et al. [144] have demonstrated using this method for synthesizing $CsPbI_3$ film from a precursor of PbI_2, CsI, and dimethylammonium iodide (DMAI) at a molar ratio of 1 : 1 : 1 in DMF. The conversion reaction is

$$DMAPbI_3 + CsI \rightarrow CsPbI_3 + DMAI \uparrow \tag{7.11}$$

During the process, the released DMAI could create pores and cracks in MHP film if it cannot be removed before the film crystallizes (Figure 7.26c,d). During annealing, the vacuum accelerates the sublimation of DMAI, accelerating the formation reaction of $CsPbI_3$, increasing the PCE from 17.26% to 20.06%.

7.7.2 Organic-gas Dosing

Solution-processed MHPs invariably contain defects, such as pinholes and voids. The post-processing morphology reconstruction of defective films is a more attractive approach for forming large-scale, high-quality MHP films. The MHP thin films are sensitive to organic gases due to their soft features. In 2015, Zhou et al. [146] firstly reported that MA gas can drive a phase and morphology transformation to $MAPbI_3$. This transformation is reversible when removed MA gas.

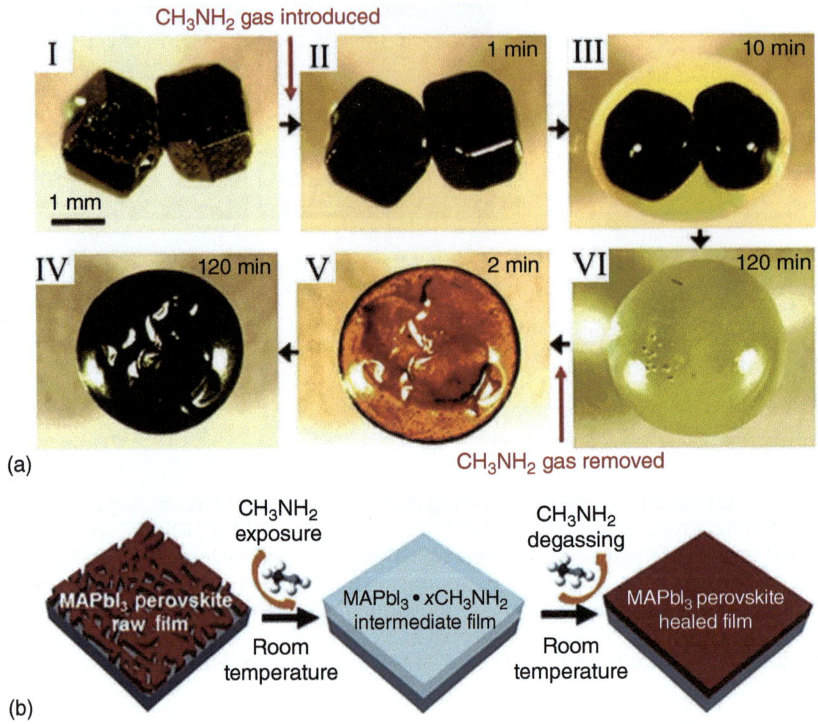

Figure 7.27 Illustration for organic gas dosing methods. (a) The transmission process of $MAPbI_3$ MHP crystals exposed to MA gas and MA gas degassing. (b) Schematic illustration of MA-induced defect healing of $MAPbI_3$ MHP thin films. Source: Reproduced with permission from Zhou et al. [146], © 2015, John Wiley & Sons.

The transformation process is shown in Figure 7.27a. When MA gas is induced, the perovskite crystals start to melt and change to the liquid state ($MAPbI_3 \cdot MA$). When MA gas is removed, the perovskite crystal begins to recrystallize, which means that the $MAPbI_3 \cdot MA$ has been converted back into $MAPbI_3$. When the phenomenon occurs in thin films, the whole process only takes a few seconds.

Furthermore, applying the phenomenon to a porous, rough $MAPbI_3$ MHP film can convert it into a fully dense, smooth film, which illustrates the reversible MA gas-induced defect-healing behavior (Figure 7.27b). Formamidine (FA) gas led to the same conversion effects on MHP layers. Zhou et al. [147] have reported that the FA gas replaced MA in $MAPbI_3$ and then converted the $MAPbI_3$ to $FAPbI_3$. When $MAPbI_3$ films are exposed to FA gas at an elevated temperature (150 °C), the cation-displacement reaction occurs, where the $MAPbI_3$ perovskite phase in the thin film is converted directly to the $FAPbI_3$ perovskite phase while preserving the desirable morphologies and microstructures of the original thin film (Figure 7.28). Furthermore, this rapid reaction is solvent-free and low reversible, which demonstrates the success of this morphology-preserving perovskite conversion approach.

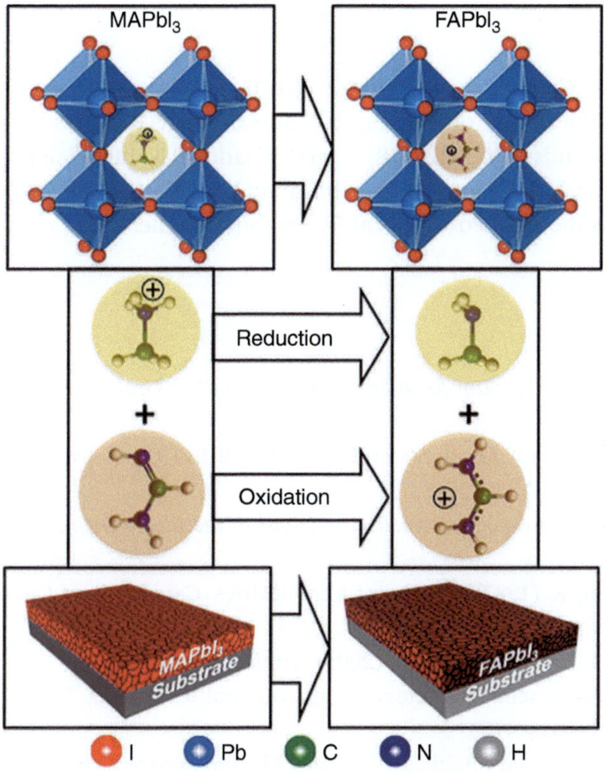

Figure 7.28 Schematic illustration of the cation displacement. The reaction between $MAPbI_3$ MHP and FA gas at 150 °C resulted in $FAPbI_3$ MHP. Source: Reproduced from Zhou et al. [147] with permission of American Chemical Society.

7.8 Summary

PSCs have revolutionized solar cells and optoelectronic research in recent years. Numerous studies have implied different fabrication methods and compositional characteristics to explore the limits of MHP materials. This chapter summarizes the fundamentals of MHP thin film and the mainstream fabrication methods from laboratory to commercialization. On the lab scale, developing new mechanistic guidance for the design of MHP materials with greater stability and superior carrier properties can assist with approaching the theoretical efficiency limits of photovoltaics and other optoelectronics. On the commercial scale, exploring high-quality, large-scale MHP thin-film fabrication methods via different methods still needs more effort. Although there are various kinds of MHP thin film synthesis methods, challenges still exist and need to be further studied, such as precise control of precursor components, scalable preparation control, and the origins of unconventional device phenomena in atomistic and microstructural terms.

Acknowledgments

Y.Z. acknowledges the Early Career Scheme (No. 22300221) and General Research Fund (No. 12302822) from the Hong Kong Research Grants Council (RGC) and the Excellent Young Scientists Funds (No. 52222318) from the National Natural Science Foundation of China (NSFC). M.H. acknowledges the Hong Kong Ph.D. Fellowship Scheme. T.D. acknowledges the RGC Postdoctoral Fellowship Scheme.

References

1 Rose, G. (1839). *De novis quibusdam fossilibus quae in montibus Uraliis inveniuntur*. typis AG Schadii.
2 Miyake, S. and Ueda, R. (1932). On polymorphic change of $BaTiO_3$. *Journal of the Physical Society of Japan* 1 (1): 32–33.
3 Wainer, E. (1946). High titania dielectrics. *Transactions of the Electrochemical Society* 89 (1): 331.
4 Cross, L. and Newnham, R. (1987). History of ferroelectrics. *Ceramics and Civilization* 3: 289–305.
5 Knutson, J.L., Martin, J.D., and Mitzi, D.B. (2005). Tuning the band gap in hybrid tin iodide perovskite semiconductors using structural templating. *Inorganic Chemistry* 44 (13): 4699–4705.
6 Green, M.A., Ho-Baillie, A., and Snaith, H.J. (2014). The emergence of perovskite solar cells. *Nature Photonics* 8 (7): 506–514.
7 Wehrenfennig, C., Eperon, G.E., Johnston, M.B. et al. (2014). High charge carrier mobilities and lifetimes in organolead trihalide perovskites. *Advanced Materials* 26 (10): 1584–1589.
8 Han, Q., Bai, Y., Liu, J. et al. (2017). Additive engineering for high-performance room-temperature-processed perovskite absorbers with micron-size grains and microsecond-range carrier lifetimes. *Energy & Environmental Science* 10 (11): 2365–2371.
9 Chondroudis, K. and Mitzi, D.B. (1999). Electroluminescence from an organic–inorganic perovskite incorporating a quaterthiophene dye within lead halide perovskite layers. *Chemistry of Materials* 11 (11): 3028–3030.
10 Mitzi, D.B., Chondroudis, K., and Kagan, C.R. (2001). Organic-inorganic electronics. *IBM Journal of Research and Development* 45 (1): 29–45.
11 Kojima, A., Teshima, K., Shirai, Y., and Miyasaka, T. (2009). Organometal halide perovskites as visible-light sensitizers for photovoltaic cells. *Journal of the American Chemical Society* 131 (17): 6050–6051.
12 Kim, H.-S., Lee, C.-R., Im, J.-H. et al. (2012). Lead iodide perovskite sensitized all-solid-state submicron thin film mesoscopic solar cell with efficiency exceeding 9%. *Scientific Reports* 2 (1): 591.
13 Zhao, Y., Ma, F., Qu, Z. et al. (2022). Inactive $(PbI_2)_2RbCl$ stabilizes perovskite films for efficient solar cells. *Science* 377 (6605): 531–534.

14 Green, M.A., Dunlop, E.D., Hohl-Ebinger, J. et al. (2022). Solar cell efficiency tables (Version 60). *Progress in Photovoltaics: Research and Applications* 30 (7): 687–701.

15 Li, Z., Li, B., Wu, X. et al. (2022). Organometallic-functionalized interfaces for highly efficient inverted perovskite solar cells. *Science* 376 (6591): 416–420.

16 Yao, M.-M., Zhang, Q., Wang, D. et al. (2022). Lead-free halide $CsAg_2I_3$ with 1D electronic structure and high stability for ultraviolet photodetector. *Advanced Functional Materials* 32 (28): 2202894.

17 Sulaman, M., Yang, S., Bukhtiar, A. et al. (2022). Hybrid bulk-heterojunction of colloidal quantum dots and mixed-halide perovskite nanocrystals for high-performance self-powered broadband photodetectors. *Advanced Functional Materials* 32 (28): 2201527.

18 Morteza Najarian, A., Vafaie, M., Johnston, A. et al. (2022). Sub-millimetre light detection and ranging using perovskites. *Nature Electronics* 5 (8): 511–518.

19 Pina, J.M., Vafaie, M., Parmar, D.H. et al. (2022). Quantum-size-effect tuning enables narrowband IR photodetection with low sunlight interference. *Nano Letters* 22 (16): 6802–6807.

20 Zhou, C., Pina, J.M., Zhu, T. et al. (2021). Quantum dot self-assembly enables low-threshold lasing. *Advanced Science* 8 (20): 2101125.

21 Liang, Y., Shang, Q., Li, M. et al. (2021). Solvent recrystallization-enabled green amplified spontaneous emissions with an ultra-low threshold from pinhole-free perovskite films. *Advanced Functional Materials* 31 (48): 2106108.

22 Li, J., Zhang, L., Chu, Z. et al. (2022). Amplified spontaneous emission with a low threshold from quasi-2D perovskite films via phase engineering and surface passivation. *Advanced Optical Materials* 10 (6): 2102563.

23 Wang, C., Dai, G., Wang, J. et al. (2022). Low-threshold blue quasi-2D perovskite laser through domain distribution control. *Nano Letters* 22 (3): 1338–1344.

24 Zhao, F., Ren, A., Li, P. et al. (2022). Toward continuous-wave pumped metal halide perovskite lasers: strategies and challenges. *ACS Nano* 16 (5): 7116–7143.

25 Privitera, A., Righetto, M., Cacialli, F., and Riede, M.K. (2021). Perspectives of organic and perovskite-based spintronics. *Advanced Optical Materials* 9 (14): 2100215.

26 Long, G., Sabatini, R., Saidaminov, M.I. et al. (2020). Chiral-perovskite optoelectronics. *Nature Reviews Materials* 5 (6): 423–439.

27 Wei, Q. and Ning, Z. (2021). Chiral perovskite spin-optoelectronics and spintronics: toward judicious design and application. *ACS Materials Letters* 3 (9): 1266–1275.

28 Saparov, B. and Mitzi, D.B. (2016). Organic–inorganic perovskites: structural versatility for functional materials design. *Chemical Reviews* 116 (7): 4558–4596.

29 Goldschmidt, V.M. (1926). Die Gesetze der Krystallochemie. *Die Naturwissenschaften* 14 (21): 477–485.

30 Weller, M.T., Weber, O.J., Henry, P.F. et al. (2015). Complete structure and cation orientation in the perovskite photovoltaic methylammonium lead iodide between 100 and 352 K. *Chemical Communications* 51 (20): 4180–4183.

31 Poglitsch, A. and Weber, D. (1987). Dynamic disorder in methylammoniumtrihalogenoplumbates (II) observed by millimeter-wave spectroscopy. *The Journal of Chemical Physics* 87 (11): 6373–6378.

32 Baikie, T., Fang, Y., Kadro, J.M. et al. (2013). Synthesis and crystal chemistry of the hybrid perovskite $(CH_3NH_3)PbI_3$ for solid-state sensitised solar cell applications. *Journal of Materials Chemistry A* 1 (18): 5628–5641.

33 Li, C., Lu, X., Ding, W. et al. (2008). Formability of ABX_3 (X = F, Cl, Br, I) halide perovskites. *Acta Crystallographica Section B* 64 (6): 702–707.

34 Manser, J.S., Christians, J.A., and Kamat, P.V. (2016). Intriguing Optoelectronic Properties of Metal Halide Perovskites. *Chemical Reviews* 116 (21): 12956–13008.

35 Mohanty, A., Swain, D., Govinda, S. et al. (2019). Phase diagram and dielectric properties of $MA_{1-x}FA_xPbI_3$. *ACS Energy Letters* 4 (9): 2045–2051.

36 Giustino, F. and Snaith, H.J. (2016). Toward lead-free perovskite solar cells. *ACS Energy Letters* 1 (6): 1233–1240.

37 Hao, F., Stoumpos, C.C., Cao, D.H. et al. (2014). Lead-free solid-state organic–inorganic halide perovskite solar cells. *Nature Photonics* 8 (6): 489–494.

38 Noel, N.K., Stranks, S.D., Abate, A. et al. (2014). Lead-free organic–inorganic tin halide perovskites for photovoltaic applications. *Energy & Environmental Science* 7 (9): 3061–3068.

39 Slavney, A.H., Hu, T., Lindenberg, A.M., and Karunadasa, H.I. (2016). A bismuth-halide double perovskite with long carrier recombination lifetime for photovoltaic applications. *Journal of the American Chemical Society* 138 (7): 2138–2141.

40 Zhao, X.-G., Yang, J.-H., Fu, Y. et al. (2017). Design of lead-free inorganic halide perovskites for solar cells via cation-transmutation. *Journal of the American Chemical Society* 139 (7): 2630–2638.

41 Zhao, X.-G., Yang, D., Sun, Y. et al. (2017). Cu–In halide perovskite solar absorbers. *Journal of the American Chemical Society* 139 (19): 6718–6725.

42 Igbari, F., Wang, Z.-K., and Liao, L.-S. (2019). Progress of lead-free halide double perovskites. *Advanced Energy Materials* 9 (12): 1803150.

43 Ju, M.-G., Chen, M., Zhou, Y. et al. (2018). Earth-abundant nontoxic titanium(IV)-based vacancy-ordered double perovskite halides with tunable 1.0 to 1.8 eV bandgaps for photovoltaic applications. *ACS Energy Letters* 3 (2): 297–304.

44 Park, B.-W., Philippe, B., Zhang, X. et al. (2015). Bismuth based hybrid perovskites $A_3Bi_2I_9$ (A: methylammonium or cesium) for solar cell application. *Advanced Materials* 27 (43): 6806–6813.

45 Zhang, Z., Li, X., Xia, X. et al. (2017). High-quality $(CH_3NH_3)_3Bi_2I_9$ film-based solar cells: pushing efficiency up to 1.64%. *The Journal of Physical Chemistry Letters* 8 (17): 4300–4307.

46 Li, X., Hoffman, J.M., and Kanatzidis, M.G. (2021). The 2D halide perovskite rulebook: how the spacer influences everything from the structure to optoelectronic device efficiency. *Chemical Reviews* 121 (4): 2230–2291.

47 Gong, J., Hao, M., Zhang, Y. et al. (2022). Layered 2D halide perovskites beyond the Ruddlesden–Popper phase: tailored interlayer chemistries for

high-performance solar cells. *Angewandte Chemie International Edition* 61 (10): e202112022.

48 Zhou, Y. and Padture, N.P. (2022). Microstructures and grain boundaries of halide perovskite thin films. In: *Perovskite Photovoltaics and Optoelectronics: From Fundamentals to Advanced Applications* (ed. T. Miyasaka), 81–105. WILEY-VCH GmbH.

49 Gottstein, G. and Shvindlerman, L.S. (2009). *Grain boundary migration in metals: thermodynamics, kinetics, applications*. CRC press.

50 Chiang, Y.-M., Birnie, D.P., and Kingery, W.D. (1997). *Physical Ceramics: Principles for Ceramic Science and Engineering*. New York: Wiley.

51 Priester, L. (2013). Precipitation at grain boundaries. In: *Precipitation at Grain Boundaries* (ed. L. Priester), 217–240. Grain Boundaries: Springer.

52 Thompson, C.V. (1990). Grain growth in thin films. *Annual Review of Materials Science* 20 (1): 245–268.

53 Zhao, Y., Cruse, K., Abdelsamie, M. et al. (2021). Synthetic approaches for thin-film halide double perovskites. *Matter* 4 (6): 1801–1831.

54 Zhou, Y., Game, O.S., Pang, S., and Padture, N.P. (2015). Microstructures of organometal trihalide perovskites for solar cells: their evolution from solutions and characterization. *The Journal of Physical Chemistry Letters* 6 (23): 4827–4839.

55 Atkinson, H.V. (1988). Overview no. 65: Theories of normal grain growth in pure single phase systems. *Acta Metallurgica* 36 (3): 469–491.

56 Giesbrecht, N., Schlipf, J., Oesinghaus, L. et al. (2016). Synthesis of perfectly oriented and micrometer-sized $MAPbBr_3$ perovskite crystals for thin-film photovoltaic applications. *ACS Energy Letters* 1 (1): 150–154.

57 Jiang, C. and Zhang, P. (2018). Crystalline orientation dependent photoresponse and heterogeneous behaviors of grain boundaries in perovskite solar cells. *Journal of Applied Physics* 123 (8): 083105.

58 Kutes, Y., Zhou, Y., Bosse, J.L. et al. (2016). Mapping the photoresponse of $CH_3NH_3PbI_3$ hybrid perovskite thin films at the nanoscale. *Nano Letters* 16 (6): 3434–3441.

59 Leblebici, S.Y., Leppert, L., Li, Y. et al. (2016). Facet-dependent photovoltaic efficiency variations in single grains of hybrid halide perovskite. *Nature Energy* 1 (8): 16093.

60 De Yoreo, J.J., Gilbert, P.U.P.A., Sommerdijk, N.A.J.M. et al. (2015). Crystallization by particle attachment in synthetic, biogenic, and geologic environments. *Science* 349 (6247): aaa6760.

61 Nielsen, A. (1964). *Kinetics of Precipitation*. Oxford: Pergamon Press.

62 Brinker, C.J. and Scherer, G.W. (2013). *Sol-Gel Science: The Physics and Chemistry of Sol-Gel Processing*. Academic press.

63 Adamson, A.W. and Gast, A.P. (1967). *Physical Chemistry of Surfaces*. New York: Interscience Publishers.

64 Ring, T.A. (1996). *Fundamentals of Ceramic Powder Processing and Synthesis*. Elsevier.

65 Stevenson, J., Sorenson, B., Subramaniam, V.H. et al. (2017). Mayer bond order as a metric of complexation effectiveness in lead halide perovskite solutions. *Chemistry of Materials* 29 (6): 2435–2444.

66 Mitzi, D.B. (2000). Organic–inorganic perovskites containing trivalent metal halide layers: the templating influence of the organic cation layer. *Inorganic Chemistry* 39 (26): 6107–6113.

67 Yan, K., Long, M., Zhang, T. et al. (2015). Hybrid halide perovskite solar cell precursors: colloidal chemistry and coordination engineering behind device processing for high efficiency. *Journal of the American Chemical Society* 137 (13): 4460–4468.

68 Stamplecoskie, K.G., Manser, J.S., and Kamat, P.V. (2015). Dual nature of the excited state in organic–inorganic lead halide perovskites. *Energy & Environmental Science* 8 (1): 208–215.

69 Hamill, J.C., Schwartz, J., and Loo, Y.-L. (2018). Influence of solvent coordination on hybrid organic–inorganic perovskite formation. *ACS Energy Letters* 3 (1): 92–97.

70 Ke, L., Luo, S., Ren, X., and Yuan, Y. (2021). Factors influencing the nucleation and crystal growth of solution-processed organic lead halide perovskites: a review. *Journal of Physics D: Applied Physics* 54 (16): 163001.

71 Guo, Y., Shoyama, K., Sato, W. et al. (2015). Chemical pathways connecting lead(II) iodide and perovskite via polymeric plumbate(II) fiber. *Journal of the American Chemical Society* 137 (50): 15907–15914.

72 Cao, J., Jing, X., Yan, J. et al. (2016). Identifying the molecular structures of intermediates for optimizing the fabrication of high-quality perovskite films. *Journal of the American Chemical Society* 138 (31): 9919–9926.

73 Hao, F., Stoumpos, C.C., Liu, Z. et al. (2014). Controllable perovskite crystallization at a gas–solid interface for hole conductor-free solar cells with steady power conversion efficiency over 10%. *Journal of the American Chemical Society* 136 (46): 16411–16419.

74 LaMer, V.K. and Dinegar, R.H. (1950). Theory, production and mechanism of formation of monodispersed hydrosols. *Journal of the American Chemical Society* 72 (11): 4847–4854.

75 Sun, Y. (2013). Controlled synthesis of colloidal silver nanoparticles in organic solutions: empirical rules for nucleation engineering. *Chemical Society Reviews* 42 (7): 2497–2511.

76 Dunlap-Shohl, W.A., Zhou, Y., Padture, N.P., and Mitzi, D.B. (2018). Synthetic approaches for halide perovskite thin films. *Chemical Reviews* 119 (5): 3193–3295.

77 Zheng, Y.C., Yang, S., Chen, X. et al. (2015). Thermal-induced volmer–weber growth behavior for planar heterojunction perovskites solar cells. *Chemistry of Materials* 27 (14): 5116–5121.

78 Kingery, W.D. (1976). Development of microstructure in ceramics. In: *Introduction to Ceramics*, 2e (ed. W.D. Kingery, H.K. Bowen, and D.R. Uhlmann), 449–468. Wiley.

79 Jeon, N.J., Noh, J.H., Kim, Y.C. et al. (2014). Solvent engineering for high-performance inorganic–organic hybrid perovskite solar cells. *Nature Materials* 13 (9): 897–903.

80 Zhou, Y., Yang, M., Wu, W. et al. (2015). Room-temperature crystallization of hybrid-perovskite thin films via solvent–solvent extraction for high-performance solar cells. *Journal of Materials Chemistry A* 3 (15): 8178–8184.

81 Nie, W., Tsai, H., Asadpour, R. et al. (2015). High-efficiency solution-processed perovskite solar cells with millimeter-scale grains. *Science* 347 (6221): 522–525.

82 Li, C., Wang, A., Xie, L. et al. (2019). Emerging alkali metal ion (Li^+, Na^+, K^+ and Rb^+) doped perovskite films for efficient solar cells: recent advances and prospects. *Journal of Materials Chemistry A* 7 (42): 24150–24163.

83 Tang, Z., Uchida, S., Bessho, T. et al. (2018). Modulations of various alkali metal cations on organometal halide perovskites and their influence on photovoltaic performance. *Nano Energy* 45: 184–192.

84 Lou, Y., Niu, Y., Yang, D. et al. (2018). Rod-shaped thiocyanate-induced abnormal band gap broadening in SCN^- doped $CsPbBr_3$ perovskite nanocrystals. *Nano Research* 11 (5): 2715–2723.

85 Li, N., Tao, S., Chen, Y. et al. (2019). Cation and anion immobilization through chemical bonding enhancement with fluorides for stable halide perovskite solar cells. *Nature Energy* 4 (5): 408–415.

86 Zhang, X., Li, L., Sun, Z., and Luo, J. (2019). Rational chemical doping of metal halide perovskites. *Chemical Society Reviews* 48 (2): 517–539.

87 Bai, S., Da, P., Li, C. et al. (2019). Planar perovskite solar cells with long-term stability using ionic liquid additives. *Nature* 571 (7764): 245–250.

88 Niu, T., Chao, L., Gao, W. et al. (2021). Ionic liquids-enabled efficient and stable perovskite photovoltaics: progress and challenges. *ACS Energy Letters* 6 (4): 1453–1479.

89 Fan, F., Zhang, Y., Hao, M. et al. (2022). Harnessing chemical functions of ionic liquids for perovskite solar cells. *Journal of Energy Chemistry* 68: 797–810.

90 Mei, A., Li, X., Liu, L. et al. (2014). A hole-conductor–free, fully printable mesoscopic perovskite solar cell with high stability. *Science* 345 (6194): 295–298.

91 Zuo, C. and Ding, L. (2021). Drop-casting to make efficient perovskite solar cells under high humidity. *Angewandte Chemie International Edition* 60 (20): 11242–11246.

92 Era, M., Hattori, T., Taira, T., and Tsutsui, T. (1997). Self-organized growth of pbi-based layered perovskite quantum well by dual-source vapor deposition. *Chemistry of Materials* 9 (1): 8–10.

93 Liu, M., Johnston, M.B., and Snaith, H.J. (2013). Efficient planar heterojunction perovskite solar cells by vapour deposition. *Nature* 501 (7467): 395–398.

94 Chen, M., Ju, M.-G., Garces, H.F. et al. (2019). Highly stable and efficient all-inorganic lead-free perovskite solar cells with native-oxide passivation. *Nature Communications* 10 (1): 16.

95 Listorti, A., Juarez-Perez, E.J., Frontera, C. et al. (2015). Effect of mesostructured layer upon crystalline properties and device performance on perovskite solar cells. *The Journal of Physical Chemistry Letters* 6 (9): 1628–1637.

96 Zhou, Y., Vasiliev, A.L., Wu, W. et al. (2015). Crystal morphologies of organolead trihalide in mesoscopic/planar perovskite solar cells. *The Journal of Physical Chemistry Letters* 6 (12): 2292–2297.

97 Mitzi, D.B., Prikas, M.T., and Chondroudis, K. (1999). Thin film deposition of organic–inorganic hybrid materials using a single source thermal ablation technique. *Chemistry of Materials* 11 (3): 542–544.

98 Li, D., Dong, G., Li, W., and Wang, L. (2015). High performance organic-inorganic perovskite-optocoupler based on low-voltage and fast response perovskite compound photodetector. *Scientific Reports* 5 (1): 7902.

99 Meggiolaro, D., Motti, S.G., Mosconi, E. et al. (2018). Iodine chemistry determines the defect tolerance of lead-halide perovskites. *Energy & Environmental Science* 11 (3): 702–713.

100 Bansode, U., Naphade, R., Game, O. et al. (2015). Hybrid perovskite films by a new variant of pulsed excimer laser deposition: a room-temperature dry process. *The Journal of Physical Chemistry C* 119 (17): 9177–9185.

101 Barraza, E.T., Dunlap-Shohl, W.A., Mitzi, D.B., and Stiff-Roberts, A.D. (2018). Deposition of methylammonium lead triiodide by resonant infrared matrix-assisted pulsed laser evaporation. *Journal of Electronic Materials* 47 (2): 917–926.

102 Dunlap-Shohl, W.A., Barraza, E.T., Barrette, A. et al. (2018). $MAPbI_3$ solar cells with absorber deposited by resonant infrared matrix-assisted pulsed laser evaporation. *ACS Energy Letters* 3 (2): 270–275.

103 Liang, K., Mitzi, D.B., and Prikas, M.T. (1998). Synthesis and characterization of organic–inorganic perovskite thin films prepared using a versatile two-step dipping technique. *Chemistry of Materials* 10 (1): 403–411.

104 Bi, D., Moon, S.-J., Häggman, L. et al. (2013). Using a two-step deposition technique to prepare perovskite ($CH_3NH_3PbI_3$) for thin film solar cells based on ZrO_2 and TiO_2 mesostructures. *RSC Advances* 3 (41): 18762–18766.

105 Burschka, J., Pellet, N., Moon, S.-J. et al. (2013). Sequential deposition as a route to high-performance perovskite-sensitized solar cells. *Nature* 499 (7458): 316–319.

106 Wu, Y., Islam, A., Yang, X. et al. (2014). Retarding the crystallization of PbI_2 for highly reproducible planar-structured perovskite solar cells via sequential deposition. *Energy & Environmental Science* 7 (9): 2934–2938.

107 Liu, T., Hu, Q., Wu, J. et al. (2016). Mesoporous PbI_2 scaffold for high-performance planar heterojunction perovskite solar cells. *Advanced Energy Materials* 6 (3): 1501890.

108 Fu, Y., Meng, F., Rowley, M.B. et al. (2015). Solution growth of single crystal methylammonium lead halide perovskite nanostructures for optoelectronic and photovoltaic applications. *Journal of the American Chemical Society* 137 (17): 5810–5818.

109 Chauhan, M., Zhong, Y., Schötz, K. et al. (2020). Investigating two-step $MAPbI_3$ thin film formation during spin coating by simultaneous in situ absorption and photoluminescence spectroscopy. *Journal of Materials Chemistry A* 8 (10): 5086–5094.

110 Hsieh, T.-Y., Huang, C.-K., Su, T.-S. et al. (2017). Crystal growth and dissolution of methylammonium lead iodide perovskite in sequential deposition: correlation between morphology evolution and photovoltaic performance. *ACS Applied Materials & Interfaces* 9 (10): 8623–8633.

111 Xiao, Z., Bi, C., Shao, Y. et al. (2014). Efficient, high yield perovskite photovoltaic devices grown by interdiffusion of solution-processed precursor stacking layers. *Energy & Environmental Science* 7 (8): 2619–2623.

112 Chen, H., Wei, Z., Zheng, X., and Yang, S. (2015). A scalable electrodeposition route to the low-cost, versatile and controllable fabrication of perovskite solar cells. *Nano Energy* 15: 216–226.

113 Koza, J.A., Hill, J.C., Demster, A.C., and Switzer, J.A. (2016). Epitaxial electrodeposition of met hylammonium lead iodide perovskites. *Chemistry of Materials* 28 (1): 399–405.

114 Chen, Q., Zhou, H., Hong, Z. et al. (2014). Planar heterojunction perovskite solar cells via vapor-assisted solution process. *Journal of the American Chemical Society* 136 (2): 622–625.

115 Li, Y., Cooper, J.K., Buonsanti, R. et al. (2015). Fabrication of planar heterojunction perovskite solar cells by controlled low-pressure vapor annealing. *The Journal of Physical Chemistry Letters* 6 (3): 493–499.

116 Chen, C.-W., Kang, H.-W., Hsiao, S.-Y. et al. (2014). Efficient and uniform planar-type perovskite solar cells by simple sequential vacuum deposition. *Advanced Materials* 26 (38): 6647–6652.

117 Li, H., Zhou, J., Tan, L. et al. (2022). Sequential vacuum-evaporated perovskite solar cells with more than 24% efficiency. *Science Advances* 8 (28): eabo7422.

118 Kim, J., Yun, J.S., Cho, Y. et al. (2017). Overcoming the challenges of large-area high-efficiency perovskite solar cells. *ACS Energy Letters* 2 (9): 1978–1984.

119 Deng, Y., Peng, E., Shao, Y. et al. (2015). Scalable fabrication of efficient organolead trihalide perovskite solar cells with doctor-bladed active layers. *Energy & Environmental Science* 8 (5): 1544–1550.

120 Zhong, Y., Munir, R., Li, J. et al. (2018). Blade-coated hybrid perovskite solar cells with efficiency > 17%: an in situ investigation. *ACS Energy Letters* 3 (5): 1078–1085.

121 Krebs, F.C. (2009). Fabrication and processing of polymer solar cells: a review of printing and coating techniques. *Solar Energy Materials & Solar Cells* 93 (4): 394–412.

122 Deng, Y., Van Brackle, C.H., Dai, X. et al. (2019). Tailoring solvent coordination for high-speed, room-temperature blading of perovskite photovoltaic films. *Science Advances* 5 (12): eaax7537.

123 Lee, D.-K., Jeong, D.-N., Ahn, T.K., and Park, N.-G. (2019). Precursor engineering for a large-area perovskite solar cell with >19% efficiency. *ACS Energy Letters* 4 (10): 2393–2401.

124 Yang, F., Dong, L., Jang, D. et al. (2021). Low temperature processed fully printed efficient planar structure carbon electrode perovskite solar cells and modules. *Advanced Energy Materials* 11 (28): 2101219.

125 Gao, L.-L., Li, C.-X., Li, C.-J., and Yang, G.-J. (2017). Large-area high-efficiency perovskite solar cells based on perovskite films dried by the multi-flow air knife method in air. *Journal of Materials Chemistry A* 5 (4): 1548–1557.

126 Whitaker, J.B., Kim, D.H., Larson Bryon, W. et al. (2018). Scalable slot-die coating of high performance perovskite solar cells. *Sustainable Energy & Fuels* 2 (11): 2442–2449.

127 He, M., Li, B., Cui, X. et al. (2017). Meniscus-assisted solution printing of large-grained perovskite films for high-efficiency solar cells. *Nature Communications* 8 (1): 16045.

128 Yang, M., Li, Z., Reese, M.O. et al. (2017). Perovskite ink with wide processing window for scalable high-efficiency solar cells. *Nature Energy* 2 (5): 17038.

129 Huang, F., Dkhissi, Y., Huang, W. et al. (2014). Gas-assisted preparation of lead iodide perovskite films consisting of a monolayer of single crystalline grains for high efficiency planar solar cells. *Nano Energy* 10: 10–18.

130 Li, X., Bi, D., Yi, C. et al. (2016). A vacuum flash assisted solution process for high-efficiency large-area perovskite solar cells. *Science* 353 (6294): 58–62.

131 Deng, Y., Zheng, X., Bai, Y. et al. (2018). Surfactant-controlled ink drying enables high-speed deposition of perovskite films for efficient photovoltaic modules. *Nature Energy* 3 (7): 560–566.

132 Bishop, J.E., Routledge, T.J., and Lidzey, D.G. (2018). Advances in spray-cast perovskite solar cells. *The Journal of Physical Chemistry Letters* 9 (8): 1977–1984.

133 Swartwout, R., Hoerantner, M.T., and Bulović, V. (2019). Scalable deposition methods for large-area production of perovskite thin films. *Energy & Environmental Materials* 2 (2): 119–145.

134 Barrows, A.T., Pearson, A.J., Kwak, C.K. et al. (2014). Efficient planar heterojunction mixed-halide perovskite solar cells deposited via spray-deposition. *Energy & Environmental Science* 7 (9): 2944–2950.

135 Das, S., Yang, B., Gu, G. et al. (2015). High-performance flexible perovskite solar cells by using a combination of ultrasonic spray-coating and low thermal budget photonic curing. *ACS Photonics* 2 (6): 680–686.

136 Bishop, J.E., Read, C.D., Smith, J.A. et al. (2020). Fully spray-coated triple-cation perovskite solar cells. *Scientific Reports* 10 (1): 6610.

137 Bishop, J.E., Smith, J.A., and Lidzey, D.G. (2020). Development of spray-coated perovskite solar cells. *ACS Applied Materials & Interfaces* 12 (43): 48237–48245.

138 Karunakaran, S.K., Arumugam, G.M., Yang, W. et al. (2019). Recent progress in inkjet-printed solar cells. *Journal of Materials Chemistry A* 7 (23): 13873–13902.

139 Eggers, H., Schackmar, F., Abzieher, T. et al. (2020). Inkjet-printed micrometer-thick perovskite solar cells with large columnar grains. *Advanced Energy Materials* 10 (6): 1903184.

140 Wei, Z., Chen, H., Yan, K., and Yang, S. (2014). Inkjet printing and instant chemical transformation of a $CH_3NH_3PbI_3$/nanocarbon electrode and interface for planar perovskite solar cells. *Angewandte Chemie International Edition* 53 (48): 13239–13243.

141 Gheno, A., Huang, Y., Bouclé, J. et al. (2018). Toward highly efficient inkjet-printed perovskite solar cells fully processed under ambient conditions and at low temperature. *Solar RRL* 2 (11): 1800191.

142 Masawa, S.M., Li, J., Zhao, C. et al. (2022). 0D/2D Mixed dimensional lead-free caesium bismuth iodide perovskite for solar cell application. *Materials* 15 (6): 2180.

143 Xiao, Z., Dong, Q., Bi, C. et al. (2014). Solvent annealing of perovskite-induced crystal growth for photovoltaic-device efficiency enhancement. *Advanced Materials* 26 (37): 6503–6509.

144 Yu, G., Jiang, K.-J., Gu, W.-M. et al. (2022). Vacuum-assisted thermal annealing of $CsPbI_3$ for highly stable and efficient inorganic perovskite solar cells. *Angewandte Chemie International Edition* 61 (27): e202203778.

145 Ke, W., Xiao, C., Wang, C. et al. (2016). Employing lead thiocyanate additive to reduce the hysteresis and boost the fill factor of planar perovskite solar cells. *Advanced Materials* 28 (26): 5214–5221.

146 Zhou, Z., Wang, Z., Zhou, Y. et al. (2015). Methylamine-gas-induced defect-healing behavior of $CH_3NH_3PbI_3$ thin films for perovskite solar cells. *Angewandte Chemie International Edition* 54 (33): 9705–9709.

147 Zhou, Y., Yang, M., Pang, S. et al. (2016). Exceptional morphology-preserving evolution of formamidinium lead triiodide perovskite thin films via organic-cation displacement. *Journal of the American Chemical Society* 138 (17): 5535–5538.

8

First Principles Atomistic Theory of Halide Perovskites

Linn Leppert

University of Twente, MESA+ Institute for Nanotechnology, Computational Chemical Physics, Drienerlolaan 5, 7500 AE Enschede, The Netherlands

8.1 Introduction: What I Talk About When I Talk About First Principles Calculations of Halide Perovskites

The structure, structural dynamics, phase transitions, the optoelectronic, charge carrier transport, and defect properties, in short, all physical and chemical consequences of the intricate, dynamical coupling of the electrons and nuclei that metal-halide perovskites consists of, are governed by quantum mechanics. In principle, this means that an *ab initio* (latin: from the beginning) understanding of these properties boils down to solving the (time-dependent) Schrödinger equation of these electrons and nuclei. In practice, finding (and storing) such solutions is impossible within the current computational paradigm due to the exponentially increasing complexity of the interacting many-particle problem.

In this chapter I will provide a brief didactic introduction to density functional theory (DFT) and Green's function-based many body perturbation theory, both widely used in physics, chemistry, and material science for approximately solving Schrödinger's equation for N-particle systems. I will discuss which approximations are applied in practical computer experiments based on these methods and describe how they can be used to calculate structural and optoelectronic properties of metal-halide perovskites. In all of this, my focus will be on conveying the most important concepts rather than providing a complete review of either the methods or the properties that can be calculated. Wherever appropriate, I will refer the interested reader to more in-depth resources.

Metal-halide perovskites are complex materials in many different respects. They are a large family of materials, as schematically illustrated in Figure 8.1, exhibiting a wide spectrum of structural and optoelectronic properties. Conventional halide perovskites – defined as crystal structures with metal-centered corner-sharing octahedra and formula ABX_3 (single perovskites) or $A_2BB'X_6$ (double

Halide Perovskite Semiconductors: Structures, Characterization, Properties, and Phenomena, First Edition. Edited by Yuanyuan Zhou and Iván Mora-Seró.

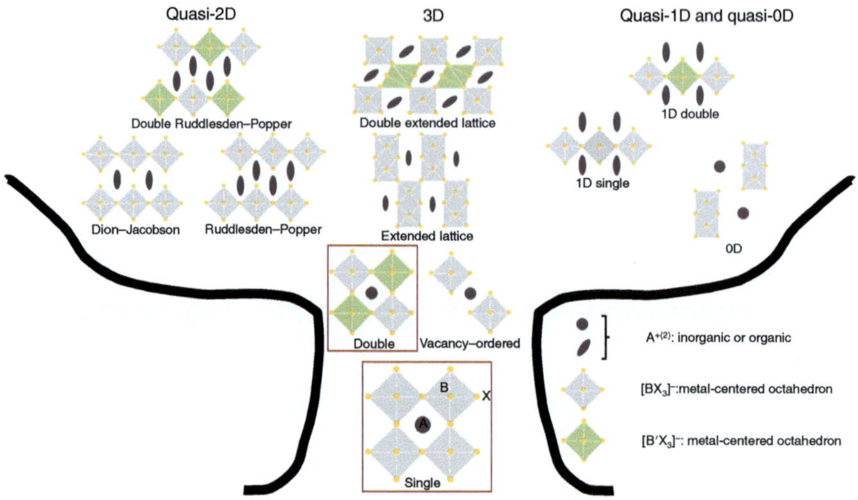

Figure 8.1 Schematic family tree of the broad family of halide perovskites, including a selection of quasi-perovskite structures with disrupted octahedral connectivity and reduced electronic dimensionality. The red boxes indicate the single ABX_3 and double $A_2BB'X_6$ perovskites that are the primary focus of this chapter.

perovskites) – allow for the realization of many hundreds of compounds through chemical substitution at the A, B, B′, and X sites. Incorporation of large A site cations leads to further structural, chemical, and phenomenological diversity, affording a variety of structural motifs with broken octahedral connectivity and reduced electronic dimensionality [1]. Both fully inorganic and organic–inorganic perovskites have elaborate structural phase diagrams as a function of temperature and pressure. Furthermore, the structural dynamics of many perovskites are governed by anharmonic effects, their lattices are soft, molecular cations can exhibit complex rotational dynamics, and halogen anions can easily migrate through the crystal lattice. The electronic structure of Pb-based and other halide perovskites that feature heavy elements is strongly affected by spin–orbit interactions; relativistic effects need to be explicitly accounted for in electronic structure calculations. Additionally, many halide perovskites feature comparably small band gaps, which are severely underestimated by standard DFT approximations.

All of this means that accurate predictions of structural and optoelectronic properties of halide perovskites call for the use of advanced first-principles techniques while also necessitating large structural models. Meeting these demands is computationally extraordinarily challenging, even on large, high-performance supercomputers. Most computational studies of halide perovskites need to strike a balance between accuracy and computational cost. The choices that are being made in practical calculations with respect to supercell size and methodology depend on the properties investigated. The following sections aim to provide an overview of how this is done in practice and in which areas care has to be taken when interpreting results from computer experiments.

8.2 Structural Properties

DFT is undoubtedly the most commonly used first-principles method for calculating the structural properties of halide perovskites. This section will therefore start with a brief introduction to DFT (Section 8.2.1), followed by an overview of the most important physical and numerical approximations in practical DFT calculations (Section 8.2.2.1). Subsequently, I will outline the central aspects of setting up and interpreting zero-temperature (static) DFT calculations of perovskite crystal structures, with a focus on the limits of these setups (Section 8.2.2.2). Section 8.2.4.1 will be devoted to the calculation of finite-temperature (dynamical) structural properties, such as phase transitions and ion migration dynamics.

8.2.1 A Short Introduction to Density Functional Theory

The foundations of DFT were laid in 1964 by Hohenberg and Kohn, who proved that the ground-state density $n(\mathbf{r})$ contains all information necessary to calculate ground-state properties of a system of N interacting electrons in the potential of M nuclei. Assuming the Born–Oppenheimer approximation, the electronic Hamiltonian of this system has the form $\hat{H} = \hat{T} + \hat{V}_{\text{ext}} + \hat{W}$, where $\hat{T} = \sum_{i=1}^{N} \hat{p}_i^2/2m$ is the kinetic energy operator, $\hat{V}_{\text{ext}} = \sum_{i=1}^{N} v_{\text{ext}}(\mathbf{r}_i)$ the interaction between the electrons and the external potential of the nuclei, and $\hat{W} = \sum_{i>j} w(\mathbf{r}_i, \mathbf{r}_j)$ is the electron–electron interaction.

The Hohenberg–Kohn theorem can be subsumed in two statements: First, for a given electron–electron interaction $\hat{W}$ there is a one-to-one mapping between the external potential v_{ext}, the (nondegenerate) ground-state wavefunction $|\Psi_0\rangle$ that one obtains by solving $\hat{H}|\Psi_0\rangle = E_0|\Psi_0\rangle$, and the ground-state density $n(\mathbf{r})$. This statement implies that $|\Psi_0\rangle$ is a unique functional of the density, and hence that every ground-state observable and in particular the ground-state energy are density functionals too. Second, using the variational principle one can obtain the exact ground-state density and energy corresponding to v_{ext} by minimizing the total energy functional

$$E[n] = F[n] + \int d^3r v_{\text{ext}}(\mathbf{r}) n(\mathbf{r}). \tag{8.1}$$

$F[n] = \langle\Psi[n]|\hat{T} + \hat{W}|\Psi[n]\rangle$ is a universal functional, independent of $v_{\text{ext}}(\mathbf{r})$. Equation (8.1) provides a simple and exact reformulation of Schrödinger's equation. However, as Hohenberg and Kohn noted already in 1964 [2]: "The major part of the complexity of the many-electron problem is associated with the determination of the universal functional $F[n]$."

The most successful approach to determining $F[n]$ was proposed in 1965 by Kohn and Sham [3], who reformulated the energy functional of a system of N interacting electrons as

$$E[n] = T_s[n] + E_H[n] + E_{\text{ext}}[n] + E_{\text{xc}}[n], \tag{8.2}$$

where T_s is the kinetic energy of a system of N non-interacting electrons, $E_H[n]$ is the classical electrostatic Hartree energy, and $E_{\text{ext}}[n]$ arises from the interaction between the electrons and the external potential v_{ext}. $E_{\text{xc}}[n]$, the exchange-correlation (xc) functional, is defined as consisting of every contribution to the total energy not included in E_{H} and T_{s}, i.e. $E_{\text{xc}} = T - T_{\text{s}} + W - E_{\text{H}}$. Here, T is the full kinetic energy of the interacting electron system, and W is the electron–electron interaction energy. The Kohn–Sham equations are obtained by minimization of Eq. (8.2) with respect to n and by realizing that the exact same ground-state density is obtained for a system of N non-interacting electrons in an effective local potential

$$v^{\text{KS}}(\mathbf{r}) = v_{\text{H}}(\mathbf{r}) + v_{\text{ext}}(\mathbf{r}) + v_{\text{xc}}(\mathbf{r}) \tag{8.3}$$

in which v_{H} is the Hartree potential and $v_{\text{xc}} = \partial E_{\text{xc}}[n]/\partial n$ is the xc potential. One can then calculate the ground-state density $n(\mathbf{r})$ of the interacting N electron system by solving the one-particle equations

$$\left(-\frac{\hbar^2}{2m}\nabla^2 + v^{\text{KS}}(\mathbf{r})\right)\varphi_i(\mathbf{r}) = \varepsilon_i\varphi_i(\mathbf{r}) \tag{8.4}$$

and summing over all occupied orbitals $\varphi_i(\mathbf{r})$

$$n(\mathbf{r}) = \sum_{i=1}^{N} |\varphi_i(\mathbf{r})|^2 \tag{8.5}$$

Equations (8.3)–(8.5) constitute the celebrated Kohn–Sham equations – an in principle exact reformulation of the full interacting problem, in which quantum-mechanical xc effects are embodied in the xc energy E_{xc}, which needs to be approximated in virtually all practical applications of DFT.

8.2.2 DFT Calculations in Practice

8.2.2.1 Approximations

How do we proceed from the Kohn–Sham equations to calculating structural properties of perovskites? Which numerical and physical approximations can be used to find an acceptable balance between accuracy and computational cost? I will distinguish between numerical/practical approximations, that are related to the implementation of Kohn–Sham DFT in electronic structure codes on the one hand. These are for example the truncation of the basis set in which the Kohn–Sham orbitals are expanded, a finite sampling of the first Brillouin zone, and the choice of pseudopotential (illustrated in Figure 8.2). On the other hand, there are physically motivated approximations, such as the choice of E_{xc}.

In terms of numerical implementation, the first step in most practical calculations of solid-state materials is to make use of the translational symmetry of the crystal structure and solve the Kohn–Sham equations with periodic boundary conditions. The Kohn–Sham orbitals $\varphi_{i\mathbf{k}}$, which now have an additional index $\mathbf{k}$ corresponding to the crystal momentum, can be expanded in a basis of plane waves

$$\varphi_{i\mathbf{k}}(\mathbf{r}) = \frac{1}{\sqrt{\Omega}}\sum_m c_{im}(\mathbf{k})\exp\left(i(\mathbf{k} + \mathbf{G}_m)\cdot\mathbf{r}\right) \tag{8.6}$$

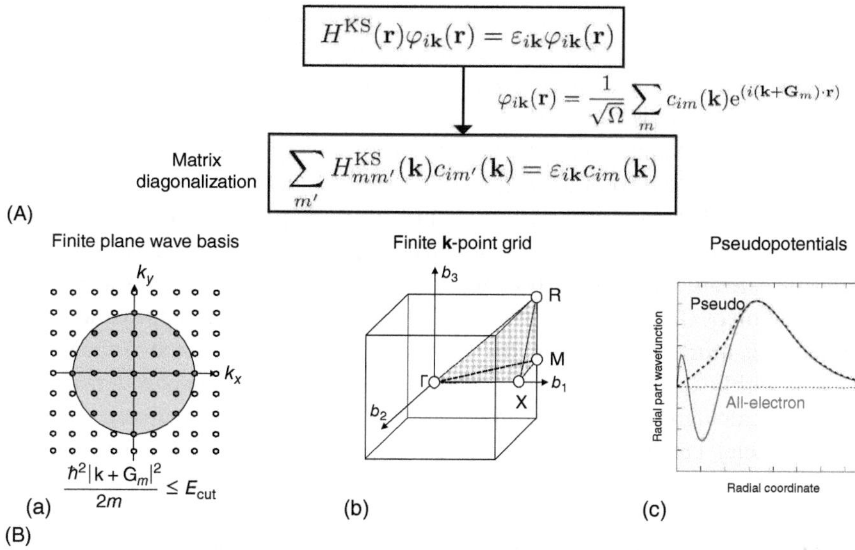

Figure 8.2 Scheme of different kinds of approximations involved in plane wave-based implementations of DFT. (A) The Kohn–Sham equations are expanded in a basis of plane waves, taking the form of a matrix eigenvalue equation that can be solved using standard matrix diagonalization schemes. (B) The most important numerical approximations include the use of a finite plane-wave basis set (a), a finite **k**-point mesh (b), and (c) the use of pseudopotentials.

where $\mathbf{G}_m$ is a reciprocal lattice vector, and c_{im} are expansion coefficients. Using a plane-wave basis set, the Kohn–Sham equations can be rewritten as matrix eigenvalue equations that can be solved for each $\mathbf{k}$:

$$\sum_{m'} H^{KS}_{mm'}(\mathbf{k})c_{im'}(\mathbf{k}) = \varepsilon_{i\mathbf{k}}c_{im}(\mathbf{k}) \tag{8.7}$$

where $H^{KS}_{mm'}$ is the Kohn–Sham Hamiltonian in the plane-wave basis. In practical calculations, the basis set is necessarily finite; only plane waves with energies below a certain cutoff energy are included in the expansion – a first convergence parameter of practical calculations – illustrated in Figure 8.2.

Another numerical choice concerns the sampling of **k**-vectors in the first Brillouin zone. Figure 8.2 shows the first Brillouin zone of a cubic perovskite lattice with the irreducible wedge spanned by the high symmetry points Γ, X, M, and R marked by the dotted gray area. Various choices for sampling the irreducible wedge are possible; in all cases, the density of the **k**-mesh needs to be carefully converged in practical calculations as well.

The third practical choice concerns the treatment of v_{ext}, the electron–nuclei potential in the Kohn–Sham equations. This term can be replaced by a so-called pseudopotential or effective core potential. Various types of pseudopotentials exist, e.g. norm-conserving pseudopotentials [4] and projector augmented wave (PAW) [5] potentials. Regardless of the details, the consequence of using a pseudopotential instead of the full external potential of the nuclei is that the low-energy

(Kohn–Sham) core orbitals are replaced with smooth functions that continuously connect to what is defined as the valence region. Pseudopotentials simplify the numerical solution of the Kohn–Sham equations because of the complex nodal structure of the orbitals close to the core (sketched in Figure 8.2 for a 3s orbital), which would necessitate many basis functions (or very dense real-space grids). Additionally, the use of pseudopotentials is physically motivated by the empirical observation that for many structural and optoelectronic properties only the valence electrons are significant, and as a pragmatic approach for taking into account relativistic effects. Since every pseudopotential is constructed with a specific xc approximation of DFT, care must be taken when mixing different approximations.

With these numerical choices, the Kohn–Sham equations can in principle be solved to within arbitrary accuracy. Knowledge of the ground-state density $n(\mathbf{r})$ then allows calculations of total energies and, via the Hellmann–Feynman theorem, forces on nuclei, i.e. the potential energy landscape. This can be used for geometry optimizations (see Section 8.2.2.2) or molecular dynamics (MD) simulations (see Section 8.2.4.1).

Next to these numerical approximations, $v_{xc}[n] = \partial E_{xc}[n]/\partial n(\mathbf{r})$, the xc potential, needs to be approximated in virtually all practical applications of DFT. In the following, I will give a very brief overview of the most relevant approximations, and refer the reader to Refs. [6] and [7] for more in-depth information. The local density approximation (LDA) is based on the idea that in a system in which the electron density varies only slowly in space, the xc energy per electron, e_{xc}, is approximately equal to e_{xc}^{hom} in the homogeneous electron gas. The exchange contribution is exactly known and given by

$$E_{x}^{LDA}[n] = -\frac{3e^2}{4}\left(\frac{3}{\pi}\right)^{1/3} \int d^3r n(\mathbf{r})^{4/3} \tag{8.8}$$

No closed expression for the correlation energy of the homogeneous electron gas is known, but highly accurate parametrizations of quantum Monte Carlo calculations [8] exist, such as the ones by Vosko et al. [9], Perdew and Zunger [10], and Perdew and Wang [11].

Generalized gradient approximations offer a systematic improvement over energies and forces calculated with the LDA by including information about the rate of spatial variation of the electron density in the form

$$E_{xc}^{GGA}[n] = \int d^3r f(n(\mathbf{r}), \nabla n(\mathbf{r})) \tag{8.9}$$

The function $f(n(\mathbf{r}), \nabla n(\mathbf{r}))$ can be constructed by satisfying exact constraints, as in the approximation by Perdew, Burke, and Ernzerhof (PBE) [12], by fitting of parameters to large sets of thermochemical data or a combination thereof.

Because of their moderate computational cost, (semi)local functionals are particularly popular for calculating ground-state properties and for large-scale applications such as high-throughput data sampling and the training of machine-learning models. However, they suffer from well-known shortcomings such as a failure to predict fundamental band gaps of solids (see Section 8.3.1.1), an inaccurate description of charge localization and distribution between subsystems [13], and a severe misrepresentation of polarizabilities, hyperpolarizabilities, and charge transfer [14, 15].

One strategy for improving over (semi)local functionals is the incorporation of exact (Fock-like) exchange (EXX)

$$E_{\mathrm{x}}^{\mathrm{EXX}} = -\frac{e^2}{2}\sum_{i,j}^{N}\int\int d^3r d^3r' \frac{\varphi_i^*(\mathbf{r})\varphi_j^*(\mathbf{r}')\varphi_j(\mathbf{r})\varphi_i(\mathbf{r}')}{|\mathbf{r}-\mathbf{r}'|} \tag{8.10}$$

However, functionals containing EXX result in a steep increase in computational cost because of their nonlocal orbital-dependence. Global hybrid functional are constructed as a combination of a fixed fraction α of EXX and semilocal exchange and correlation. The parameter α is typically determined by fitting to extensive molecular sets [16] or by virtue of the adiabatic connection formula [17], such as in the global hybrid PBE0 in which $\alpha = 0.25$ and the semilocal contribution is described by PBE exchange [18, 19].

The concept of global hybrid functionals is generalized by allowing EXX and semilocal exchange to dominate the total xc energy at different electron–electron distances r. So-called range-separated hybrid (RSH) functionals separate the Coulomb interaction W into a long-range and a short-range term via

$$\frac{1}{r} = \frac{\alpha + \beta\mathrm{erf}(\gamma r)}{r} + \frac{1-[\alpha + \beta\mathrm{erf}(\gamma r)]}{r} \tag{8.11}$$

where α is the fraction of short-range EXX, $\alpha + \beta$ the fraction of long-range EXX, and γ is a parameter that determines at which length scale semilocal and EXX dominate, respectively. One advantage of RSH functionals over global hybrids is that they allow for an xc potential with the correct asymptotic behavior, by requiring $\alpha + \beta = 1$ for atoms, molecules and clusters [20], and $\alpha + \beta = 1/\epsilon_\infty$ for solids [21], where ϵ_∞ is the orientationally averaged high-frequency dielectric constant, to ensure an on-average correct description of long-range dielectric screening effects. The only free parameter left then is γ, which can be optimally tuned to satisfy exact criteria, such as the ionization potential theorem of DFT [22].

Finally, I want to mention meta-generalized gradient approximations (GGAs) as another well-established avenue for overcoming the shortcomings of (semi)local functionals. Meta-GGAs depend on the kinetic energy density

$$\tau(\mathbf{r}) = \frac{\hbar^2}{2m}\sum_{i=1}^{N}|\nabla\varphi_i(\mathbf{r})|^2 \tag{8.12}$$

and sometimes terms $\nabla^2 n(\mathbf{r})$. Meta-GGAs developed in recent years such as the SCAN [23] and TASK [24] functionals, have focused on satisfying a large number of exact constraints of the exact xc functional, and allow for a more accurate prediction of total energies and improved band gaps at a computational cost comparable to that of semilocal xc approximations.

For xc functionals that depend explicitly on the orbitals such as global and RSH functionals and meta-GGAs, the calculation of the xc potential is not as straightforward as the simple functional derivative of E_{xc} with respect to the electron density introduced in Section 8.2.1 (see Ref. [7] for an instructive overview on orbital-dependent functionals in DFT). In practice, many implementations of hybrid and other orbital-dependent functionals are rooted in the so-called generalized Kohn–Sham framework, in which the potential in the single-particle equations is allowed to be nonlocal:

$$\left[-\frac{\hbar^2}{2m}\nabla^2 + v_{\mathrm{H}}(\mathbf{r}) + v_{\mathrm{ext}}(\mathbf{r})\right]\varphi_i^{\mathrm{gKS}}(\mathbf{r}) + \int d^3r' v_{\mathrm{xc}}(\mathbf{r},\mathbf{r}')\varphi_i^{\mathrm{gKS}}(\mathbf{r}') = \varepsilon_i^{\mathrm{gKS}}\varphi_i^{\mathrm{gKS}}(\mathbf{r}) \tag{8.13}$$

This has particularly important consequences for the calculation of fundamental band gaps (see Section 8.3.1.1).

8.2.2.2 Calculations of Structural Properties

The ingredients described in Section 8.2.2.1 are sufficient for calculating total energies and, by virtue of the (generalized) Kohn–Sham equations, electronic structures. To obtain information about structural parameters such as equilibrium geometries, but also lattice dynamics, such as phonon modes, we additionally need a method for calculating forces acting on the atomic nuclei.

In our derivation of the Kohn–Sham equations in Section 8.2.1, we had assumed that the coordinates $\mathbf{R} = \{\mathbf{R}_I\}$ of the nuclei only appear as parameters in the Kohn–Sham equations (Born–Oppenheimer approximation). The Born–Oppenheimer potential energy surface, sometimes also called the clamped-ion energy, corresponds to the Kohn–Sham ground-state energy (Eq. (8.2)) plus the electrostatic interaction energy between different nuclei $E_N(\mathbf{R})$. The equilibrium geometry of a system is given by the condition that the forces on each nucleus must vanish, i.e.

$$\mathbf{F}_I = -\frac{\partial E(\mathbf{R})}{\partial \mathbf{R}_I} = 0. \tag{8.14}$$

Similarly, vibrational frequencies ω can be calculated by finding the eigenvalues of the Hessian of the total energy

$$\det\left|\frac{1}{\sqrt{M_I M_J}}\frac{\partial^2 E(\mathbf{R})}{\partial \mathbf{R}_I \partial \mathbf{R}_J} - \omega^2\right| = 0 \tag{8.15}$$

where M_I is the mass of nucleus I. These first and second derivatives can be calculated with the help of the Hellmann–Feynman theorem, which connects the derivative of the energy with respect to a parameter λ to the expectation value of the derivative of the Hamiltonian with respect to λ:

$$\frac{\partial E_\lambda}{\partial \lambda} = \left\langle \Psi_\lambda \left| \frac{\partial H_\lambda}{\partial \lambda} \right| \Psi_\lambda \right\rangle \tag{8.16}$$

where E_λ is solution of the eigenvalue equation $H_\lambda \Psi_\lambda = E_\lambda \Psi_\lambda$.

In the context of the Born–Oppenheimer approximation, the Hellmann–Feynman theorem can be used to connect the first and second derivatives of the total energy with respect to the nuclear coordinates to the variational properties of the Kohn–Sham Hamiltonian H^{KS}. The forces $\mathbf{F}_I$ acting on the nuclei can then be written as

$$\mathbf{F}_I = -\frac{\partial E(\mathbf{R})}{\partial \mathbf{R}_I} = -\int d^3r\, n(\mathbf{r})\frac{\partial v_{\mathrm{ext}}(\mathbf{r})}{\partial \mathbf{R}_I} - \frac{\partial E_N(\mathbf{R})}{\partial \mathbf{R}_I} \tag{8.17}$$

where $n(\mathbf{r})$ is the ground-state density corresponding to the nuclear configuration $\mathbf{R}$.

Similarly, the Hessian is calculated as

$$\frac{\partial^2 E(\mathbf{R})}{\partial \mathbf{R}_I \mathbf{R}_J} = \int \frac{\partial n(\mathbf{r})}{\partial \mathbf{R}_J} \frac{\partial v_{\text{ext}}(\mathbf{r})}{\partial \mathbf{R}_I} d^3r + \int n(\mathbf{r}) \frac{\partial^2 v_{\text{ext}}(\mathbf{r})}{\partial \mathbf{R}_I \mathbf{R}_J} d^3r + \frac{\partial^2 E_N(\mathbf{R})}{\partial \mathbf{R}_I \mathbf{R}_J} \tag{8.18}$$

The first term of Eq. (8.18) contains the linear response of the ground-state density to a distortion of the nuclear geometry, which can be calculated from density functional perturbation theory (DFPT). In DFPT, the Kohn–Sham equations are linearized with respect to a change in the Kohn–Sham orbitals, ground-state density and Kohn–Sham potential variation (for example due to lattice distortions). This leads to a set of self-consistent equations for the perturbed system, the solution of which allows for the calculation of phonon frequencies, phonon dispersion relations, and related properties. For a review of the method with a particular focus on phonons, the reader is referred to the excellent review article by Baroni et al. [25].

Alternatively, the frequencies of selected phonon modes can be calculated from the energy differences and/or forces produced by finite periodic displacements of atoms in a supercell approach (frozen phonon approach), or from molecular dynamics simulations, which will be discussed in more detail in Section 8.2.4.1.

8.2.3 Zero-Temperature Calculations for Halide Perovskites

When comparing calculated structural properties to experiment, it is important to setup calculations based on appropriate structural models. This is true for all materials classes, but particularly for halide perovskites with their often intricate interplay between the rotational dynamics of the molecular cation and distortions and tilts of the metal-halide sublattice. The former are not accounted for in zero-temperature DFT calculations, which can lead to problems in the interpretation of temperature-dependent structural and optoelectronic properties. Additionally, some properties of halide perovskites can be strongly affected by large anharmonic displacements, which are not included in DFPT or frozen phonon calculations (see Section 8.2.4.1). Therefore, the low-temperature phase should be chosen for validating structural properties extracted from zero-temperature calculations in comparison with experiment.

As an example, consider the low-temperature structure of $CH_3NH_3PbI_3$ ($MAPbI_3$) with *Pnma* space group symmetry (shown in Figure 8.3a) featuring corner-sharing PbI_6 octahedra with $a^-a^-b^+$ octahedral tilt pattern in Glazer notation. At low temperature, the methylammonium (MA) molecules show a preferential orientation, which is well-understood based on X-ray diffraction and neutron-scattering experiments [26–28]. For this structural model, a DFT-based geometry optimization can be used to understand, which xc functional describes structural parameters such as lattice parameters and unit cell volumes best. Table 8.1 shows lattice parameters of orthorhombic $MAPbI_3$ calculated with different DFT approximations in comparison with experiment. Trends agree qualitatively well with observations for other solids: the LDA tends to underestimate lattice parameters by ~2% and PBE overestimates them by around the same amount. The comparison also shows that details of the computational

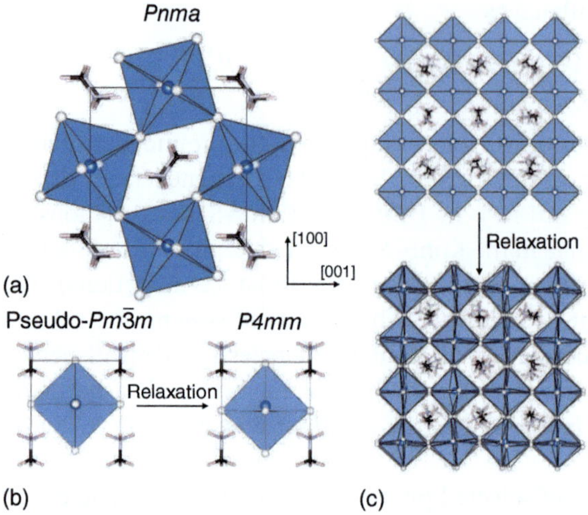

Figure 8.3 Representation of different structural models of $MAPbI_3$. Pb atoms are represented by blue spheres and are at the center of each octahedron, I atoms are shown in white, C atoms in black, N atoms in light blue, and H atoms in pink. (a) Low-temperature structure with *Pnma* symmetry, (b) primitive unit cell of cubic $MAPbI_3$ with pseudo-$Pm\bar{3}m$ symmetry, which relaxes into a non-centrosymmetric *P4mm* symmetry upon geometry optimization. (c) $3 \times 3 \times 3$ formula unit structural model with randomly oriented MA molecules before and after relaxation.

setup, such as constraints on the symmetry and the convergence parameters discussed in Section 8.2.2.1, play a decisive role in determining the results. Dispersion interactions have been reported to be crucial for an accurate description of structural properties of halide perovskites [29], and indeed the dispersion-corrected functional optB86b-vdW results in excellent agreement with experiment for orthorhombic $MAPbI_3$ [28]. Whether dispersion-corrected functionals describe the structural properties of the high-temperature phases of $MAPbI_3$ equally well is still a matter of debate at the time of writing [30, 31].

For the all-inorganic perovskite $CsPbI_3$, the low-temperature phase is a "yellow" non-perovskite structure. However, an orthorhombic structure with *Pbnm* symmetry can be stabilized; the lattice parameters collected in Table 8.1 refer to that structure. For this material, it is interesting to note that the LDA, PBE, and the meta-GGA SCAN all overestimate the a lattice parameter, while the b and c parameters follow the usual trends. Overall, the SCAN functional seems to describe structural parameters of $CsPbI_3$ better than standard (semi)local functionals, as expected based on its performance for other solids [32].

In principle, zero-temperature geometry optimizations can also be carried out for the high temperature phases of halide perovskites, such as the tetragonal *I4/mcm* and the cubic $Pm\bar{3}m$ phase of $MAPbI_3$. However, for these elevated-temperature phases, the choice of structural model is crucial and highly nontrivial for two reasons. First, the structural dynamics of halide perovskites can be highly anharmonic, i.e. the assumption that atoms oscillate around their equilibrium positions harmonically breaks down (see Section 8.2.4.1). The rotational dynamics of molecular cations

Table 8.1 Lattice parameters of orthorhombic (low-temperature) phases of $MAPbI_3$ and $CsPbI_3$.

Material	*a* (Å)	*b* (Å)	*c* (Å)
Pnma $MAPbI_3$			
LDA[a)]	8.68	8.32	12.39
PBE[a)]	9.23	8.62	12.88
optB86b-vdW[a)]	8.83	8.57	12.65
PBE[b)]	8.94	8.94	12.59
PBEsol[b)]	8.67	8.67	12.34
optB86b-vdW[c)]	8.85	8.56	12.69
Exp[c)]	8.81	8.56	12.59
Exp[d)]	8.84	8.56	12.58
Exp[e)]	8.86	8.58	12.62
Pbnm $CsPbI_3$[f)]			
LDA	8.94	7.95	12.22
PBE	9.10	8.73	12.64
SCAN	9.02	8.52	12.52
Exp	8.85	8.62	12.50
Exp	8.86	8.58	12.47
Exp	8.86	8.86	12.48

a) Wang et al. [33], no symmetry constraints, VASP with PAW pseudopotentials, 520 eV plane wave cutoff, $4 \times 4 \times 4$ **k**-points.
b) Ong et al. [34], $a = b$ fixed, VASP with PAW pseudopotentials, 500 eV plane wave cutoff, $3 \times 2 \times 3$ **k**-points.
c) Whitfield et al. [28], no symmetry constraints, VASP with PAW pseudopotentials, 900 eV plane wave cutoff, $7 \times 5 \times 7$ **k**-points.
d) Baikie et al. [27].
e) Poglitsch and Weber [26].
f) Kaczkowski and Płowaś-Korus [35] and references therein, no symmetry constraint, VASP with PAW pseudopotentials, 800 eV plane wave cutoff, $6 \times 6 \times 5$ **k**-points.

at elevated temperatures is anharmonic too, and strongly coupled to structural distortions of the metal-halide sublattice. Second, the cubic phase of perovskites can exhibit nonthermal, i.e. static distortions, which are local in nature and can therefore not readily be detected in X-ray diffraction measurements.[1] To describe the electronic structure and excited states of these on-average cubic phases, structural models featuring a distribution of low-symmetry octahedral tilts, rotations, and ionic displacements, dubbed *polymorphic networks*, may be more appropriate than a single cubic unit cell [38].

1 Local deviations from the on-average space group of a crystal can, for example, be determined by measuring the pair-distribution function [36, 37].

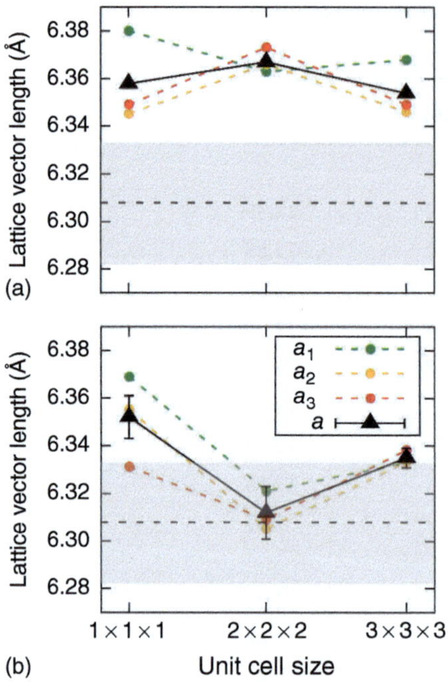

Figure 8.4 Pseudocubic lattice parameters a_1, a_2, and a_3 as functions of unit cell size. The shaded gray area indicates the range of experimental lattice parameters for the high-temperature phase of $MAPbI_3$. (a) With MA cations oriented along the (001) direction. (b) With random MA orientations (see text for details).

In Figure 8.3b,c, the problem of choosing a representative structural model for $MAPbI_3$ is illustrated for the cubic phase. This phase has on average $Pm\bar{3}m$ symmetry, a space group that can in principle be represented by a single formula unit cell as shown in Figure 8.3b. However, in such a unit cell, no matter, which orientation of the organic cation is chosen, an artificially ordered structure is generated when periodic boundary conditions are applied. A geometry optimization of the structure leads to a large off-centering motion of the Pb ion and overall *P4mm* symmetry, in disagreement with experiment [26, 39]. A more realistic structural model can be built by using a larger supercell and distributing the MA molecular orientations randomly as shown in Figure 8.3c.[2] Relaxing this structure also leads to characteristic octahedral tilts, but the average structure is closer to cubic symmetry than a single unit cell.

The same effect is also illustrated in Figure 8.4, which shows the pseudocubic lattice vectors a_1, a_2, and a_3 as a function of unit cell size for a structural model with molecular orientation parallel to the (001) direction (a) and one in which the MA orientations are generated randomly (b). The structural optimizations were carried out using the VASP code with PAW potentials, a plane wave energy cutoff of 440 eV, and a **k**-mesh of $8 \times 8 \times 8$ points for the single formula unit cell, and progressively fewer points for the larger unit cells. We used the PBE functional with Tkatchenko–Scheffler dispersion corrections [40]. In the structure with parallel orientation, the lattice parameters are clearly overestimated with respect to experiment. Both the lattice parameters and the magnitude of Pb off-centering in (001) direction

2 Note that a static model of the $Pm\bar{3}m$ phase of perovskites should ideally consist of an even number of unit cell repetitions to allow for single-mode octahedral rotations.

depend only little on the unit cell size, and are consistent with *P4mm* symmetry. For the systems with random molecular orientations, we generated three $(1 \times 1 \times 1)$, ten $(2 \times 2 \times 2)$, and six $(3 \times 3 \times 3)$ sets of structures. The data shown in the bottom panel of Figure 8.4 is the average over the lattice parameters of each of these sets of structures. The standard deviation is shown as an error bar of the average lattice parameter a. The results illustrate that the agreement with experiment improves for larger unit cell sizes and the difference between the three lattice parameters decreases. This is because with increasing unit cell size the importance of specific MA orientations decreases. In the limit of an infinitely large supercell, the *average* structure would have $Pm\bar{3}m$ symmetry, even though locally structural distortions and specific MA orientations can lead to sizeable deviations from the idealized cubic symmetry.

These observations are not only relevant for the determination of structural properties, but also for calculations of optoelectronic properties, which are highly sensitive to the choice of structural model (see Section 8.3.1.4). More realistic approaches for static models of the high-temperature structure of $MAPbI_3$ and other perovskites have been proposed [38, 41]. However, for a full understanding of temperature-dependent structural properties of halide perovskites, nuclear dynamics have to be taken into account explicitly in simulations.

8.2.4 Structural Dynamics

8.2.4.1 Molecular Dynamics: From Classical Force Fields to DFT Accuracy

Structural dynamics of materials can be modeled with molecular dynamics (MD) simulations. Here, I will only cover simulations in which the nuclei are considered as classical objects that move according to Newton's equations. Furthermore, the Born–Oppenheimer approximation is applied, which assumes that the dynamics of electrons and nuclei can be decoupled because of the significantly larger mass of the nuclei. Born–Oppenheimer MD simulations start with an initial set of positions and velocities of each atom. The latter can for example be chosen such that the mean kinetic energy corresponds to a chosen temperature T. In the next step, the forces $\mathbf{F}_I$ on each atom are calculated, which are then used to integrate Newton's equation of motion, e.g. by using a Verlet algorithm, in which the position of an atom $\mathbf{R}_I$ at a time $t + \Delta t$ is calculated as $\mathbf{R}_I(t + \Delta t) = 2\mathbf{R}_I(t) - \mathbf{R}_I(t - \Delta t) + \mathbf{F}_I(t)/M_I \Delta t^2$, where M_I is the mass of atom I and Δt is the time step of the molecular dynamics simulation. Such a simulation samples a microcanonical, i.e. NVE ensemble. Constant temperature, i.e. canonical or NVT ensembles, can be simulated by modifying the equation of motion with a thermostat – a functional expression that amounts to acceleration and deceleration of the nuclei so that the average temperature of the simulation remains constant. A number of different thermostats exist, for a more in-depth discussion of their specifics and implementation, see, for example Ref. [42].

The energies and forces between atoms can be calculated with DFT as outlined in Section 8.2.2.2 using the Hellmann–Feynman theorem. The accuracy of the simulation is then only restricted by the validity of the Born–Oppenheimer approximation and the numerical and physical approximations described in Section 8.2.2.1. In practice, DFT scales too steeply with the number of atoms to allow for sufficiently long simulations of systems with more than a couple dozen atoms for many properties of interest.

On the other end of the accuracy spectrum are molecular dynamics simulations based on classical force fields. Force fields rely on the (semi-)empirical parametrization of a functional expression for the potential energy surface. This expression usually contains bond lengths, bond angles, and other structural parameters, which are fitted to experimental or computational data sets. Classical force fields allow for molecular dynamics simulations that scale linearly with the number of atoms; they can be applied to systems with millions of atoms and simulation times can be very long. However, they are only reliable for simulations of systems similar to the data set used for the parametrization, and important processes such as bond breaking, polarization, and charge transfer cannot readily be described. Machine-learning schemes that generate classical force fields based on DFT (or other first principles) data, offer a way out of this catch.

Machine-learning applications have recently exploded in material science, physics, quantum chemistry, and related fields. They range from material discovery and crystal structure prediction to the construction of new DFT approximations and the entire replacement of the first principles approach as I described it in Section 8.1. I refer the reader to a recent review by Schmidt et al. for an overview of the state-of-the-art of these applications [43]. For the generation of machine-learned force fields, the potential energy is written as a sum of *local* atomic energies E_I, corresponding to each atom I in the system. Machine-learning methods then *learn* expressions for these local energies (and sometimes additionally forces and stress tensors) based on suitably chosen descriptors. These descriptors are functions that represent the local chemical environment surrounding each atom. A compact review of these methods can be found in Ref. [44]. Machine-learning force fields allow for the simulation of the dynamics of large and complex systems on long time scales with DFT accuracy, providing atomistic insights into hitherto inaccessible material properties, such as the simulation of phase transitions of $MAPbI_3$ [45] and other halide perovskites [46]. For example, machine-learned force fields were used to show that the first-order orthorhombic to tetragonal phase transition of $MAPbI_3$ is uniquely related to the *unfreezing* of the MA cation's rotational degrees of freedom, which are absent in fully inorganic perovskites with a similar progression of structural phase transitions [45].

8.2.4.2 Perovskites and the Breakdown of the Harmonic Approximation

The harmonic approximation assumes that the potential energy of a crystal lattice is proportional to $\Delta\mathbf{R}_I(t)^2$, where $\Delta\mathbf{R}_I(t)$ are the time-dependent displacements of the nuclei out of their equilibrium position $\mathbf{R}_I^0$. Because of the translational symmetry of the crystal lattice, these displacements are quantized. Their collective behavior, i.e. the quantized vibrations of the lattice are called phonons. In the harmonic approximation, phonons are independent vibrational modes, i.e. normal modes, of the lattice. For some halide perovskites, this approximation breaks down [47].

Signatures of strong anharmonic effects have been observed experimentally [48, 49] and in numerical simulations [46, 50–52] in both organic–inorganic and fully inorganic perovskites. In first principles simulations, anharmonic effects can be determined by extracting vibrational density of states from molecular dynamics simulations. The vibrational density of states of atom I is the Fourier transform of the velocity autocorrelation function

$$f_I(t) = \frac{\langle \mathbf{V}_I(t')\mathbf{V}_I(t'')\rangle}{\langle \mathbf{V}_I(t'')\mathbf{V}_I(t'')\rangle} \tag{8.19}$$

where $\mathbf{V}_I(t)$ is the velocity of atom I at time $t = t' - t''$, and $\langle\cdot\rangle$ is a thermal average that has to be taken over all unit cells of the crystal. The total vibrational density of states is then obtained by summing over the individual atomic contributions to the Fourier transform of $f_I(t)$, weighted with the mass of atom I:

$$F(\omega) = \sum_I \sqrt{M_I} f_I(\omega) \tag{8.20}$$

A reciprocal space representation, containing information about contributions from all accessible reciprocal space vectors $\mathbf{k}$, can be obtained by Fourier transforming the mass-weighted velocity field into reciprocal space. For a perfectly harmonic solid, the resulting $\mathbf{k}$-resolved vibrational density of states, should be identical to a phonon band structure as obtained from, for example, DFPT (see Section 8.2.2.2), implying that the lattice dynamics can be decomposed into a set of independent harmonic oscillators. It is this assumption that does not necessarily hold for halide perovskites, as shown by explicit calculation of vibrational density of states based on molecular dynamics, for example of $CsPbBr_3$ [46, 50], and the double perovskite $Cs_2AgBiBr_6$ [52]. The anharmonic nature of these crystals is particularly relevant for thermal transport properties, phase stability, and other thermodynamical properties. But it also has important consequences for electron–phonon and exciton–phonon interactions and the description of charge carrier transport. However, a full first-principles treatment of these effects in the anharmonic limit remains elusive.

8.2.4.3 A Primer on Ion Migration

Halide perovskites are mixed electronic-ionic conductors. Ion migration, in particular of halogen ions, is also relevant for material degradation, phase separation, and material growth processes. In principle, molecular dynamics simulations as described in Section 8.2.4.1 can be used to directly simulate the movement of ions through the crystal lattice. However, this requires very large simulation cells and long simulation times [53].

An alternative method for gaining atomistic insight into ion migration kinetics, is the calculation of the activation energy of relevant ion migration processes. The activation energy E_a is the sum of the defect formation energy E^f and the energy barrier for ion migration, defined as the energy difference between the highest energy transition state between equilibrium initial and final states (Figure 8.5c). Transition states correspond to saddle points in the potential energy landscape, and various ways exist to identify their structure and energy. For a review of methods, the reader is referred to Ref. [54]. Here, I want to mention the nudged elastic band (NEB) method for finding the minimum energy path between a known initial and final state of a transition, because this method has been used extensively for calculating the activation energies of ion migration in halide perovskites [55–58]. In the NEB method, $N - 1$ *images*, i.e. structure snapshots characterized by a set of atomic coordinates $\{\mathbf{R}_i\}$

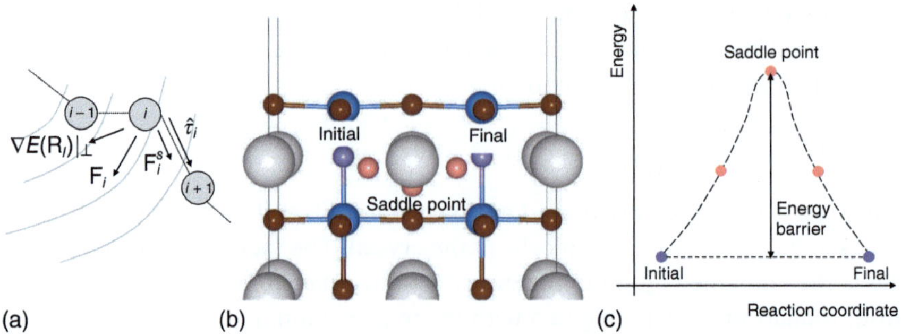

Figure 8.5 (a) Forces on image i in the NEB method. Source: Figure adapted from Ref. [54]. (b) Minimum energy path for Br migration at surface of cubic $CsPbBr_3$. The migrating Br ion is depicted in color, the initial and final position in purple and pictures along the minimum energy path in pink, whereas all other Br ions are brown. (c) Schematic of the total energy along the reaction coordinate. The energy barrier for ion migration is the energy difference between a saddle point of the potential energy surface, the transition state, and the energy of the initial/final state.

$(i = 0, 1, 2, \ldots, N)$, between the initial $(\{\mathbf{R}_0\})$ and final state $(\{\mathbf{R}_N\})$ are constructed. These images are connected by spring forces of the form

$$\mathbf{F}_i^{\mathrm{s}} = k(|\mathbf{R}_{i+1} - \mathbf{R}_i| - |\mathbf{R}_i - \mathbf{R}_{i-1}|)\hat{\tau}_i \tag{8.21}$$

parallel to the local tangent $\hat{\tau}_i$, and characterized by a spring constant k, a parameter of the calculation. These spring forces are what turns the collection of structure snapshots into an elastic band. The total force acting on each image is the sum of the spring force along the local tangent and the true force perpendicular to the local tangent

$$\mathbf{F}_i = \mathbf{F}_i^{\mathrm{s}} - \nabla E(\mathbf{R}_i)|_{\perp} \tag{8.22}$$

as illustrated in Figure 8.5a. Minimization of $\mathbf{F}_i$ in principle leads the elastic band onto the minimum energy path. If the path contains a sufficient number of images, the saddle point between the initial and final state can be determined with arbitrary accuracy. In practice, one can use a so-called "climbing image" modification of the NEB method to guarantee that the saddle point is found with a limited number of images. For that purpose, the spring forces acting on the image with the highest energy are turned off, so that this image feels the full force due to the potential with the component along the elastic band inverted, and is guaranteed to "climb up" to the saddle point [59]. An exemplary setup used for an NEB calculation of Br migration at a surface of $CsPbBr_3$ is shown in Figure 8.5b,c [58].

The formation energy of a defect X in charge state q is defined as

$$E^{\mathrm{f}}(X^q) = E(X^q) - E(\text{pristine}) - \sum_i n_i\mu_i + qE_{\mathrm{F}} + E_{\mathrm{corr}} \tag{8.23}$$

where $E(X^q)$ is the total energy of the system with defect, E(pristine) is the total energy of the pristine system without defect, n_i is the number of atoms of type i that have been added to $(n_i > 0)$ or removed from $(n_i < 0)$ the system to form the defect, μ_i are the chemical potentials of these species, E_{F} is the Fermi energy, and

E_{corr} are correction terms that account for the finite size of the supercell and finite **k**-point sampling. Reference [60] by Freysoldt et al. provides an excellent review on first principles calculations for point defects in solids.

In principle, both the energy barrier for ion migration and the defect formation energy can be calculated with DFT. For halide perovskites, particular care must be taken in three different areas: 1. Choice of the xc functional, 2. choice of structural model, and 3. anharmonic effects.

1. **Choice of xc functional**: The choice of functional is particularly relevant for the calculation of defect formation energies. Energy barriers are expected to be less sensitive because they rely on total energy differences between equivalent supercells. Therefore, xc functionals that yield reliable total energies for halide perovskites are sufficient, and are not expected to lead to results significantly different from computationally more demanding approximations. This was explicitly shown for $MAPbI_3$ by Meggiolare et al., who compared energy barrier calculations with PBE and the screened hybrid functional HSE06 for $MAPbI_3$ and found energy differences of only ~0.1 eV. The same is not true for the calculation of defect transition levels, because they are calculated with respect to the band edges and therefore strongly depend on the quality of the chosen xc approximation (see Section 8.3.1.4).
2. The choice of structural model is particularly relevant for any calculations of high-temperature structures, even more so in the case of halide perovskites with molecular A-site cations. As described in Section 8.2.2.2, the intricate coupling between the organic and the inorganic sublattices, and the rotational–vibrational degrees of freedom of the A-site cation can lead to artificial distortions, introducing systematic errors in computed ion migration and defect formation energies.
3. Anharmonic effects are expected to lower energy barriers for ion migration, but the magnitude of this effect in various halide perovskites has been unexplored to date. In the case of defect formation energies, large-scale molecular dynamics simulations based on DFT have shown for the case of $CsPbBr_3$ that at room temperature, Br vacancy defect levels can oscillate by as much as 1 eV on the picosecond time scale [61]. In how far these dynamical defect properties affect carrier transport, and other optoelectronic properties of halide perovskite, has yet to be determined.

8.3 Optoelectronic Properties

In the following discussion of optoelectronic properties of halide perovskites, I will refer to two types of electronic excitations: Charged excitations, in which an N electron system becomes an $N + 1$ or $N - 1$ system as a consequence of an excitation through a photon. For example, the band structure and band gap of a solid are properties that are related to charged excitations, and can be measured in (inverse) photoemission experiments. Section 8.3.1 is intended to help the reader to interpret the results of band structure calculations based on DFT and

Green's function-based many-body perturbation theory approaches. Furthermore, I will attempt an overview of the most important qualitative and quantitative properties of halide perovskite band structures in Section 8.3.1.4. The second type of excitation is that in which the electron number remains unchanged (neutral excitations), as measured in optical absorption. Neutral excitations can be calculated using the Bethe–Salpeter Equation (BSE) approach in the framework of Green's function-based many-body perturbation theory (Section 8.3.2.1). Results for halide perovskites with a focus on the description of bound excitons will be discussed in Section 8.3.2.2.

8.3.1 Electronic Band Structures

8.3.1.1 What Can DFT Tell Us About Band Gaps of Solids?

The band gap E_{gap} of a solid is an excited state property because its definition is linked to the removal/addition of an electron from the neutral (N-electron) solid (E_{gap} is the unbound limit of a series of bound excitations – excitons). However, E_{gap} is defined as the difference between the ionization energy $I(N)$ and the electron affinity $A(N)$, and can therefore be calculated from ground-state energy differences

$$E_{\text{gap}}(N) = I(N) - A(N) = [E(N-1) - E(N)] - [E(N) - E(N+1)] \tag{8.24}$$

where $N-1$ and $N+1$ refer to the system with one electron removed and added, respectively (see for example Ref. [7] for an in-depth discussion). Equation (8.24) implies that E_{gap} can be determined from three ground-state DFT calculations of the total energies of the $N-1$, N, and $N+1$-electron systems.[3] In practice, one sometimes sees the energy difference between the energy of the highest occupied (called ε_{VBM} in the following) and the lowest unoccupied (ε_{CBM}) electronic state, equated with E_{gap}. When ε_{CBM} and ε_{VBM} stem from a Kohn–Sham DFT calculation, this equality is fundamentally untrue. The difference between E_{gap} and the Kohn–Sham band gap $E_{\text{gap}}^{\text{KS}} = \varepsilon_{\text{CBM}} - \varepsilon_{\text{VBM}}$ has been identified as the constant by which the Kohn–Sham potential jumps when an electron is added to the N-electron system – the derivative discontinuity Δ_{xc} [62–64]. It is important to stress that the failure of Kohn–Sham DFT to predict fundamental band gaps, is not a problem of individual xc approximations (e.g. because semilocal functionals do not possess a derivative discontinuity). Even the exact functional and approximations that contain a derivative discontinuity would not result in $E_{\text{gap}} = E_{\text{gap}}^{\text{KS}}$, because $E_{\text{gap}}^{\text{KS}}$ does not include Δ_{xc} [65].

An equivalent way of looking at this is through the lens of the ionization potential theorem of DFT, which states that $\varepsilon_{\text{VBM}} = -I(N)$ when calculated with the exact Kohn–Sham potential [66]. This relation is reminiscent of Koopman's theorem of Hartree–Fock theory. It is, however, a much stronger statement, since it refers to the *relaxed* vertical ionization energy, whereas Koopman's theorem only holds under the assumption that the orbitals of the N and the $N-1$ electron system are identical (*unrelaxed* ionization energy). In DFT, ε_{CBM} is not related to the electron affinity

3 This method is sometimes referred to as the Δ-SCF (self-consistent field) method.

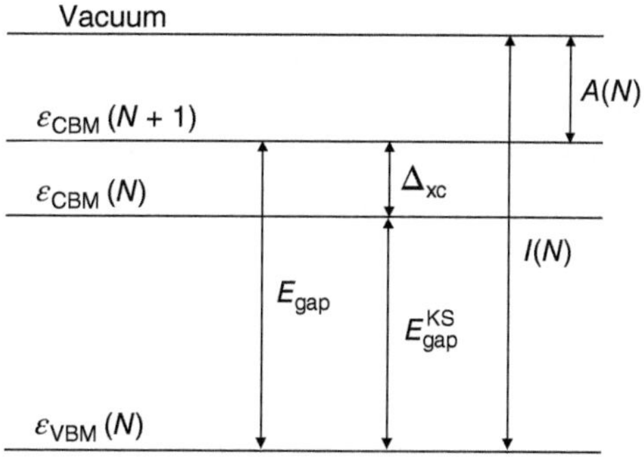

Figure 8.6 Schematic picture of energy eigenvalues from Kohn–Sham DFT and their relation to the N electron ionization energy, electron affinity, and the fundamental band gap. Source: Figure adapted from Ref. [67].

of the N electron system in the same way, precisely because of the existence of the derivative discontinuity in the Kohn–Sham potential. These relations are schematically presented in Figure 8.6.

Hybrid functionals and meta-GGAs can lead to band gaps in better agreement with experiment when they are implemented in a generalized Kohn–Sham framework [68]. As shown in Eq. (8.13), in generalized Kohn–Sham theory the xc potential is an integral operator instead of a multiplicative potential. In other words, while in Kohn–Sham theory all electrons *see* the same *mean-field* potential, this is not the case in the generalized Kohn–Sham approach. As a consequence, the derivative discontinuity is (partially) incorporated in the generalized Kohn–Sham band gap. Generalized Kohn–Sham band gaps, calculated from eigenvalue energy differences, can in principle agree with exact fundamental band gaps (see for example Ref. [69] and references therein). The degree to which they do will depend on the chosen xc approximation and on the material in question (see Section 8.3.1.4).

8.3.1.2 A Short Introduction to the GW Approach

The GW approach is an approximation to an exact set of coupled equations that were derived by Lars Hedin in 1965 [70], and is considered to be the state-of-the-art for calculating band structures and band gaps of solids. Hedin's equations can be derived through a perturbation expansion of the single particle Green's function G.

$$G(\mathbf{r}, t, \mathbf{r}', t') = -i\langle \Psi_0^N | \hat{T}\{\hat{\psi}(\mathbf{r}, t)\hat{\psi}^\dagger(\mathbf{r}', t')\} | \Psi_0^N \rangle \tag{8.25}$$

which corresponds to the probability amplitude that a particle created at space-time point $\mathbf{r}, t$ is destroyed at space-time point $\mathbf{r}', t'$. Particle creation and destruction are mathematically described by a product of time-ordered field operators $\hat{\psi}(\mathbf{r}, t)$ acting on the N-particle ground state wavefunction Ψ_0^N, where $\hat{T}$ is the time-ordering operator. The poles of this Green's function correspond to electron addition ($\epsilon_j = E(N+1, j) - E(N)$) and removal ($\varepsilon_j = E(N) - E(N-1, j)$) energies to and from a bound initial state j. This property of the single particle Green's function can be seen in the Lehmann representation of G:

$$G(\mathbf{r}, \mathbf{r}', \omega) = \sum_j \frac{g_j(\mathbf{r}) g_j^*(\mathbf{r}')}{\omega - \varepsilon_j + i\eta \cdot \mathrm{sgn}(\varepsilon_j - \varepsilon_\mathrm{F})} \tag{8.26}$$

Here, the $g_j(\mathbf{r})$ are transition amplitudes between N and $N \pm 1$-particle states with $g_j(\mathbf{r}) = \langle \Psi_j^{N-1} | \hat{\psi}(\mathbf{r}) | \Psi_0^N \rangle$ for $\varepsilon_j < \varepsilon_F$ and $\langle \Psi_0^N | \hat{\psi}(\mathbf{r}) | \Psi_j^{N+1} \rangle$ for $\varepsilon_j \geq \varepsilon_F$. Hedin's equations offer a route for determining this object in practice. They connect the single-particle Green's function with the reducible electronic self-energy Σ, the screened Coulomb interaction W, the irreducible polarizability χ and the so-called vertex function Γ, a three-particle object. The GW approximation neglects the vertex in the expression for the self energy leading to the following set of coupled integro-differential equations:

$$G(1,2) = G_0(1,2) + \int d3d4 G_0(1,3)\Sigma(3,4)G(4,2) \tag{8.27}$$

$$\chi_0 = -iG(1,2)G(2,1) \tag{8.28}$$

$$W(1,2) = v(1,2) + \int d3d4 v(1,3)\chi_0(3,4)W(4,2) \tag{8.29}$$

$$\Sigma(1,2) = iG(1,2)W(1^+,2) \tag{8.30}$$

In the notation $1 = (r_1, t_1)$, a single number represents a space-time point, and the + superscript 1^+ corresponds to $t_1 + \eta$, an infinitesimally small increase in the time coordinate t_1. G_0 is the zeroth-order Green's function that can be constructed using the eigensystem obtained through self-consistent solution of the Kohn–Sham (Eqs. (8.3)–(8.5)), generalized Kohn–Sham (Eq. (8.13)), or other single-particle equations:

$$G_0(\mathbf{r},\mathbf{r}',\omega) = \sum_j \frac{\varphi_j(\mathbf{r})\varphi_j^*(\mathbf{r}')}{\omega - \varepsilon_j + i\eta \cdot \mathrm{sgn}(\varepsilon_j - \varepsilon_F)} \tag{8.31}$$

Based on an initial guess for G_0, Eqs. (8.27)–(8.30) can in principle be solved self-consistently. However, this is prohibitively expensive for most materials of practical interest.

When electron addition or removal processes can be described by a single-particle like excitation, they are called quasiparticle (QP) excitations, and can be associated with a single-particle state ϕ_j^{QP} and an energy $\varepsilon_j^{\mathrm{QP}}$. Within this picture, the interacting single-particle Green's function becomes

$$G(\mathbf{r},\mathbf{r}',\omega) = \sum_j \frac{\varphi_j^{\mathrm{QP}}(\mathbf{r})\varphi_j^{*\mathrm{QP}}(\mathbf{r}')}{\omega - \varepsilon_j^{\mathrm{QP}} + i\eta \cdot \mathrm{sgn}(\varepsilon_j^{\mathrm{QP}} - \varepsilon_F)} \tag{8.32}$$

Together with the Dyson equation (8.27), Eqs. (8.31) and (8.32) can be used to derive effective single-particle equations with a structure resembling the Hartree–Fock and generalized Kohn–Sham equations:

$$\left[-\frac{\hbar^2}{2m}\nabla^2 + v_{\mathrm{H}}(\mathbf{r}) + v_{\mathrm{ext}}(\mathbf{r})\right]\varphi_j^{\mathrm{QP}}(\mathbf{r}) + \int d^3r'\Sigma(\mathbf{r},\mathbf{r}',\varepsilon_j^{\mathrm{QP}})\varphi_j^{\mathrm{QP}}(\mathbf{r}') = \varepsilon_j^{\mathrm{QP}}\varphi_j^{\mathrm{QP}}(\mathbf{r}) \tag{8.33}$$

Equation (8.33) is a useful starting point for computationally efficient approximations to fully self-consistent GW. In lowest-order perturbation theory, dubbed G_0W_0, one approximates the QP states with (generalized) Kohn–Sham eigenfunctions and obtains from (8.33) and (8.13) a set of equations

$$\varepsilon_j^{\mathrm{QP}} = \varepsilon_j^{\mathrm{gKS}} + \langle \varphi_j^{\mathrm{gKS}} | \Sigma(\varepsilon_j^{\mathrm{QP}}) - v_{\mathrm{xc}} | \varphi_j^{\mathrm{gKS}} \rangle \tag{8.34}$$

The self-energy Σ is constructed from G_0 (Eq. (8.31)) and

$$W_0(\mathbf{r}, \mathbf{r}', \omega) = \int d^3r'' \epsilon^{-1}(\mathbf{r}, \mathbf{r}'', \omega) v(\mathbf{r}'', \mathbf{r}') \tag{8.35}$$

in the random phase approximation (RPA). Here $v(\mathbf{r}, \mathbf{r}')$ is the bare Coulomb interaction, and the dielectric function ϵ

$$\epsilon(\mathbf{r}, \mathbf{r}', \omega) = \delta(\mathbf{r}, \mathbf{r}') - \int d^3r'' v(\mathbf{r}, \mathbf{r}'') \chi_0(\mathbf{r}'', \mathbf{r}', \omega) \tag{8.36}$$

is constructed from (generalized) Kohn–Sham eigenfunctions and eigenvalues as well.

For more information regarding the GW approach, I refer the reader to an excellent review article by Golze et al. [71]. At this point, I only note that various forms of (partial) self-consistency have been introduced that allow for a treatment of QP effects beyond the G_0W_0 approximation. The simplest and computationally least expensive method consists of using the eigenvalues obtained by solving Eq. (8.34) to recalculate G and W until convergence is reached (eigenvalue self-consistency). Quasiparticle self-consistent GW (QPGW) introduced by Kotani et al. additionally includes an update in the QP states, by determining the variationally optimal generalized Kohn–Sham potential for a given Σ [72]. Fully self-consistent GW calculations are computationally expensive and relatively rare.

8.3.1.3 The Band Structure of Halide Perovskites: A Tight-Binding Perspective

The band structure of a perfectly cubic ($Pm\bar{3}m$) ABX_3 perovskites with B = Pb, Sn is characterized by a direct band gap at the R point of the Brillouin zone (corresponding to reciprocal space coordinates of $(\frac{2\pi}{a}, \frac{2\pi}{a}, \frac{2\pi}{a})$, where a is the lattice parameter) and highly dispersive conduction and valence bands. In the literature on these materials, one often reads that their conduction band is derived from metal-p orbitals and the valence band from metal-s and halogen-p orbitals. Remembering Section 8.2.2.1, in which the Kohn–Sham equations were derived in a basis of plane waves, one might wonder how to reconcile a model of the electronic structure in terms of completely delocalized electrons with that of orbitals localized at atomic sites. In practical terms, information about the orbital *character* of a band can be extracted from first principles calculations, by projecting the (generalized) Kohn–Sham eigenfunctions in the plane wave basis onto spherical harmonics in a specified radius centered around each ion in the lattice. Naturally, the *partial charges* and *projected density of states* that can be calculated with this approach depend on the chosen ionic radii, and therefore have to be interpreted with care.

A second first principles perspective on expressing the electronic structure of periodic solids in terms of localized wavefunctions is the Wannier representation, in which the Bloch wavefunctions are *localized* through a set of unitary matrix transformations. In practice this can be achieved by a localization procedure introduced by Marzari and Vanderbilt that minimizes the total spread of the Wannier functions in real space and yields so-called *maximally localized Wannier functions* (MLWF) [73]. MLWFs have been used extensively, for example for understanding chemical bonds and the electronic structure of solids. They are also conceptually related to the modern theory of polarization [74], and can be used for constructing model Hamiltonians for phenomena ranging from electron dynamics to magnetic interactions in solids [75].

A related semi-empirical approach is the tight-binding method, in which instead of deriving localized orbitals directly from first principles results, a model Hamiltonian is constructed in terms of atomic orbitals at different lattice sites, and the interactions between them. A tight-binding model for cubic ABX_3 perovskites was first proposed by Boyer-Richard et al. [76]. The model takes into account 16 basis functions if spin–orbit coupling (SOC) is disregarded. With SOC, the basis size has to be doubled, and the tight-binding Hamiltonian contains an additional term proportional to $\hat{\mathbf{L}} \cdot \hat{\mathbf{S}}$, where $\hat{\mathbf{L}}$ is the orbital angular momentum, and $\hat{\mathbf{S}}$ is the spin angular momentum operator. The basis functions are the s and p orbitals at the metal and halogen sites, respectively. Electronic contributions from the A site are not taken into account because in all ABX_3 halide perovskites synthesized to date, the electronic states associated with the molecular cations can be found deep in the valence and conduction band, respectively, and therefore do not contribute directly to the electronic band structure close to the band gap or optical absorption in the visible region.

In the Slater–Koster approach to tight-binding [77], the matrix elements of the tight-binding Hamiltonian are parameters that can be fitted to results from first principles calculations such as DFT. The diagonal elements are on-site interaction terms, whereas the off-diagonal elements contain the hopping interaction parameter and a geometrical, **k**-dependent function that depends on the symmetry of the crystal structure. Note that the number of parameters needed in the model depends on the atomic orbitals in the basis and their relative positions with respect to each other. This is illustrated schematically in Figure 8.7a, which shows possible interactions between s and p orbitals on B and X lattice sites in a cubic lattice, resulting in four different interaction parameters V_{ss}, V_{sp}, $V_{pp\sigma}$, and $V_{pp\pi}$. Figure 8.7b shows the resulting tight-binding band structure of cubic $MAPbI_3$ with parameters adapted from Ref. [76] with and without SOC.

Similar approaches can be used to describe the electronic structure of more complex halide perovskites, such as the quaternary halide double perovskites $A_2BB'X_6$. Tight binding models of these materials are more elaborate than for the single cubic perovskites, because the primitive unit cell of the double perovskite lattice is rhombohedral, the atomic orbital basis is at least twice as large, and many relevant materials feature d-orbital derived valence bands, leading to additional parameters. Nonetheless, a qualitative model capable of predicting the position of valence and conduction band edges in reciprocal space was put forward by Slavney et al. [78]. The model shows that by deliberate combination of B and B′

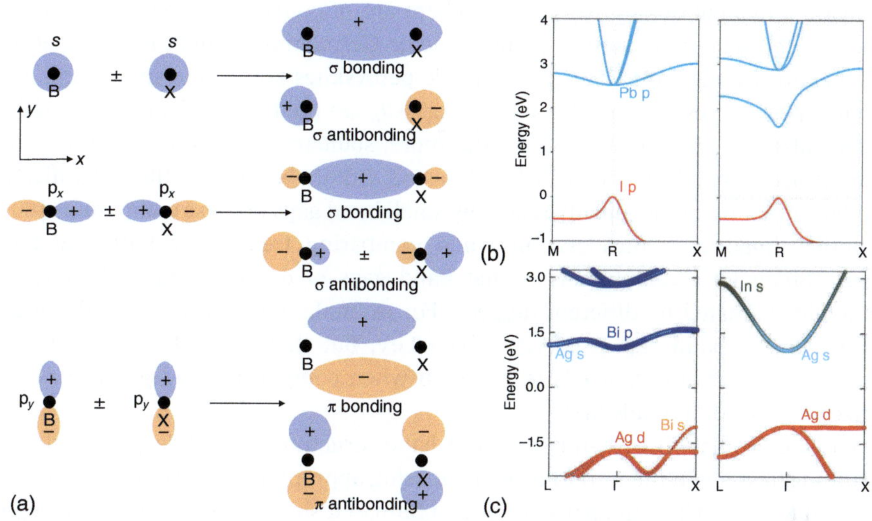

Figure 8.7 (a) Schematic of σ and π bonding and antibonding nearest-neighbor interactions arising from interactions of B and X site atomic s and p orbitals in a cubic lattice. In this model, interactions between p_z orbitals are equivalent to those of p_y orbitals, s–p interactions are not drawn, and all other interaction terms vanish. (b) Tight binding band structure of cubic $MAPbI_3$ without (left) and with (right) SOC using a model and parameters. Source: Adapted from [76]. (c) G_0W_0@PBE band structures of double perovskites $Cs_2AgBiCl_6$ and $Cs_2AgInCl_6$. The *orbital character* of the bands, calculated by projection onto spherical harmonics, is shown in color.

site metals, the double perovskite stoichiometry allows for the realization of a vast variety of electronic structures, in which the positions and dispersion of the band edges are primarily determined by the frontier orbitals of the B and B′ site ions. This is exemplified in Figure 8.7c, which shows *GW* band structures and the orbital character of the band edges for the two double perovskites $Cs_2AgBiCl_6$ and $Cs_2AgInCl_6$. Although these two materials only differ at the B′ site, they have qualitatively very different band structures featuring the conduction band minimum and valence band maximum at different high-symmetry points, and very different band dispersion.

8.3.1.4 Toward Predictive Band Structure Calculations for Halide Perovskites

While tight-binding models are useful for obtaining a qualitative understanding of the electronic structure of materials, predicting optoelectronic properties of materials with quantitative accuracy requires the use of first-principles methods.[4] The limits of standard DFT approximations for the calculation of band gaps were highlighted in Section 8.3.1.1. Nonetheless, it is good computational practice to start investigations of the electronic structure of new materials using standard xc functionals both for reasons of computational efficiency and because they usually

---

4 Data from first principles calculations can of course be used to fed tight-binding, or other semiempirical methods, or train machine-learning models.

provide a reliable qualitative picture. There are of course prominent exceptions, such as the seemingly simple semiconductor Germanium, which is predicted to be metallic by standard DFT. In the halide perovskites family, the small band gap double perovskites $Cs_2AgTlBr_6$ and $Cs_2AgTlCl_6$ are predicted to be (semi)metallic by standard xc functionals [79]. Furthermore, standard approximations favor the emergence of band inversion in $MAPbI_3$ and $CsPbI_3$ because of their significant underestimation of the already relatively small band gaps of these compounds [80]. Another important consideration when comparing trends based on standard Kohn–Sham DFT calculations is that band gaps of different materials tend to be underestimated by different degrees. For example, the PBE functional underestimates the band gap of $MAPbI_3$ by ~1 eV, but that of $MAPbBr_3$ by almost 2 eV, meaning that care needs to be taken when comparing band gap trends for mixed-halide compounds [81].

For the prediction of band gaps and band structures of halide perovskites, I want to stress again the importance of choosing appropriate structural models (see Section 8.2.2.2). The band gaps of Pb- and Sn-based single perovskites are strongly blueshifted by B-site off-center distortions and octahedral tilts and rotations. For example, the band gap of $MAPbI_3$ as calculated within the computational setup described in Section 8.2.2.2, i.e. using the PBE xc functional and the PAW formalism as implemented in Vienna Ab-initio simulation package (VASP), varies by several 100 meV depending on the chosen orientation of the molecular cations within the Pb-I sublattice and the size of the supercell. In this context it is important to realize that photoabsorption or -emission experiments from which band gaps can be predicted *see* an average structure. Electron–phonon interactions lead to a renormalization of the band gap, but care has to be taken in how their effect is incorporated in electronic structure calculations [82, 83].

For many halide perovskites, the choice of structural model not only affects the magnitude of the band gap, but also other qualitative features of the band structure. A prominent example is the emergence of a strong Rashba/Dresselhaus splitting in the band structure of metal-halide perovskites with strong spin–orbit interactions and a non-centrosymmetric structure. This splitting of the energy bands in reciprocal space arises as a consequence of polar distortions such as B-site off-center distortions in combination with sizeable spin–orbit interactions [84–87]. While there are ongoing discussions in the perovskite community regarding the magnitude and significance of the Rashba effect in $MAPbI_3$ [39, 88, 89], it is clear that in calculations its relative size in conduction and valence band, and the spin texture of the energy bands depend sensitively on the nature and direction of distortions present in the chosen structural model [90, 91].

The predictions of standard (semi)local xc approximations are significantly improved by hybrid functionals when implemented in a generalized Kohn–Sham framework (see Section 8.3.1.1). Global and RSH functionals can be parametrized based on first-principles considerations, and reproduce experimental band gaps well, as shown for a wide range of solids [92, 93] including some all-inorganic halide perovskites [94].

The GW approach (Section 8.3.1.2) is considered to be the state-of-the-art for band structure and band gap predictions, and as starting point for the calculation of optical properties within the BSE approach (Section 8.3.2.1). The first G_0W_0 calculations for halide perovskites were reported in 2014 by Umari et al., who used an implementation in which SOC was only included in the construction of the zeroth-order Green's function G_0, and not in the screened Coulomb interaction W_0. The reported band gaps for the tetragonal *I4/mcm* phases of $MAPbI_3$ and $MASnI_3$ were 1.67 and 1.10 eV, respectively, in almost perfect agreement with experimental results [95]. Including SOC also in the construction of W_0 reduces the G_0W_0 band gap as shown for $MAPbI_3$ by Brivio et al., albeit with a different structural model [96]. In the same study, it was also shown that iterating *G* and *W* in the form of a QP self-consistent approach [72] resolves the band gap underestimation of G_0W_0 and leads to excellent agreement with experiment. The importance of (partial) self-consistency in *GW* has been confirmed by further calculations of band gaps and band structures of $MAPbI_3$ and $CsPbI_3$ [97–99].

The G_0W_0 approach is well-known to suffer from a strong dependence on the (generalized) Kohn–Sham eigensystem used to construct G_0 and W_0. This was shown for a range of single and double halide perovskites with band gaps ranging between 1 and ~3.5 eV in Ref. [81]. Particular care has to be taken for halide perovskites with small band gaps, for example due to strong SOC or a large band dispersion. For these materials, for example $MAPbI_3$, $MAPbBr_3$, $CsSnBr_3$, and $Cs_2AgTlBr_6$, standard xc approximation lead to *overscreening* [100], and either hybrid functional starting points or *GW* self-consistency may be necessary [81].

These examples show, that while accurate band structures and band gaps can be calculated within the *GW* and the (generalized) Kohn–Sham method, there is (at the time of writing) no computationally inexpensive, predictive *one size fits all* approach for band gaps of halide perovskites. Both *GW* self-consistency and tuning schemes for global and RSH functionals are computationally demanding (although they scale differently with system size) and remain impractical for high-throughput applications. While electronic structure methods like DFT are powerful tools for material design and predictions [79, 101–103], validation through experimental results remains indispensable.

8.3.2 Optical Properties

In this section, I will briefly introduce the BSE approach, and its applications to the excited-state structure of halide perovskites. A detailed introduction to the method can be found in Ref. [104]. Note, that time-dependent DFT is well-suited for calculating optical properties of solids too, when based on appropriate xc kernels [105, 106], but has only rarely been employed for the calculation of optical properties of halide perovskites at the time of writing [107].

8.3.2.1 A Short Introduction to the Bethe–Salpeter Equation Approach

The BSE is a perturbative approach for calculating the two-particle correlation function $iL(1,2;1',2') = -G(1,2;1',2') + G(1,2)G(1',2')$, where $G(1,2;1',2')$ is the two-particle Green's function [108, 109]. It describes the propagation of an excited electron–hole pair or exciton:

$$L(1,2;1',2') = L_0(1,2;1',2') + \int d3d4d5d6L_0(1,4;1',3)K(3,5;4,6)L(6,2;5,2') \tag{8.37}$$

where $L_0(1,4;1',3) = G(1,3)G(4,1')$ is the noninteracting correlation function and the kernel is defined as

$$K(3,5;4,6) = v(3,6)\delta(3,4)\delta(5,6) - W(3^+,4)\delta(3,6)\delta(4,5) \tag{8.38}$$

in the GW approximation. From this, one can derive a matrix eigenvalue equation, where it is commonly assumed that the screened Coulomb interaction is frequency-independent, i.e. that $W(1,2) \approx W(\mathbf{r},\mathbf{r}',t)\delta(t-t')$ is local in time. For calculations of excitons in solids, one typically makes the Tamm–Dancoff approximation. The equation to solve for each exciton state S then becomes

$$(\varepsilon_a^{\text{QP}} - \varepsilon_i^{\text{QP}})X_{ia}^S + \sum_{jb}\langle ia|K|jb\rangle = \Omega^S X_{ia}^S \tag{8.39}$$

where Ω^S is the excitation energy and X_{ia}^S the corresponding exciton wavefunction, and the electron–hole interaction kernel expressed in the product state basis of occupied and unoccupied orbitals is

$$\langle ia|K|jb\rangle = v_{ijab} - W_{iajb} \tag{8.40}$$

The eigenvalue difference in Eq. (8.39) is typically based on QP energies calculated within the GW approach. From the solution of Eq. (8.39), the imaginary part of the macroscopic transverse dielectric function can be calculated as

$$\epsilon_2(\omega) = \frac{16\pi e^2}{\omega^2}\sum_S |\mathbf{e}\cdot\langle 0|\mathbf{v}|S\rangle|^2\delta(\omega-\Omega_S) \tag{8.41}$$

where $\mathbf{e}$ is the polarization vector of light and $\mathbf{v}$ is the velocity operator. Exciton binding energies can be calculated by comparing Eq. (8.41) with

$$\epsilon_2^{\text{IP}}(\omega) = \frac{16\pi e^2}{\omega^2}\sum_{ia} |\mathbf{e}\cdot\langle i|\mathbf{v}|a\rangle|^2\delta(\omega-(\varepsilon_a^{\text{QP}}-\epsilon_i^{\text{QP}})) \tag{8.42}$$

in the independent particle (IP) approximation, in which excitations correspond to vertical transitions between independent electron and hole QP states. In crystals, local field effects, i.e. off-diagonal matrix elements of the dielectric matrix $\epsilon_{\mathbf{G},\mathbf{G}'}(\omega)$, can suppress the dielectric function. They can be calculated by determining $\epsilon_2^{\text{IP}}(\omega)$ from the head of the inverse dielectric matrix $\epsilon^{-1}_{\mathbf{G},\mathbf{G}'}(\omega)$, where $\mathbf{G}$ are reciprocal lattice vectors.

8.3.2.2 Neutral Excitations in Halide Perovskites

The calculation of neutral excitations of halide perovskites within the GW + BSE approach is challenging because of the problems of structural complexity and band gap underestimation described in the previous sections. Furthermore, $MAPbI_3$ and similar perovskites feature highly dispersive valence and conduction bands.

Therefore, excitons in these materials are spatially strongly delocalized and need to be calculated on very dense **k**-point grids, for example using model dielectric functions or patched-sampling methods [110–113].

The results of these calculations on $MAPbI_3$ and other single halide perovskites show three things: First, excitons are well-described by a hydrogenic Wannier–Mott model. This means that exciton binding energies follow $E_X = \mu/2\varepsilon_\infty^2 \cdot 1/n^2$ (in atomic units), where μ is the effective mass of the electron–hole pair, calculated as the second derivative of the GW energy of conduction and valence band with respect to the wave vector, ε_∞ is the optical dielectric constant, and n is the principal quantum number. Moreover, the exciton wavefunctions X_{ia}^S from Eq. (8.39) are also accurately described by the hydrogenic model. This is illustrated for several inorganic single perovskites in Figure 8.8a [112]. Second, the exciton binding energies overestimate experiment by up to a factor of 3. This overestimation has been attributed to polaronic [110] and phonon screening effects [111]. However,

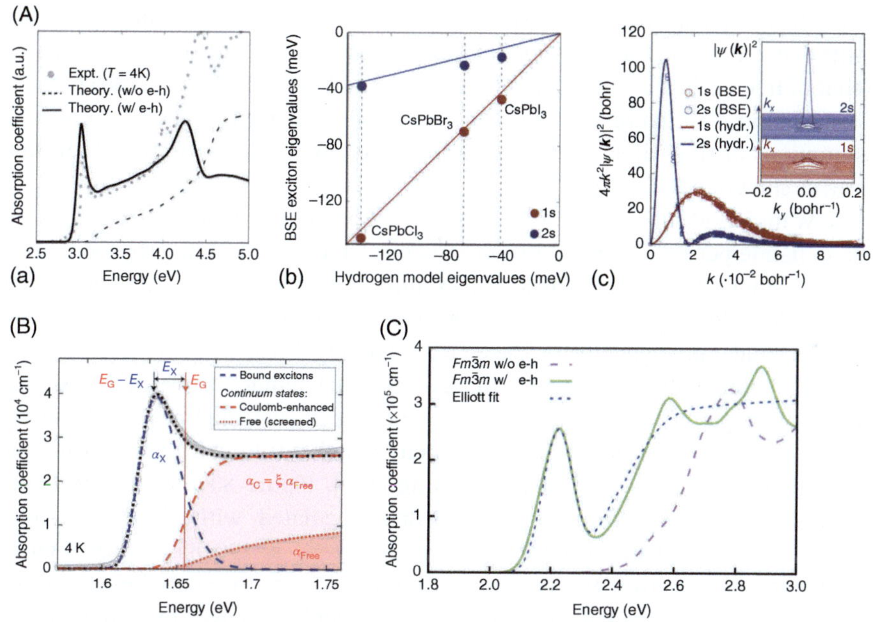

Figure 8.8 (A) (a) Linear optical absorption spectrum of $CsPbCl_3$ calculated with the G_0W_0 + BSE approach (continuous line), random phase approximation (dashed line), and from experiment (gray dots), (b) exciton binding energies predicted from G_0W_0 + BSE (filled circles) and the Wannier–Mott model (lines), (c) exciton radial probability density (main) and probability of localization (inset) in reciprocal space, calculated from G_0W_0 + BSE and the Wannier–Mott model. Source: Reproduced from [112]/with permission of American Physical Society. (B) Experimental absorption spectrum of $MAPbI_3$ taken at 4 K. The black dotted line is a fit to the data based on Elliott's theory as described in the main text. Arrows indicate the position of the band gap (E_G) and the value of the exciton binding energy (E_X). Source: [89]/Springer Nature/Licensed under CC BY 4.0. (C) Linear optical absorption spectrum of $Fm\bar{3}m$ $Cs_2AgBiBr_6$ calculated with the G_0W_0 + BSE approach (green continuous line), the random phase approximation (purple dashed line), and the Elliott model (blue dashed line).

including phonon screening explicitly accounts for only up to ~20% of the discrepancy with experimental results. Polaronic effects, i.e. the interactions of bare electrons and holes with phonons, are expected to play a significant role in halide perovskites. However, a full first-principles treatment of their effect on excitons is not available at the time of writing. Third, despite the importance of electron–phonon and exciton–phonon coupling, the lineshape of the excitonic feature calculated within the GW + BSE approach as described in Section 8.3.2.1, i.e. without taking these effects into account, is in very good agreement with experiment (see Figure 8.8a) [112].

Experimentally, exciton binding energies are frequently extracted from absorption spectra using the Elliott model [114]. This model assumes that only one pair of (parabolic and isotropic) valence and conduction bands is involved in optical transitions, and that the electric dipole transition matrix element is independent of the photon energy. The absorption coefficient in Elliott theory is then a sum of the absorption from bound excitons, α_X and electron–hole continuum states α_C, where α_X has the form of a Rydberg-like series at energies $-E_X$ below the band gap E_G (see Figure 8.8b for an illustration [89]). Continuum absorption is enhanced as compared to that of free excited electrons and holes (denoted by α_{free} in Figure 8.8b) because of the Coulomb attraction between the electron–hole pair [114]. For $MAPbI_3$, the assumptions of the Elliott model are fulfilled for energies close to the absorption onset, as shown in Ref. [89].

The Wannier–Mott and Elliott models need to be modified, or even fail, for excitons in halide perovskites with greater structural and chemical heterogeneity than found in single ABX_3 perovskites. For example, excitons in quasi-2D perovskites with Ruddlesden–Popper and Dion–Jacobson structures are strongly affected by the spatial heterogeneity of dielectric screening [115, 116], as known from other 2D semiconductors [117, 118]. But also 3D halide perovskites, like the halide double perovskite $Cs_2AgBiBr_6$, may significantly deviate from the Wannier–Mott model due to their chemical heterogeneity, which results in a highly anisotropic electronic structure and dielectric screening [119]. Figure 8.8c shows the linear optical absorption spectrum of $Cs_2AgBiBr_6$, calculated with the G_0W_0 + BSE approach. The calculated exciton binding energy E_X is 170 meV, in qualitative agreement with experimental reports, but neither E_X nor the spatial extent of the exciton are well-described by the Wannier–Mott model. Moreover, by fitting the first principles absorption spectrum with Elliott's formula for optical absorption, one obtains an exciton binding energy 35% higher than the first principles result [119].

8.4 Concluding Remarks: First Person Singular

I hope that this chapter could convince the reader – you – of the unique power of using first-principles numerical modeling techniques for understanding a class of materials as complex and diverse as the halide perovskites. I find that DFT and Green's function-based approaches are not only useful but also exceptionally beautiful and elegant in their simplicity. They reduce the intractable many-body problem of quantum mechanics to one that can not only be solved and stored (on any computer with reasonable hardware), but also interpreted in terms of familiar concepts of one-electron quantum mechanics and physical chemistry. Undoubtedly, this is one of the many reasons for the success of these methods, particularly DFT, in a wide variety of scientific fields. I chose the examples highlighted throughout this chapter to demonstrate the ability of these methods to provide an atomistic understanding of structural and optoelectronic properties of halide perovskites, to help interpret experimental findings and to disentangle disparate physical effects. However, they were also meant to showcase the danger of using computational setups that are blind to the fundamental and practical limits of the theories and methods they rely on. Therefore, despite methodological advancements, ever-increasing computational capabilities of high-performance computing resources, and the recent surge of machine-learning-based methods, researchers – yes, you – are still key to setting up, validating, and interpreting smart first principles computer experiments.

Acknowledgments

I would like to thank S. Krach for her work on size-dependent properties of perovskites with disordered A sites, J. Schiphorst for his implementation of a tight-binding model for halide perovskites, R.-I. Biega for the double perovskite band structures and absorption spectrum shown in Figures 8.7 and 8.8 and proof-reading of this chapter, and M. Filip, M. Bokdam, S. Reyes-Lillo, and J. Neaton for instructive and inspiring discussions.

References

1 Smith, M.D., Crace, E.J., Jaffe, A., and Karunadasa, H.I. (2018). The diversity of layered halide perovskites. *Annual Review of Materials Research* 48 (1): 111.
2 Hohenberg, P. and Kohn, W. (1964). Inhomogeneous electron gas. *Physical Review* 136 (3): B864.
3 Kohn, W. and Sham, L.J. (1965). Self-consistent equations including exchange and correlation effects. *Physical Review* 385: A1133.
4 Hamann, D.R., Schlüter, M., and Chiang, C. (1979). Norm-conserving pseudopotentials. *Physical Review Letters* 43 (20): 1494.
5 Blöchl, P.E. (1994). Projector augmented wave method. *Physical Review B* 50 (24): 17953–17979.
6 Kraisler, E. (2020). Asymptotic behavior of the exchange-correlation energy density and the Kohn–Sham potential in density functional theory: exact results and strategy for approximations. *Israel Journal of Chemistry* 2 (2): 1–18.
7 Kümmel, S. and Kronik, L. (2008). Orbital-dependent density functionals: theory and applications. *Reviews of Modern Physics* 80 (1): 3.
8 Ceperley, D.M. and Alder, B.J. (1980). Ground state of the electron gas by a stochastic method. *Physical Review Letters* 45 (7): 566.

9 Vosko, S.H., Wilk, L., and Nusair, M. (1980). Accurate spin-dependent electron liquid correlation energies for local spin density calculations: a critical analysis. *Canadian Journal of Physics* 58 (8): 1200–1211.

10 Perdew, J.P. and Zunger, A. (1981). Self-interaction correction to density-functional approximations for many-electron systems. *Physical Review B* 23 (10): 5048–5079.

11 Perdew, J.P. and Wang, Y. (1992). Accurate and simple analytic representation of the electron-gas correlation energy. *Physical Review B* 45 (23): 13244.

12 Perdew, J.P., Burke, K., and Ernzerhof, M. (1996). Generalized gradient approximation made simple. *Physical Review Letters* 77 (18): 3865–3868.

13 Mori-Sánchez, P., Cohen, A., and Yang, W. (2008). Localization and delocalization errors in density functional theory and implications for band-gap prediction. *Physical Review Letters* 100 (14): 146401.

14 Dreuw, A. and Head-Gordon, M. (2004). Failure of time-dependent density functional theory for long-range charge-transfer excited states: the zincbacteriochlorin-bacteriochlorin and bacteriochlorophyll-spheroidene complexes. *Journal of the American Chemical Society* 126 (12): 4007–4016.

15 Cohen, A.J., Mori-Sánchez, P., and Yang, W. (2008). Insights into current limitations of density functional theory. *Science* 321 (5890): 792.

16 Stephens, P.J., Devlin, F.J., Chabalowski, C.F., and Frisch, M.J. (1994). Ab initio calculation of vibrational absorption and circular dichroism spectra using density functional force fields. *Journal of Physical Chemistry* 98 (45): 11623.

17 Becke, A.D. (1993). A new mixing of Hartree-Fock and local density-functional theories. *Journal of Chemical Physics* 98 (2): 1372.

18 Adamo, C. and Barone, V. (1999). Toward reliable density functional methods without adjustable parameters: the PBE0 model. *Journal of Chemical Physics* 110 (13): 6158.

19 Perdew, J.P., Ernzerhof, M., and Burke, K. (1996). Rationale for mixing exact exchange with density functional approximations. *Journal of Chemical Physics* 105 (22): 9982–9985.

20 Kronik, L., Stein, T., Refaely-Abramson, S., and Baer, R. (2012). Excitation gaps of finite-sized systems from optimally tuned range-separated hybrid functionals. *Journal of Chemical Theory and Computation* 8 (5): 1515–1531.

21 Refaely-Abramson, S., Sharifzadeh, S., Jain, M. et al. (2013). Gap renormalization of molecular crystals from density-functional theory. *Physical Review B* 88 (8): 081204.

22 Stein, T., Eisenberg, H., Kronik, L., and Baer, R. (2010). Fundamental gaps in finite systems from eigenvalues of a generalized Kohn–Sham method. *Physical Review Letters* 105 (26): 266802.

23 Sun, J., Ruzsinszky, A., and Perdew, J. (2015). Strongly constrained and appropriately normed semilocal density functional. *Physical Review Letters* 115 (3): 036402.

24 Aschebrock, T. and Kümmel, S. (2019). Ultranonlocality and accurate band gaps from a meta-generalized gradient approximation. *Physical Review Research* 1 (3): 033082.

25 Baroni, S., de Gironcoli, S., Dal Corso, A., and Giannozzi, P. (2001). Phonons and related crystal properties from density-functional perturbation theory. *Reviews of Modern Physics* 73: 515.

26 Poglitsch, A. and Weber, D. (1987). Dynamic disorder in methylammoniumtrihalogenoplumbates (II) observed by millimeter-wave spectroscopy. *Journal of Chemical Physics* 87 (11): 6373–6378.

27 Baikie, T., Fang, Y., Kadro, J.M. et al. (2013). Synthesis and crystal chemistry of the hybrid perovskite $(CH_3NH_3)PbI_3$ for solid-state sensitised solar cell applications. *Journal of Materials Chemistry A* 1 (18): 5628.

28 Whitfield, P.S., Herron, N., Guise, W.E. et al. (2016). Structures, phase transitions and tricritical behavior of the hybrid perovskite methyl ammonium lead iodide. *Scientific Reports* 6: 35685.

29 Egger, D.A. and Kronik, L. (2014). Role of dispersive interactions in determining structural properties of organic–inorganic halide perovskites: insights from first-principles calculations. *Journal of Physical Chemistry Letters* 5: 2728–2733.

30 Lahnsteiner, J., Kresse, G., Heinen, J., and Bokdam, M. (2018). Finite-temperature structure of the $MAPbI_3$ perovskite: comparing density functional approximations and force fields to experiment. *Physical Review Materials* 2 (7): 073604.

31 Beck, H., Gehrmann, C., and Egger, D.A. (2019). Structure and binding in halide perovskites: analysis of static and dynamic effects from dispersion-corrected density functional theory. *APL Materials* 7 (2): 021108.

32 Isaacs, E.B. and Wolverton, C. (2018). Performance of the strongly constrained and appropriately normed density functional for solid-state materials. *Physical Review Materials* 2 (6): 063801.

33 Wang, Y., Gould, T., Dobson, J.F. et al. (2014). Density functional theory analysis of structural and electronic properties of orthorhombic perovskite $CH_3NH_3PbI_3$. *Physical Chemistry Chemical Physics* 16 (4): 1424–1429.

34 Ong, K.P., Goh, T.W., Xu, Q., and Huan, A. (2015). Mechanical origin of the structural phase transition in methylammonium lead iodide $CH_3NH_3PbI_3$. *Journal of Physical Chemistry Letters* 6 (4): 681–685.

35 Kaczkowski, J. and Płowaś-Korus, I. (2021). The vibrational and thermodynamic properties of $CsPbI_3$ polymorphs: an improved description based on the SCAN meta-GGA functional. *Journal of Physical Chemistry Letters* 12 (28): 6613–6621.

36 Zhao, X.G., Dalpian, G.M., Wang, Z., and Zunger, A. (2020). Polymorphous nature of cubic halide perovskites. *Physical Review B* 101 (15): 155137.

37 Girdzis, S.P., Lin, Y., Leppert, L. et al. (2021). Revealing local disorder in a silver-bismuth halide perovskite upon compression. *Journal of Physical Chemistry Letters* 12 (1): 532–536.

38 Zhao, X.G., Wang, Z., Malyi, O.I., and Zunger, A. (2021). Effect of static local distortions vs. dynamic motions on the stability and band gaps of cubic oxide and halide perovskites. *Materials Today* 49: 107–122.

39 Frohna, K., Deshpande, T., Harter, J. et al. (2018). Inversion symmetry and bulk Rashba effect in methylammonium lead iodide perovskite single crystals. *Nature Communications* 9 (1): 1829.

40 Tkatchenko, A. and Scheffler, M. (2009). Accurate molecular van der Waals interactions from ground-state electron density and free-atom reference data. *Physical Review Letters* 102 (7): 073005.

41 Kotiuga, M., Halilov, S., Kozinsky, B. et al. (2021). A microscopic picture of paraelectric perovskites from structural prototypes. *arXiv*, 2107.04628:117. http://arxiv.org/abs/2107.04628.

42 Frenkel, D. and Smits, B. (2002). *Understanding Molecular Simulation*. Elsevier.

43 Schmidt, J., Marques, M.R.G., Botti, S., and Marques, M.A.L. (2019). Recent advances and applications of machine learning in solid-state materials science. *npj Computational Materials* 5 (1): 83.

44 Behler, J. (2016). Perspective: machine learning potentials for atomistic simulations. *Journal of Chemical Physics* 145: 170901.

45 Jinnouchi, R., Lahnsteiner, J., Karsai, F. et al. (2019). Phase transitions of hybrid perovskites simulated by machine-learning force fields trained on the fly with Bayesian inference. *Physical Review Letters* 122 (22): 225701.

46 Lahnsteiner, J. and Bokdam, M. (2022). Anharmonic lattice dynamics in large thermodynamic ensembles with machine-learning force fields: $CsPbBr_3$ a phonon-liquid with Cs rattlers. *Physical Review B* 105: 024302.

47 Menahem, M., Benshalom, N., Asher, M. et al. (2023). Disorder origin of Raman scattering in perovskite single crystals. *Physical Review Materials* 7 (4). https://doi.org/10.1103/physrevmaterials.7.044602.

48 Guo, P., Xia, Y., Gong, J. et al. (2017). Polar fluctuations in metal halide perovskites uncovered by acoustic phonon anomalies. *ACS Energy Letters* 2 (10): 2463–2469.

49 Yaffe, O., Guo, Y., Tan, L.Z. et al. (2017). Local polar fluctuations in lead halide perovskite crystals. *Physical Review Letters* 118 (13): 136001.

50 Simoncelli, M., Marzari, N., and Mauri, F. (2019). Unified theory of thermal transport in crystals and glasses. *Nature Physics* 15 (8): 809–813.

51 Gehrmann, C. and Egger, D.A. (2019). Dynamic shortening of disorder potentials in anharmonic halide perovskites. *Nature Communications* 10 (1): 3141.

52 Klarbring, J., Hellman, O., Abrikosov, I.A., and Simak, S.I. (2020). Anharmonicity and ultralow thermal conductivity in lead-free halide double perovskites. *Physical Review Letters* 125 (4): 45701.

53 Pols, M., Brouwers, V., Calero, S., and Tao, S. (2023). How fast do defects migrate in halide perovskites: insights from on-the-fly machine-learned force fields. *Chemical Communications* 59 (31): 4660–4663. https://doi.org/10.1039/d3cc00953j.

54 Sheppard, D., Terrell, R., and Henkelman, G. (2008). Optimization methods for finding minimum energy paths. *Journal of Chemical Physics* 128 (13): 134106.

55 Egger, D.A., Kronik, L., and Rappe, A.M. (2015). Theory of hydrogen migration in organic–inorganic halide perovskites. *Angewandte Chemie International Edition* 54: 12437–12441.

56 Oranskaia, A., Yin, J., Bakr, O.M. et al. (2018). Halogen migration in hybrid perovskites: the organic cation matters. *Journal of Physical Chemistry Letters* 9 (18): 5474–5480.

57 Meggiolaro, D., Mosconi, E., and De Angelis, F. (2019). Formation of surface defects dominates ion migration in lead-halide perovskites. *ACS Energy Letters* 4 (3): 779–785. https://doi.org/10.1021/acsenergylett.9b00247.

58 Biega, R.-I. and Leppert, L. (2021). Halogen vacancy migration at surfaces of $CsPbBr_3$ perovskites: insights from density functional theory. *Journal of Physics: Energy* 3: 034017.

59 Henkelman, G., Uberuaga, B.P., and Jónsson, H. (2000). Climbing image nudged elastic band method for finding saddle points and minimum energy paths. *Journal of Chemical Physics* 113 (22): 9901–9904.

60 Freysoldt, C., Grabowski, B., Hickel, T. et al. (2014). First-principles calculations for point defects in solids. *Reviews of Modern Physics* 86 (1): 253–305.

61 Cohen, A.V., Egger, D.A., Rappe, A.M., and Kronik, L. (2019). Breakdown of the static picture of defect energetics in halide perovskites: the case of the Br vacancy in $CsPbBr_3$. *Journal of Physical Chemistry Letters* 10 (16): 4490–4498.

62 Perdew, J.P., Parr, R.G., Levy, M. et al. (1982). Density-functional theory for fractional particle number: derivative discontinuities of the energy. *Physical Review Letters* 4 (23): 1691.

63 Perdew, J.P. and Levy, M. (1983). Physical content of the exact Kohn–Sham orbital energies: band gaps and derivative discontinuities. *Physical Review Letters* 51 (6): 1884.

64 Sham, L.J. and Schlüter, M. (1983). Density-functional theory of the energy gap. *Physical Review Letters* 51 (20): 1888.

65 Aouina, A., Gatti, M., Chen, S. et al. (2023). Accurate Kohn–Sham auxiliary system from the ground-state density of solids. *Physical Review B* 107 (19). https://doi.org/10.1103/physrevb.107.195123.

66 Janak, J.F. (1978). Proof that $\delta E/\delta n_i = \varepsilon_i$ in density-functional theory. *Physical Review B* 18: 7165.

67 Capelle, K. (2006). A bird's-eye view of density-functional theory. *Brazilian Journal of Physics* 36: 1318.

68 Seidl, A., Görling, A., Vogl, P. et al. (1996). Generalized Kohn–Sham schemes and the band-gap problem. *Physical Review B* 53 (7): 3764–3774. https://doi.org/10.1103/physrevb.53.3764.

69 Perdew, J.P., Yang, W., Burke, K. et al. (2017). Understanding band gaps of solids in generalized Kohn–Sham theory. *Proceedings of the National Academy of Sciences of the United States of America* 114 (11): 2801–2806.

70 Hedin, L. (1965). New method for calculating the one-particle Green's function with application to the electron–gas problem. *Physical Review* 139 (3A): A796–A823. https://doi.org/10.1103/physrev.139.a796.

71 Golze, D., Dvorak, M., and Rinke, P. (2019). The GW compendium: a practical guide to theoretical photoemission spectroscopy. *Frontiers in Chemistry* 7: 377.

72 Kotani, T., van Schilfgaarde, M., and Faleev, S. (2007). Quasiparticle self-consistent GW method: a basis for the independent-particle approximation. *Physical Review B* 76 (16): 165106.

73 Marzari, N. and Vanderbilt, D. (1997). Maximally localized generalized Wannier functions for composite energy bands. *Physical Review B* 56 (20): 12847–12865.

74 King–Smith, R.D. and Vanderbilt, D. (1993). Theory of polarization of crystalline solids. *Physical Review B* 47 (3): 1651–1654. https://doi.org/10.1103/physrevb.47.1651.

75 Marzari, N., Mostofi, A.A., Yates, J.R. et al. (2012). Maximally localized Wannier functions: theory and applications. *Reviews of Modern Physics* 84 (4): 1419–1475. https://doi.org/10.1103/revmodphys.84.1419.

76 Boyer-Richard, S., Katan, C., Traoré, B. et al. (2016). Symmetry-based tight binding modeling of halide perovskite semiconductors. *Journal of Physical Chemistry Letters* 7: 3833–3840.

77 Slater, J.C. and Koster, G.F. (1954). Simplified LCAO method for the periodic potential problem. *Physical Review* 94 (6): 1498.

78 Slavney, A.H., Connor, B.A., Leppert, L., and Karunadasa, H.I. (2019). A pencil-and-paper method for elucidating halide double perovskite band structures. *Chemical Science* 10 (48): 11041.

79 Slavney, A.H., Leppert, L., Saldivar Valdes, A. et al. (2018). Small-band-gap halide double perovskites. *Angewandte Chemie International Edition* 57 (39): 12765–12770.

80 Jin, H., Im, J., and Freeman, A.J. (2012). Topological insulator phase in halide perovskite structures. *Physical Review B* 86 (12): 121102.

81 Leppert, L., Rangel, T., and Neaton, J.B. (2019). Towards predictive band gaps for halide perovskites: lessons from one-shot and eigenvalue self-consistent GW. *Physical Review Materials* 3 (10): 103803.

82 Giustino, F. (2016). Electron-phonon interactions from first principles. *Reviews of Modern Physics* 89: 1–64.

83 Wiktor, J., Rothlisberger, U., and Pasquarello, A. (2017). Predictive determination of band gaps of inorganic halide perovskites. *Journal of Physical Chemistry Letters* 8 (22): 5507–5512.

84 Rashba, E. (1960). Properties of semiconductors with an extremum loop. 1. Cyclotron and combinational resonance in a magnetic field perpendicular to the plane of the loop. *Soviet Physics-Solid State* 2: 1109–1122.

85 Dresselhaus, G. (1955). Spin-orbit coupling effects in zinc blende structures. *Physical Review B* 100: 580–586.

86 Kim, M., Im, J., Freeman, A.J. et al. (2014). Switchable S = 1/2 and J = 1/2 Rashba bands in ferroelectric halide perovskites. *Proceedings of the National Academy of Sciences of the United States of America* 111: 6900–6904.

87 Kepenekian, M. and Even, J. (2017). Rashba and Dresselhaus couplings in halide perovskites: accomplishments and opportunities for spintronics and spin–orbitronics. *Journal of Physical Chemistry Letters* 8: 3362–3370.

88 Etienne, T., Mosconi, E., and De Angelis, F. (2016). Dynamical origin of the Rashba effect in organohalide lead perovskites: a key to suppressed carrier recombination in perovskite solar cells? *Journal of Physical Chemistry Letters* 7: 1638–1645.

89 Davies, C.L., Filip, M.R., Patel, J.B. et al. (2018). Bimolecular recombination in methylammonium lead triiodide perovskite is an inverse absorption process. *Nature Communications* 9 (1): 293.

90 Zheng, F., Tan, L.Z., Liu, S., and Rappe, A.M. (2015). Rashba spin-orbit coupling enhanced carrier lifetime in $CH_3NH_3PbI_3$. *Nano Letters* 15: 7794–7800.

91 Leppert, L., Reyes-Lillo, S.E., and Neaton, J.B. (2016). Electric field and strain induced Rashba effect in hybrid halide perovskites. *Journal of Physical Chemistry Letters* 7: 3683–3689.

92 Skone, J.H., Govoni, M., and Galli, G. (2014). Self-consistent hybrid functional for condensed systems. *Physical Review B* 89 (19): 195112.

93 Wing, D., Ohad, G., Haber, J.B. et al. (2020). Band gaps of crystalline solids from Wannier-localization based optimal tuning of a screened range-separated hybrid functional. *Proceedings of the National Academy of Sciences of the United States of America* 118 (34): e2104556118.

94 Bischoff, T., Wiktor, J., Chen, W., and Pasquarello, A. (2019). Nonempirical hybrid functionals for band gaps of inorganic metal-halide perovskites. *Physical Review Materials* 3 (12): 123802.

95 Umari, P., Mosconi, E., and De Angelis, F. (2014). Relativistic GW calculations on $CH_3NH_3PbI_3$ and $CH_3NH_3SnI_3$ perovskites for solar cell applications. *Scientific Reports* 4: 4467.

96 Brivio, F., Butler, K.T., Walsh, A., and van Schilfgaarde, M. (2014). Relativistic quasiparticle self-consistent electronic structure of hybrid halide perovskite photovoltaic absorbers. *Physical Review B* 89 (15): 155204.

97 Filip, M.R., Eperon, G.E., Snaith, H.J., and Giustino, F. (2014). Steric engineering of metal-halide perovskites with tunable optical band gaps. *Nature Communications* 5: 5757.

98 Filip, M.R., Verdi, C., and Giustino, F. (2015). GW band structures and carrier effective masses of $CH_3NH_3PbI_3$ and hypothetical perovskites of the type $APbI_3$: A = NH_4, PH_4, AsH_4 and SbH_4. *Journal of Physical Chemistry C* 119: 25209–25219.

99 Poncé, S., Schlipf, M., and Giustino, F. (2019). Origin of low carrier mobilities in halide perovskites. *ACS Energy Letters* 4 (2): 456–463.

100 Blase, X., Attaccalite, C., and Olevano, V. (2011). First-principles GW calculations for fullerenes, porphyrins, phtalocyanine, and other molecules of interest for organic photovoltaic applications. *Physical Review B* 83 (11): 115103.

101 Volonakis, G., Filip, M.R., Haghighirad, A.A. et al. (2016). Lead-free halide double perovskites via heterovalent substitution of noble metals. *Journal of Physical Chemistry Letters* 7: 1254–1259.

102 Volonakis, G., Haghighirad, A.A., Milot, R.L. et al. (2017). $Cs_2InAgCl_6$: a new lead-free halide double perovskite with direct band gap. *Journal of Physical Chemistry Letters* 8: 772–778.

103 Volonakis, G., Sakai, N., Snaith, H.J., and Giustino, F. (2019). Oxide analogs of halide perovskites and the new semiconductor Ba_2AgIO_6. *Journal of Physical Chemistry Letters* 10 (8): 1722–1728.

104 Strinati, G. (1988). Application of the Green's functions method to the study of the optical properties of semiconductors. *La Rivista del Nuovo Cimento* 11 (12): 1–86.

105 Onida, G., Reining, L., and Rubio, A. (2002). Electronic excitations: density-functional versus many-body Green's-function approaches. *Reviews of Modern Physics* 74: 601.

106 Wing, D., Haber, J.B., Noff, R. et al. (2019). Comparing time-dependent density functional theory with many-body perturbation theory for semiconductors: screened range-separated hybrids and the GW plus Bethe-Salpeter approach. *Physical Review Materials* 3 (6): 064603.

107 Sun, J., Yang, J., and Ullrich, C.A. (2020). Low-cost alternatives to the Bethe-Salpeter equation: towards simple hybrid functionals for excitonic effects in solids. *Physical Review Research* 2 (1): 013091.

108 Rohlfing, M. and Louie, S.G. (1998). Electron-hole excitations in semiconductors and insulators. *Physical Review Letters* 81 (11): 2312–2315.

109 Rohlfing, M. and Louie, S.G. (2000). Electron-hole excitations and optical spectra from first principles. *Physical Review B* 62 (8): 4927.

110 Bokdam, M., Sander, T., Stroppa, A. et al. (2016). Role of polar phonons in the photo excited state of metal halide perovskites. *Scientific Reports* 6 (I): 28618.

111 Umari, P., Mosconi, E., and De Angelis, F. (2018). Infrared dielectric screening determines the low exciton binding energy of metal-halide perovskites. *Journal of Physical Chemistry Letters* 9 (3): 620–627.

112 Filip, M.R., Haber, J.B., and Neaton, J.B. (2021). Phonon screening of excitons in semiconductors: the case of halide perovskites hybrid organic–inorganic halide perovskite. *Physical Review Letters* 127: 067401.

113 Da Jornada, F.H., Qiu, D.Y., and Louie, S.G. (2017). Nonuniform sampling schemes of the Brillouin zone for many-electron perturbation-theory calculations in reduced dimensionality. *Physical Review B* 95 (3): 035109.

114 Elliott, R.J. (1957). Intensity of optical absorption by excitons. *Physical Review* 108 (6): 1384–1389.

115 Blancon, J.C., Stier, A.V., Tsai, H. et al. (2018). Scaling law for excitons in 2D perovskite quantum wells. *Nature Communications* 9 (1): 2254.

116 Cho, Y. and Berkelbach, T.C. (2019). Optical properties of layered hybrid organic–inorganic halide perovskites: a tight-binding GW-BSE study. *Journal of Physical Chemistry Letters* 10 (20): 6189–6196.

117 Qiu, D.Y., Da Jornada, F.H., and Louie, S.G. (2013). Optical spectrum of MoS_2: many-body effects and diversity of exciton states. *Physical Review Letters* 111: 216805.

118 Berkelbach, T.C., Hybertsen, M.S., and Reichman, D.R. (2013). Theory of neutral and charged excitons in monolayer transition metal dichalcogenides. *Physical Review B* 88 (4): 045318.

119 Biega, R.-I., Filip, M.R., Leppert, L., and Neaton, J.B. (2021). Chemically localized resonant excitons in silver–pnictogen halide double perovskites. *Journal of Physical Chemistry Letters* 12 (8): 2057–2063.

9

Comparing the Charge Dynamics in $MAPbBr_3$ and $MAPbI_3$ Using Microwave Photoconductance Measurements

Tom J. Savenije, Jiashang Zhao, and Valentina M. Caselli

Delft University of Technology, Faculty of Applied Sciences, Department of Chemical Engineering, van der Maasweg 9, 2629 HZ Delft, The Netherlands

This chapter is built up as follows: Section 9.1 comprises a short comprehensive overview regarding the electrodeless time-resolved microwave conductivity (TRMC) technique. Section 9.2 introduces a kinetic model to obtain quantitative data from the TRMC results, including charge carrier mobilities and trap densities. Section 9.3 presents the TRMC measurements on $MAPbI_3$ and $MAPbBr_3$, including analysis of the decay kinetics. The found mobilities will be discussed and compared to other reported values determined by, e.g. terahertz spectroscopy. In Section 9.4, TRMC measurements on both metal halide perovskites (MHPs) supplied with different CSTLs are presented. The CSTLs comprise Spiro-OMeTAD as a hole transport layer or C_{60} as electron transport layer. Possible implications of the mobilities and trap densities, which determine the charge carrier diffusion lengths, on the charge collection and performance of a corresponding cell will be discussed.

9.1 Time-Resolved Microwave Conductivity

The TRMC technique can be used to study the dynamics of photo-induced charge carriers in semiconductor materials with low background conductivities [1–5]. This technique is based on the interaction between the electric field component of the microwaves (GHz regime) and mobile carriers (see Figure 9.1). Hence, the photoconductivity can be determined without contacting the semiconductor with electrodes, thereby avoiding interfacial effects or undesired chemical reactions between the MHP and the metallic electrodes [6, 7].

In general, if photo excitation of a material results in the generation of free, mobile charge carriers, the conductivity of this material changes. The electrical conductivity, σ scales with the concentration of free charge carriers, n and their mobility μ according to:

$$\sigma = e\sum_i n_i\mu_i \tag{9.1}$$

Halide Perovskite Semiconductors: Structures, Characterization, Properties, and Phenomena, First Edition. Edited by Yuanyuan Zhou and Iván Mora-Seró.

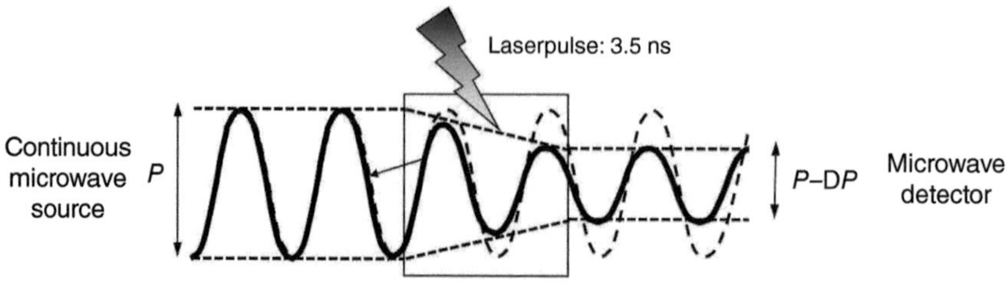

Figure 9.1 Representation of photo-induced time-resolved microwave conductivity (TRMC) technique. In essence, the MHP is optically excited by a short laser pulse, yielding excess charge carriers. These carriers interact with microwaves, resulting in a reduction of the microwave power, which is measured on nanosecond timescales. TRMC is used to measure the photoconductance, which scales with the time-dependent concentration n and mobility μ of free electrons and holes. The sinusoidal (dashed) line represents the magnitude of the microwave electric field as it passes through the (excited) MHP.

where e is the elementary charge. With the TRMC technique, the integrated change in σ over the complete film thickness, L, is measured, which yields a change in conductance, ΔG:

$$\Delta G = \beta \int_0^L \Delta\sigma(z)\mathrm{d}z \tag{9.2}$$

Here β is the ratio between the wide and small inner walls of the microwave guide. Hence, ΔG is proportional to the product of the total number of photo-induced free charges and their mobility. Absorption of microwaves by the free carriers reduces the microwave power on the detector. The normalized reduction in microwave power $(\Delta P(t))/P$ is related to $\Delta G(t)$ by:

$$\frac{\Delta P(t)}{P} = -K\Delta G(t) \tag{9.3}$$

where K is the sensitivity factor, which can be quantified as described previously [5, 8].

Figure 9.2 shows a representation of the TRMC setup. Photo-excitation is realized by laser pulses of 3–5 ns FWHM with a tunable wavelength at a repetition rate of 10 Hz. Metallic, neutral-density filters with different optical densities are

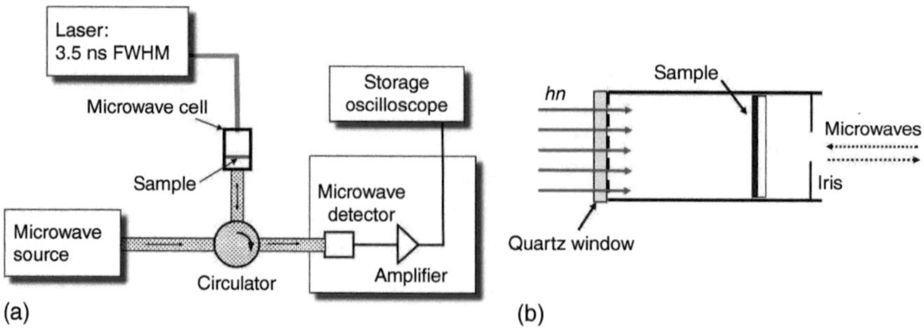

Figure 9.2 Schematic representation of (a) the TRMC set-up and (b) the TRMC cell. Source: Adapted from Savenije et al. [3].

applied to vary the photon fluence between 10^9 and 10^{15} photons cm^{-2} per pulse. A voltage-controlled oscillator is utilized to generate the microwaves with a frequency in the 10 GHz range. The sample of interest is placed in a microwave cell that ends with a metal grating (see Figure 9.2b), which largely transmits the laser power but fully reflects the microwaves. A quartz window is glued on top of the grating to seal the cell and avoid exposure of the sample to air. In the presence of an iris (see Figure 9.2b) the sensitivity factor, K is much higher (c. 40 times) than for an iris free microwave cell. However, this sensitivity gain leads to an increase in the response time from c. 1 to 18 ns. A microwave diode detector converts the microwave power into a transient voltage, which is recorded using a digital oscilloscope.

If the sensitivity factor K is known, ΔG can be quantitatively obtained from the measured $\Delta P(t)/P$ using Eq. (9.3). Then, the TRMC signal can be expressed in the remaining two unknown parameters: i.e. the mobility, μ and the charge carrier generation yield, φ. If every absorbed photon creates a single positive and a single negative charge carrier, Eq. (9.1) simplifies to:

$$\sigma = en(\mu_e + \mu_h) \tag{9.4}$$

in which n is the concentration, μ_e is the electron mobility, and μ_h is the hole mobility. The yield can be defined as

$$\varphi = \frac{Ln}{F_A I_0} \tag{9.5}$$

in which I_0 is the intensity of the laser in photons/pulse/unit area and F_A is the fraction of light absorbed at the excitation wavelength. By combining Eqs. (9.2), (9.4), and (9.5), the product of yield and mobility can be obtained from ΔG_{max}

$$\varphi(\mu_e + \mu_h) = \frac{Ln}{F_A I_0}\frac{\Delta\sigma}{en} = \frac{L}{F_A I_0}\frac{\Delta G_{max}}{e\beta L} = \frac{\Delta G_{max}}{F_A I_0 e\beta} \tag{9.6}$$

Expressing the TRMC signal in $\varphi(\mu_e + \mu_h)$ products enables us to directly compare TRMC measurements of different samples.

In contrast to DC techniques, such as time-of-flight or field effect transistor measurements, charges do not undergo a net displacement in the photo-active layer during a TRMC measurement. In fact, due to the low electric field strength of the microwaves in the cell and the rapid oscillating direction of the electric field, the drift distances are relatively small, i.e. in the nanometer regime. This implies that transport over grain boundaries does not limit mobilities unless the grain size becomes <100 nm [9]. For techniques in which excess carriers have to cross grain boundaries, reported mobilities are typically at least an order of magnitude less [10]. An extensive discussion over the latter is reported for thin films of $MAPbI_3$ by Levine et al. [11].

Note that both electrons and holes' contributions to the photoconductance depend on their mobilities. This is in contrast with more frequently used time-resolved photoluminescence (TRPL) measurements, which specifically detect radiative recombination events. Therefore, the decays obtained with TRMC cannot be a priori compared to TRPL signals. Instead, TRMC gives complementary insights to TRPL since it displays the recombination of all free mobile charge carriers, both radiative and nonradiative. Therefore, TRMC is especially useful for studying the recombination kinetics in MHPs showing poor light outcoupling and substantial reabsorption of emitted photons, such as macroscopic MHP crystals or MHPs supplied with a transport material, where most recombination is nonradiative.

9.2 Global Modeling of TRMC Data

In order to extract quantitative data out of TRMC measurements, the generic kinetic model, as detailed in Figure 9.3, can be used [12, 13]. This model accounts for different recombination pathways of photo-excited electrons and holes and assumes homogeneous generation and decay of charges, which can be experimentally realized by using an excitation wavelength close to the absorption onset. In view of the relatively small exciton binding energy of most MHPs compared to thermal energy [14], electron–hole pairs dissociate into free conduction band (CB) electrons and valence band (VB) holes.

Formation of charge carriers is represented by the generation term, G_C and takes into account the temporal profile and total light intensity of the laser pulse. In general, perovskite semiconductors are often unintentionally doped due to impurities and defect states in the crystal lattice. This leads to additional CB electrons (n-type) or VB holes (p-type) already present before photoexcitation. In case of a p-type doped perovskite, as shown in Figure 9.3, the concentration of dark carriers is represented by p_0. Note that p_0 does not contribute to the photoconductance since the TRMC technique is an AC technique; however, p_0 affects the recombination rates.

The following set of coupled differential equations implements the processes shown in Figure 9.3 and describes n_e (Eq. (9.7)), n_h (Eq. (9.8)), and n_t (Eq. (9.9)) as

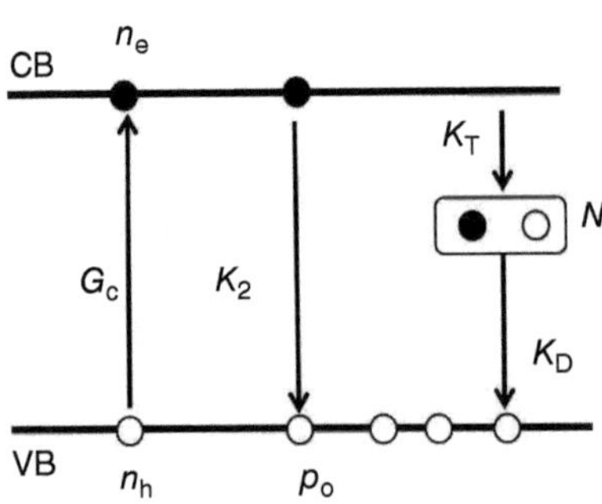

Figure 9.3 Kinetic model of processes occurring upon photoexcitation of a (unintentionally) p-type semiconductor perovskite. Here, G_C represents the photoexcitation of electrons (closed circles) from the valence band (VB) to the conduction band (CB). The CB electrons can recombine with holes (open circles) in the VB via k_2. In competition with k_2, electrons can be immobilized in intraband gap trap states with total densities of N_T via k_T. Finally, the trapped electrons can recombine with holes from the VB via k_D. In the case of an (unintentionally) doped MHP, there will be additional holes (p_0, p-type) on top of the photogenerated holes. Note that this fully mathematical model also holds for an n-type MHP with trap states for holes.

functions of time. The rate constants for band-to-band electron–hole recombination, trap filling, and trap emptying are represented by k_2, k_T, and k_D, respectively.

$$\frac{dn_e}{dt} = G_C - k_2 n_e(n_h + p_0) - k_T n_e(N_T - n_t) \tag{9.7}$$

$$\frac{dn_h}{dt} = G_C - k_2 n_e(n_h + p_0) - k_D n_t(n_h + p_0) \tag{9.8}$$

$$\frac{dn_t}{dt} = k_T n_e(N_T - n_t) - k_D n_t(n_h + p_0) \tag{9.9}$$

Solving the equations using numerical methods yields the time-dependent concentrations n_e, n_h and n_t. The change in photoconductance as function of time is calculated from the product of charge carrier concentrations and mobilities according to:

$$\Delta G(t) = e(n_e(t)\mu_e + n_h(t)\mu_h)\beta L \tag{9.10}$$

Here, the trapped charge carriers, n_t are assumed to be immobile and do not contribute to ΔG. The mobilities are assumed to be constant within the time window of the measurement and independent of the charge density. Finally, a convolution is applied to take into account the instrumental response function of the setup [5]. To obtain accurate values for all kinetic parameters describing a specific MHP, a set of TRMC traces induced by widely varying laser intensities (preferably more than two orders of magnitude) should be recorded. Fits to these TRMC traces vary only by adapting the laser intensity, which is incorporated in the generation term, G_C. By following this global approach, accurate quantification of all parameters is feasible.

9.3 TRMC Measurements on $MAPbI_3$ and $MAPbBr_3$

Results from TRMC measurements on MHPs are provided in Figure 9.4a,b for $MAPbI_3$ and $MAPbBr_3$, respectively. The photoconductance, normalized for the number of absorbed photons, is shown as function of time on photoexcitation using incident light intensities ranging over at least two orders of magnitude. The rise of the signal at $t = 0$ originates from the formation of mobile excess charges; the maximum signal height reflects the product of charge carrier generation yield and mobility. For $MAPbI_3$ at higher intensities, the decay kinetics become increasingly faster, which is typically observed in a regime where higher-order recombination dominates. Apart from this, the maximum signal somewhat reduces due to the fact that pulse recombination occurs. At very low excitation densities, the number of generated excess carriers is below the trap density. For these excitation intensities, the band-to-band second-order recombination process is not the dominant decay process any more due to the rapid trapping of the minority carriers. Note that at these intensities, the counter charges are responsible for the observed TRMC signal, as these are still mobile. Most notably, for $MAPbBr_3$, we observe very different decay kinetics, featuring a very fast, sharp decrease followed by a long-lived almost

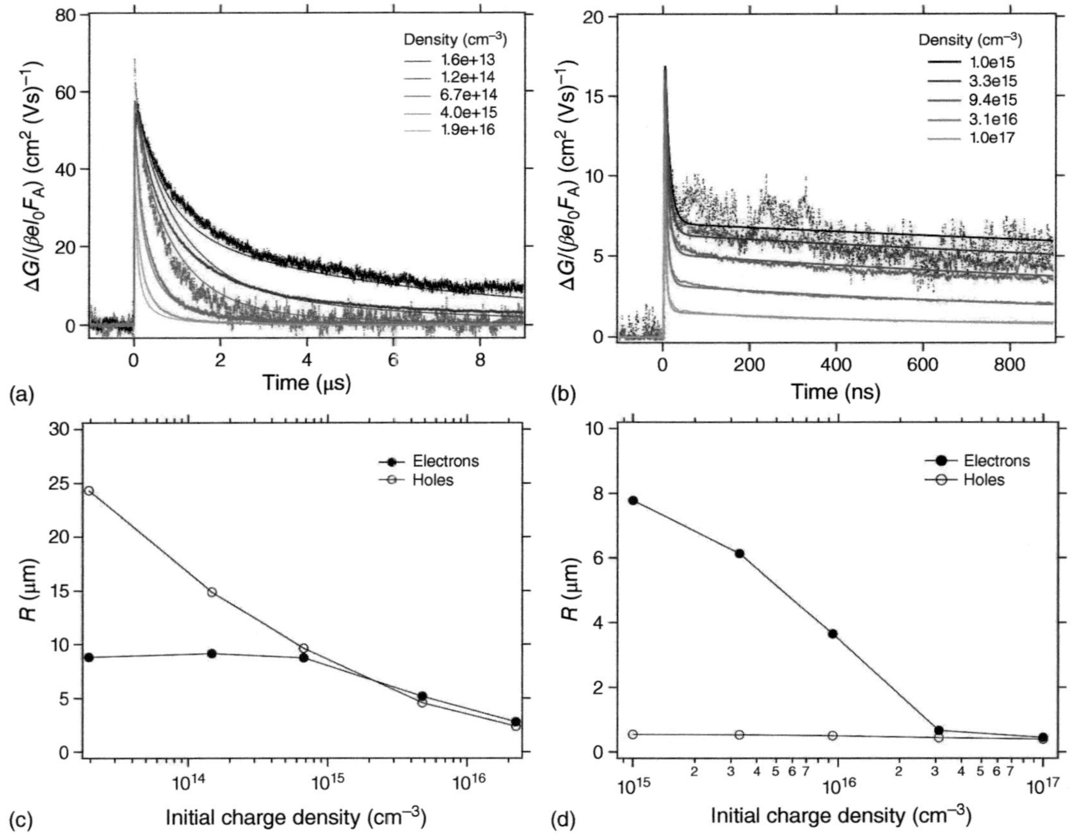

Figure 9.4 Experimental TRMC traces for (a) $MAPbI_3$ and (b) $MAPbBr_3$ recorded at wavelengths of $\lambda = 650$ and 500 nm, respectively. The full lines are calculated by numerically solving differential Eqs. (9.7)–(9.9) and converting the time-dependent concentration curves to the light intensity normalized ΔG traces. Note the difference in time scale! Panels (c) and (d) show the calculated charge carrier diffusion lengths on basis of charge carrier lifetime ($t_{1/2}$) and mobility for different laser intensities. Source: Adapted from Zhao et al. [15].

Table 9.1 Kinetic parameters used for fits to the TRMC traces for bare $MAPbI_3$ and $MAPbBr_3$ layers.

	$MAPbI_3$	$MAPbBr_3$
k_2 ($\times 10^{-9}$ cm^3 s^{-1})	1.2	5.5
k_T ($\times 10^{-9}$ cm^3 s^{-1})	8	2
k_D ($\times 10^{-10}$ cm^3 s^{-1})	0.5	0.45
N_T ($\times 10^{15}$ cm^{-3})	0.2	35
p_0/n_0 ($\times 10^{15}$ cm^{-3})	0.02	3.5
μ_e (cm^2 (Vs)$^{-1}$)	75	9.5
μ_h (cm^2 (Vs)$^{-1}$)	50	9.5

constant signal. In analogy with $MAPbI_3$, at higher intensities, the normalized TRMC becomes increasingly smaller.

Using the kinetic model as described in Section 9.2, the TRMC traces for both MHPs are fitted using a single set of kinetic parameters, and results are collected in Table 9.1. Note that for obtaining the different TRMC traces, only the magnitude of the generation term, G_C is altered according to the incident intensity. Despite the very different decay kinetics for both MHPs, we see that the model describes the TRMC traces for both MHPs rather well. Using this approach, it is feasible to obtain accurate values for all kinetic parameters describing the charge carrier dynamics. Some prominent dissimilarities between both MHPs can be observed, including mobilities, band-to-band recombination rate constants, and values for the trap and background density. For $MAPbI_3$ layers mobilities between 50 and 100 cm^2 (Vs)$^{-1}$ are close to previously reported values measured using TRMC or optically pump terahertz probe measurements (OPTP) [11, 16, 17]. From Table 9.1 there is a clear difference between the mobilities of $MAPbI_3$ and $MAPbBr_3$. The mobility according to the classical Drude model is given by

$$\mu = \frac{e\tau}{m^*} \tag{9.11}$$

in which τ is the scattering time and m^* the effective mass. From first principles calculations or magneto-optical studies, the effective masses of the charge carriers have been derived, which show similar values for electrons and holes of $0.10m_0$ to $0.2m_0$ for both materials [18–20]. With respect to the scattering time, τ it has been argued that the interactions between the longitudinal phonon modes are higher for bromide perovskites than iodide perovskites. The resulting shorter scattering times for $MAPbBr_3$ provide an explanation for the differences in the observed mobilities [21]. The values for k_2 for $MAPbI_3$ are very close to previously reported numbers [22]. The higher value in k_2 for $MAPbBr_3$ might find its origin in surface enhanced second-order recombination, as recently deduced [15].

Next, we discuss the large difference in trap densities found in $MAPbI_3$ and $MAPbBr_3$. The density of the latter is more than two orders of magnitude larger. Same is true for the number of dark carriers. The number of traps has of course major effects on the charge dynamics, which is evident from the TRMC decay kinetics shown in Figure 9.4a,b. From the parameters, we can calculate the lifetime of a carrier in the conduction or valence band. Knowing the half life, $\tau_{1/2}$ and the mobilities enables us to get an idea about the charge carrier diffusion length, R using:

$$R = \sqrt{D\tau_{1/2}} \tag{9.12}$$

The diffusion coefficient, D can be derived from the charge mobility obtained by fitting the TRMC traces and using the Einstein–Smoluchowski equation, $D = kT\mu/e$. As shown in Figure 9.4c, R for the minority carriers in $MAPbI_3$ is very small in comparison to that of the majority carriers. At excitation intensities yielding a charge carrier concentration just below N_T we observe that the values of R for both electrons and holes are similar and maximal. At higher densities, R becomes smaller for both types of carriers due to enhanced band-to-band recombination. Thus, the electron and hole diffusion lengths are most balanced at charge densities just above the trap density, which is desirable for optimum collection of both electrons and holes in an operating solar cell.

In Figure 9.4d, the R values are provided for $MAPbBr_3$. In particular for the minority carriers, we see very short values, which can be explained by the fact that minority carriers get trapped on very short time scales. Only at very high laser intensities do both diffusion lengths become similar. However, at these laser intensities, rapid band-to-band recombination is also operative due to the correspondingly high concentrations. Hence, generally substantially shorter values of R are found for $MAPbBr_3$. In summary, we conclude that due to smaller mobility values as well as the substantially larger concentration of deep trap states, the charge carrier diffusion lengths, in particular for the minority carriers in $MAPbBr_3$, are much smaller than those in $MAPbI_3$.

9.4 TRMC Measurements on $MAPbI_3$ and $MAPbBr_3$ with Charge Selective Contacts

While in the previous part we focused on mobilities and trap states, in this part we want to elucidate the types of carriers and traps responsible for the observed charge carrier dynamics. Since TRMC does not differentiate between holes and electrons and both carriers contribute to the photo conductance according to Eq. (9.1), other specific methods need to be invoked. To reveal which type of carrier is the minority carrier for both layers, we deposited a CSTL on top of the MHP. As electron-accepting layer, we used C_{60}, while for hole transport layer, Spiro-OMeTAD was used (see diagrams at top of Figure 9.5). Figure 9.5 shows the effect of introducing the CSTL on $MAPbI_3$ and $MAPbBr_3$, together with the bare MHP, using identical laser wavelengths and intensities. The signal of the bare MHP layer is normalized to 1, and the bilayer is normalized with the same factor. As discussed previously, two important aspects can be noticed [15, 23, 24]. First, the signal for all bilayers drops quickly below the signal of the bare perovskite layer. Obviously, the

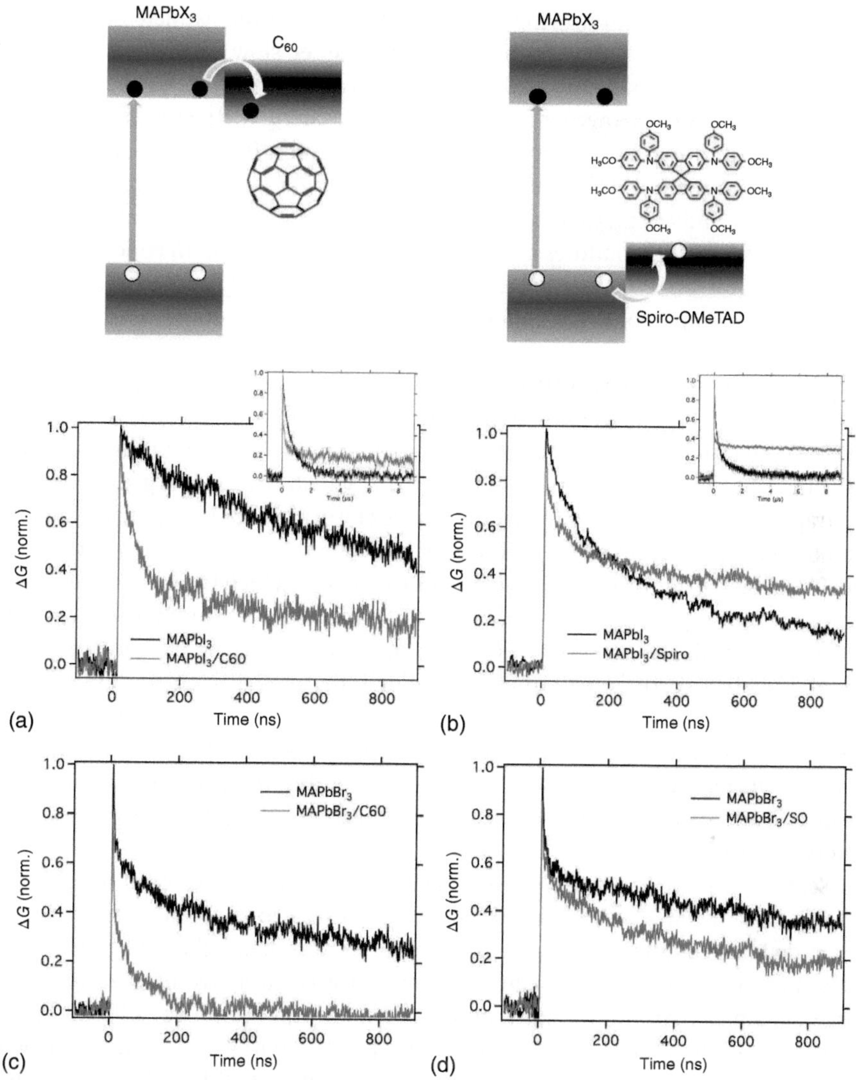

Figure 9.5 Top panels illustrate the charge extraction process: For left panels, the CSTL is C_{60}, while for right panels; the CSTL is Spiro-OMeTAD. (a) and (b) show the TRMC traces recorded at initial charge carrier densities of $5-10\times10^{14}\,cm^{-3}$ per pulse for single and double layers comprising $MAPbI_3$ with excitation at $\lambda = 650$. (c) and (d) are the traces for $MAPbBr_3$ with excitation at $\lambda = 500\,nm$. The signal of the bare MHP layer is normalized to 1, while the signal of the bilayer is normalized with the same factor. Insets show TRMC traces on 10 ms timescale. Source: Adapted from Zhao et al. [15].

CSTLs efficiently extract parts of the carriers. Since the mobilities for electrons in C_{60} and for holes in spiro-OMeTAD are both smaller than in perovskite, the overall TRMC signal lowers. Thus, it is important to realize that the predominant part of the TRMC signal is due to counter charges remaining in the perovskite. Interestingly, for $MAPbI_3$-based bilayers, both insets show that the charge carrier lifetimes extend over many microseconds, implying that recombination of the extracted carrier by the CSTL with the remaining counter charge is rather slow. If full extraction of either electrons or holes from a CSTL occurs, it is feasible to assess the mobility of the counter charge. However, since, apart from charge extraction also charge trapping and band-to-band recombination also occur simultaneously, extraction is most likely not complete. However, we do see that similar long-lived signals appear for both CSTLs, indicating that extraction for both transport layers takes place and that the charge carrier mobilities are close to being balanced [25]. In conclusion, we can state that for $MAPbI_3$, both charge carriers are efficiently extracted and their mobilities are comparable. The latter is in line with the results of the fits to the bare layers.

Finally, we observe a very different behavior for the bilayers with $MAPbBr_3$. For $MAPbBr_3/C_{60}$, the signal drops quickly to almost zero, while for the $MAPbBr_3$/spiro-OMeTAD the differences with the bare layer are limited. Combining this information with the results from the fits from Section 3, which revealed large amounts of deep trap states, we can conclude that these states form traps for holes and the majority mobile carriers are electrons. This can be rationalized as follows: Optical excitation of $MAPbBr_3$ leads to generation of excess carriers, in which the holes are quickly trapped. These trapped holes are not able to be extracted by the Spiro-OMeTAD, while the C_{60} rapidly extracts the electrons. In the latter bilayer, we conclude that no highly mobile carriers remain, since the mobility in C_{60} is $<1\,cm^2\,(Vs)^{-1}$ and the corresponding holes are rapidly trapped in $MAPbBr_3$, explaining the fast decaying signal. These observations can explain the typically poor performances observed for solar cells based on $MAPbBr_3$. Hence by combining the perovskite layer with transport layers specific information can be obtained regarding the type of minority carriers and deep traps.

In Sections 9.1 and 9.2, we presented the basics of the time-resolved microwave technique and the analysis of decay kinetics. TRMC is a contactless technique, and thus, it does not require prolonged exposure to electric fields that can potentially lead to ion migration in perovskites, making it a very attractive method to study charge dynamics [26]. Additionally, photogeneration can occur over a broad range of intensities in the 1-sun equivalent regime [11]. In Sections 9.3 and 9.4, we applied this technique to $MAPbI_3$ and $MAPbBr_3$ with and without CSTLs. Analysis of the fits on the bare layers recorded using various light intensities yields the kinetic parameters describing the charge carrier dynamics. From the bilayers, specific information regarding the electron and hole mobility can be obtained, as well as the sign of the carriers. While $MAPbI_3$ can be regarded as highly intrinsic with doping densities $p_0 < 10^{14}\,cm^{-3}$, $MAPbBr_3$ behaves as unintentional n-type doped with $n_0 > \times 10^{15}\,cm^{-3}$. Furthermore, the high number of deep trap states makes the charge carrier diffusion length of the minority carriers in $MAPbBr_3$ small, $<0.5\,\mu m$, which is typically not sufficient for optimal solar cell performance. In this chapter, we have shown that the TRMC technique is able to reveal the charge carrier dynamics on timescales relevant for optoelectronic devices. The versatile and electrodeless nature of the technique allows screening of perovskite layers with differences in material composition or deposition method.

Acknowledgement

This work was supported by the Netherlands Organization for Scientific Research (NWO) under the Echo grant number: 712.014.007.

References

1 Kroeze, J.E., Savenije, T.J., Vermeulen, M.J.W., and Warman, J.M. (2003). *The Journal of Physical Chemistry B* 107: 7696.
2 Warman, J.M., De Haas, M.P., Dicker, G. et al. (2004). *Chemistry of Materials* 16: 4600.
3 Savenije, T.J., Ferguson, A.J., Kopidakis, N., and Rumbles, G. (2013). *Journal of Physical Chemistry C* 117: 24085.
4 Oga, H., Saeki, A., Ogomi, Y. et al. (2014). *Journal of the American Chemical Society* 136: 13818.
5 Hutter, E.M., Savenije, T.J., and Ponseca, C.S. Jr. (2017). *Perovskite Solar Cells Principle, Materials. Devices*, 179–232. World Scientific Publishing Co. Pte. Ltd.
6 Kerner, R.A., Schulz, P., Christians, J.A. et al. (2019). *APL Materials* 7. https://doi .org/10.1063/1.5083812.
7 Zhao, L., Kerner, R.A., Xiao, Z. et al. (2016). *ACS Energy Letters* 1: 595.
8 Reid, O.G., Moore, D.T., Li, Z. et al. (2017). *Journal of Physics D: Applied Physics* 50. https://doi.org/10.1088/1361-6463/aa9559.
9 Muscarella, L.A., Hutter, E.M., Sanchez, S. et al. (2019). *Journal of Physical Chemistry Letters* 10: 6010.
10 Haber, K.S. and Albrecht, A.C. (1984). *The Journal of Physical Chemistry* 88: 6025.
11 Levine, I., Gupta, S., Bera, A. et al. (2018). *Journal of Applied Physics* 124. https://doi.org/10.1063/1.5037637.
12 Stranks, S.D., Burlakov, V.M., Leijtens, T. et al. (2014). *Physical Review Applied* 2: 1.
13 Hutter, E.M., Eperon, G.E., Stranks, S.D., and Savenije, T.J. (2015). *Journal of Physical Chemistry Letters* 6: 3082.
14 Miyata, A., Mitioglu, A., Plochocka, P. et al. (2015). *Nature Physics* 11: 582.
15 Zhao, J., Caselli, V.M., Bus, M. et al. (2021). *ACS Applied Materials and Interfaces* 13: 16309.
16 Brenes, R., Guo, D., Osherov, A. et al. (2017). *Joule* 1: 155.
17 Herz, L.M. (2017). *ACS Energy Letters* 2: 1539.
18 Brivio, F., Butler, K.T., Walsh, A., and Van Schilfgaarde, M. (2014). *Physical Review B: Condensed Matter and Materials Physics* 89: 1.
19 Galkowski, K., Mitioglu, A., Miyata, A. et al. (2016). *Energy & Environmental Science* 9: 962.
20 Whalley, L.D., Frost, J.M., Jung, Y.K., and Walsh, A. (2017). *The Journal of Chemical Physics* 146. https://doi.org/10.1063/1.4984964.

21 Wright, A.D., Verdi, C., Milot, R.L. et al. (2016). *Nature Communications* 7. https://doi.org/10.1038/ncomms11755.

22 Davies, C.L., Filip, M.R., Patel, J.B. et al. (2018). *Nature Communications* 9: 1.

23 Hutter, E.M., Hofman, J.J., Petrus, M.L. et al. (2017). *Advanced Energy Materials* 7. https://doi.org/10.1002/aenm.201602349.

24 Caselli, V.M., Wei, Z., Ackermans, M.M. et al. (2020). *ACS Energy Letters* 5: 3821.

25 Ponseca, C.S., Savenije, T.J., Abdellah, M. et al. (2014). *Journal of the American Chemical Society* 136: 5189.

26 Futscher, M.H., Lee, J.M., McGovern, L. et al. (2019). *Materials Horizons* 6: 1497.

10

Hot Carriers in Halide Perovskites

Jia Wei Melvin Lim[1,2], Yue Wang[1], and Tze Chien Sum[1]

[1] *Nanyang Technological University, School of Physical and Mathematical Sciences, Division of Physics and Applied Physics, 21 Nanyang Link, Singapore 637371, Singapore*
[2] *Nanyang Technological University, Energy Research Institute @ NTU (ERI@N), Interdisciplinary Graduate School, Singapore 637371, Singapore*

10.1 Introduction

The well-reported exemplary optoelectronic properties of halide perovskites have propelled their use as excellent solar absorbers in perovskite solar cells (PSCs). Current record power conversion efficiencies (PCE) of PSCs have exceeded 25% in just over a decades' worth of research and are approaching the Shockley–Queisser (SQ) theoretical limit [1]. The maximum PCE of a single junction solar cell, according to detailed balance calculations, is shown in Figure 10.1a. For a standard single-junction PSC with an absorber bandgap of ~1.6 eV that is typical of a standard PSC absorber composition, this limit is ~31%. With efficiencies just a few percent shy of the SQ limit, recent advancements in improving PCEs have decelerated considerably, with efforts to optimize the absorber composition and morphology as well as engineering of the hole and electron transport layers (HTLs and ETLs) seemingly well-explored and exhausted.

Consequently, more attention has been drawn to exploring other advanced photovoltaic concepts to harness even more energy from the Sun in recent years. Figure 10.1b shows the various loss channels present in a single junction solar cell with an absorber bandgap of 1.6 eV and their respective contributions indicated on the standard AM1.5G solar spectrum. Undoubtedly, the largest source of losses is due to unabsorbed photons of energy less than the absorber bandgap, which could be harnessed in concepts such as tandem and intermediate band solar cells. The focus of our discussion in this chapter concerns the second-highest loss channel, which is attributed to the cooling loss of high energy photons from the Sun; that is, for photons with energy $E > E_g$. The energy from this loss channel could be harnessed through the novel concepts of the *hot carrier solar cell* (HCSC) and *multiple exciton generation solar cell* (MEGSC), which will be discussed in detail in Sections 10.4 and 10.8, respectively.

Halide Perovskite Semiconductors: Structures, Characterization, Properties, and Phenomena, First Edition. Edited by Yuanyuan Zhou and Iván Mora-Seró.

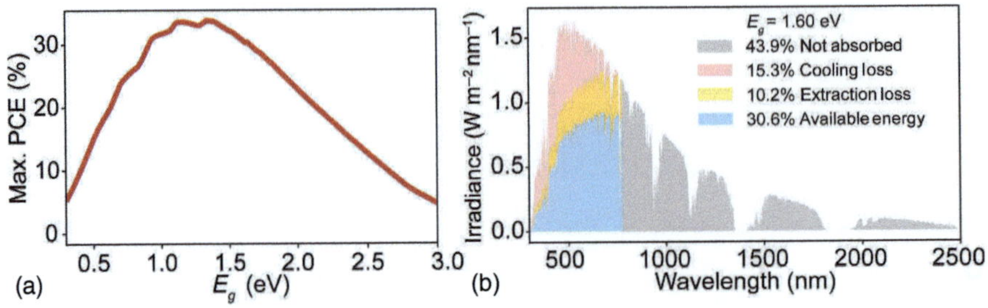

Figure 10.1 Shockley–Queisser (SQ) limit and the losses in single junction solar cells. (a) Maximum power conversion efficiencies (PCE) versus bandgap (E_g) for a single junction solar cell under 1 Sun illumination calculated according to the SQ detailed balance limit. (b) The various contributions to losses limiting the PCE of a single junction solar cell of $E_g = 1.60$ eV.

10.1.1 Potential of Perovskites for Next-Generation Photovoltaics

The potential of halide perovskites for use in novel next-generation photovoltaic concepts was first highlighted by Xing et al., who reported slow hot hole cooling in the archetypal $CH_3NH_3PbI_3$ ($MAPbI_3$) halide perovskite (Figure 10.2) [2]. Using transient absorption (TA) spectroscopy with energy-selective excitations as seen in Figure 10.2a,b, the authors unveiled a relatively slow hot carrier cooling time of ~0.4 ps. A slowing of the band edge photobleaching (PB) rise time dynamics, accompanied by a decay in the higher energy PB, was seen with the higher energy pump (Figure 10.2a) compared to the lower energy one seen in Figure 10.2b. By studying

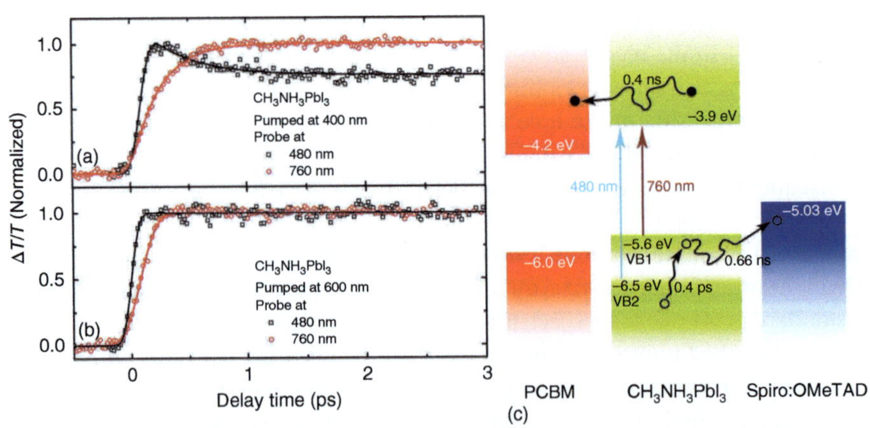

Figure 10.2 Slow hot hole cooling is revealed by transient absorption (TA) spectroscopy. (a, b) Early time dynamics of the high (480 nm, black squares) and low energy (760 nm, red circles) photobleaching peaks in the TA spectra of $CH_3NH_3PbI_3$ ($MAPbI_3$) pumped at (a) 400 nm and (b) 600 nm. (c) Schematic depicting energy alignment of the conduction and valence bands in $MAPbI_3$, PCBM and Spiro:OMeTAD. The arrows depict the transitions that give rise to the dynamics in (a, b). Source: Reproduced from Xing et al. [2]/with permission of American Association for the Advancement of Science.

the dynamics with different extractor layers, the authors constructed the energy band alignments of $MAPbI_3$ seen in Figure 10.2c and assigned the ∼0.4 ps lifetime to the cooling of the hot holes. A subsequent work also further confirmed the hot electron cooling time to be similar, about ∼0.4 ps [3]. Since then, numerous studies have reported similar slow-hot carrier cooling properties in other halide perovskite compositions and even in nanostructures.

The hot carrier cooling time of halide perovskites is longer than that of most organic semiconductors, whose cooling time is typically on the scale of ∼100 fs. Such slow hot carrier cooling time is a prerequisite for both HCSCs and MEGSCs, which are both novel next-generation photovoltaic concepts that were proposed to transcend the SQ limit. This slow hot carrier cooling time is required for hot charge extraction in HCSCs, while the cooling time should be sufficiently slow for the multi-exciton generation process to occur efficiently in MEGSCs. Hence, the interest in halide perovskites for next-generation photovoltaic concepts was invigorated given their exceptional prospects for these technologies.

10.2 Hot Carrier Cooling Mechanisms

Hot carriers refer to photogenerated charge carriers in a material that possess energy larger than the bandgap energy of the material. In general, these hot carriers will rapidly "cool" by losing the excess energy through various processes until they reach the band edge. The cooling process of hot carriers is generally complicated, with multiple competing processes taking place simultaneously. However, it is possible to separate these processes by the timescales on which they dominate, which gives rise to the different stages of the whole cooling process. A schematic of these main processes is depicted in Figure 10.3 below, and the following paragraphs describe each stage in detail.

Carrier Thermalization (10–100 fs): Within the first few femtoseconds after photoexcitation, the population of non-thermally excited carriers undergoes carrier thermalization to form a quasi-thermalized carrier distribution obeying Fermi

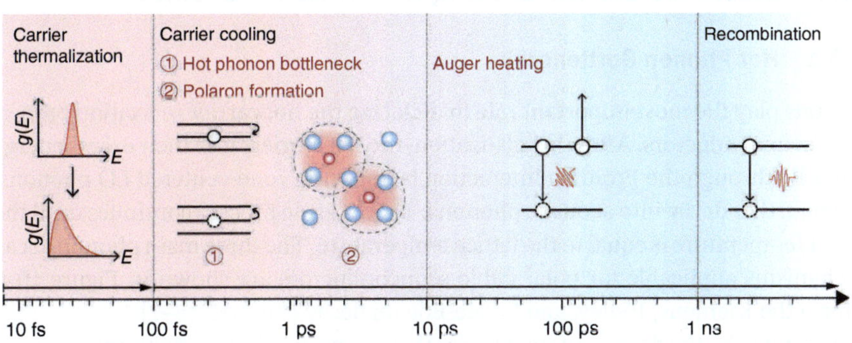

Figure 10.3 Schematic of the timescale of the main ultrafast processes involved in hot carrier cooling of halide perovskites.

statistics with a characteristic carrier temperature (T_{eh}). This process is mainly dominated by carrier-carrier scattering events, where the excess energy is rapidly distributed among the nonequilibrium carrier population. In this initial stage of carrier cooling, the energies of the carriers are still conserved. The carrier thermalization rates in halide perovskites are strongly dependent on both the excitation energy and carrier density, where the rates increase with both quantities, indicating that carrier-carrier scattering is the main mechanism in this process [4]. Typically, this process is completed within 100 fs in halide perovskites.

Lattice thermalization (100 fs to >1 ns): Phonons come into play in this second stage of cooling, where the excess energies of the hot carriers are lost to the crystal lattice through the emission of phonons. The main mechanism responsible for this stage in polar semiconductors like halide perovskites is carrier-phonon scattering through the Fröhlich interaction, in which hot carriers lose energy to the lattice by emission of longitudinal optical (LO) phonons. LO phonon emission continues until the carrier temperature equals the lattice temperature. Hence, this process persists until the carriers have fully thermalized to the lattice temperature and their eventual recombination in the nanosecond timescale. This is the main energy loss channel for excess energies in hot carriers, and most of the mechanisms behind the slow hot carrier cooling process in halide perovskites pertain to this stage of carrier cooling. They will be discussed in detail in a later part of this chapter.

Polaron formation (100–1000 fs): In polar semiconductors, the emitted LO phonons can couple with carriers to form polarons, which are electrons or holes dressed by LO phonons that move together in the lattice as a large quasiparticle. Polarons exhibit different transport properties as compared to the bare carriers and are also thought to be responsible for certain observed slow cooling phenomena of hot carriers in halide perovskites. The properties of the polaron in semiconductors depend on the Fröhlich coupling constant (α) that measures the strength of the electron–phonon coupling. The timescale of polaron formation is similar to the LO phonon lifetimes in halide perovskites.

10.3 Slow Hot Carrier Cooling in Halide Perovskites

10.3.1 Hot Phonon Bottleneck

Phonons play the most important role in dictating the hot carrier relaxation properties of semiconductors. After thermalization, the hot carriers lose their excess energy primarily through the Fröhlich interaction by emitting zone-centered LO phonons, which further decay into acoustic phonons. This cascade process continues until the carrier temperature is equal to the lattice temperature. The three main phonon decay mechanisms applicable for polar cubic semiconductors are shown in Figure 10.4, namely the Klemens, Ridley, and Vallée-Bogani decay channels [5–7].

As denoted by the decay equation under their respective illustrations, the Klemens decay channel involves the decay of a LO phonon into two daughter longitudinal acoustic (LA) phonons. The Ridley channel involves the decay of a LO phonon into

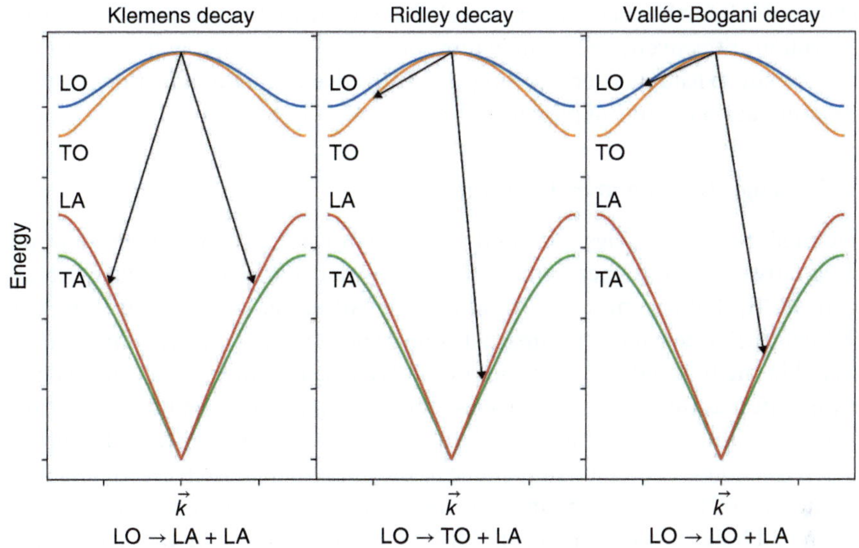

Figure 10.4 Main LO phonon decay channels in polar semiconductors. LO phonons decay into a combination of optical and/or acoustic daughter phonons through the Klemens, Ridley and Vallée-Bogani decay channels. The arrows illustrate possible transitions for each mechanism and are not drawn to scale. In the actual process, both energy and momentum must be conserved.

a transverse optical (TO) and a LA phonon. Lastly, the Vallée-Bogani decay channel involves the decay of a LO phonon into another LO phonon and a LA phonon. We note here that the main decay channel responsible for carrier cooling in halide perovskite materials is through the Klemens channel for the following reasons. Firstly, the Ridley channel is less efficient for halide perovskites, which possess highly symmetric crystal structures. This causes the LO and TO branches to be essentially overlapping, and hence its inefficiency to dissipate excess energies. On the other hand, the Vallée-Bogani channel is expected to be less efficient because of the relatively small LO phonon energy of halide perovskites, which is on the order of ~10 meV for lead iodide perovskites and ~20 meV for lead bromide perovskites. This mechanism was found to be the dominant channel for GaAs, which has larger LO phonon energy (~36 meV). We emphasize that while these channels are less efficient, it does not mean that they do not participate in LO phonon relaxation. The observed phonon dynamics is expected to be an ensemble average of all three decay channels, with the Klemens channel being the main channel dictating the energy loss to the lattice.

A hot phonon bottleneck *effect* was found to be present in halide perovskites and was used to explain the relatively slow hot carrier cooling in halide perovskites [2, 8]. Later, it was found that this bottleneck effect was due to the impediment of the Klemens decay channel caused by a large gap between the LO and LA phonon branches, effectively causing a large phonon bandgap that slows LO phonon relaxation [9]. Furthermore, the small LO phonon energy of halide perovskites leads to a large build-up of LO phonons at the lowest LO phonon branch, which can both inhibit further LO phonon emission and increase the likelihood of phonon reabsorption. Effectively, the combination of small LO phonon energy and large phonon bandgap leads to a lengthening of the effective LO phonon lifetime, causing the manifestation of this hot phonon bottleneck effect in halide perovskites.

10.3.2 Auger Heating of Hot Carriers

At elevated carrier densities, a second stage of hot carrier cooling emerges in the cooling curves of halide perovskites. This was later attributed to Auger heating (Figure 10.5a) that becomes significant at high carrier densities of $\sim 10^{19}$ cm^{-3}. At such high carrier densities, multiparticle effects become more likely, which can be observed in the band edge PB kinetics in the TA spectra. Multiparticle effects generally manifest as the appearance of a fast decay component with increasing carrier densities. It was found that this second stage of hot carrier cooling has a lifetime of tens of picoseconds, which is comparable to the nonradiative Auger recombination (AR) lifetime of halide perovskites. In this inter-band Auger heating process, an electron and hole pair recombine, and the energy is transferred nonradiatively to a third carrier, thereby reheating it above the band edge. This process is depicted in the schematic in Figure 10.5b. This further lengthens the lifetime of the hot carriers by tens of picoseconds, enabling them to remain above the lattice temperature for longer periods. The combination of the hot phonon bottleneck effect and Auger heating effectively lengthens the lifetime of hot carriers in halide perovskites, as seen in Figure 10.5a. Without the participation of these effects (magenta dashed line), the hot carriers would cool within $\sim$1 ps.

10.3.3 Large Polaron Formation

At low carrier densities below the Mott density ($<10^{18}$ cm^{-3}), long-lived energetic hot carriers can still be observed at carrier densities as low as $\sim 10^{16}$ cm^{-3}, which was

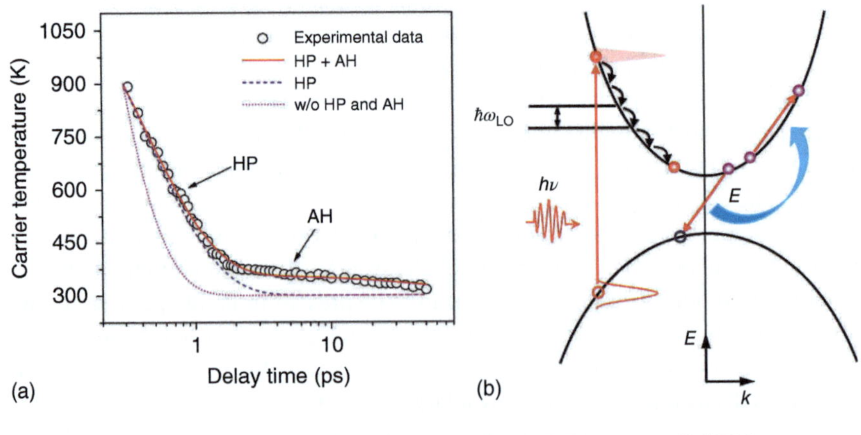

Figure 10.5 Auger heating in halide perovskites. (a) Cooling curve of $MAPbI_3$ post-excitation at 2.48 eV. A second stage of carrier cooling manifests and is attributed to Auger heating (AH). (b) Schematic depicting the mechanism of AH. The energy released by one electron–hole pair re-excites a third carrier. Source: Fu et al. [9]/Springer Nature/Licensed under CC BY 4.0.

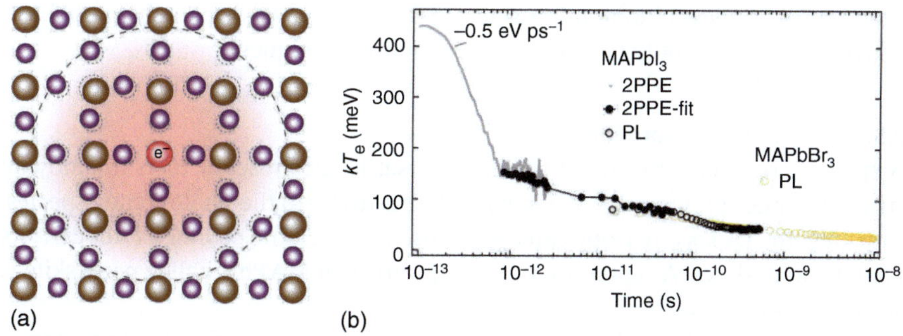

Figure 10.6 Large polarons in halide perovskites. (a) Illustration of a polaron. The presence of an electron localized at a lattice site distorts the surrounding lattice and creates a polarized cloud of ions. (b) Carrier excess energies as a function of time post-excitation in $MAPbI_3$. The carrier cooling rate slows considerably following the formation of polarons. Source: Reproduced from Joshi et al. [10]/with permission of John Wiley & Sons.

attributed to the large polaron screening of hot carriers [10]. As seen in Figure 10.6a, the presence of a charge-carrier (hole or electron) in a polar semiconductor distorts the surrounding lattice, attracting and repelling oppositely charged ions, giving rise to a polarized cloud of ions that moves together with the charge carrier. This is equivalent to a charge carrier dressed by phonons and can be thought of as a quasi-particle called the polaron. The term large polaron refers to the Fröhlich polaron, whose size is larger than the lattice constant. In $MAPbI_3$, the calculated polaron radius is ~25 Å, much larger than the unit cell [11].

Notably, the dressing of charge carriers with the polarized cloud of ions (phonons) screens the charge-carriers and reduces subsequent phonon scattering rate. After polaron formation, the cooling of hot carriers slows down considerably due to this screening effect. As seen in Figure 10.6b, the excess energies of carriers in $MAPbI_3$ initially decay rapidly in the first ~0.5 ps. Thereafter, the cooling rate slows considerably after polaron formation sets in, and the carriers can remain ~100 meV above the bandgap energy into the nanosecond timescale.

One stark difference between this mechanism and those discussed earlier is that this mechanism only dominates at low carrier densities (~10^{16} to 10^{18} cm^{-3}). Moreover, the hot carrier population decreases with increasing carrier density in this range, which is the complete opposite trend to that expected when considering the hot phonon bottleneck mechanism. This is due to the destabilization of the large polarons, where they start to overlap at increasing carrier densities, leading to less effective screening of the hot carriers. For even larger carrier densities beyond the Mott density, the carriers exist as ionized plasmas, and the hot carrier cooling is mainly governed by the hot phonon bottleneck and Auger heating effects.

10.3.4 Spectroscopic Signature of Hot Carriers

Ultrafast spectroscopy is the most common type of technique used to study the cooling of hot carriers in halide perovskites since the typical time resolution of ultrafast techniques is sufficient to study the main stages of hot carrier cooling following carrier thermalization. The two most widely employed ultrafast techniques are ultrafast

TA and time-resolved photoluminescence (TRPL). Here, we discuss the manifestations of the signature of hot carriers in these main techniques.

10.3.4.1 Transient Absorption

TA spectroscopy remains one of the most versatile spectroscopy tools in studying carrier dynamics of materials due to its high sensitivity to the photoexcited species. In this regard, TA spectroscopy can also be used to track the cooling of hot carriers in halide perovskites. A representative TA spectrum of $MAPbI_3$ halide perovskite is shown in Figure 10.7a.

The main photophysical processes that are related to carrier cooling happen within the first 1–2 ps, which is the focus of the discussion here. As can be observed, at this moderate carrier density, the main PB peak that is centered around the bandgap energy of the perovskite grows to its maximum within ~1 ps. Furthermore, there is an induced absorption band at energies below the bandgap value, which decays with the concomitant redshifting of the main PB peak. This redshift is highlighted in the magnified view of the contour plot in Figure 10.7a, and it can be attributed to the competing effects of the Burstein–Moss and band gap renormalization (BGR) phenomena. The former induces a blueshift of the bandgap, while the latter induces a redshift of the bandgap. At early time delays, the BGR effect dominates, causing an overall observable redshift. On the other hand, the TA spectra on the high-energy side of the PB peak exhibits an obvious broadening; this part of the TA spectra is widely known as the "high energy tail" by the hot carrier community. Figure 10.7b shows time slices of the TA spectra in panel an at several time delays, where the high-energy tail is much more clearly visualized. As can be seen, the high-energy tail decays within the 1.4 ps time window and is attributed to the presence of hot carriers that possess energies larger than the bandgap energy of the material. This tail becomes more obvious at high fluences. The decay of

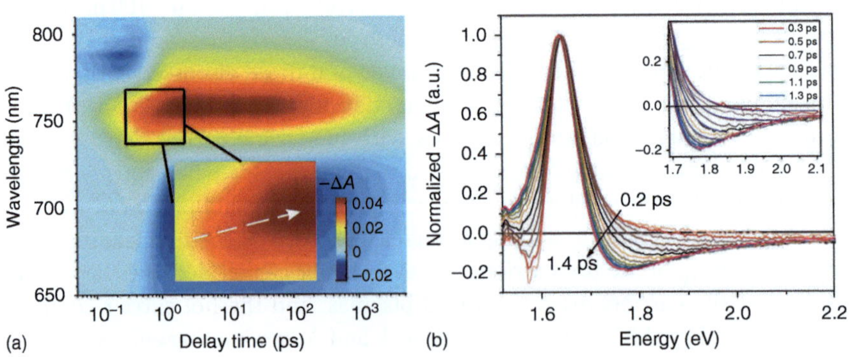

Figure 10.7 Spectroscopic signature of hot carriers in halide perovskites. (a) Contour plot of the TA spectra of $MAPbI_3$ excited above bandgap at moderate carrier density. Obvious broadening and redshift of the main photobleach peak are seen at early time delays, which is clearly visible in the magnified view. (b) Corresponding TA spectra in the first 1.4 ps showing the dynamics of the high energy region over time. Source: Fu et al. [9]/Springer Nature/Licensed under CC BY 4.0.

this broadening coincides with the simultaneous growth of the main PB peak and characterizes the hot carrier cooling rate of the carriers.

Typically, there are two main methods of characterizing the cooling of hot carriers using TA spectroscopy. The first and most straightforward method is to study the ground state bleach (GSB) dynamics in the TA spectra of halide perovskites. After high-energy above-bandgap pump excitation, the carriers cool and repopulate the band edge in a timescale comparable to the hot carrier cooling time. Hence, the rise time of the GSB kinetics can be used as a measure of the cooling time of the hot carriers, where a lengthening of the GSB rise time can be observed for increasing pump energy and/or fluences. The rise time is typically extracted by fitting the kinetics with a rising exponential function that is convolved with the instrument response function (IRF). As a word of caution, the IRF should be shorter than the total rise dynamics to ensure accuracy in the values of the extracted rise times.

The second method of characterizing the hot carrier dynamics from the TA spectra of halide perovskites is through the construction of the cooling curves, which is the time dependence of the carrier temperature (T_c) following excitation. As mentioned earlier, in the early-time TA spectra of halide perovskites, the high-energy side of the main PB peak exhibits an obvious broadening that implies the presence of hot carriers. By fitting the TA spectra at several time delays, either partially by approximating with a Maxwell–Boltzmann (MB) distribution function [8], or a full spectrum fit [12], the carrier temperatures can be extracted and used to construct the cooling curves. The MB fitting method remains the most widely used due to its simplicity and low computation requirement. In this method, when the condition $E \gg E_F$ holds (E_F is the quasi-Fermi level splitting), the quasi-thermalized Fermi–Dirac distribution of hot carriers can be approximated by the MB distribution function to fit the high-energy side of the main bleaching peak in the TA spectra as:

$$\Delta A(E) \propto A_0(E) \exp \frac{E_F - E}{k_b T_c} \tag{10.1}$$

where $\Delta A(E)$ is the high energy part of the TA spectra, $A_0(E)$ is the linear absorbance, k_b is the Boltzmann constant, and T_c is the carrier temperature. The cooling curves, as shown in Figure 10.5a, can be constructed, and the stages of hot carrier cooling can be distinguished. Typically, the cooling curve can be fitted with a biexponential decay function with two lifetimes corresponding to the two stages of hot carrier cooling. It is worth noting that this method cannot distinguish the individual carrier temperatures of the electron and hole; they are assumed to be equal. This is reasonable given the roughly equal effective masses of the electron and holes in halide perovskites. Later, it was found that this MB fitting method overestimates the actual carrier distribution temperature even though the dynamics of the cooling curves seem to be largely unaffected [12]. When accuracy in the carrier temperatures is crucial, a full-spectrum fit approach should be used, albeit with increased computation time and workload:

$$\Delta A = (1 - f_e(E, E_F, T_c))^2 \times A_{exc} - A_0 \tag{10.2}$$

where ΔA is the change in absorption, f_e is the Fermi–Dirac distribution function, and A_{exc}, A_0 are the absorption spectra after and before excitation, respectively. It is important to model A accurately and not treat it like a constant density of states. Normally, the Elliott equation is sufficient for describing A with reasonable accuracy within the narrow window required for fitting. Using this method, the entire TA spectra can be reproduced unambiguously without any discrepancies arising from the choice of fitting region when using the MB distribution fitting method, and the quantities T_c and E_F can be accurately determined from the fitting.

10.3.4.2 Fluorescence-Based Techniques

Fluorescence-based techniques such as TRPL or ultrafast photoluminescence (PL) techniques can also be used as a more direct approach to study the hot carrier cooling dynamics. Under moderate to high pump intensities, the PL peak of halide perovskites exhibits an obvious blueshift and broadens toward the high energy side. As seen in Figure 10.8a, the main PL peak in the TRPL spectra of $FASnI_3$ exhibits obvious broadening toward the high energy side. This can be attributed to the radiative emission from hot carriers with energies above the bandgap energy of the material. Gradually, this broadening subsides, depending on the hot carrier cooling rate. In Figure 10.8b, the PL intensity monitored at different parts of the PL peak shows slightly different kinetics. Notably, the time trace of the PL intensity at a higher energy of 1.65 eV that is in the region of the high energy broadening exhibits an additional faster component that can be attributed to the hot PL emission and has a much shorter lifetime than the main PL peak. The extracted carrier temperatures (Figure 10.8c) from the PL spectra also show similar dynamics, confirming the nature of the high energy emission to be from hot carriers.

Moreover, hot PL emission can also be observed in halide perovskites under continuous-wave (CW) excitation at sufficiently high intensities. As seen in Figure 10.8d, the CW PL spectra of $FAPbI_3$ thin film exhibit similar behavior to that under pulsed excitation conditions. The high-energy broadening and blueshift of the main PL peak can also be observed when the excitation fluence is varied between 0.5 and 40 W cm^{-2}. This observation confirms that the hot carriers can be sustained even under continuous illumination conditions.

Like conventional group III–V semiconductors, the generalized Planck equation can also be used to describe the emission profile of halide perovskites:

$$I_{PL}(E) = \frac{2\pi}{h^3c^2}\frac{E^2A_0(E)}{\exp\left(\frac{E-\Delta\mu_{eh}}{k_bT_{eh}}\right)-1} \tag{10.3}$$

where I_{PL} is the photoluminescence intensity, h and c the Planck constant and speed of light, $A_0(E)$, $\Delta\mu_{eh}$ and T_{eh} are the absorption spectrum, quasi-Fermi splitting, and the carrier temperature respectively. Similar to the generalized model for the TA spectra (Eq. (10.2)), an appropriate model for $A(E)$ should be used, in which the Eliott equation is usually sufficient to describe it within the fitting window [14]. Sometimes, the MB distribution form of Eq. (10.3) is also used as an approximation to fit the high energy region of the PL peak, for energies greater than the quasi-fermi level:

$$I_{PL}(E) \propto A_0(E)\exp\left(\frac{\Delta\mu_{eh}-E}{k_bT_{eh}}\right) \tag{10.4}$$

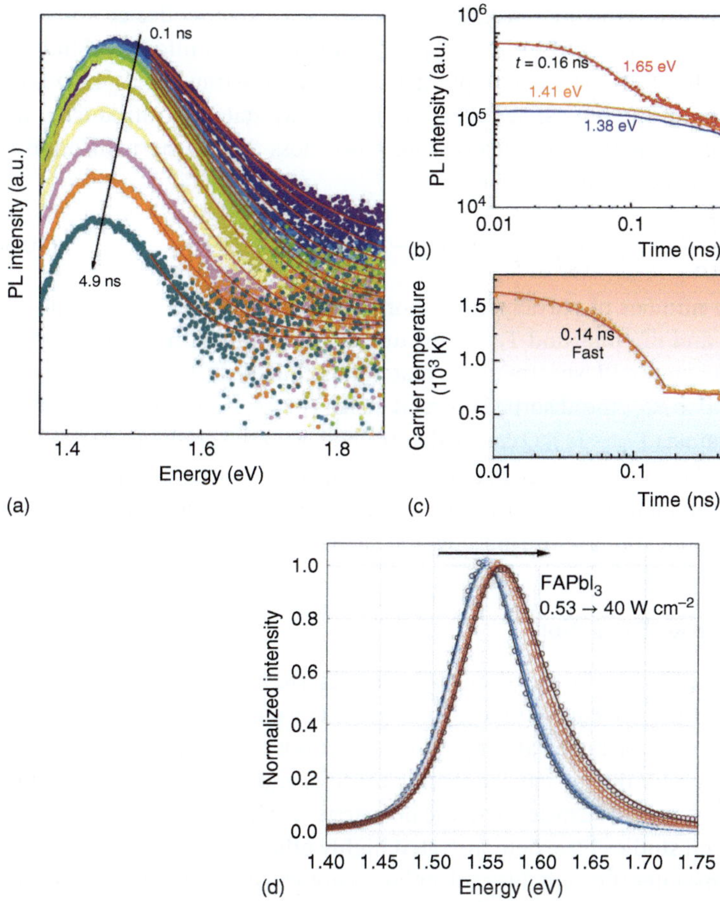

Figure 10.8 Spectroscopic signature of hot carriers in the photoluminescence (PL) spectra of halide perovskites. (a) Time-resolved photoluminescence spectra of $FASnI_3$ thin film at several time delays. (b) The time trace of the PL intensity at several energies. A second component is visible for the time trace at the high energy shoulder of the PL peak. (c) The corresponding extracted carrier temperatures versus delay time. (d) PL spectra of $FAPbI_3$ thin film illuminated under continuous-wave excitation. Obvious broadening toward the high-energy side and blueshift are observed at high excitation densities. Source: (a–c) Fang et al. [13]/Springer Nature/Licensed under CC BY 4.0. (d) Reproduced from Lim et al. [14]/with permission of American Chemical Society.

However, we again stress that this may compromise the accuracy of the obtained absolute values of T_{eh}. When accuracy in these values is crucial, the full form (Eq. (10.3)) should be used. By fitting the PL spectra at several time delays post-excitation, the cooling curves can also be constructed to track the cooling of the hot carriers. Furthermore, the power-dependent steady-state photoluminescence (SSPL) spectra, which is the time-averaged TRPL spectra, can be collected and fitted with Eq. (10.3) gives the trend of how T_{eh} evolves with pump intensity. Studying the evolution of the SSPL spectra with pump fluence and the trend of T_{eh} allows for the quantification of the hot carrier effects in the steady-state. By introducing the thermalization coefficient Q, the steady-state power loss due to thermalization to the lattice (P_{th}) can be quantified [15]:

$$P_{th} = \frac{n_0 E_{LO}}{\tau_{LO}} \exp\left(-\frac{E_{LO}}{k_b T_{eh}}\right) = Q(T_{eh} - T_L)\exp\left(-\frac{E_{LO}}{k_b T_{eh}}\right) \tag{10.5}$$

where n_0 is the number of participating phonon modes, E_{LO} and τ_{LO} are the LO phonon energy and lifetime, and T_L is the lattice temperature. Under open-circuit conditions (as when the PL spectra of the material is collected), P_{th} can be approximated to be equal to P_{abs}, the absorbed power density. Hence, by plotting the quantity $P_{abs} \exp \frac{E_{LO}}{k_b T_{eh}}$ against ($T_{eh} - T_L$), Q can be determined from the gradient of the linear fit. Q represents the power loss due to thermalization per unit area per unit Kelvin of temperature of the material above the lattice temperature. Therefore, a low Q value is desirable for applications requiring slow hot carrier cooling.

10.3.5 Hot Carrier Extraction

One approach to utilizing the excess energy of hot carriers is to extract them before they cool fully to the band edge. This can be achieved by pairing the perovskite layer with a suitable extractor material. Importantly, this hot carrier extraction process competes directly with the hot carrier cooling process, and therefore, the extraction rate of hot carriers should be faster than the hot carrier cooling rate for their efficient extraction. Several studies have demonstrated highly efficient hot carrier extraction from halide perovskites into suitable molecular acceptors such as bathophenanthroline (BPhen) and bathocuproine (BCP) [16–18]. Traditionally, the verification of successful hot carrier extraction is done by studying the TA dynamics with and without the carrier extraction layer. When successful hot carrier extraction occurs, a reduction in the magnitude of the TA spectra, especially on the high energy side, will be observed. This is also accompanied by a faster build-up of the band edge kinetics, indicating faster hot carrier cooling with the extraction layer. The cooling curves will also show a difference, mainly in the reduction of carrier temperatures and faster overall cooling dynamics with the presence of the hot carrier extraction layer. On the other hand, hot carrier extraction can also be studied with other techniques such as pump–push–probe (PPP) spectroscopy. Briefly, the PPP technique is a variant of TA spectroscopy, with the addition of a third "push" pulse that is usually set to arrive at a later delay, after the carriers have fully cooled to the band edge. This push pulse re-excites a portion of the cold carriers above the band edge, and their subsequent dynamics are characterized by the hot carrier cooling rate. Lim et al. demonstrated hot carrier extraction from $MAPbI_3$ into BPhen, a hot electron acceptor molecule, using the PPP technique [17]. A reduction of the TA band edge kinetics with the arrival of the push pulse was observed, and the corresponding change in TA signal, that is, the change due to the push pulse, is long-lived and does not fully recover, implying a loss in band edge carriers in $MAPbI_3$ and validating hot carrier extraction into BPhen. Furthermore, by varying the energy of the push pulse, the authors also found that a Schottky barrier forms at the interface between the two materials that needs to be overcome for the successful extraction of hot carriers, which has great implications in the search for suitable hot carrier acceptors.

10.4 Utilizing Hot Carriers in Halide Perovskites

10.4.1 Hot Carrier Solar Cell

Undoubtedly, the quest to fabricate a working HCSC has been the main motivation behind decades of hot carrier research stemming from conventional group III–V semiconductors, and now in halide perovskites. The theoretical maximum PCE of a single-junction solar cell is limited by the SQ limit of ~33.7%. One of the main loss channels that limits the PCE is the cooling loss of carriers that absorb above-bandgap energy solar photons. This excess energy is lost to the lattice in the form of phonons, and harvesting these excess energies pushes the maximum PCE to ~66%. Hence, a photovoltaic device that could harvest both the hot and cold carriers is called a HCSC and is one of the most promising strategies to overcome the SQ limit. A schematic of the HCSC is given in Figure 10.9. Apart from having the hole transport layer (HTL) and electron transport layer (ETL), energy selective contacts (ESCs) for the holes and

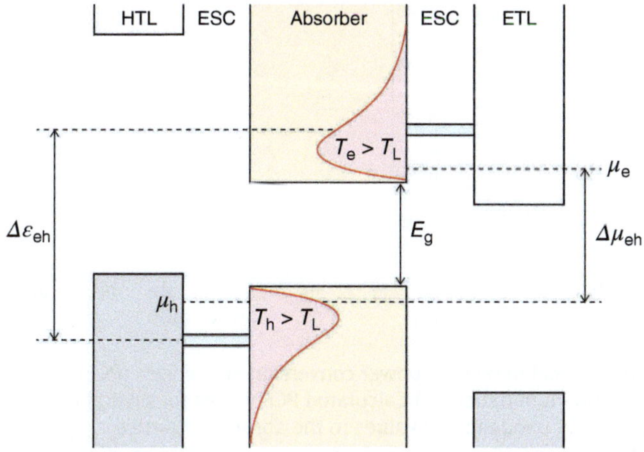

Figure 10.9 Working principle of a hot carrier solar cell. With appropriate energy-selective contacts (ESCs) positioned above the band edges of the absorber, the voltage of the cell can be enhanced.

electrons positioned at energy levels above the respective band edges are needed to extract the hot carriers.

The voltage of the cell is enhanced with the extraction of hot carriers, which is above the limit of a conventional solar cell that extracts only cold carriers. The external voltage of the HCSC is given by:

$$eV_{\text{ext}} = \Delta\mu_{\text{eh}}\left(\frac{T_{\text{L}}}{T_{\text{eh}}}\right) + \Delta\varepsilon_{\text{eh}}\left(1 - \frac{T_{\text{L}}}{T_{\text{eh}}}\right) \tag{10.6}$$

where e is the elementary charge, V_{ext} is the external voltage of the cell, T_{L} and T_{eh} are the lattice and carrier temperatures respectively, and $\Delta\varepsilon_{\text{eh}}$ is the separation of the ESC energy levels. Assuming the hot carriers remain above the lattice temperature, i.e. $T_{\text{eh}} > T_{\text{L}}$, the external voltage of the cell can be enhanced according to Eq. (10.6). When $T_{\text{eh}} = T_{\text{L}}$, Eq. (10.6) reduces to the conventional solar cell case, where V_{ext} is the limit impose by the quasi-Fermi level splitting of the cold carrier distributions in the cell. Figure 10.10a shows the theoretical maximum PCE values of a HCSC under 1 Sun solar illumination calculated for several values of T_{eh} as a function of the bandgap value of the absorber (E_{g}).

The calculated PCE values generally show an increasing trend with T_{eh}, which is mainly due to the enhancement of the V_{ext} of the cell according to Eq. (10.6). The increments are larger for lower values of E_{g}, due to an increasing proportion of solar irradiation that generates hot carriers as the bandgap value decreases. While it may seem that maximizing the value of T_{eh} gives the best PCE values, one must consider that this case assumes that the hot carriers remain at these temperatures indefinitely, which is unlikely in practical conditions. In an actual device, there will be energy lost to the surrounding lattice due to the temperature difference between T_{eh} and the lattice temperature T_{L}. This should be considered as an additional power

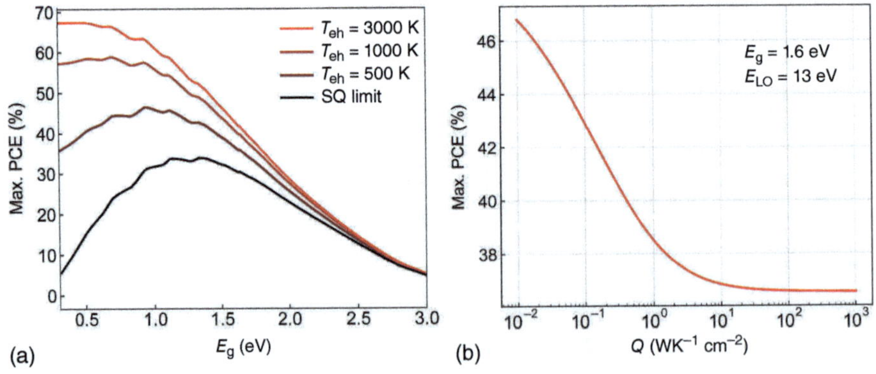

Figure 10.10 Calculated theoretical maximum power conversion efficiencies (PCE) for a hot carrier solar cell under 1 Sun illumination. (a) Calculated PCE values for several carrier temperatures (T_{eh}) as a function of bandgap (E_{g}) values in the absence of lattice thermalization effects. (b) Calculated PCE values with lattice thermalization considered for a cell with $E_{\text{g}} = 1.6$ eV and LO phonon energy, $E_{\text{LO}} = 13$ meV under 10 000 Sun illumination as a function of thermalization coefficient (Q). Source: (b) Lim et al. [14]/with permission of American Chemical Society.

loss term due to lattice thermalization, P_{th}. The total output power density (P) of the HCSC is then given by:

$$P = P_{abs} - P_{em} - P_{th} \tag{10.7}$$

P_{abs} is the total absorbed power density and P_{em} is the power loss due to recombination. P_{th} is given by Eq. (10.5) discussed previously, which increases when the temperature difference between the hot carriers and the lattice, ΔT, increases. This imposes an additional constraint on T_{eh}, and consequently the optimal value of T_{eh} that maximizes the PCE need not be as large, depending on the value of the thermalization coefficient Q. The calculated PCE values for a HCSC with $E_g = 1.6\,\text{eV}$ and $E_{LO} = 13\,\text{meV}$ under 10 000 Sun illumination is shown in Figure 10.10b. As can be seen, when lattice thermalization is considered, the maximum PCE values have an inverse relationship with Q; lower values of Q allows for larger values of T_{eh} that further increases the PCE. When Q is large, the carriers are considered fully thermalized, and this corresponds to the SQ limit PCE values. Hence, low values of Q are desirable for HCSC applications. It was found that halide perovskites possess Q values less than 1, which are the lowest values for any material system [14]. This highlights halide perovskites as attractive candidates for use as potential HCSC absorbers.

10.4.2 Toward the Realization of Perovskite Hot Carrier Solar Cells

The concept of an HCSC is certainly an attractive approach to overcoming the constraints of conventional single junction solar cells. While some work has demonstrated the proof-of-concept of an HCSC, the actual realization of a working HCSC device with actual improvements over traditional solar cells remains a great challenge. Here, we discuss the challenges and considerations in their realization with reference to seminal works in the perovskite literature.

10.4.2.1 Cooling Loss to the Lattice

The first and most significant obstacle to overcome when considering actual HCSCs is sustaining the hot carriers at elevated temperatures under steady-state solar illumination. While there are abundant studies detailing the slow hot carrier cooling properties in halide perovskites, one must note that these studies were conducted under intense, usually monochromatic, high-energy pump illumination. These phenomena were also reported at the ultrafast timescale, while reports detailing the manifestations of these slow hot carrier cooling effects under steady-state conditions are scarce. Under solar illumination, for which the AM1.5G spectrum is a widely accepted standard (Figure 10.1b), the spectral intensity of the high-energy photons is far lower than those used in spectroscopy experiments. Since these hot carrier effects are pump intensity dependent, it is expected that such slow hot carrier cooling phenomena would not be as obvious under solar illumination. Moreover, the high carrier temperatures observed in spectroscopy experiments typically decay rapidly in the first few picoseconds due to strong LO phonon coupling in these polar halide perovskite semiconductors. Under steady-state conditions, the average carrier temperature should be much lower considering the pump fluences and timescales involved. While studies have shown that hot carrier temperatures can be maintained above the lattice temperature under steady-state conditions, they are typically an order of magnitude lower than the carrier temperatures observed in the first few picoseconds in ultrafast experiments [14]. Most significantly, detailed balance calculations for HCSC efficiencies (Figure 10.10a) do not consider this thermalization loss. When considering thermalization losses, the projected maximum PCEs are reduced significantly, much so under 1 Sun illumination. The concept of the thermalization coefficient (Q) introduced earlier is useful in quantifying thermalization losses under steady-state conditions, where smaller values of Q imply lower losses to thermalization. The dependence of the maximum PCE of an HCSC with Q and solar concentration is shown in Figure 10.11. We note that only solar concentrations of 100 Suns and above are shown, as the PCE gains below 100 Suns would require $Q < 10^{-3}$ which seems physically impractical.

From Figure 10.11, we can observe that the maximum PCE values increase with smaller values of Q, while for larger Q values the PCE reduces and tapers off toward the corresponding SQ limit efficiencies that correspond to a fully thermalized carrier distribution. It is obvious that a decrease in the solar concentration requires lower values of Q to achieve any PCE enhancements from hot carriers over the SQ limit values. The reported Q values for halide perovskite thin films are in the range of 0.26–0.66 W K^{-1} cm^{-2} [14]. This translates to somewhat noticeable enhancements above 1000 Suns illumination, but rather minute enhancements are forecasted under 100 Suns solar illumination. While these values are typically applicable to concentrator photovoltaics, possible thermal and photo degradation of the material and the longevity of the device under such intense illuminations should be considered. Notably, the Q values of halide perovskites are an order of magnitude lower than that of conventional semiconductors, which have typical values of Q above 1. Further reduction in the Q values through various engineering strategies, such as making use of quantum confinement that slows the carrier

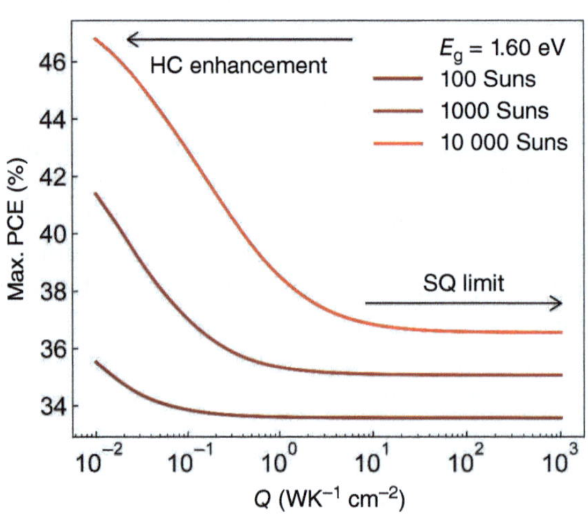

Figure 10.11 Variation of the projected maximum power conversion efficiency of an HCSC on the thermalization coefficient (Q) at different solar concentrations for an HCSC with E_g = 1.60 eV. Source: Reproduced from Lim et al. [14]/with permission of American Chemical Society.

cooling rate further, could make the PCE gains from hot carriers more attainable under lower solar concentrations.

10.4.2.2 Energy Selective Contacts

Another challenge in the realization of a practical HCSC is the search for appropriate ESCs. Several studies have undoubtedly demonstrated efficient hot carrier extraction from halide perovskites into suitable molecular forms, as discussed previously. However, another consideration in this regard is whether these ESCs are *energy-selective.* Being energy selective requires the energy bandwidth of the ESC to be as narrow as possible. So far, there has been a lack of studies investigating this aspect in the perovskite literature. This requirement is equally important because carrier cooling within the ESC will occur and deplete any potential excess energy if the energy bandwidth of the ESC is not sufficiently narrow. The added complication of Schottky barriers forming at the interfaces proves that the search for such an appropriate ESC with a low barrier for hot carrier extraction is indeed a tough one. Some strategies explored in the conventional semiconductor community include concepts such as tunneling barriers in quantum wells and minibands in quantum dots.

10.4.2.3 Loss of Cold Carriers

The last challenge in realizing a working HCSC is harvesting not only the hot but also the cold carriers. In an ideal HCSC, majority of the carriers in the junction are hot, such that the cold carriers only make up a small minority. In this case, just focusing on extracting the HCs alone through their appropriate ESCs with the simple straightforward device architecture (Figure 10.9) will lead to higher PCE values through V_{oc} enhancement. However, as discussed before, in realistic situations, when thermalization to the surrounding lattice is considered, the HC temperature is reduced drastically, hence making this scenario highly unlikely. Therefore, practical HCSC devices should try to efficiently extract not only the hot carriers but also the cold ones, such that enhancements over a standard solar cell can be attained. Thus, it appears that an alternative device architecture needs to be developed for next-generation photovoltaics that can harvest both hot and cold carriers.

So far, the experimental realization of a HCSC is still a proof of concept that only considers the extraction of hot carriers, and these devices typically have efficiencies even lower than those of their conventional counterparts. It seems that in the quest to achieve a working HCSC, attention is so focused on the hot carriers that the cold carriers are completely neglected. One of the main reasons for the low PCE in these devices is the poor electrical properties of the ESCs; organic molecules commonly used to demonstrate hot carrier extraction suffer from low carrier mobility and high resistance, while quantum dots (QDs) themselves also suffer from the same issues. The situation is worse for colloidal nanocrystals (NCs)/QDs, which are stabilized by long-chain insulating ligands. While advancements in ligand exchange and chemistry have been made, the main issue lies in their stability when deposited as QD solids or films. Lastly, it may seem that the standard solar cell architecture (such as that in Figure 10.9) needs to be carefully rethought out for next-generation photovoltaics to harvest both hot and cold carriers.

An alternative solution is to use concentrated solar illumination such that the fraction of hot carriers will be increased in the carrier bath. In this way, the phonon bottleneck effects should be more obvious, and significant reheating of carriers may be expected [19]. It might then be possible to maintain a bath of cold carriers that are constantly being replenished, and the reheating of these cold carriers will also help sustain the HC population. This way, extracting only the HCs at a higher energy level will lead to an increased V_{oc} and thus the PCE of the solar cells.

With further studies focused on engineering suitable device architecture and identifying potential ESCs, perhaps these issues may be surmountable in the future.

10.5 Multiple Exciton Generation

Multiple exciton generation (MEG), also known as carrier multiplication (CM) or impact ionization (II, in bulk materials), refers to a process of generating multiple excitons (≥2 electron–hole pairs) from the excitation of one high energy photon [20]. This process is analogous to singlet fission (SF) that occurs in organic molecules [21]. MEG is considered another promising approach for overcoming the SQ limit in a single junction solar cell [22].

In a typical photoexcitation scenario, absorption of a single photon leads to the generation of only a single electron–hole pair, or exciton. The excess photon energy above bandgap (E_g) cannot be extracted as electrical energy and would eventually dissipate in the form of heat via lattice vibrations. As shown in Figure 10.12a, an excited carrier with excess kinetic energy that corresponds to $\Delta E = h\nu - E_g$ relaxes to the band edge and the process is known as *hot carrier cooling* or *thermal relaxation* as previously discussed [23]. However, in QDs or NCs, the hot carrier cooling time is prolonged because of the quantum confinement-induced discrete electronic states. Thus, hot carriers bearing energies of at least twice or larger multiples of E_g could pass the excess energy to another carrier and excite it across the bandgap (Figure 10.12b). Thus, the absorption of a single photon bearing twice or multiple times the bandgap energy leads to more than one photo-induced carrier pair

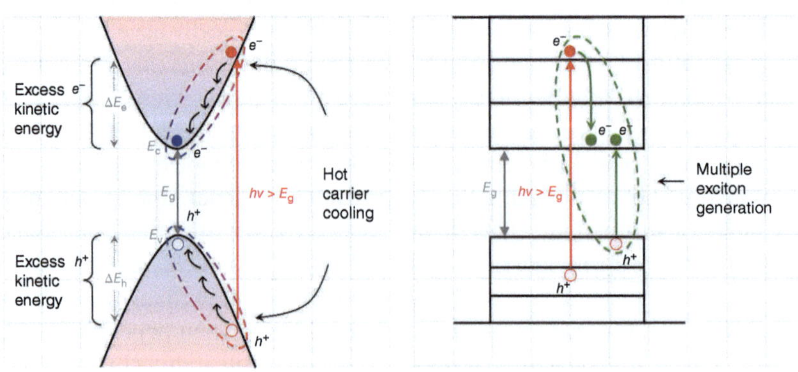

(a) (b)

Figure 10.12 Schematic diagram of (a) hot carrier cooling (HCC) and (b) multiple exciton generation (MEG).

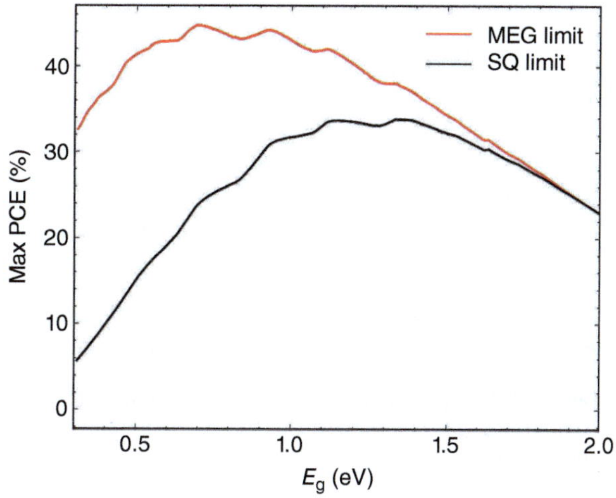

Figure 10.13 Maximum PCE values versus bandgap energy of a MEG solar cell for the case of ideal MEG (the MEG limit) as compared to a conventional single junction solar cell limited by the Shockley–Queisser (SQ) limit.

or exciton, in a process that is called *multiple exciton generation* (MEG), carrier multiplication (CM) or impact ionization (II) [20]. According to theoretical calculations, MEG/CM/II provides the possibility of increasing the PCE of a single-junction solar cell from the SQ limit of 33.7% up to 44.4% (Figure 10.13) [20].

The impact ionization process was originally identified in bulk semiconductors in 1953 [24]. Bulk materials such as Si, Ge, PbTe, PbS, and PbSe have been thoroughly investigated for decades. Then, in 2004, Schaller and Klimov made the first demonstration of MEG in PbSe NCs [25]. Whereafter, effective MEG has been discovered in a variety of halide perovskite materials, paving the path for the next-generation solar cells [24, 26, 27].

10.5.1 MEG Metrics

The *MEG quantum yield* (QY), *MEG efficiency* (η), and *MEG threshold* (E_{th}) are the essential metrics for assessing the MEG process in materials. The MEG QY, also known as MEG quantum efficiency (QE), represents the quantity of excitons produced per absorbed photon. The ideal MEG QY/QE is a step-like rise in the QE at energies corresponding to each integer multiple of E_g (i.e. $h\nu/E_g$), as illustrated in Figure 10.14b by the purple curve. However, in real-world situations, this ideal step-like behavior is usually not observed, but rather a gradual increase in the QE with increasing photon energies is obtained, such as the different curves corresponding to the different values of $\eta = 33\%$, 50%, and 99% in Figure 10.14b. The curves are calculated according to the model proposed by Beard et al. [28], which is schematically represented by Figure 10.14a.

In this model, the competition between the MEG rate (k_{MEG}) and hot carrier cooling rate (k_{cool}) is considered. As seen in the schematic, a high-energy photon can excite a single exciton with large initial excess energy into an excited single exciton state n_1^*. This high energy single exciton can either cool to the relaxed single exciton state n_1 or, if energy conservation permits, leads to the formation of an excited biexciton state n_2^*, which can also cool into the relaxed biexciton state n_2 or further produce a trion state n_3^* if energy conservation permits, and so on. This cascade process continues until the energy of the final excitons is less than twice the bandgap energy,

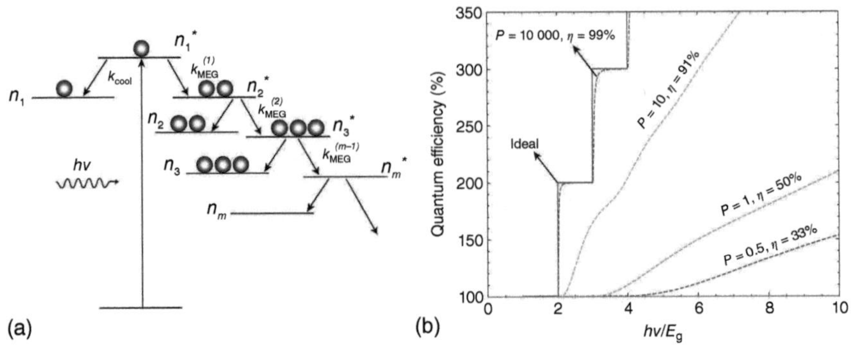

Figure 10.14 (a) Schematic of the cascade model for describing exciton and multiexciton populations. (b) Quantum efficiencies of the photon to exciton conversion for different P and η values.

consequently cooling to the band edge. Here, we denote the populations of the hot and relaxed m^{th} multiexciton states as n_m^* and n_m, respectively, and their rate of formation is denoted by $k_{\text{MEG}}^{(m-1)}$. By solving for the single and multiexciton populations, the final QY can be represented by:

$$\text{QY} = \sum_{j=1}^{m} \frac{k_{\text{cool}} \prod_{i=1}^{j} k_{\text{MEG}}^{(i=1)}}{\prod_{i=1}^{j} \left(k_{\text{cool}} + k_{\text{MEG}}^{(i)}\right)} \tag{10.8}$$

To evaluate the QY, a relation between the rates k_{MEG} and k_{cool} needs to be established, which can be expressed as [29]:

$$k_{\text{MEG}} = k_{\text{cool}} P \left(\frac{h\nu - h\nu_{\text{th}}}{h\nu_{\text{th}}}\right)^{s} \tag{10.9}$$

where $E_{\text{th}} = h\nu_{\text{th}}$ is the MEG threshold energy, P is a constant characterizing the competition between carrier cooling and MEG, and s is an exponent that varies between 2 and 5, where a value of two corresponds to the case of an ideal semiconductor. For halide perovskites, the value of s is found to be close to 2 [30]. The MEG efficiency η can then be expressed as:

$$\eta = \frac{P}{1+P} \tag{10.10}$$

where a larger value of P will lead to more efficient MEG. Moreover, the value of P also describes whether the onset of MEG is gradual or sharp, like in the ideal case. A value of $P < 1$ leads to a gradual, linear-like increase, whereas larger values lead to a sharper increase at the threshold energies with a more step-like behavior, as seen in Figure 10.14b. Furthermore, the threshold energy to produce m excitons $E_{\text{th}}^{(m)}$ can be expressed as:

$$E_{\text{th}}^{(m)} = E_{\text{g}} \left(1 + \frac{m}{\eta}\right) \tag{10.11}$$

Notably, if one assumes that k_{MEG} for all multiexciton states are equal, which is a reasonable approximation, then Eq. (10.8) simplifies to:

$$\text{QY} = \sum_{j=1}^{m} \frac{k_{\text{cool}} k_{\text{MEG}}^{j}}{(k_{\text{cool}} + k_{\text{MEG}})^{j}} \tag{10.12}$$

Using Eqs. (10.9) and (10.12), the QY for each value of $h\nu/E_g$ can be computed as shown in Figure 10.14b. Furthermore, the QY values evaluated from data obtained from experiments can be fitted with Eq. (10.12) to evaluate the key MEG metrics for the material. These values are typically evaluated using time-resolved spectroscopy techniques, with TA spectroscopy being the most common. The fingerprint of the MEG signal in NCs is a rapid fast decay with enhanced amplitude attributed to the AR of multiple excitons in a single NC or QD at very low pump fluence. The spectroscopic signature of MEG will be discussed in detail in Section 10.7.4. Aside from TA, TRPL, terahertz (THz) spectroscopy and quasi continuous wave (CW) spectroscopy are also approaches for identifying MEG. The corresponding signatures are: multiexciton PL dynamics (TRPL), enhanced far infrared absorption of multi-excitons (THz) as well as a red shift and variations in line shape of PL caused by the multi-excitons (CW) [31].

10.6 Multiple Exciton Generation Mechanisms

10.6.1 The Debate Over the MEG Threshold and MEG Mechanism

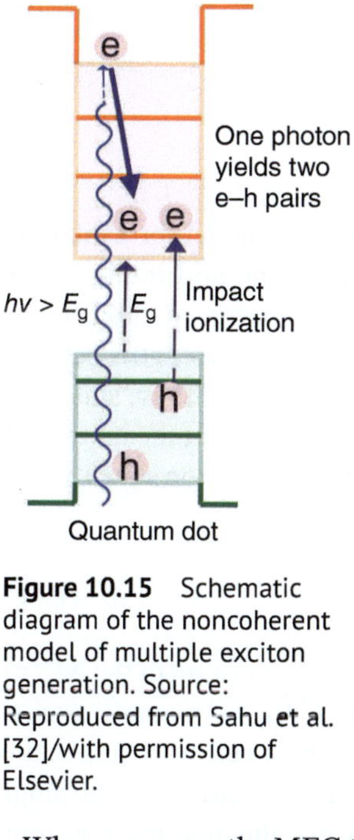

Figure 10.15 Schematic diagram of the noncoherent model of multiple exciton generation. Source: Reproduced from Sahu et al. [32]/with permission of Elsevier.

In general, two main mechanisms are hypothesized in the literature for the MEG process. The first is based on noncoherent impact ionization, whereas the second utilizes coherent superposition of multi-exciton and single exciton states to represent MEG [31–35]. The MEG thresholds, according to each model, are found to be $3E_g$ and $2E_g$ respectively. Figure 10.15 shows a schematic for the noncoherent model of MEG [32]. According to this model, after absorbing photons with energy greater than the bandgap (i.e. $h\nu > E_g$), the generated hot carriers possessing excess energies are dispersed based on their effective masses (m_e and m_h for electrons and holes, respectively). At least one of the carriers should have $h\nu > E_g$ to create additional electron–hole pairs. The MEG threshold energy (E_{th} or $\hbar\omega_{CM}$) can be simply expressed as follows:

$$E_{th} = \left(2 + \frac{m_e}{m_h}\right) E_g \tag{10.13}$$

When $m_e \approx m_h$, the MEG threshold is around $3E_g$, but if $m_e \ll m_h$, the MEG barrier can be reduced to $2E_g$ (Figure 10.16) [34]. In this case, the impact ionization (MEG) process competes with hot carrier cooling.

In conventional lead-based semiconductors such as lead selenide (PbSe) NCs for instance, the MEG threshold is approximately $2.9E_g$ [34]. In cadmium selenide (CdSe), because the ratio m_e/m_h is ~0.17, the MEG threshold can be decreased to ~$2.17E_g$ [35]. In indium arsenide (InAs), because its m_e is substantially less than m_h, the threshold can approach ~$2E_g$ [35].

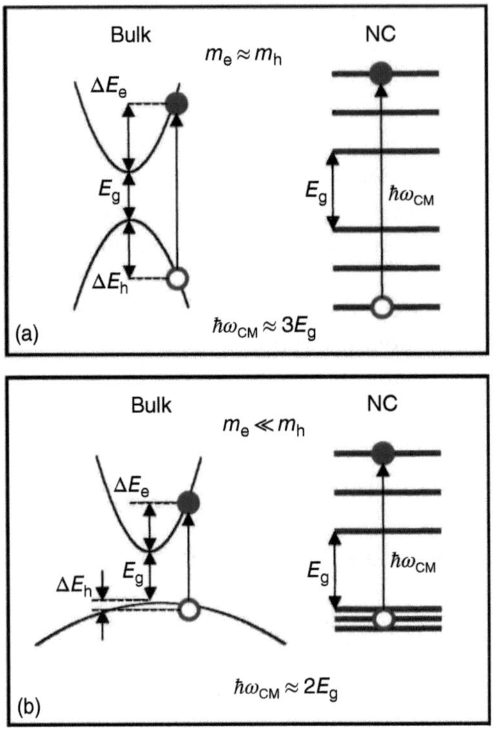

Figure 10.16 Schematic diagram of variation of MEG threshold E_{th} when carrier effective mass are (a) similar ($m_e \approx m_h$; $E_{th} \approx 3E_g$) or (b) different ($m_e \ll m_h$; $E_{th} \approx 2E_g$). Source: Reproduced from Schaller et al. [34]/with permission of AIP Publishing.

Figure 10.17 depicts the schematic diagram of the second model for explaining MEG, which uses the idea of coherent superposition of single and multi-exciton states [32]. The model employs a time-dependent density matrix method, which allows for the simultaneous consideration of arbitrary coupling strengths between single and multi-exciton states as well as varied dephasing rates for these states and a short pulse excitation of the NCs [31]. In the diagram, W_c represent the matrix element of the Coulomb interaction between the single and multi-exciton states, which can be expressed as follows:

$$W_c = \left\langle \Psi_{1Ex} \mid \frac{e^2}{\varepsilon r} \mid \Psi_{nEx} \right\rangle \tag{10.14}$$

where Ψ_{1Ex} is the initial single exciton state, ε is the dielectric constant of the NC, r is the distance between the electronic articles in the NC and Ψ_{nEx} is the multi-exciton state. $W_c/\hbar$ indicates the rate of Coulomb coupling between the single and multi-exciton states. γ_1 and γ_2 represent the rate of energy relaxation for the single excitons and multi-excitons, respectively. For efficient MEG, the quantity $1/\gamma_1$ must be lower than both $W_c/\hbar$ and $1/\gamma_2$ (i.e. $\gamma_1 \ll \gamma_2$). The much slower $1/\gamma_1$ than $1/\gamma_2$ is due to the asymmetric charge distribution in the excited biexcitons which enhances electron–phonon coupling. Thus, a biexciton is created from a single photon due to the favorable dephasing of the coherence for excited biexcitons.

V is the overall decay rate from the excited single exciton state to the ground state. It addresses the various processes that are responsible for MEG in NCs. The decay

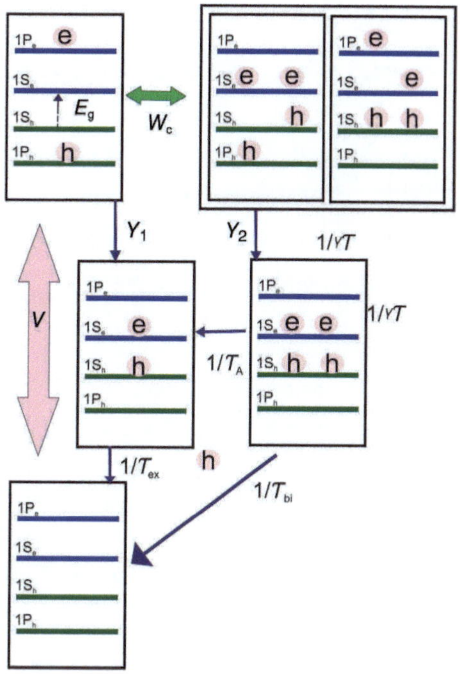

Figure 10.17 Schematic diagram of coherent model of multiple exciton generation. Source: Reproduced from Sahu et al. [32]/with permission of Elsevier.

rates of relaxed single excitons and biexcitons to the ground state are given by $1/\Gamma_{ex}$ and $1/\Gamma_{bi}$, which is much longer than $1/\gamma_1$ and $1/\gamma_2$. The ratio of the population of ground biexcitons to the single excitons (N_{bi}/N_{ex}) can be used to calculate the MEG efficiency as:

$$\frac{N_{bi}}{N_{ex}} = (\gamma_2/\gamma_1)P_{1-2} \tag{10.15}$$

where P_{1-2} is the probability of the excited biexciton state populating the single-exciton state generated by a photon through Coulomb interaction, which is always smaller than 1. For efficient MEG, $\gamma_1 \ll \gamma_2$ must be fulfilled. The increase of γ_2/γ_1 always leads to an increase in MEG QY. This model can also explain the MEG threshold at $2E_g$.

Apart from these two mechanisms, Schaller and Klimov proposed a third MEG mechanism based on the direct multiplication generation via a virtual exciton state to account for MEG in CdSe and PbSe NCs, which have extremely short τ_{MEG} (50–200 fs) and cannot be explained either by the non-coherent or coherent models [36].

10.6.2 Underlying Mechanism of the Efficient MEG in Perovskite

Thus far, the mechanism behind MEG is still inconclusive for halide perovskite materials. The MEG thresholds for halide perovskites can often approach ~$2E_g$, as discussed in detail in the subsequent chapters [24, 26, 27, 30]. The widely accepted reason is the slow hot carrier cooling rate ($1/\tau_{cool}$) in perovskite materials that

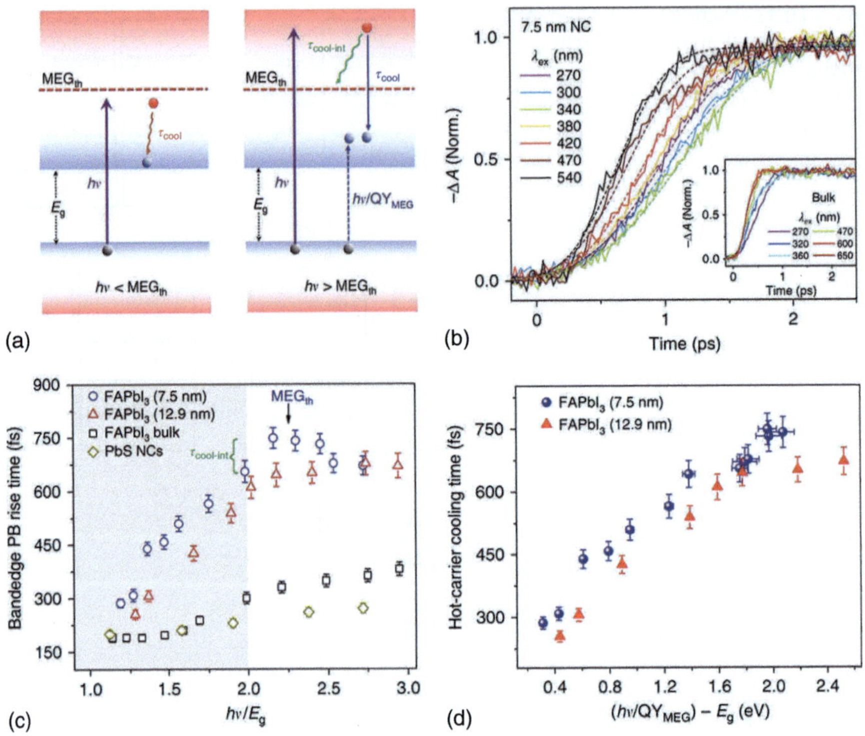

Figure 10.18 Slow HCC in perovskite NCs. (a) Schematic illustration for HCC and MEG under photo energy below/above MEG threshold. (b) PB rise time in perovskite NCs and bulk films (inset). (c) PB rise time as a function of $h\nu/E_g$ in different materials. (d) HCC time as a function of $\left(\frac{h\nu}{\mathrm{QY_{MEG}}} - E_g\right)$ in different materials. Source: Li et al. [26]/Springer Nature/

allows MEG to proceed more favorably, as seen in Figure 10.18a. As seen in the schematic, $\tau_{\text{cool-int}}$, which refers to the interval time it takes to cool down from the initial photoexcited state to the state of the MEG threshold, is larger than the MEG rate such that $1/\tau_{\text{cool-int}} < 1/\tau_{\text{MEG}}$ for efficient MEG to take place.

The total cooling time τ_{cool} can be calculated from the PB rise time (Figure 10.18b). Thus, the $\tau_{\text{cool-int}}$ is computed as $\tau_{(\text{cool}\,h\nu>2E_g)} - \tau_{(\text{cool}\,h\nu=2E_g)}$ (Figure 10.18c). The $\tau_{\text{cool-int}}$ of 7.5 nm $FAPbI_3$ NCs is ~90 ± 5 fs [26, 30]. Thus, its τ_{MEG} is less than 90 fs. An intriguing phenomenon is that τ_{cool} increases with $h\nu/E_g$ in $FAPbI_3$ bulk materials, while the τ_{cool} in 7.5 nm $FAPbI_3$ NCs drops slightly beyond MEG threshold (i.e. when $h\nu/E_g$ reaches 2.5, the $\tau_{\text{cool}\,\frac{h\nu}{E_g}>2.5}$ is roughly the same as $\tau_{\text{cool}\,\frac{h\nu}{E_g}=2.0}$, indicating an efficient MEG process). This can be attributed to the inverse Auger process caused by MEG that competes with hot carrier cooling. The MEG rate can be increased with better carrier Coulomb coupling in smaller perovskite NCs, allowing for a more effective MEG. When compared to 12.9 nm NCs, the 7.5 nm perovskite NCs have a longer HCC time and a more efficient MEG (Figure 10.18d).

Another proposed explanation for the $2E_g$ MEG threshold in perovskite materials is a possible unbalanced distribution of surplus energy between electrons and holes. If the second conduction or valence band (CB2 or VB2) energy is close to $2E_g$, the MEG threshold might be close to $2E_g$ [27]. The asymmetric transitions may render quantum confinement unnecessary. While this theory is verified in Pb chalcogenide materials, further investigations are needed for perovskites.

10.6.3 Controversy and Pitfalls Over Photocharging and Artifactual MEG Signal

One of the biggest issues in MEG research is the large disparity in MEG QY and MEG efficiencies in some early investigations. These were later found to be mainly due to photocharging artifacts [20, 35, 37–39]. Figure 10.19a illustrates the relaxation processes of excitons photogenerated by high-energy photons in relaxed and photocharged QDs. The unexcited or relaxed QD in the upper left panel is labeled as n_0. A hot exciton is generated in a QD when it is excited by a photon with energy greater than E_g at a rate of γ_{abs} into the excited state denoted by n^*. The following two situations can occur. In the first situation without MEG, the hot exciton cools to the band edge with rate γ_{cool} and the QD relaxes into the state n_1 that possesses a single cold exciton. In the presence of MEG, the high energy photon can generate multi-excitons at a rate of γ_{MEG} and the QD transits into the state n_2. The QD possessing multi-excitons can then relax into the single exciton state n_1 via AR of the multi-excitons with rate γ_{AR}. However, complications can arise when the QD undergoes photocharging that results in a long-lived charged state denoted by n_T in the top right of Figure 10.19a.

Notably, this state has a lifetime greater than the time interval between laser pulses, which is the basis behind this artifact. When this photocharged QD absorbs a photon from a subsequent laser pulse, a hot trion is generated that subsequently cools with rate γ_{cool} and the trion-containing QD moves into the state n_{2T}. The QD possessing trions can then relax back to the state n_T via a nonradiative AR process with rate γ_{trion} that has dynamics comparable to multi-exciton recombination (characterized by γ_{AR}). The similar dynamics of MEG and AR in trions can lead to the misidentification of MEG due to this photocharging artifact. Such photocharging artifacts may raise the perceived MEG QY or reduce the perceived MEG threshold below its true value. This is illustrated in Figure 10.19b, where the open data points corresponding to the case with photocharging all show an increased perceived QY compared to the situation without photocharging (solid data points). In the worst-case scenario, this artifact could lead to the false identification of MEG when, in fact, no MEG has occurred.

For colloidal NCs or QDs, the photocharging effects can be mitigated by sufficiently stirring or inducing a continuous flow of the probed samples within the excitation volume, thus preventing the formation of trions. Solid-state samples such as bulk films can be placed on a sample stage that can be continuously moved perpendicularly to the laser beams to mitigate the photocharging effects.

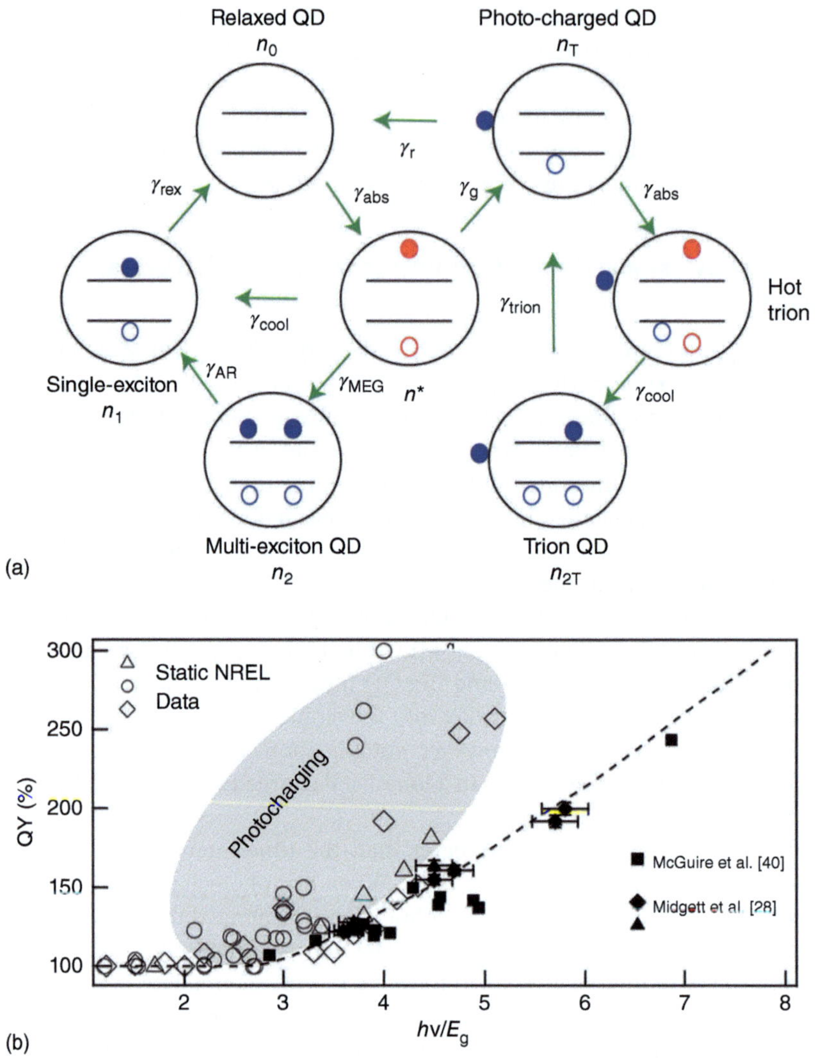

Figure 10.19 (a) Relaxation pathways of excitons photogenerated by high-energy photons in relaxed QDs (left) and photo charged QDs (right). (b) MEG QY calculated using TA data under static (gray shadow, open symbol) and stirring (solid symbol) conditions. Source: Reproduced from Beard et al. [37]/with permission of American Chemical Society.

Figure 10.20 illustrates how photocharging can influence the determination of MEG metrics in spectroscopy measurements [40]. The main problem stems from the use of the ratio of early to late-time signal amplitudes (A/B) to characterize the MEG effect. The signal amplitude is proportional to the average number of photogenerated excitons $\langle N_{abs} \rangle$ in a QD that follows Poissonian statistics [41], where the probability of a QD having N excitons is $P(N) = \exp(-\langle N_0 \rangle)$. In the absence of photocharging artifacts, for example, when the colloidal sample is stirred, the ratio A/B tends to 1, with decreasing $\langle N_{abs} \rangle$ when the QD is excited below the MEG threshold

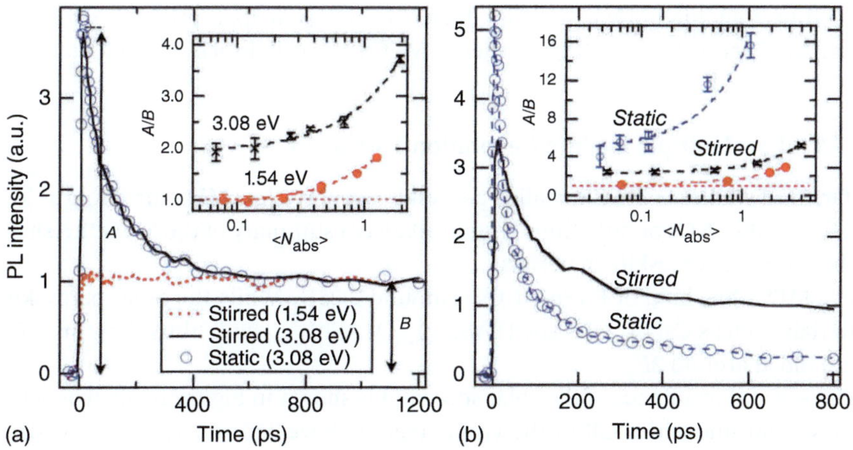

Figure 10.20 The effect of stirring to mitigate photocharging artifacts in PL dynamics of PbSe QDs. (a, b) PL dynamics collected with 1.54 eV (red) and 3.08 eV (black) excitation energies under static (blue open circles) and stirred (red dotted and solid black lines) conditions. The laser fluence used corresponds to the initial numbers of photogenerated excitons, $\langle N_{abs} \rangle$, of 2.0 and 1.4 for (a) and (b), respectively. Source: Reproduced from McGuire et al. [40]/with permission of American Chemical Society.

(red traces). When the sample is excited above the MEG threshold (black traces), a fast decay component manifests in the PL decay trace, resulting in a ratio A/B approaching 2 at low $\langle N_{abs} \rangle$.

However, for a static sample with photocharging (blue open circles), the nonradiative AR process in charged trions may show an increase in the multi-exciton decay component, leading to an exaggerated ratio of A/B. Figure 10.20a depicts the case when this photocharging effect is not too significant, as can be observed by the fact that the difference in the early time amplitude A between the static and stirred cases is not too large. However, in the case of Figure 10.20b, where photocharging effects are more significant, the difference in A is very large, and the ratio A/B can approach up to ~4 at low $\langle N_{abs} \rangle$, that leads to an overestimation of important MEG metrics like the MEG QY. Importantly, photocharging artifacts will influence the interpretation of MEG determined not only from PL spectroscopy but also TA spectroscopy, since the widely used method to observe MEG in the latter also relies on the ratio of the amplitudes A/B in a similar way. Thus, careful consideration and mitigation of photocharging effects are vital for accurate determination of MEG metrics.

10.7 Efficient Multiple Exciton Generation in Halide Perovskites

Since 2018, MEG studies have been conducted on various perovskite materials such as formamidinium lead iodide ($FAPbI_3$) NCs, caesium lead iodide ($CsPbI_3$) NCs, formamidinium lead-tin iodide ($FAPb_{1-x}Sn_xI_3$) NCs, and formamidinium methylammonium lead-tin iodide ($FA_{0.6}MA_{0.4}Pb_{0.4}Sn_{0.6}I_3$) bulk films. Their low

MEG threshold, high MEG efficiency, high MEG QY, and appropriate E_g reveal the potential of perovskites for high efficiency MEG solar cell device implementation.

10.7.1 Low Multiple Exciton Generation Threshold

So far, most studies of MEG in halide perovskites report a low MEG threshold as low as $2E_g$ in either NCs or bulk films. Figure 10.21 is a summary of the MEG threshold of the reported perovskite materials.

The MEG threshold of $FAPbI_3$ NCs is around $2.25E_g$, while the other perovskite materials such as $CsPbI_3$ NCs and $(FASnI_3)_{0.6}(MAPbI_3)_{0.4}$ bulk films show an MEG threshold at around $2E_g$.

The spectral irradiance of the solar spectrum is shown in Figure 10.22, where the highest light intensities fall in the visible region above 350 nm. Thus, a low MEG threshold is advantageous for the development of high efficiency MEG PSCs because MEG may occur in these ideal regions of the solar spectrum that have relatively high intensities (>350 nm) given that the bandgap of the material is narrow enough. Large MEG thresholds that correspond to wavelengths below 350 nm are not desirable

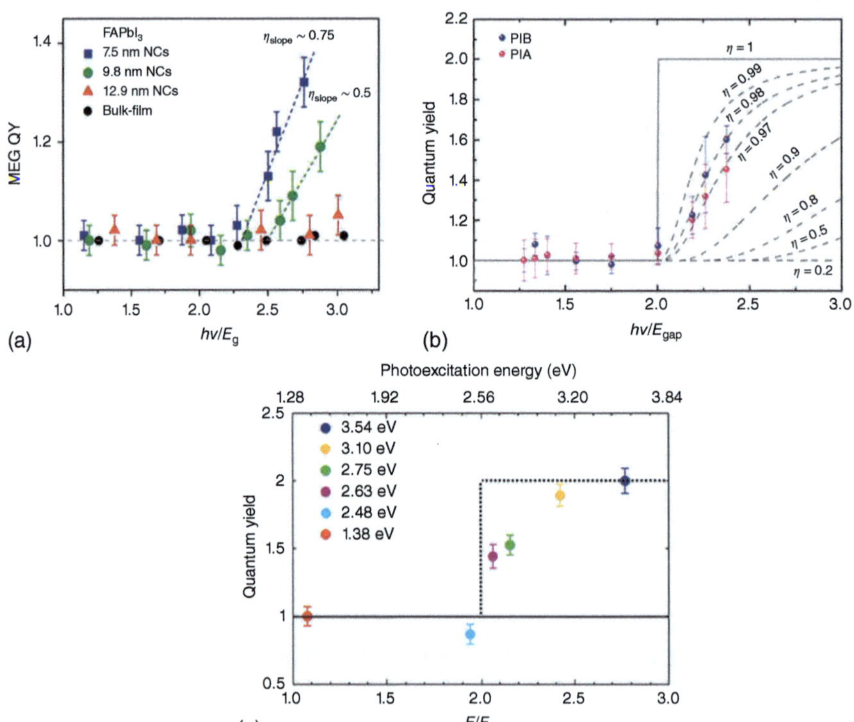

Figure 10.21 MEG QY as a function of $h\nu/E_g$ of (a) $FAPbI_3$ NCs, (b) $CsPbI_3$ NCs and (c) $(FASnI_3)_{0.6}(MAPbI_3)_{0.4}$ bulk films. Source: (a) Li et al. [26]/Springer Nature/Licensed under CC BY 4.0. (b) de Weerd et al. [24]/Springer Nature/Licensed under CC BY 4.0. (c) Reproduced from Maiti et al. [27]/with permission of American Chemical Society.

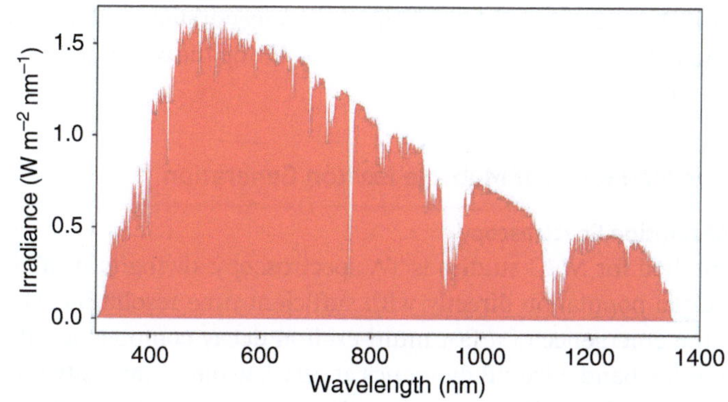

Figure 10.22 The spectral irradiance of the AM1.5G solar spectrum in the region of 300–1400 nm.

because the light intensity in this region is relatively low, and thus any enhancement from MEG is unlikely to contribute significantly to PCE enhancements of the solar cell. As a result, a perovskite material with a narrow bandgap and a low MEG threshold, such as Pb/Sn mixed perovskite with a bandgap that lies in the 1000 nm region and an MEG threshold of $2E_g$, could be a promising candidate for MEG PSCs.

10.7.2 High Multiple Exciton Generation Efficiency

Halide perovskite QDs outperform typical inorganic QDs in terms of their MEG efficiency η. As indicated in Section 10.5.1, the value of η can be determined from the slope of MEG QY to $h\nu/E_g$ or from modelling, and the maximum value is 1 (100%) that corresponds to the case where the QY approaches 200%. In PbS NCs, for instance, the MEG threshold is $3E_g$ with $\eta \sim 0.4$ [26]. The MEG efficiency for halide perovskite materials, as illustrated in Figure 10.21, can reach 75% or even 98%, which is close to the ideal case with of $\eta = 100\%$. The steeper the slope, the greater the MEG efficiency, and the closer the MEG behavior of the material is to the ideal step-like increase. A higher MEG efficiency would mean greater MEG QY at low light intensities, which benefits the performance of MEG PSCs.

10.7.3 Large Multiple Exciton Generation Quantum Yield

Perovskite materials can reach a high MEG QY of ~1.6 at 2 or $3E_g$, as seen in Figure 10.21. For Pb/Sn mixed perovskite, the MEG QY can reach up to 2 at $2.8E_g$. The high MEG QY, low MEG threshold and high MEG efficiency make halide perovskites promising materials for MEGSCs. The MEG QY is typically evaluated by time-resolved spectroscopy according to the following equation:

$$R_{pop} = \frac{A}{B} = \frac{\delta \langle N_0 \rangle \mathrm{QY}}{1 - \exp(-\langle N_0 \rangle)} \tag{10.16}$$

where R_{pop} is the ratio A/B of the early time initial signal amplitude A and late time signal amplitude B from time-resolved experiments, δ is the decay of the single exciton population that can be estimated in the absence of MEG, when the QY is assumed to be 100%, and $\langle N_0 \rangle$ is the average number of initial photogenerated excitons per NC or QD [26, 31]. Thus, the MEG QY can be retrieved by fitting the dependence of R_{pop} against $\langle N_0 \rangle$ with Eq. (10.16).

10.7.4 Spectroscopic Signatures of Multiple Exciton Generation

10.7.4.1 Transient Absorption Spectroscopy

The most common method for MEG studies is TA spectroscopy, owing to its ability to monitor the exciton population directly with sufficient time resolution. The signature of MEG is the emergence of a fast multi-exciton decay component with increased amplitude in the band edge PB dynamics at *very low* pump intensity (i.e. at fluences where $\langle N_0 \rangle \ll 1$). Here, $\langle N_0 \rangle$ represents the average number of initially photogenerated excitons per NC by the pump pulse and is determined by the function: $\langle N_0 \rangle = J_0 \sigma_{\text{p}}$, where σ_{p} is the absorption cross-section of the material at the pump wavelength and J_0 is the intensity of the photon flux [41].

In the absence of MEG, each QD absorbs one photon, which generates a single exciton at very low $\langle N_0 \rangle$. These single excitons possess a relatively slow single exponential decay that lasts for several to tens of nanoseconds, as can be observed in the trace collected $h\nu = 1.51E_{\text{g}}$ in Figure 10.23a [26]. In contrast, at high pump energies, MEG can occur, and multiple excitons in a single QD can be generated from a single photon. These multiple excitons will recombine much faster through

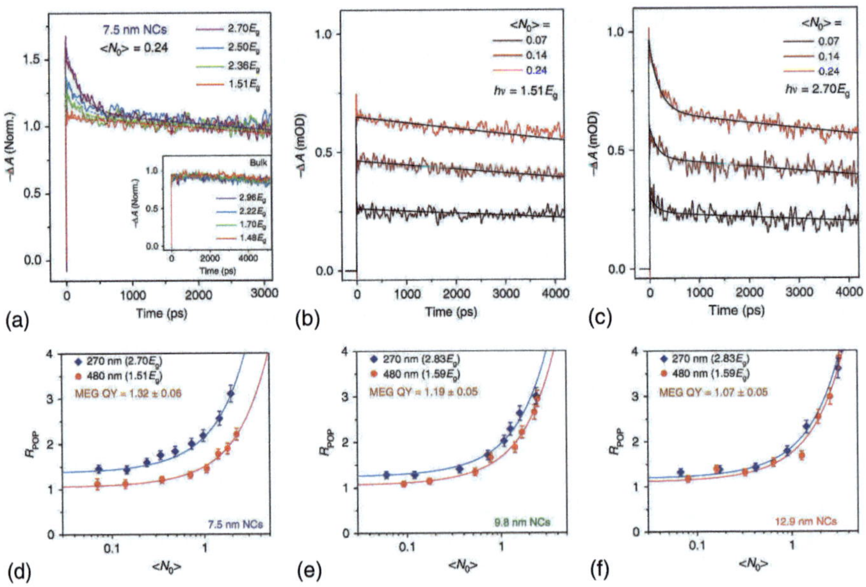

Figure 10.23 MEG signature from TA dynamics. (a) TA dynamics in NCs and bulk (inset) $FAPbI_3$ materials at different pump energies. (b) TA dynamics in $FAPbI_3$ NCs with different pump fluences excited at pump energy below MEG threshold ($1.51E_{\text{g}}$). (c) TA dynamics in $FAPbI_3$ NCs with different pump fluences excited at pump energy above MEG threshold ($2.70E_{\text{g}}$). (d–f) R_{pop} as a function of $\langle N_0 \rangle$ in $FAPbI_3$ NCs with different sizes. Source: Li et al. [26]/Springer Nature/Licensed under CC BY 4.0.

AR process, which leads to the appearance of a fast decay component, resulting in a multiexciton lifetime in the PB dynamics that can be seen in Figure 10.23a,c for $h\nu > 2E_g$ [41]. The additional fast decay component measured under low pump intensity and high pump energy that is otherwise absent in the PB dynamics collected with pump energy below the MEG threshold is considered the fingerprint of MEG [24]. In addition, since the TA signal amplitude is proportional to the exciton population, MEG will lead to an increased initial signal amplitude (A) relative to the situation without MEG. This leads to $R_{pop} > 1$ with MEG even at low $\langle N_0 \rangle$, as observed in Figure 10.23d–f for the data collected above the MEG threshold (blue squares). On the contrary, for the data collected under the MEG threshold (red squares), R_{pop} tends to 1 at low $\langle N_0 \rangle$ indicating the absence of MEG effects. This most widely used method is straightforward but susceptible to the photocharging effects discussed in the earlier section. Thus, due care must be taken to mitigate any effects of photocharging the samples, which would otherwise influence the interpretation of MEG.

Furthermore, by performing the same analysis as in Figure 10.23 at several pump energies to retrieve the corresponding MEG QY values, the dependence of the MEG QY on $h\nu/E_g$ can be plotted, from which a linear fit of the region with MEG could be performed to retrieve the MEG efficiency η and the MEG threshold energy. This is more clearly illustrated in Figure 10.24, where the plots of the MEG QY versus energy for different sized $FAPbI_3$ NCs are shown in panel (a), with the same analysis performed on PbS NCs plotted in panel (b). The linear fits of the MEG QY in panel a yielded the value of η that can be as large as 0.75 for smaller sized $FAPbI_3$ NCs and the extrapolated MEG threshold is $\sim 2.25E_g$.

The variation of absolute value of initial amplitude $|\Delta A_{max}|$, which is proportional to the exciton population, can likewise be employed as an MEG signature. A larger $|\Delta A_{max}|$ at the same absorbed photon fluence indicates the generation of additional excitons. By fitting the slope of $|\Delta A_{max}|$ as a function of fluence (defined as the absorbed photons per unit area), the MEG QY can also be approximately calculated.

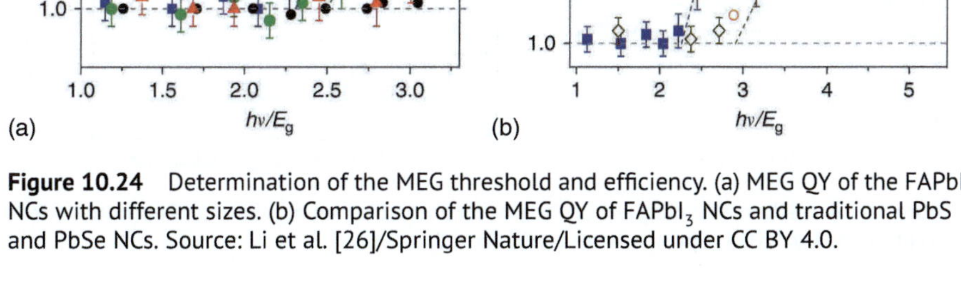

Figure 10.24 Determination of the MEG threshold and efficiency. (a) MEG QY of the $FAPbI_3$ NCs with different sizes. (b) Comparison of the MEG QY of $FAPbI_3$ NCs and traditional PbS and PbSe NCs. Source: Li et al. [26]/Springer Nature/Licensed under CC BY 4.0.

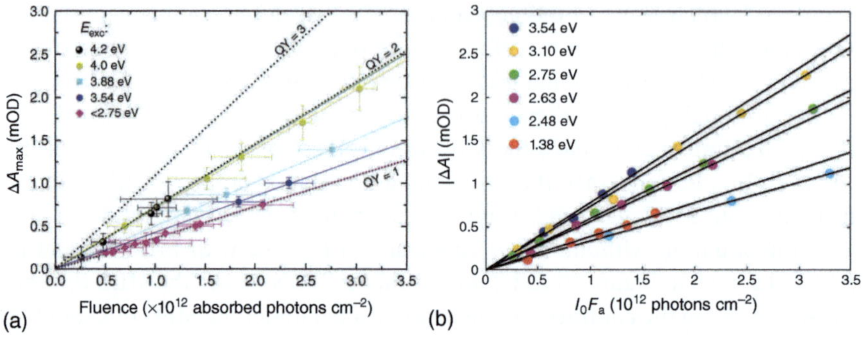

Figure 10.25 $|\Delta A_{max}|$ as a function of pump fluence in $CsPbI_3$ NCs and Pb/Sn-mixed perovskite films. Source: (a) de Weerd et al. [24]/Springer Nature/Licensed under CC BY 4.0. (b) Reproduced from Maiti et al. [27]/with permission of American Chemical Society.

At low pump fluence, $|\Delta A_{max}|$ increases linearly with pump fluence and a linear fit can be applied to obtain the gradient. Assuming that the QY is unity in the case without MEG, the MEG QY can be estimated as the ratio of the gradient for the case in the presence of MEG to the gradient of the case without MEG. For example, Figure 10.25 shows the data for $CsPbI_3$ NCs, where the gradient for excitation <2.75 eV is defined for the case without MEG. For the data collected under 4.0 eV excitation, this gradient is twice that of the previous and hence the MEG QY could be estimated as 2, where for the same photon fluence, the signal amplitude $|\Delta A_{max}|$ is doubled compared to the case without MEG.

10.7.4.2 Photocurrent-Based Techniques

Another way for detecting MEG is the transient photocurrent (TPC) technique shown in Figure 10.26 which was used to study MEG in PbSe QD films [32, 42]. A fast oscilloscope is used to monitor the dynamics of photocurrent generated following ultrafast laser excitation in a photoconductive switch. The photocurrent's temporal evolution can be described by:

$$j(h\nu, t) = q(h\nu)eE[n_e(t)\mu_e(t) + n_h(t)\mu_h(t)] \tag{10.17}$$

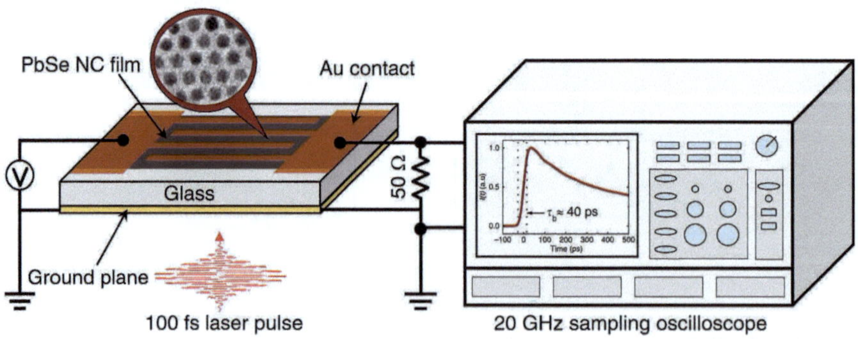

Figure 10.26 Schematic diagram of the photoconductive switch. Source: Reproduced from Gao et al. [42], © 2015, Springer Nature/CC BY 4.0.

where $j(h\nu, t)$ is the current density, q is the photon-to-exciton conversion, n is the photo generated carrier density, and μ is the carrier mobility. The subscripts e and h refer to electrons and holes. Assuming that the respective quantities for electrons and holes are similar in the QDs, the equation can be summarized as follows:

$$j(h\nu, t) = 2q(h\nu)e\mu(t)En_{\mathrm{QD}}\langle N_t\rangle \tag{10.18}$$

It is apparent that μ and n_{QD} are essential to photocurrent dynamics. The observed photocurrent dynamics is similar to the TA dynamics discussed earlier, and the presence of multiple excitons at early time similarly manifests as a rapid short decay component. The average number of the electron–holes pairs in each QD ($\langle N_x\rangle$) can be quantitatively expressed as:

$$\langle N_x\rangle = (2p_2 + p_1)/(p_1 + p_2) \tag{10.19}$$

where p_1 and p_2 are the fraction of single exciton and biexciton, respectively. In absence of MEG, $\langle N_x\rangle$ remains 1 when fluence approaches 0, which is consistent with TA measurement. The single exciton dynamics can be expressed as:

$$j(t) \propto u_x = (1-f) + fe^{-t/\tau_x} \tag{10.20}$$

The corresponding $j(t)$ described using p_1 and p_2 is stated as:

$$j(t) \propto (p_1 + p_2)u_x + p_2u_{xx} = (p_1 + p_2)\left[(1-f) + fe^{-t/\tau_x}\right] + p_2e^{-t/\tau_{xx}} \tag{10.21}$$

By fitting Eqs. (10.19) and (10.21), the MEG QY can be calculated using the photocurrent dynamics.

Photocurrent behavior under continuous illumination from a mercury (Hg) lamp with various band pass filters was also used to study MEG in NCs [30]. The authors observed an internal quantum efficiency (IQE) larger than 100% in a photodetector under a 4 V bias voltage illuminated above the MEG threshold at 254 nm. However, it should be noted that interpretation of MEG behavior from photocurrent experiments may not be equivalent to the situation in a solar cell device. This is because the photodetectors in photocurrent experiments are typically measured under an applied bias voltage, and thus the measured external quantum efficiency (EQE) values can easily exceed 100% [43]. On the contrary, solar cell devices are measured under zero bias. Thus, an observed IQE that is slightly larger than 100% measured under bias voltage in a photodetector may not necessarily be equal to the same situation measured under zero bias in a solar cell device. An EQE and IQE larger than 100% in solar cell devices measured from the incident-photon-to-current efficiency (IPCE) or calculated from J_{sc} is yet to be demonstrated in perovskite devices. Nevertheless, the observation of a rise in IQE under illumination above the MEG threshold could still be used as a signature of MEG as a relative comparison, but not as an absolute value.

10.8 Utilizing Multiple Exciton Generation in Halide Perovskites

Despite several spectroscopic studies on MEG, the ultimate goal of the MEG study is its real-world applicability in actual solar cell devices. The crucial evidence for the existence of multiple excitons generated in the devices is an EQE and IQE surpassing 100% under *zero bias*. The EQE is commonly measured by the IPCE, and IQE is normally computed as EQE/(1 – R), where R is the reflectance of the entire device and 1 – R represents the absorption of the whole device. IQE can also be calculated using optical modeling by using a combination of the measured refractive index n and extinction coefficient k for each layer to obtain the modeled reflectance and verify the measured IQE. Optical modeling can also be used to extract the absorptance of the sole active layer which can then be used to calculate the IQE of the absorbing layer only. However, the accuracy of optical modeling must be considered both in the n and k modeling process and in the overall device modeling. To date, a demonstration of EQE or IQE exceeding 100% in PSCs has yet to be reported. Here, our discussion primarily references the pioneering works on PbSe QD based MEGSCs to guide the direction and implementation of future perovskite-based MEGSCs.

10.8.1 Multiple Exciton Generation Solar Cells

The PbSe QD MEGSCs with an EQE and IQE surpassing 100% were reported by Semonin, Nozik, and Beard et al. in 2011 [44]. The EQE, which is measured under 1 Sun condition at zero bias represents the ratio of free carriers collected by the external circuit to the number of incident photons at each wavelength, whereas the IQE considers the reflected photons that do not reach the absorbing layer and should be greater than EQE. Thus, in the absence of MEG, the IQE and even more so, the EQE, should not surpass 100%. The observation of an EQE exceeding 100% can be seen as a definite proof of MEG enhancement in solar cell devices. Using both 1,2-ethanedithiol (EDT) and hydrazine for ligand change during QD film deposition, solar cell devices with a structure of ITO/ZnO/QDs/Au were prepared. The EQE of the devices based on QDs with three different sizes was measured, and the IQE was calculated according to the reflectance measured experimentally and simulated by modeling (Figure 10.27). With the increase in QD size, the bandgap becomes smaller and the peak EQE and IQE become larger. The highest EQE of 106 ± 3% and highest IQE of 130% were achieved by the 5.6 nm QD (E_g = 0.72 eV) at 3.44 eV. With the application of an antireflection coating, the peak EQE can even reach 114 ± 1% (a). By normalizing the photon energy to the bandgap and plotting the QY as a function of $h\nu/E_g$, it was observed that the onset of MEG in the three QDs remains the same, which is ~3E_g (Figure 10.28b,c). Although the PCE of the PbSe MEG solar cell devices is relatively low (~2–4.5%), this study provides a proof of concept for a MEGSC and guides the design for future practical MEGSCs that could potentially possess PCEs beyond the SQ limit.

The ligand length, surface passivation, and passivation degrees of inorganic PbSe QDs may impact MEG efficiency and MEG observation. The MEG threshold, on the

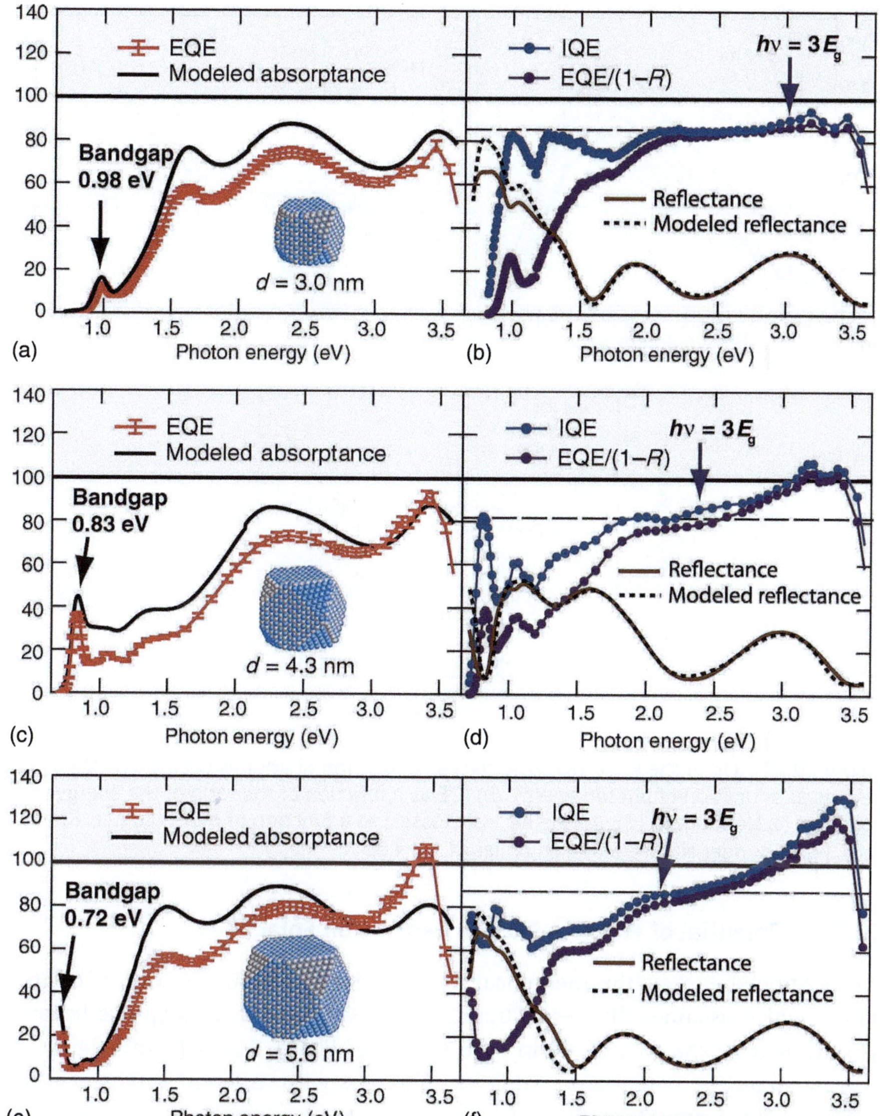

Figure 10.27 EQE and IQE of PbSe QDs with different sizes. (a, c, e) EQE and modeled absorptance of the QDs. (b, d, f) Modeled IQE and the IQE calculated with $1 - R$ of the QDs. Source: Reproduced from Lohse et al. [45]/with permission of American Chemical Society.

other hand, does not seem to vary with QD size. While the 3 nm QD possesses the greatest degree of quantum confinement, the 5.6 nm QD with the smallest bandgap exhibits the most significant MEG enhancement, perhaps because of its smaller bandgap. Hence, a balance between quantum confinement for increased MEG efficiency and a narrow bandgap for practical MEG applications must be similarly considered for perovskite MEGSCs.

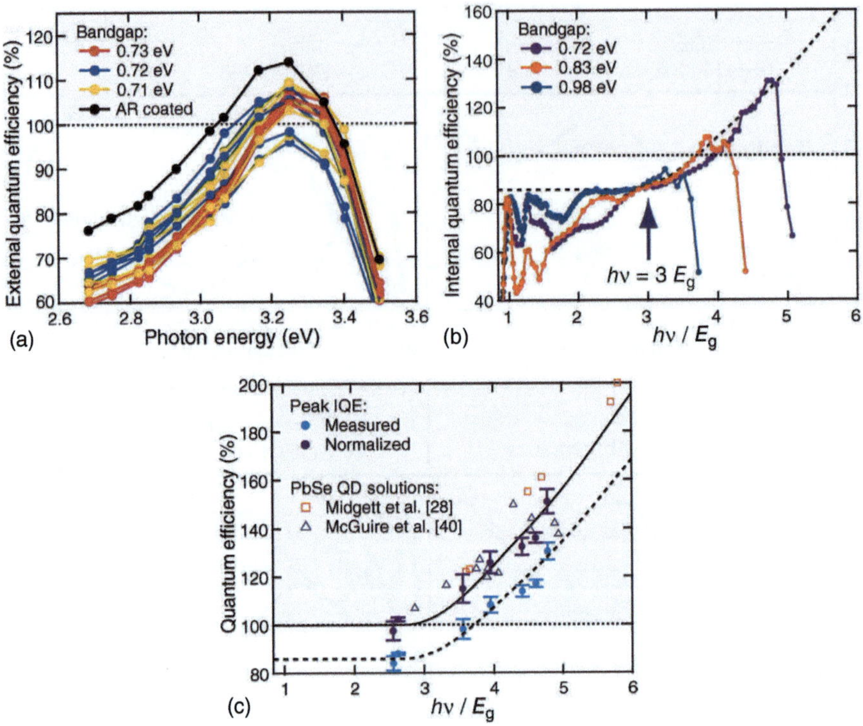

Figure 10.28 QE of the PbSe QD solar cells as a function of photon energy. (a) EQE of 18 devices as a function of photon energy. (b) IQE as a function of the ratio of the energy to bandgap. (c) Normalized IQE (to the intrinsic losses) as a function of $h\nu/E_g$. Source: Rhee et al. [46]/Springer Nature/Licensed under CC BY 4.0.

10.8.2 Potential of Multiple Exciton Generation Solar cells

As mentioned earlier, the theoretical maximum PCE attainable for a MEGSC is 44.4%, which assumes the ideal situation of MEG QY with the step-like behavior as indicated by the grey curve in both panels of Figure 10.29. In this figure, we also illustrate the potential of MEG for materials with different values of P that characterizes their MEG behavior according to Eqs. (10.8)–(10.12) discussed previously.

As can be seen, for larger values of P (and thus η), the QE behaves increasingly more like the ideal case, which translates to increasing PCE enhancements in an MEGSC. In all cases, it can be observed that PCE enhancements over the SQ limit can be achieved, with the larger enhancements corresponding to narrower values of bandgap energies, once again highlighting the importance of choosing an optimal narrow bandgap that aligns with an optimal region of the solar spectrum that has considerable intensities. The main prerequisites for materials for MEGSC device applications are a suitable bandgap, small MEG threshold, high MEG efficiency, and high MEG QY. Other considerations for MEGSCs include having a high transport ability of the electron and hole transport layers to prevent carrier saturation and a

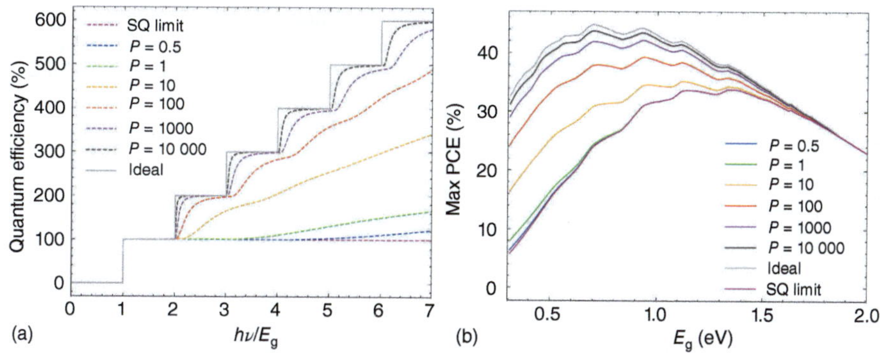

Figure 10.29 The potential of MEG solar cells. (a) Illustration of the quantum efficiency (QE) of the photon to exciton conversion for different values of *P*. (b) The corresponding calculated maximum power conversion efficiencies for a MEGSC.

high transmittance of the inactive layer before the absorbing layer. Furthermore, the absorbance of indium tin oxide (ITO) and fluorine-doped tin oxide (FTO) coated glass, which are the commonly used transparent conductive electrodes in PSCs, in the <350 nm range. Therefore, materials with MEG thresholds below 350 nm are unlikely to be suitable for efficient MEGSC implementation because of competing absorption from the ITO or FTO layers. Perovskite materials that possess narrow bandgap (~1.24 eV), slow HCC, low MEG threshold, and high MEG efficiencies are thus promising candidates for realizing high-efficiency MEGSCs for breaking the SQ limit.

10.9 Conclusion and Outlook

The slow hot carrier cooling properties of halide perovskites have both enabled high carrier temperatures to persist comparatively longer than other semiconductors and facilitated the highly efficient MEG process with high MEG QY in these materials. Such a material that could simultaneously exhibit both high carrier temperatures and efficient MEG makes halide perovskites one of the best candidates for next-generation solar cells that could break the SQ efficiency limit. Furthermore, we envisage a halide perovskite solar cell device that could exploit both HC and MEG effects to gain efficiency enhancements that are greater than harnessing either effect alone. Such a device would have a J–V characteristic such as that shown in Figure 10.30. The MEG enhancement would increase the J_{sc} while the enhancement from HCs will increase the V_{oc} of the cell. Both enhancements, when combined, could lead to even larger PCE of the cell than if only considering one effect alone. Of course, competition from one effect over the other must also be taken into consideration for a compromise and the right balance. Further studies are warranted.

With further efforts channeled toward developing a suitable device architecture that could exploit both effects simultaneously and searching for suitable extraction layers with high mobilities, perhaps such a highly efficient next-generation solar

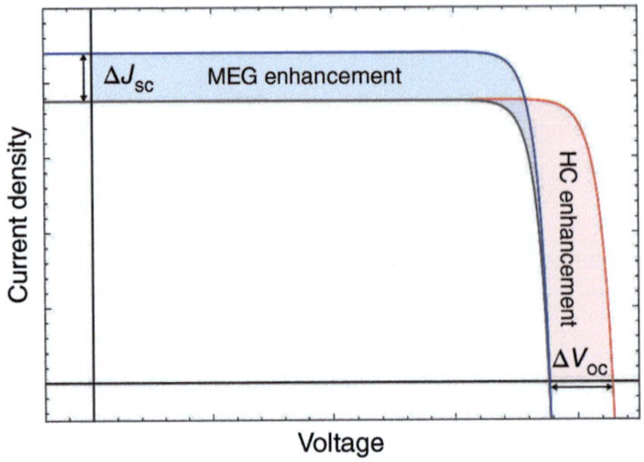

Figure 10.30 Simulated *JV* curve for a hypothetical solar cell that could harness both HC and MEG effects for efficiency enhancements.

cell device based on halide perovskites could one day be realized in the near future to help humanity combat and reverse the effects of climate change.

References

1 Shockley, W. and Queisser, H.J. (1961). Detailed balance limit of efficiency of p-n junction solar cells. *Journal of Applied Physics* 32 (3): 510–519.

2 Xing, G., Mathews, N., Sun, S. et al. (2013). Long-range balanced electron-and hole-transport lengths in organic–inorganic $CH_3NH_3PbI_3$. *Science* 342 (6156): 344–347.

3 Sum, T.C., Mathews, N., Xing, G. et al. (2016). Spectral features and charge dynamics of lead halide perovskites: origins and interpretations. *Accounts of Chemical Research* 49 (2): 294–302.

4 Richter, J.M., Branchi, F., Valduga de Almeida Camargo, F. et al. (2017). Ultrafast carrier thermalization in lead iodide perovskite probed with two-dimensional electronic spectroscopy. *Nature Communications* 8 (1): 1–7.

5 Klemens, P. (1966). Anharmonic decay of optical phonons. *Physical Review* 148 (2): 845.

6 Ridley, B. (1989). Electron scattering by confined LO polar phonons in a quantum well. *Physical Review B* 39 (8): 5282.

7 Vallée, F. and Bogani, F. (1991). Coherent time-resolved investigation of LO-phonon dynamics in GaAs. *Physical Review B* 43 (14): 12049.

8 Yang, Y., Ostrowski, D.P., France, R.M. et al. (2016). Observation of a hot-phonon bottleneck in lead-iodide perovskites. *Nature Photonics* 10 (1): 53.

9 Fu, J., Xu, Q., Han, G. et al. (2017). Hot carrier cooling mechanisms in halide perovskites. *Nature Communications* 8 (1): 1300.

10 Joshi, P.P., Maehrlein, S.F., and Zhu, X. (2019). Dynamic screening and slow cooling of hot carriers in lead halide perovskites. *Advanced Materials* 31 (47): 1803054.

11 Frost, J.M., Whalley, L.D., and Walsh, A. (2017). Slow cooling of hot polarons in halide perovskite solar cells. *ACS Energy Letters* 2 (12): 2647–2652.

12 Lim, J.W.M., Giovanni, D., Righetto, M. et al. (2020). Hot carriers in halide perovskites: how hot truly? *The Journal of Physical Chemistry Letters* 11 (7): 2743–2750.

13 Fang, H.-H., Adjokatse, S., Shao, S. et al. (2018). Long-lived hot-carrier light emission and large blue shift in formamidinium tin triiodide perovskites. *Nature Communications* 9 (1): 243.

14 Lim, J.W.M., Wang, Y., Fu, J. et al. (2022). Spotlight on hot carriers in halide perovskite luminescence. *ACS Energy Letters* 7: 749–756.

15 Le Bris, A., Lombez, L., Laribi, S. et al. (2012). Thermalisation rate study of GaSb-based heterostructures by continuous wave photoluminescence and their potential as hot carrier solar cell absorbers. *Energy & Environmental Science* 5 (3): 6225–6232.

16 Li, M., Bhaumik, S., Goh, T.W. et al. (2017). Slow cooling and highly efficient extraction of hot carriers in colloidal perovskite nanocrystals. *Nature Communications* 8: 14350.

17 Lim, S.S., Giovanni, D., Zhang, Q. et al. (2019). Hot carrier extraction in $CH_3NH_3PbI_3$ unveiled by pump-push-probe spectroscopy. *Science Advances* 5 (11): eaax3620.

18 Ramesh, S., Giovanni, D., Righetto, M. et al. (2022). Tailoring the energy manifold of quasi-two-dimensional perovskites for efficient carrier extraction. *Advanced Energy Materials* 12: 2103556.

19 Yang, J., Wen, X., Xia, H. et al. (2017). Acoustic-optical phonon up-conversion and hot-phonon bottleneck in lead-halide perovskites. *Nature Communications* 8: 14120.

20 Pietryga, J.M., Park, Y.S., Lim, J. et al. (2016). Spectroscopic and device aspects of nanocrystal quantum dots. *Chemical Reviews* 116 (18): 10513–10622.

21 Nozik, A.J. (2021). Quantization effects in semiconductor nanostructures and singlet fission in molecular chromophores for photovoltaics and solar fuels. *Chemical Physics Reviews* 2 (2): 021305.

22 Li, M. et al. (2019). Slow hot-carrier cooling in halide perovskites: prospects for hot-carrier solar cells. *Advanced Materials* 31 (47): e1802486.

23 Nozik, A.J. (2001). Spectroscopy and hot electron relaxation dynamics in semiconductor quantum wells and quantum dots. *Annual Review of Physical Chemistry* 52: 193–231.

24 de Weerd, C., Gomez, L., Capretti, A. et al. (2018). Efficient carrier multiplication in $CsPbI_3$ perovskite nanocrystals. *Nature Communications* 9 (1): 4199.

25 Schaller, R.D. and Klimov, V.I. (2004). High efficiency carrier multiplication in PbSe nanocrystals: implications for solar energy conversion. *Physical Review Letters* 92 (18): 186601.

26 Li, M., Begum, R., Fu, J. et al. (2018). Low threshold and efficient multiple exciton generation in halide perovskite nanocrystals. *Nature Communications* 9 (1): 4197.

27 Maiti, S., Ferro, S., Poonia, D. et al. (2020). Efficient carrier multiplication in low band gap mixed Sn/Pb halide perovskites. *Journal of Physical Chemistry Letters* 11 (15): 6146–6149.

28 Beard, M.C., Midgett, A.G., Hanna, M.C. et al. (2010). Comparing multiple exciton generation in quantum dots to impact ionization in bulk semiconductors: implications for enhancement of solar energy conversion. *Nano Letters* 10 (8): 3019–3027.

29 Ridley, B.K. (2013). *Quantum Processes in Semiconductors*. Oxford University Press.

30 Chen, Y., Yin, J., Wei, Q. et al. (2022). Multiple exciton generation in tin–lead halide perovskite nanocrystals for photocurrent quantum efficiency enhancement. *Nature Photonics* 16, 7: 485–490.

31 Nozik, A.J. (2008). Multiple exciton generation in semiconductor quantum dots. *Chemical Physics Letters* 457 (1–3): 3–11.

32 Sahu, A., Garg, A., and Dixit, A. (2020). A review on quantum dot sensitized solar cells: past, present and future towards carrier multiplication with a possibility for higher efficiency. *Solar Energy* 203: 210–239.

33 Beard, M.C., Knutsen, K.P., Yu, P. et al. (2007). Multiple exciton generation in colloidal silicon nanocrystals. *Nano Letters* 7 (8): 2506–2512.

34 Schaller, R.D., Petruska, M.A., and Klimov, V.I. (2005). Effect of electronic structure on carrier multiplication efficiency: comparative study of PbSe and CdSe nanocrystals. *Applied Physics Letters* 87 (25): 253102.

35 Kershaw, S.V. and Rogach, A.L. (2017). Carrier multiplication mechanisms and competing processes in colloidal semiconductor nanostructures. *Materials (Basel)* 10 (9): 1095.

36 Schaller, R.D., Agranovich, V.M., and Klimov, V.I. (2005). High-efficiency carrier multiplication through direct photogeneration of multi-excitons via virtual single-exciton states. *Nature Physics* 1 (3): 189–194.

37 Beard, M.C. (2011). Multiple exciton generation in semiconductor quantum dots. *Journal of Physical Chemistry Letters* 2 (11): 1282–1288.

38 Smith, C. and Binks, D. (2013). Multiple exciton generation in colloidal nanocrystals. *Nanomaterials (Basel)* 4 (1): 19–45.

39 Tyagi, P. and Kambhampati, P. (2011). False multiple exciton recombination and multiple exciton generation signals in semiconductor quantum dots arise from surface charge trapping. *The Journal of Chemical Physics* 134 (9): 094706.

40 McGuire, J.A., Sykora, M., Joo, J. et al. (2010). Apparent versus true carrier multiplication yields in semiconductor nanocrystals. *Nano Letters* 10 (6): 2049–2057.

41 Klimov, V.I. (2007). Spectral and dynamical properties of multiexcitons in semiconductor nanocrystals. *Annual Review of Physical Chemistry* 58: 635–673.

42 Gao, J., Fidler, A.F., and Klimov, V.I. (2015). Carrier multiplication detected through transient photocurrent in device-grade films of lead selenide quantum dots. *Nature Communications* 6: 8185.

43 Li, L., Zhang, F., Wang, J. et al. (2015). Achieving EQE of 16,700% in P3HT:PC_{71}BM based photodetectors by trap-assisted photomultiplication. *Scientific Reports* 5: 9181.

44 Semonin, O.E., Luther, J.M., Choi, S. et al. (2011). Peak external photocurrent quantum efficiency exceeding 100% via MEG in a quantum dot solar cell. *Science* 334 (6062): 1530–1533.

45 Lohse, S.E. and Murphy, C.J. (2012). Applications of colloidal inorganic nanoparticles: from medicine to energy. *Journal of the American Chemical Society* 134 (38): 15607–15620.

46 Rhee, J.H., Chung, C.-C., and Diau, E.W.-G. (2013). A perspective of mesoscopic solar cells based on metal chalcogenide quantum dots and organometal-halide perovskites. *NPG Asia Materials* 5 (10): e68.

11

Ionic Transport in Perovskite Semiconductors

Wenke Zhou[1], *Yicheng Zhao*[2,3], *and Qing Zhao*[4]

[1]*Army Engineering University of PLA, State Key of Laboratory for Disaster Prevention and Mitigation of Explosion and Impact, Nanjing 210007, China*
[2]*University of Electronic Science and Technology of China, Chengdu 611731, China*
[3]*Helmholtz-Institute Erlangen-Nürnberg (HI-ERN), Erlangen, Germany*
[4]*Peking University, State Key Lab for Mesoscopic Physics and Frontiers Science Center for Nano-optoelectronics, School of Physics, Beijing 100871, China*

Electrical conduction has two basic modes: one based on electrons' motion, the other on ions' motion. The corresponding conductivity is a measure of how efficiently it conducts an electric current. Fine-tuning of a material's conductivity, including both electronic conductivity and ionic conductivity, is considered to be a primarily developing strategy in modern technologies, such as metal-based electrocatalysis, semiconductor-based microchips, ceramic oxides-based fuel cells, metal oxides-based lithium-ion batteries, electroencephalography-based, and brain-computer interfaces. Thus, a deep understanding of a material's conductivity is of vital importance.

In the following chapters, we will be focusing on the ionic conductivity in metal-halide perovskite (MHP) semiconductors, considering that mobile ions have a huge impact on the electronic properties of MHPs. We will start with the theory of ionic transport and then discuss the extrinsic and intrinsic factors that impact ionic transport. We will close by specifying the detrimental effects of ionic accumulation that would play a role in the performance of MHPs function devices.

11.1 Theoretical Basis of Ionic Transport

In solid-state thin films, ionic transport is normally mediated by defects *via* hopping process, in bulk or at boundary, which is described below [1]:

$$\sigma_{\mathrm{i}} = \frac{{Z_{\mathrm{i}}}^2 e^2 N_{\mathrm{A}} C_{\mathrm{v}} D_{\mathrm{v}}}{k_{\mathrm{B}} T V_{\mathrm{m}}} \tag{11.1}$$

where σ_{i} is the ionic conductivity, Z_{i} is the ionic charge, N_{A} is Avogadro's constant, C_{v} is the vacancy concentration, D_{v} is the vacancy diffusion coefficient, V_{m} is the

Halide Perovskite Semiconductors: Structures, Characterization, Properties, and Phenomena, First Edition. Edited by Yuanyuan Zhou and Iván Mora-Seró.

molar volume of perovskite, k_B is Boltzmann constant, and T is temperature. At low temperatures, the vacancy concentration C_v is a constant due to the low thermal energy; however, at high temperatures, the vacancy concentration will increase exponentially with temperature, which is dependent on the formation energy of Schottky defects. The diffusion coefficient D_v is also a function of temperature:

$$D_V = D_V^0 \exp\left(-\frac{E_a}{k_B T}\right) \tag{11.2}$$

where E_a is the activation energy for vacancy diffusion and D_v^0 is a constant. Thus, at low temperatures, the product of σ_i and T has an exponential relationship with E_a. At high temperatures, the scenario becomes more complicated due to the change in defect concentration, as demonstrated in Eq. (11.3) for a chemical formula of ABX_3:

$$[V_A][V_B][V_X]^3 = K_S = K_S{}^0 \exp\left(-\frac{\Delta H_S}{k_B T}\right) \tag{11.3}$$

where K_s is the Schottky constant, ΔH_s is the enthalpy change for the above reaction and K_s^0 is a constant. The electroneutrality requires:

$$[V_A] = [V_B] = \frac{1}{3}[V_X] \tag{11.4}$$

By combining Eqs. (11.3) and (11.4), the concentration of vacancies in perovskite at one temperature can be written:

$$C_X = [V_X] = \left(9K_s{}^0\right)^{(1/5)} \exp\left(-\frac{\Delta H_S}{5k_B T}\right) \tag{11.5}$$

Finally, we have:

$$\sigma_i T = \frac{Z_i{}^2 e^2 N_A C_{v0} D_0}{k_B V_m} \exp\left(-\frac{\Delta H_S}{5k_B T} - \frac{E_a}{k_B T}\right) = \sigma_0 \exp\left(-\frac{E_a{}^{\text{eff}}}{k_B T}\right) \tag{11.6}$$

where $E_a{}^{\text{eff}} = \frac{\Delta H_S}{5} - E_a$ corresponds to the effective activation energy. The E_a value is sensitive to ionic radius, hopping distance, and local crystal structure, which are closely correlated with the quality of the polycrystalline perovskite film. Generally, ions migrate faster in crystals with more and bigger available interstitial sites, smaller and less charged ions, and smaller jumping distances.

11.2 Characterizations of Ionic Transport

In solid-state thin films, the ion migration can be characterized by optical, electrical, and electrochemical methods. Therefore, a proper electric field is required to induce the motion of ions, which, in most cases, demands two electrodes that can in-situ apply electric field to the sample. The electrodes can be deposited on the sample to construct either a vertical structure or a lateral structure, as shown below in Figure 11.1:

To study the ionic transport, the lateral structure is highly recommended due to the following reasons: First, most mobile ions in thin films are concentrated at the

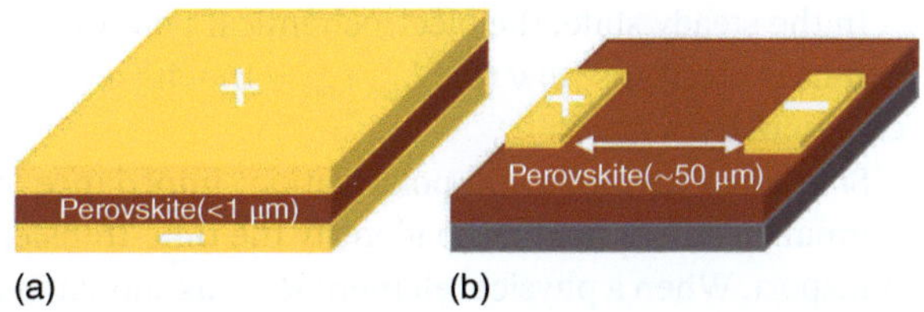

Figure 11.1 Two basic structures of MHPS-based electronic devices. (a) Vertical structure. (b) Lateral structure.

surface or grain boundary, making the lateral structure involve a larger ratio of ionic transport than that with vertical structure; second, a lateral structure enables in-situ optical characterizations during the electric poling process since there is no coverage on perovskite layer; third, external stimuli, such as light illumination and gas absorption, can be easily imposed on the sample surface to examine its impact on ionic transport. Besides, few characterization methods with a nanoscale resolution can be applied to the study of ion migration in vertical structure, especially when the film thickness is below the resolution limit of most optical microscopes (e.g. <1 μm).

For electrical characterization of ionic transport, galvanostatic DC polarization measurement (Figure 11.2) is a simple but effective tool to identify the ratio of ionic conductivity to electronic conductivity.

In this measurement, a weak current is applied to the sample to induce the motion of both ions and electrons reversibly without causing any degradation of the sample. Upon switching on the current, the voltage instantaneously reaches a finite value, V_0, and then it progressively increases up to the saturation value V_S. Such a transient behavior is typical for stoichiometric polarization, which stems from the electrodes that block the mobile ions. In this way, an ionic concentration gradient builds up to balance the net charge at the interface. The instantaneous value V_0 by turning on the current corresponds to the total electrical resistance resulting from both the electronic and the ionic transport [2].

$$V_0 = i\frac{R_{\text{ion}}R_{\text{ele}}}{R_{\text{ion}} + R_{\text{ele}}} \tag{11.7}$$

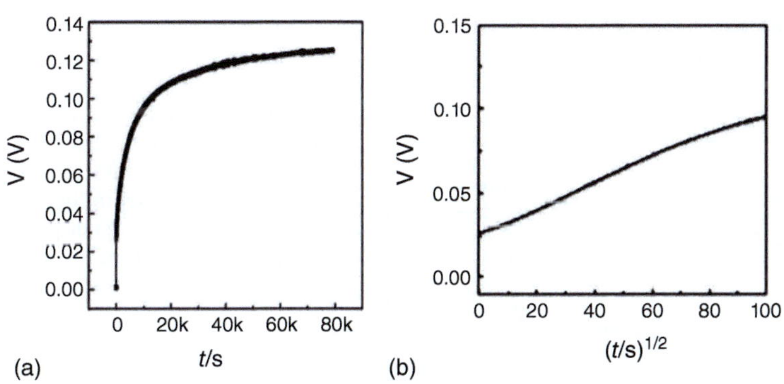

Figure 11.2 (a) DC polarization curve for a symmetric Graphite/$MAPbI_3$/Graphite cell (thickness 0.6 mm) measured at 30 °C in air flow by applying a current of 2 nA. (b) Voltage versus square root of time up to 104 seconds. Source: Reproduced from Yang et al. [2]/with permission of John Wiley & Sons.

In the steady state, the electrochemical potential gradient of the ions disappears, and only electrons flow ($i = i_{\mathrm{ele}}$, $i_{\mathrm{ion}} = 0$) with $V_{\mathrm{s}} = iR_{\mathrm{ele}}$, finally giving us the ionic conductivity.

Similar to galvanostatic polarization, impedance spectroscopy analysis is also a normal electrical method to identify the time-transient behavior of ionic/electronic transport. When a physical element R_1C_1 as shown in Figure 11.3 is perturbed with a step voltage, the electric current decays with exponential shape defined by the characteristic time.

$$\tau = R_1C_1 \tag{11.8}$$

If the perturbation is ac voltage, the relation of AC voltage to current is a complex impedance (Figure 11.4).

$$Z(\omega) = \frac{\mathrm{d}U}{\mathrm{d}I} = \mathrm{Re}(Z(\omega)) + i\mathrm{Im}(Z(\omega)) \tag{11.9}$$

Briefly, in ordinary ionic conductors, the main aspects of the capacitance include dielectric capacitance, chemical capacitance, depletion layer at contact barrier (or possibly accumulation or inversion capacitances), and electrode polarization [4]. Under the bias voltage, the accumulation of ions at the contact interface produces a capacitance due to surface space charge that responds in the low-frequency range of the measurement. This capacitance occurs at a short distance from the contact and is independent of the active film thickness, which can be defined as ion Debye length

$$L_{\mathrm{D}} = \frac{C^{\delta}}{\varepsilon} = \sqrt{\frac{\varepsilon\varepsilon_0 k_{\mathrm{B}}T}{q^2N}} \tag{11.10}$$

Here, N accounts for the density of ionic charges.

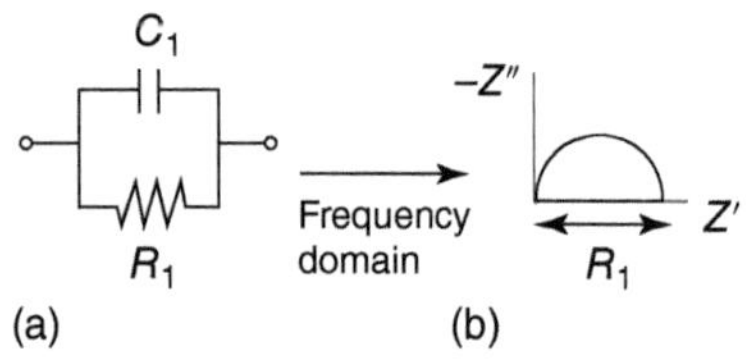

Figure 11.3 (a) Parallel RC equivalent circuit. (b) The experimental in the signature frequency domain. Source: Reproduced with permission from Maier [3]. © 2007, Springer Nature.

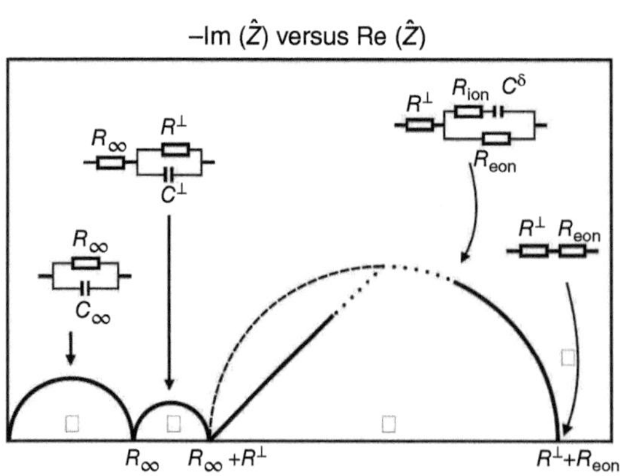

Figure 11.4 Nyquist plot of the complex impedance at which the mixed conductor is contacted by electrodes. Source: Reproduced with permission from Maier [3]. © 2007, Springer Nature.

For optical characterizations, optical imaging, photoluminescence (PL), time-resolved photoluminescence (TRPL), and photothermal induced resonance (PTIR) microscopy [5] are powerful tools. Especially for PTIR, it uses a pulsed wavelength-tunable IR laser for excitation and atomic force microscopy (AFM) tip operating in contact mode as a local detector of IR spectra and maps with a resolution as high as tens of nanometers. Because of the specificity of IR spectroscopy to the chemical composition, PTIR allows mapping the distribution of given chemical species (or chemical groups) and has enabled the characterization of diverse samples, including plasmonic nanomaterials, metal-organic frameworks, polymers, and, very recently, MHPs.

For electrochemical characterizations, a solid-state electrochemical cell is used to identify the mobile ion species by monitoring the chemical reaction at various interfaces.

11.3 Mobile Ions in Perovskite Film Under Electric Field

The crystal structure of MHPs provides many tunnels for ions to migrate in the crystal, especially at the surface and grain boundary [6–10]. Some first-principles calculations on the E_a of different ions through Schottky defects reveal a basic relationship in $MAPbI_3$ materials: $E_a(I) < E_a(MA) < E_a(Pb)$. That is, I^- moves along the I–I edge of the $[PbI_6]^{4-}$ octahedron with the lowest activation energy of 0.58 eV (Figure 11.5a). MA^+ migrates through the unit cell face, surrounded by four I^- ions with a higher activation energy of 0.84 eV (Figure 11.5b); and Pb^{2+} ions move along <110> direction of the unit cell with the highest migration energy barrier of 2.31 eV [11]. Further considering Eq. (11.4), V_I should be the dominant defect in perovskite that makes the conductivity of V_I dominate the ionic conductivity in MHPs [12, 13]. Depending on the migration path and mechanism, different theoretical works could produce different results. For example, smaller E_a values are obtained to be 0.08, 0.80, and 0.46 eV for I^-, Pb^{2+}, and MA^+ ions, respectively. In another work, proton can exist in the MHPs due to a deprotonation process, and it has lower activation

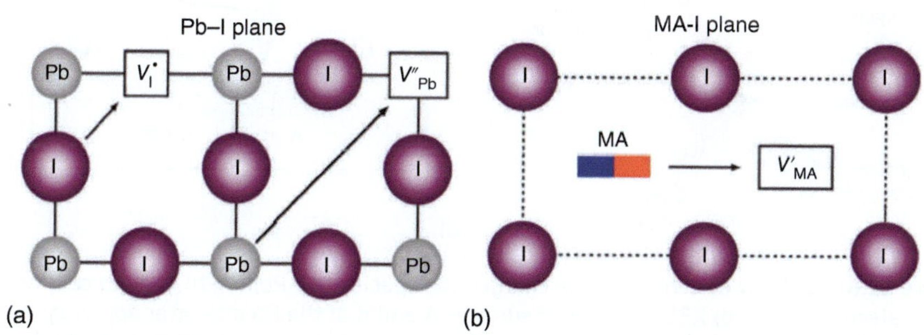

Figure 11.5 (a) Schematic illustration of the I^- transport mechanisms involving conventional vacancy hopping between neighboring positions. (b) MA^+ migration into a neighboring vacant A-site cage involving motion normal to the unit cell face composed of four iodide ions [11]. Source: Eames et al. [11]/Springer Nature/Licensed under CC BY 4.0.

energy of 0.07 eV via quantum-tunneling mechanism [14]. It should be noted that most theoretical analysis is based on the bulk, not surface or grain boundary. The latter would be more valuable to guide experiments.

To find the direct evidence of ionic transport, Yang et al. set up a solid-state electrochemical cell, Pb|$MAPbI_3$|AgI|Ag (Figure 11.6a), consisting of one electronic electrode (Pb electrode taken as "+") and one ionic electrode (AgI electrode taken as "-"), to investigate the chemical effects at various interfaces upon long-term charge transference [2]. Combining with scanning electron microscopy (SEM) micrograph, energy dispersive X-ray spectroscopy (EDX), and X-ray diffraction (XRD) characterization (Figure 11.6b), PbI_2 was detected at the Pb|$MAPbI_3$ interface. It demonstrated that I^- can easily migrate through $MAPbI_3$ film and provided strong and compelling evidence for the presence of electric field-induced ionic transport in $MAPbI_3$.

Then Yuan et al. applied PTIR microscopy to map the spatial distribution of MA^+ before and after electric poling on a lateral $MAPbI_3$ device [5] (Figure 11.6c). After exerting a moderate electric field of 1.6 V μm^{-1} for 100–200 seconds, PTIR signal intensity indicated a redistribution of MA^+ in the perovskite film, robustly proving the electromigration of MA^+. Then, via Kelvin probe force microscopy (KPFM) characterization (Figure 11.6d), Yuan et al. observed a reversible electric-field-driven

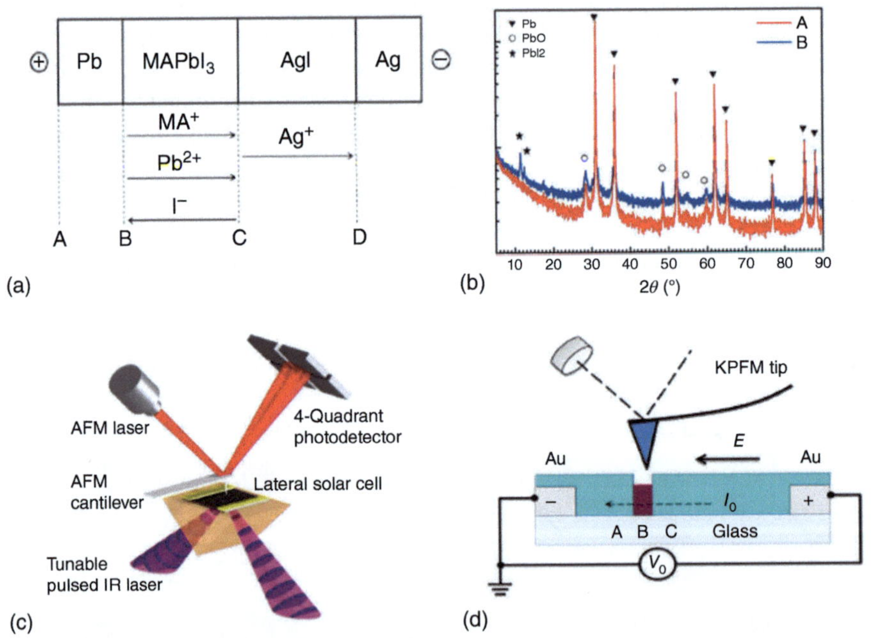

Figure 11.6 (a) Flow directions of the charged ion species in a Pb|$MAPbI_3$|AgI|Ag cell under electrical bias. (b) XRD patterns of surfaces A and B of the Pb disk after applying a DC of 10 nA for a week. Source: Reproduced from Yang et al. [2]/with permission of John Wiley & Sons. (c) Schematic illustration of the PTIR measurement on $MAPbI_3$ film. Source: Reproduced from Yuan et al. [5]/with permission of John Wiley & Sons. (d) Schematic illustration of the KPFM measurement around the PbI_2 thread. Source: Reproduced with permission from Yuan et al. [15]. © 2015, Wiley-VCH.

conversion between $MAPbI_3$ and PbI_2 at an elevated temperature of 330 K [15], which could be explained by a massive migration of MA^+ and I^-.

Besides I^- and MA^+, the proton migration was proven to exist by cryogenic galvanostatic characterization between 17 and 297 K [16]. And the details will be discussed in the following Section 11.4.3.

11.4 The Factors Affecting Ionic Transport in Perovskites

In the works reported by several groups, the relaxation time of ionic transport in perovskite film varied several orders of magnitude [17], even under the same electric poling condition. Crystal quality and surrounding environments could be the invisible hand, regulating ionic transport behavior. Therefore, figuring out which factors affect ionic transport and then obtaining a clear map of MHPs' operational mechanism is the ticket to improved device stability.

11.4.1 Moisture

In 2015, Leijtens et al. investigated how the water molecules affect the ionic motion in $MAPbI_3$ film [18]. Under 30% humidity, by applying an electric field around $0.06\ V\ \mu m^{-1}$, the bare perovskite film exhibits an extremely quick and irreversible degradation (Figure 11.7a). Then, Zhao et al. found that, when coated with a PMMA layer, the $MAPbI_3$ film slowly, over many hours, turned into a yellow phase. In addition, the film shows no obvious change in vacuum condition [19] (Figure 11.7b). In typical MHPs, the build-in field is around $1\ V\ \mu m^{-1}$, which is two orders of magnitude higher than that used in Leijtens's experiment. The above experimental results strongly suggest that humidity can enhance ion migration. The possible mechanism for moisture-enhanced ionic transport can be described as follows: Once the water molecules (H_2O) absorb on the surface of perovskite crystal, hydrogen bonding will form between H_2O and organic cation (MA^+ or FA^+). The external disturbance may reduce the interaction between MA^+/FA^+ and the surrounding $[PbI_6]^{4-}$ octahedron, lowering the energy barrier for MA^+/FA^+ migration [18]. Therefore, applying an inert environment in the fabrication process, in combination with strict encapsulation, is vital to isolate the MHPs from water molecules for better stability.

11.4.2 Light Illumination

Perovskite devices can maintain their original performance after several months' storage in darkness in an inert gas environment; however, they exhibit a fast decay under light [20–22]. To get further insight into how light affects perovskite film, Zhao et al. investigated the variation of perovskite morphology under different light intensities via an $Au/MAPbI_3/Au$ device structure by optical microscope [20]. After applying electric poling, they found that, when the illumination intensities

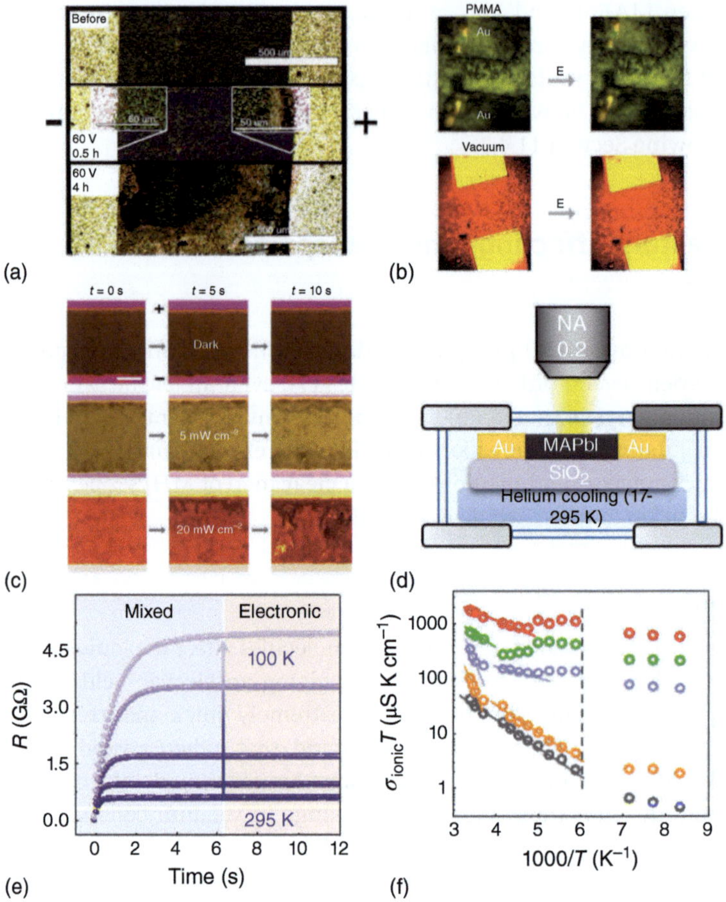

Figure 11.7 (a) Microscope images of the lateral different stages of electric field-induced degradation in ~30% humidity. Source: Reproduced with permission from Leijtens et al. [18]. © 2015, John Wiley & Sons. (b) Optical images change after 1 minute of high-field poling ($1\,V\,\mu m^{-1}$) for the perovskite film with PMMA coating in ambient air and without any coating in a vacuum. Source: Reproduced with permission from Zhao et al. [19]. © 2016, American Chemical Society. (c) Optical dynamic images of perovskite film under electric poling and illumination of various intensities in ambient air at room temperature. The scale bar represents 20 μm. (d) The apparatus used in cryogenic galvanostatic measurement, with a helium cooling system. (e) Five typical galvanostatic curves were recorded at different temperatures under a 20 pA current and $1\,mW\,cm^{-2}$ illumination. (f) Ionic conductivity multiplied by temperature (from 17 to 295 K), $\sigma_{ion}T$, as a function of $1000/T$ under various illumination intensities (gray: $0\,mW\,cm^{-2}$; Orange: $0.05\,mW\,cm^{-2}$; purple: $1\,mW\,cm^{-2}$; green: $5\,mW\,cm^{-2}$; red: $20\,mW\,cm^{-2}$). The dashed line indicates the phase transition temperature. Source: Reproduced with permission from Zhao et al. [20]. © 2016, Springer Nature.

increased, a growing number of pinholes and dendrites (PbI_2) emerged in perovskite film (Figure 11.7c). After comparing the element distribution in perovskite film before and after electric poling, a possible mechanism could be deduced: under the electric field and stronger light, massive I^- migrated toward the cathode and volatilized as I_2; MA^+ evaporated away in the form of CH_3NH_2 gas during the migration process from cathode to anode, and the $MAPbI_3$ gradually turned into PbI_2 phase. That is to say, light can enhance ion migration.

Then, the next question is how light enhances ionic transport in $MAPbI_3$ film. One prediction is that light can stimulate the attempt frequencies of ionic transport [23–25] and another is that photo-crystal interaction can reduce the energy barrier for ionic transport [26, 27]. To quantitatively evaluate the activation energy for ionic transport, Zhao et al. conducted cryogenic galvanostatic measurements to obtain the temperature-dependent ionic conductivities under different light intensities, with a tiny current, on the $Au/MAPbI_3/Au$ with a helium cooling system (Figure 11.7d). First, the current was switched from zero to 20 pA. Then they measured the voltage of the structure as a function of time [28, 29] (Figure 11.7e). Considering Eq. (11.6), the fitting results (Figure 11.7f) indicated that, with stronger light illumination, the activation energy for ionic transport was significantly reduced. That is to say, ions are easier to migrate under light. Similarly, deQuilettes et al. reported the light-induced PL "brightening" and halide redistribution of perovskite film [30]; Xing et al. also found light-promoted ionic transport with reduced energy barrier for ionic transport [31]; and Kim et al. discovered that the ionic conductivity of $MAPbI_3$ was enhanced by several orders of magnitude under light excitation [32].

Then by carefully analyzing the zoomed-in view of temperature-dependent ionic conductance, the activation energy region shows two separate parts (Figure 11.7f): E_{a1} ($T > 250$ K) and E_{a2} ($180\text{ K} < T < 250$ K). Published results for E_a for I^- and MA^+ varied from 0.1–0.8 eV. For H^+, the activation energy was predicted to be ~0.17 eV or even smaller, if considering nuclear quantum tunneling effect. Thus, the E_{a1}, ranging from 0.14 to 0.82 eV, corresponds to I^- or MA^+, whereas the E_{a2}, from 0.06 to 0.13 eV, can only be assigned to H^+.

Besides, the dynamic process of ionic transport could be analyzed by extracting the temperature-dependent relaxation time (kinetic constant) [16] (Figure 11.8a). Below 80 K, the $\ln(K_{ion})$ of $MAPbI_3$ exhibits a T-independent behavior (Figure 11.8b), which should be related to the deep quantum tunneling of the ions rather than conventional classical hopping. Owing to the light mass of H, H^+ is the most likely ion with which this tunneling can happen [14]. Further, when the target perovskite changes to $CsPbI_2Br$ without H, the $\ln(K_{ion})$ ascends until the lowest temperature (Figure 11.8c). Thus, for $CsPbI_2Br$, no quantum tunneling existed. Therefore, both experimental and theoretical work powerfully indicate that the horizontal lines below 80 K in Figure 11.8b are intrinsic to H^+ migration in $MAPbI_3$.

11.4.3 Perovskite Composition

Understanding which ions are responsible for the light-enhanced ionic transport is of great significance since it will provide clues for the developed perovskite materials with increased stability for device applications. Therefore, the target perovskite for light excitation was changed to $MAPbBr_3$, $MAPb(I_{0.9}Br_{0.1})_3$, and $CsPbI_2Br$ by Zhou et al. [33]. In $MAPbBr_3$ and $MAPb(I_{0.9}Br_{0.1})_3$ films, plane dendritic structures were still observed under stronger light, indicating halide anions are not the culprits for

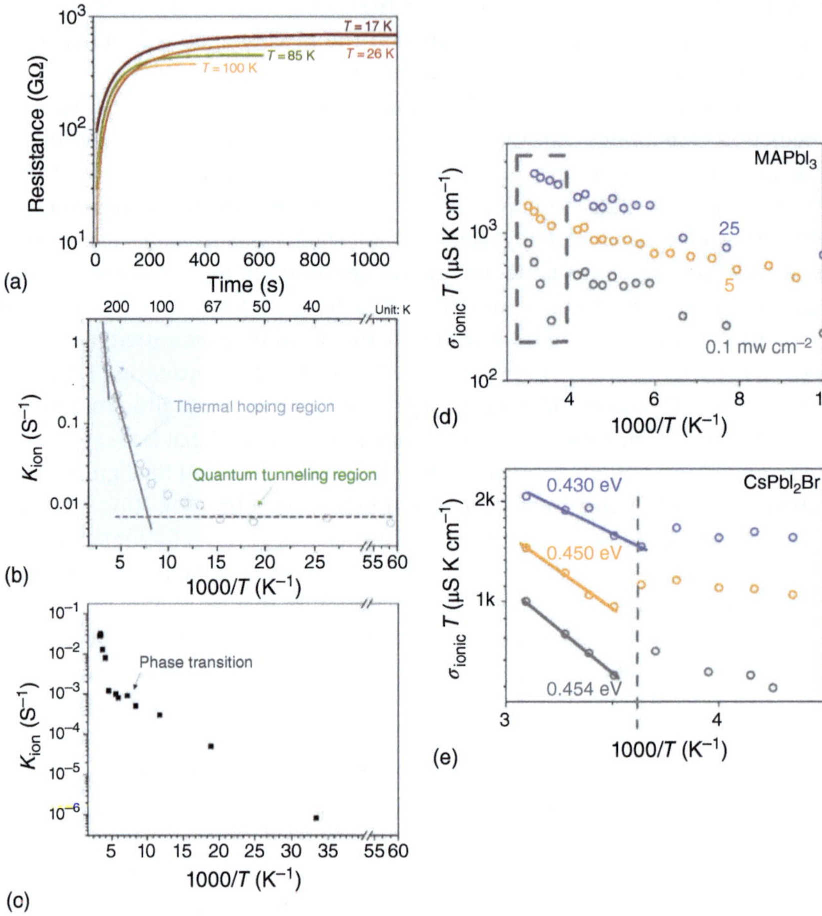

Figure 11.8 (a) Galvanostatic curves of the ionic conductance at temperatures (17–100 K) for the $MAPbI_3$ sample, from which the ionic kinetic constant is derived from exponential fitting for these curves. (b) The ionic kinetic constants versus T for the $MAPbI_3$ perovskite sample. (c) The ionic kinetic constants versus T for $CsPbI_2Br$ perovskite sample. Source: Reproduced with permission from Feng et al. [16]. © 2018, American Chemical Society. (d, e) Ionic conductivity of $MAPbI_3$ films (d) and $CsPbI_2Br$ films (e) and, $\sigma_{ion}T$, under different light intensities (gray: 0.1 mW cm^{-2}; orange: 5 mW cm^{-2}; purple: 25 mW cm^{-2}). Source: Reproduced with permission from Zhou et al. [33]. © 2017, American Chemical Society.

light-enhanced ion migration. However, with $CsPbI_2Br$ in the same structure, they found that even when the light intensities increased to 25 mW cm^{-2}, the perovskite film showed no obvious change and no light-induced phase segregation, which is consistent with results by Beal et al. [34] and implies the suppressed ionic transport by Cs substitution. Then, the activation energy for ionic transport in $CsPbI_2Br$ and $MAPbI_3$ was extracted by galvanostatic measurement (Figure 11.8d and e). In comparison with $MAPbI_3$, the barrier for ionic transport of Cs-perovskite remains constant (0.45 eV) under growing light intensities, indicating the light-independent ionic transport in all-inorganic perovskite film. This finding strongly implies the key role of organic cation in the complicated interaction between photoexcitation and ionic transport since organic cation-based perovskite film demonstrates light-enhanced ionic transport. Therefore, Cs substitution is an effective method to suppress the cation migration [35–37].

Kim et al. found that bromide substitution could strongly weaken the ionic conductivity enhancement in $APbI_3$ (A = MA^+, FA^+, or Cs^+) under illumination [38]; however, the variations of A-site cations make a trivial impact on light effects. In MHPs, ion migration is not simply equal to the movement of charged defects. Rather, there is a dynamic interaction between carriers, traps, and mobile ions, all of which could be the factor affecting ionic conductivities. For a perovskite film with larger quantity of vacant defects, the interstitial ions could move and fill them at the initial stage of the ionic transport process, during which the "healing effects" [30] will reduce the number of mobile ions and then lower the ionic conductivities. Additionally, the defect state densities of iodine- and bromide-based perovskite films may exhibit significant variation, on which the ionic conductivities heavily depend. Therefore, we can just jump to the conclusion that if the bromide-based perovskite contains higher amount of vacant defect state densities, more charged ions could compensate for them and reduce the ionic conductivities temporarily. Although ions can heal part of defects at the early stage of migration, prolonged ionic transport can cause irreversible changes in perovskite with a large number of defect states near the electrodes, indicated by quenched fluorescence, and eventually cause cells to become unstable.

11.4.4 Grain Boundary

In solid-state films, more defect states mean a higher possibility for ionic transport. Apart from bulk point defects, grain boundaries (GBs) could act as pathways for ionic transport due to the high densities of lattice dislocation [39], grain disorientation [40], excess volume [41], and wrong bonds at GBs [42]. Since the state-of-the-art MHPs are fabricated by low-temperate solution process (<150 °C), the target film has large amounts of GBs [43]. Although the defect states located at GBs may not act as deep-level traps, they could still boost the hopping-dominated ionic transport.

Shao et al. first discovered a faster ionic transport at GBs [44] via comparing the extent of current hysteresis at GBs and in the grains of polycrystalline $MAPbI_3$ films (Figure 11.9a–c). Then by comparing the extent of hysteresis in perovskite film with different grain sizes, they found that smaller grains could induce severer I–V hysteresis (Figure 11.9d–f) and lead to a higher contribution of ionic conductance to the entire conductivities in galvanostatic characterization. Quantitatively, the perovskite film with a larger grain size (around 1 μm) exhibited a 0.50 eV (under darkness) and 0.14 eV (under light) activation energy for ionic transport (Figure 11.9g), and in comparison, the value reduced to 0.27 and 0.08 eV for perovskite with smaller grains (300 nm) (Figure 11.9h), respectively [31]. Sparklingly, the activation energy for ionic transport in a single crystal is around 1.10 eV in darkness and 0.47 eV under light (Figure 11.9i). All the above results reveal that GBs can facilitate ionic transport;

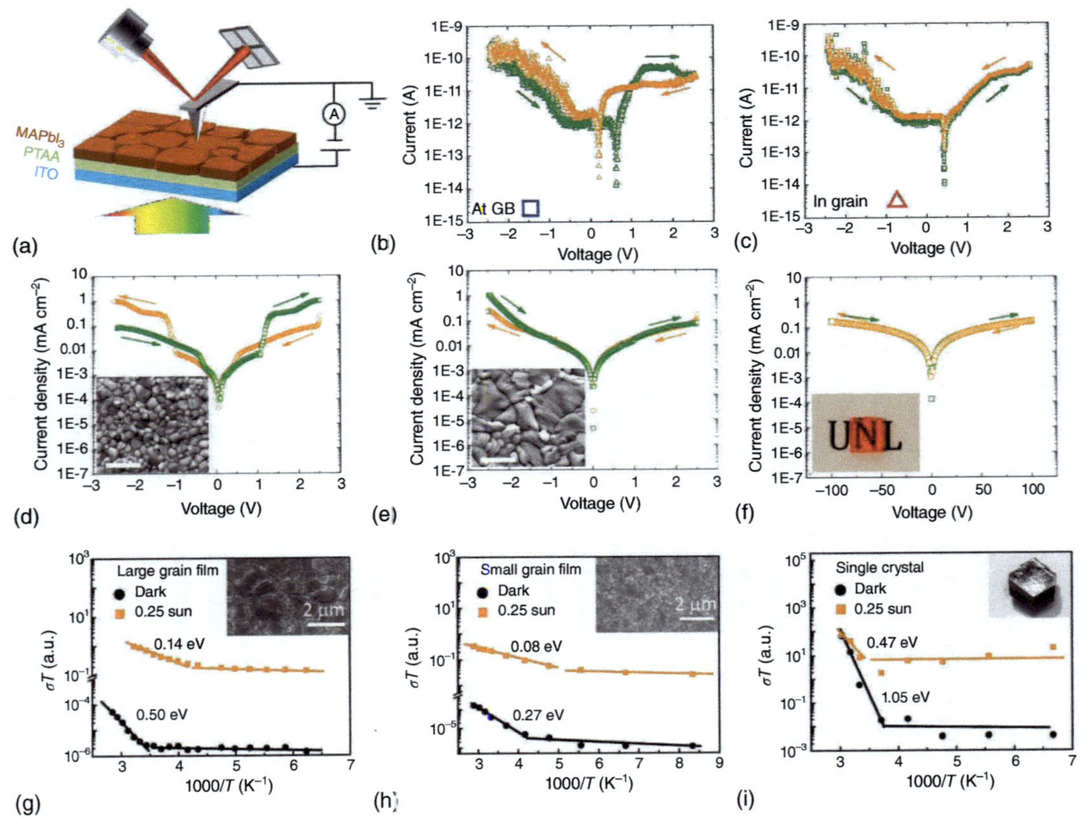

Figure 11.9 (a) The c-AFM measurement set up to study the ionic transport associated local current at GBs and on grains. (b) Local dark current measured at GBs. (c) Local dark current measured in grain. (d–f) Grain size-dependent ionic transport in perovskite materials measured at macroscopic level. (d) small grain size, (e) large grain size, (f) single crystal. The scale bar of SEM image is 2 μm. Source: Reproduced with permission from Shao et al. [44]. © 2016, Royal Society of Chemistry. (g–i) The temperature-dependent conductivity of a large-grain film (g), a small-grain film (h), and single crystal (i). The inset pictures of Figure 11.9g–i are SEM images of tested samples. Source: Reproduced with permission from Xing et al. [31]. © 2016, Royal Society of Chemistry.

therefore, increasing the grain size and passivating defect states at GBs may lead to improved device stability.

11.4.5 Lattice Strain

Due to the formation of GBs and thermal lattice expansion [45, 46] in solid-state crystal, compressive stress and tensile strain are widely distributed in the out-of- and in-plane directions of perovskite crystal [47]. Although subsequent thermal annealing can lead to Ostwald ripening [48], which could partly release the strains and reduce GB's densities. However, the residue strains, growing on substrates, can reach 50 MPa [49]. Zhao et al. qualitatively and quantitatively studied the impact of strain on the stability of $MAPbI_3$ film [50]. First, they curved the film into a convex and a concave shape to enhance and reduce the lattice strain, respectively. After 500 hours of illumination (50 mW cm^{-2}), a large area of the convex film turned yellow, which is verified to be PbI_2 through XRD measurement. In contrast, the concave film was highly stable without any PbI_2 appearance.

Then, the mechanism of strain-accelerated degradation was discovered to be related to ionic transport, given that the perovskite was covered with a thin polystyrene layer to prevent moisture. From temperature-dependent conductance measurement, the convex, flat, and concave $MAPbI_3$ films show gradually enlarged activation energies, of which are 0.29, 0.39, and 0.53 eV under dark and 0.046, 0.074, and 0.083 eV under 25 mW cm^{-2}, respectively. The convex film shows the smallest activation energy, which means the most severe ionic transport. The results also agree well with the light-enhanced ionic transport by showing much smaller activation energy under light compared to darkness. Additionally, Saidaminov et al. demonstrated that local lattice strain could facilitate the formation of vacancy defects and be detrimental to device stability. Therefore, strain-accelerated ionic transport is responsible for the faster degradation in convex film and sheds light on a path to improve perovskite stability. As to the reason for strain-enhanced ionic transport, we can explain the scenarios like this: When the perovskite crystal sustains tensile strain, the crystal structure would be stretched; and a higher density of lattice distortion/mismatch would generate in crystal; additionally, the GBs may split to form "cracks." All those possibilities would bring in wider channels for ions to move and enhance the extent of ionic transport.

It is worth mentioning that Chen et al. discovered a 1% response of compressive strain in a $MAPbI_3$ single crystal under an electric field of 3.7 V μm^{-1} [51]. Based on the second-order relationship between strain response and applied voltage (S~M 2), the electromechanical response is caused by electrostriction, not piezoelectric effect, for which there is a first-order coupling. Then, after carefully analyzing several possible mechanisms, like anharmonic displacement of positive/negative ions, Maxwell stress, and MA^+ cation ordering, ion/defect migration was considered to be responsible for the electrostrictive response. Finally, they estimated that about 0.01%–0.03% compressive strain could be generated in a typical perovskite solar cell with a built-in field around 0.3–0.65 V μm^{-1}. Due to the cooling process of $MAPbI_3$ crystal formation, mismatched thermal expansion can lead to residual tensile strain (about 0.47%) in the in-plane direction. Thus, the electrostriction can result in a larger lattice in the in-plane direction, which will then reduce the tensile strain and improve the device's stability.

11.5 The Impact of Ionic Transport on Perovskite Films and Devices

Ions could migrate in perovskite film at an electric field of $0.3\,V\,\mu m^{-1}$ or even smaller [15, 52], and cause switchable photovoltaic effect in a lateral device [5], with a relatively low activation energy [14]. Therefore, revealing the mechanism between device degradation and ionic transport is highly needed for fixing intrinsic instability before moving toward highly stable MHPs.

11.5.1 Phase Segregation

In 2015, Hoke et al. have observed reversible light-induced phase segregation in mixed-halide $MAPb(I_xBr_{1-x})_3$ [53]. The perovskite film with bromide concentrations exceeding 20% showed a PL redshift from 1.9 to 1.68 eV in less than one minute, under illumination lower than one sun ($15\,mW\,cm^{-2}$) at room temperature. Combined with XRD and absorption characterization, $MAPb(I_xBr_{1-x})_3$ was found to split into I-rich and Br-rich domains; in addition, after five minutes' storage in darkness, the PL spectra recover to their original position, indicating different stable states for perovskite under light and dark. Furthermore, Beal et al. found a similar shift of PL spectra in $CsPb(Br_xI_{1-x})_3$ for $0.4 < x < 1$ [34]. The I-rich component perovskite, with a lower bandgap, could act as charge-trapping defects and lead to the reduction of open circuit voltage (V_{oc}) in MHPs. Given that the observed second timescale for the PL shift is close to the ionic transport, the phase segregation is thought to be related to ionic transport, which needs more evidence.

Later, Zhao et al. used galvanostatic measurement with in-situ PL scan and real-time optical microscopy recording, to monitor the ionic transport in $MAPbI_3$ film [19] (Figure 11.10a). This characterization was realized in a lateral setup ($Au/MAPbI_3/Au$), upon switching the current from zero to a constant small value (Figure 11.10b). During electric poling, they found that the PL intensity decreased in several seconds both at anode (Figure 11.10c) and cathode (Figure 11.10d) after the current switched on, illustrating that defect states increased across the perovskite film. Besides, an obvious blue shift around 10 nm occurred near the anode, which may have arisen from the band-filling effect induced by the iodine vacancies with n-doping [54, 55]. Then, in I–Br mixed perovskite system, PL peak position showed around 30 nm of red shift at anode and 50 nm of blue shift at cathode [56], giving strong evidence for halide redistribution.

Additionally, DFT calculation results showed that, after electric poling, I trimers can form in the I-poor region and Pb dimers in the I-rich area [19] (Figure 11.10e–g), introducing deep-level defects around 0.3 eV below the conduction band minimum. Therefore, ionic transport-induced trap states can act as nonradiative recombination

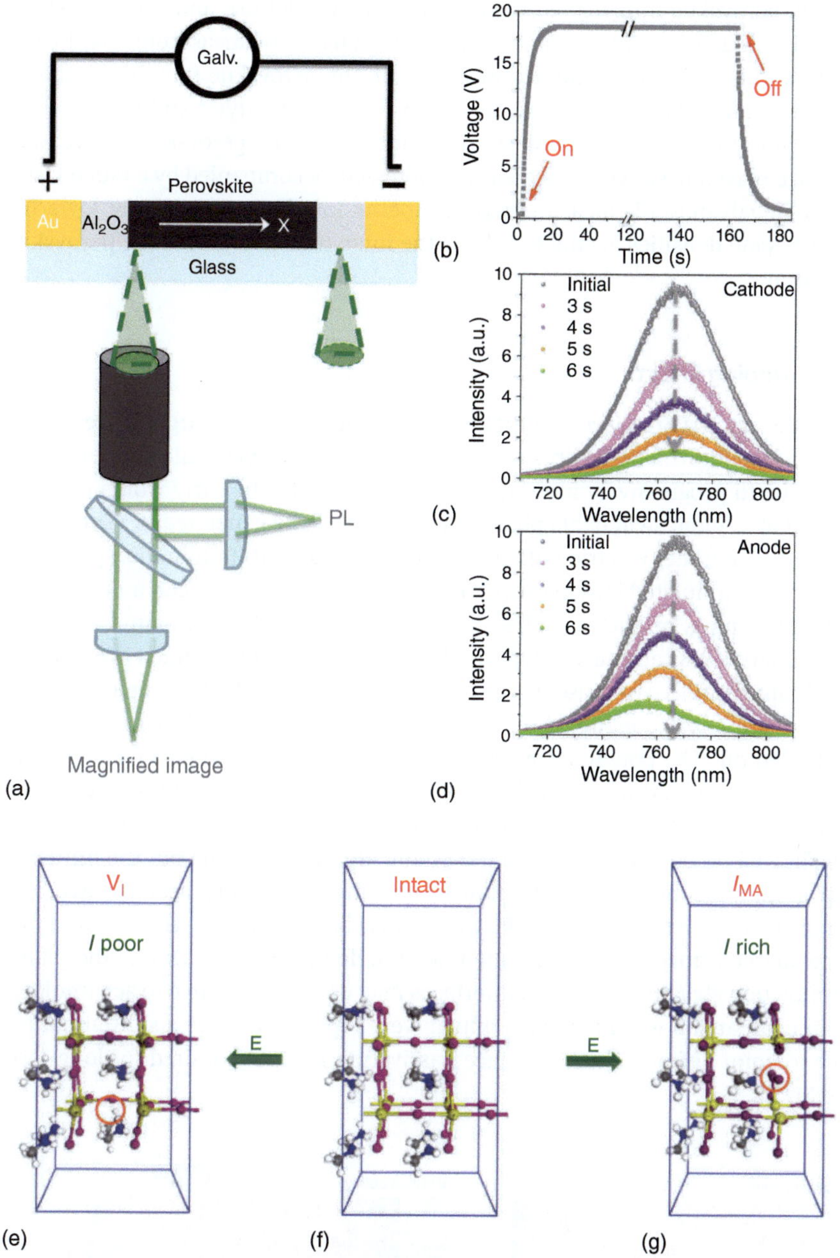

Figure 11.10 (a) Schematic illustration of the in-situ PL characterization with galvanostatic measurement. (b) Voltage–time (V–t) curve for a fixed direct current at 10 nA. The on and off indicate the current switches on and off, respectively. (c) PL spectra show a significant decrease at the positive pole (cathode) after electric poling at 1 V μm^{-1} for 6 seconds. (d) PL spectra show a significant decrease at the negative pole (anode) after electric poling at 1 V μm^{-1} for 6 seconds, accompanied by a strong blue shift from ~765 to 757 nm. (e–g) Atomic structure of two-dimensional perovskite $(CH_3NH_3)_3Pb_2I_7$ viewed along the direction perpendicular to the z-axis. Color scheme: Pb: gold; I: purple; C: dark gray; N: blue; H: white. (f). Red circles in (e) and (g) represent the I trimer and Pb dimer, respectively. Source: Reproduced with permission from Zhao et al. [19]. © 2016, American Chemical Society.

centers in perovskite film, decrease the charge carrier lifetime, and cause solar cell degradation. Furthermore, through liquid-state ^{1}H NMR measurements, PCBM was found to have a strong interaction with I^- in the perovskite film due to charge transfer to a buckyball and the van der Waals force with butyric and methyl ester. Thus, to grab mobile ions, PCBM was introduced into the perovskite device, and the voltage-driven hysteresis was greatly suppressed, accompanied by a much faster response in galvanostatic measurements for perovskite film. Accordingly, lowering the defect state densities is vital to inhibit the ionic transport inside the perovskite bulk crystal.

11.5.2 Doping Effects

Besides the photoactive layer, interface plays a vital role in charge carrier transportation in photovoltaic devices [57–63], and migrated ions could influence this process due to progressively interfacial accumulation. In the early works reported by Xiao et al., a giant switchable photovoltaic effect can be generated in a symmetrical solar device under electric poling ($<1\ V\ \mu m^{-1}$) and light illumination [27]. This means that accumulated ions in the perovskite near the electrodes can induce p-type and n-type doping, respectively. Theoretically, positively charged iodine vacancies or MA^+ interstitials can cause n-type doping, while negatively charged MA^+ vacancies or I^- interstitials can cause p-type doping [8, 64]. To verify this scenario, KPFM characterization was conducted on an Au/Perovskite device with part of Au electrode peeled off. The potential images indicated that, when the positively charged ions move toward cathode, it will result in a p–i–n structure, and a reverse bias can flip it to n–i–p (Figure 11.11a and b) [65]. During this process, surface topography showed no obvious change in the poling areas, thus ruling out the influence of topography on the surface potential characterization. Thus, accumulated ions could bring in defect states at interface and cause band-bending, which would hinder the charge transportation and attenuate the device stability. Due to the planar heterojunction structure of MHPs, interfaces may be more sensitive to accumulated ions, in comparison to bulk crystal, which were reported to have a substantial tolerance to point defects. Thus, interface passivation is highly desired for long-term stable MHPs.

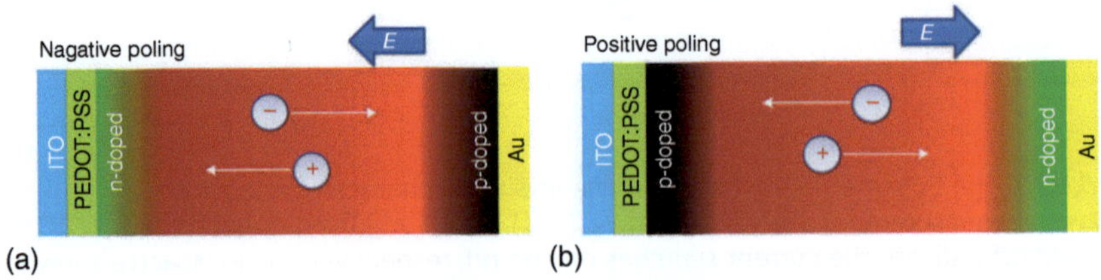

Figure 11.11 (a, b) Schematics of ion drift in perovskite during positive and negative poling, respectively, shows that accumulated ions in the perovskite near the electrodes induce p-doing and n-doping. Source: Reproduced with permission from Xiao et al. [65]. © 2014, Springer Nature.

11.5.3 SCLC and TFT Devices

Space charge-limited current (SCLC) characterization is one of the most common methods for extracting the carrier mobility and trap densities of semiconductors and has been widely published in halide-perovskite literature. Ideally, SCLC measurements consist of a dark JV curve measurement of a single carrier device with symmetric ohmic contacts on either side of a semiconductor. When the JV curve is plotted on a log–log scale, several regimes can be identified: First a low-voltage regime with a slope $\left[\frac{\mathrm{d}\log(J)}{\mathrm{d}\log(V)}\right]$ of 1, followed by a regime with a high slope (>2) due to trap filling (if any), and, finally, at high voltage, the so-called SCLC regime with a slope of 2. By applying the Mott–Gurney fitting equation [66]

$$J = \frac{9}{8}\varepsilon\mu\frac{(V - V_{\mathrm{BI}})^2}{L^3} \tag{11.11}$$

We can extract the mobility value, where J is the current density, V and V_{BI} are the applied and built-in voltages, respectively, ε is the dielectric constant, L is the thickness, and μ is the mobility. The trap density can be extracted from the so-called trap-filled-limit voltage:

$$V_{\mathrm{tfl}} = \frac{qn_{\mathrm{t}}L^2}{2\varepsilon} \tag{11.12}$$

However, their applicability to halide perovskites is not compatible, due to the ionic nature of those materials. The mobile ions raise notorious J–V hysteresis curves in the SCLC measurement and make it impossible to extract reliable trap density values using V_{tfl}. The ions at the interface can result in the formation of a thin PbI_2 layer with a small injection barrier (≈0.2–0.3 eV) next to the electrode, which indeed complicates the use of the backward scan to gain better accuracy of the trap density.

In the perovskite-based thin-film transistor, there is also an obvious hysteresis in the forward and backward JV scans at room temperature. The ions migrate under the electric field and then accumulate at the perovskite/dielectric interface, which partly screens the grid electric field. This will greatly affect the modulation of grid and reduce the field-effect mobility, which finally makes it tedious to obtain a reliable room-temperature operated TFT.

Besides, the mobile ions could significantly influence the determination of devices' characteristics and lifetimes based on IS results. First, under illumination, there is a very large increase in the low-frequency (from 10^{-1} to 10^{-2} Hz) capacitance in comparison with dark capacitance at room temperature. Although some authors claimed that under illumination, holes accumulate in the front of the hole-transport layer as a consequence of the unbalanced charge transport in the device. However, at the long time scale (microseconds to seconds), the ion accumulation at the contacts will be responsible for such behavior. The accumulation of ions at the contact interface produces a capacitance due to the surface space charge that responds in the low-frequency range in ordinary ionic conductors, within an extension equivalent to the Debye length. This capacitance happens at a short distance from the contact and is dependent on the active film thickness. Second, the cell device shows a transition point in the capacitance-frequency spectroscopy and negative imaginary impedance below 1 kHz under a small poling electric field (~−0.5 V) after long-term operating. Those phenomena represent a negative capacitance in the solar device, which is closely correlated with the sub-band-gap defects originated from the migrated ionic vacancies or interstitials.

11.5.4 Degradation in Functional Devices

Due to the soft organic nature of Spiro-OMeTAD, part of the accumulated ions can pass through the perovskite/HTL interface and then penetrate HTL layer. Zhao et al. systemically investigated the mobile MA^+ induced degradation of organic HTL [67]. Initially, they measured XRD patterns of perovskite films in fresh and degraded devices (operation for eight minutes in air with ~35% humidity), which showed no appearance of excess PbI_2, indicating that negligible decomposition occurs in the perovskite films. Then by replacing the Spiro-OMeTAD layer in the degraded cells with a fresh film, the cell efficiency increases from 4.9% (in the degraded cells) to 11.2% (in the recovered cells), with an initial PCE of 16.1% in the fresh cells. This efficiency recovery means that the degradation of the HTL layer should be responsible for part of the efficiency loss. Then, time-of-flight secondary-ion mass spectrometry (ToF-SIMS) clearly showed that, in the degraded devices, there was a conspicuous overlap between the MA^+ and Spiro distributions (Figure 11.12b), much larger than that in the fresh cell (Figure 11.12a). This gave strong evidence for MA^+ penetration into HTL, which can be explained by the MA^+ migration under electric field and illumination (Figure 11.12c). Similarly, in the inverted devices with p–i–n alignment structures, Wang et al. found that both MA^+ and I^- ions diffuse through the PCBM layer and accumulate at the Ag internal surface in N_2 under a standard test temperature of 85 °C [68] (Figure 11.12d and e). Massive loss of MAI at the GBs leads to the fusing of the $MAPbI_3$ grains, resulting in the generation of large PbI_2 gaps with low conductivity. Under the driving force of forming AgI, MA^+, and I^- ions are first released from the GBs of perovskite layer, and the MA^+ and I^- ions from the bulk grains migrate to reconstruct the GBs, inducing more defects in both the thin films and the $MAPbI_3$/PCBM interface. Consequently, the efficiency suffers from the deteriorated $MAPbI_3$ thin film, poor electron extraction, and the generation of AgI barrier. To improve the device resistance against MA^+/I^- penetration, applying inorganic hole transport material is an option, considering the robust covalent bonds in the inorganic crystal; in addition, an ion-blocking layer could be inserted between perovskite/transport layer interfaces.

11.6 Summary and Outlook

Generally, the operational stability of typical MHPs cannot meet commercial standards right now. Solving the instability problem is the first step toward the commercialization of MHPs [69]. As concluded in Sections 11.5, ionic transport could generate defect states in perovskite bulk crystal, induce interfacial band doping,

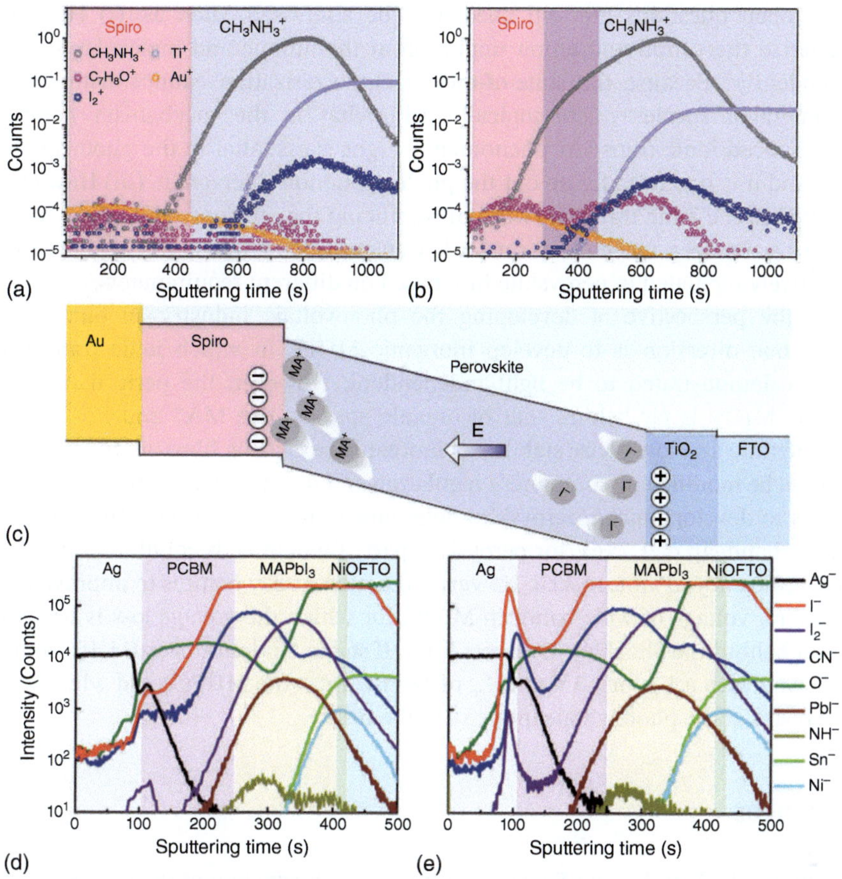

Figure 11.12 (a, b) Counts versus sputtering time by ToF-SIMS for different ions in the Au/HTL/perovskite/TiO_2/FTO architecture in fresh cells (a) and degraded devices (b) with over 60% efficiency loss. (c) Schematic of the electric field distribution in MHPs and the corresponding direction of ionic transport. Source: Reproduced with permission from Zhao et al. [67]. © 2016, American Chemical Society. ToF-SIMS elemental depth profiles (d) before and (e) after a thermal treatment at 85 °C for 24 hours in the Ag/PCBM/$MAPbI_3$/NiO/FTO architecture. The depth profiles of Ag^-, I^-, and CN^- after different temperatures of thermal treatment. Source: Yuan et al. [15]/Reproduced with permission of John Wiley & Sons.

and destroy the organic hole transport layer, which intrinsically influences the solar cell device stability. Since the photoactive, organic-cationic, solution-processed, and polycrystalline nature of perovskite materials, light illumination/organic cations/grain boundaries/tensile strain have been found to affect migration to some degree and, further, affect the stability of MHPs. Accordingly, various strategies could be developed to suppress ion migration toward highly stable MHPs [70]. For the perovskite layer, one can lower defect state densities, enlarge grain size, and increase cesium content; and for carrier transport materials, an ion-blocking layer may benefit.

Many open questions are still needed to be addressed, such as: (i) How to characterize the cation and anion migration at the atomic and molecular scales independently? Because the state-of-the-art characterization cannot distinguish such migrated species microscopically. (ii) What is the mechanism behind light-enhanced ionic transport phenomena? Light is essential to the photovoltaic device, and it is urgent to figure out the photon–phonon interaction. (iii) How can you suppress the ionic transport without sacrificing the efficiency of MHPs? After uncovering the deep mechanism for ionic transport, ionic transport is believed to be effectively regulated in perovskite film based on different requirements.

From the perspective of developing the photovoltaic industry, in our point of view, one direction is to develop inorganic MHPs, in which ionic transport has been demonstrated to be light independent. However, the performance of inorganic MHPs is far behind that of organic species, like MA^+ and FA^+. Poor carrier lifetime and low phase stability of inorganic perovskite films are the origins, which can be modified by component regulation and interface passivation. Another option is to develop silicon/perovskite heterojunction tandem solar cells with an optimum bandgap of 1.75 eV for perovskite part. Tandem cells could improve the device's efficiency to around 30%. However, the present key point is to improve the open-circuit voltage of wide-bandgap MHPs, for which the voltage loss is beyond 0.6 V, far behind the Shockley–Queisser limit. If stable tandem silicon/MHPs could be achieved with a V_{oc} of 1.3 V and J_{sc} of 18 mA cm^{-2}, the MHPs could win a seat on the board of the photovoltaic industry in the future.

References

1 Mizusaki, J., Arai, K., and Fueki, K. (1983). Ionic conduction of the perovskite-type halides. *Solid State Ionics* 11: 203–211.

2 Yang, T.-Y., Gregori, G., Pellet, N. et al. (2015). The significance of ion conduction in a hybrid organic–inorganic lead-iodide-based perovskite photosensitizer. *Angewandte Chemie, International Edition* 54: 7905–7910.

3 Maier, J. (2007). Solid state electrochemistry II: devices and techniques. In: *Modern Aspects of Electrochemistry*, vol. 41 (ed. C. Vayenas, R.E. White, and M.E. Gambpa-Aldeco), 1–128.

4 Lunkenheimer, P., Bobnar, V., Pronin, A.V. et al. (2002). Origin of apparent colossal dielectric xonstants. *Physical Review B* 66: 052105.

5 Yuan, Y., Chae, J., Shao, Y. et al. (2015). Photovoltaic switching mechanism in lateral structure hybrid perovskite solar cells. *Advanced Energy Materials* 5: 1500615.

6 Yan, Y. and Wei, S.-H. (2008). Doping asymmetry in wide-bandgap semiconductors: origins and solutions. *Physica Status Solidi B* 245: 641–652.

7 Wei, S.-H. (2004). Overcoming the doping bottleneck in semiconductors. *Computational Materials Science* 30: 337–348.

8 Yin, W.-J., Shi, T., and Yan, Y. (2014). Unusual defect physics in $CH_3NH_3PbI_3$ perovskite solar cell absorber. *Applied Physics Letters* 104: 063903.

9 Wang, Q., Shao, Y., Xie, H. et al. (2014). Qualifying composition dependent P and N self-doping in $CH_3NH_3PbI_3$. *Applied Physics Letters* 105: 163508.

10 Shi, T., Yin, W.-J., Hong, F. et al. (2015). Unipolar self-doping behavior in perovskite $CH_3NH_3PbBr_3$. *Applied Physics Letters* 106: 103902.

11 Eames, C., Frost, J.M., Barnes, P.R.F. et al. (2015). Ionic transport in hybrid lead iodide perovskite solar cells. *Nature Communications* 6: 7497.

12 Chen, B., Yang, M., Priya, S., and Zhu, K. (2016). Origin of J–V hysteresis in perovskite solar cells. *Journal of Physical Chemistry Letters* 7: 905–917.

13 Haruyama, J., Sodeyama, K., Han, L., and Tateyama, Y. (2015). First-principles study of ion diffusion in perovskite solar cell sensitizers. *Journal of the American Chemical Society* 137: 10048–10051.

14 Egger, D.A., Kronik, L., and Rappe, A.M. (2015). Theory of hydrogen migration in organic–inorganic halide perovskites. *Angewandte Chemie, International Edition* 54: 12437–12441.

15 Yuan, Y., Wang, Q., Shao, Y. et al. (2016). Electric-field-driven reversible conversion between methylammonium lead triiodide perovskites and lead iodide at elevated temperatures. *Advanced Energy Materials* 6: 1501803.

16 Feng, Y., Zhao, Y., Zhou, W.-K. et al. (2018). Proton migration in hybrid lead iodide perovskites: from classical hopping to deep quantum tunneling. *Journal of Physical Chemistry Letters* 9: 6536–6543.

17 Pockett, A., Eperon, G.E., Sakai, N. et al. (2017). Microseconds, milliseconds and seconds: deconvoluting the dynamic behaviour of planar perovskite solar cells. *Physical Chemistry Chemical Physics* 19: 5959–5970.

18 Leijtens, T. et al. (2015). Mapping electric field-induced switchable poling and structural degradation in hybrid lead halide perovskite thin films. *Advanced Energy Materials* 5: 1500962.

19 Zhao, Y., Zhou, W., Ma, W. et al. (2016). Correlations between immobilizing ions and suppressing hysteresis in perovskite solar cells. *ACS Energy Letters* 1: 266–272.

20 Zhao, Y.-C., Zhou, W.-K., Zhou, X. et al. (2017). Quantification of light-enhanced ionic transport in lead iodide perovskite thin films and its solar cell applications. *Light Science and Applications* 6: e16243–e16243.

21 Domanski, K., Alharbi, E.A., Hagfeldt, A. et al. (2018). Systematic investigation of the impact of operation conditions on the degradation behaviour of perovskite solar cells. *Nature Energy* 3: 61–67.

22 Bryant, D., Aristidou, N., Pont, S. et al. (2016). Light and oxygen induced degradation limits the operational stability of methylammonium lead triiodide perovskite solar cells. *Energy & Environmental Science* 9: 1655–1660.

23 Morimoto, T., Nagai, M., Minowa, Y. et al. (2019). Microscopic ion migration in solid electrolytes revealed by terahertz time-domain spectroscopy. *Nature Communications* 10: 2662.

24 León, C., Rivera, A., Várez, A. et al. (2001). Origin of constant loss in ionic conductors. *Physical Review Letters* 86: 1279–1282.

25 Koettgen, J., Zacherle, T., Grieshammer, S., and Martin, M. (2017). Ab initio calculation of the attempt frequency of oxygen diffusion in pure and samarium doped ceria. *Physical Chemistry Chemical Physics* 19: 9957–9973.

26 Jones, A., Slater, P.R., and Islam, M.S. (2008). Local defect structures and ion transport mechanisms in the oxygen-excess apatite $La_{9.67}(SiO_4)_6O_{2.5}$. *Chemistry of Materials* 20: 5055–5060.

27 Islam, M.S., Driscoll, D.J., Fisher, C.A.J., and Slater, P.R. (2005). Atomic-scale investigation of defects, dopants, and lithium transport in the $LiFePO_4$ olivine-type battery material. *Chemistry of Materials* 17: 5085–5092.

28 Yokota, I. (1961). On the theory of mixed conduction with special reference to conduction in silver sulfide group semiconductors. *Journal of the Physical Society of Japan* 16: 2213–2223.

29 Yokota, I. (1953). On the electrical conductivity of cuprous sulfide: a diffusion theory. *Journal of the Physical Society of Japan* 8: 595–602.

30 de Quilettes, D.W., Zhang, W., Burlakov, V.M. et al. (2016). Photo-induced halide redistribution in organic–inorganic perovskite films. *Nature Communications* 7: 11683.

31 Xing, J., Wang, Q., Dong, Q. et al. (2016). Ultrafast ion migration in hybrid perovskite polycrystalline thin films under light and suppression in single crystals. *Physical Chemistry Chemical Physics* 18: 30484–30490.

32 Kim, G.Y., Senocrate, A., Yang, T.-Y. et al. (2018). Large tunable photoeffect on ion conduction in halide perovskites and implications for photodecomposition. *Nature Materials* 17: 445–449.

33 Zhou, W., Zhao, Y., Zhou, X. et al. (2017). Light-independent ionic transport in inorganic perovskite and ultrastable Cs-based perovskite solar cells. *Journal of Physical Chemistry Letters* 8: 4122–4128.

34 Beal, R.E., Slotcavage, D.J., Leijtens, T. et al. (2016). Cesium lead halide perovskites with improved stability for tandem solar cells. *Journal of Physical Chemistry Letters* 7: 746–751.

35 Saliba, M. et al. (2016). Cesium-containing triple cation perovskite solar cells: improved stability, reproducibility and high efficiency. *Energy & Environmental Science* 9: 1989–1997.

36 Zhao, Y. et al. (2018). Perovskite seeding growth of formamidinium-lead-iodide-based perovskites for efficient and stable solar cells. *Nature Communications* 9: 1607.

37 Li, Q., Zhao, Y., Fu, R. et al. (2018). Efficient perovskite solar cells fabricated through CsCl-enhanced PbI_2 precursor via sequential deposition. *Advanced Materials* 30: 1803095.

38 Kim, G.Y., Senocrate, A., Wang, Y. et al. (2021). Photo-effect on ion transport in mixed cation and halide perovskites and implications for photo-demixing. *Angewandte Chemie, International Edition* 60: 820–826.

39 Read, W.T. and Shockley, W. (1950). Dislocation models of crystal grain boundaries. *Physics Review* 78: 275–289.

40 Brandon, D.G. (1966). The structure of high-angle grain boundaries. *Acta Metallurgica* 14: 1479–1484.

41 Pond, R.C. (1979). The measurement of excess volume at grain boundaries using transmission electron microscopy. *Journal of Microscopy* 116: 105–114.

42 Lambrecht, W.R.L., Lee, C.H., Methfessel, M. et al. (1990). Wrong bonds at compound semiconductor grain boundaries. *MRS Proceedings* 209: 667.

43 Long, R., Liu, J., and Prezhdo, O.V. (2016). Unravelling the effects of grain boundary and chemical doping on electron–hole recombination in $CH_3NH_3PbI_3$ perovskite by time-domain atomistic simulation. *Journal of the American Chemical Society* 138: 3884–3890.

44 Shao, Y. et al. (2016). Grain boundary dominated ion migration in polycrystalline organic–inorganic halide perovskite films. *Energy & Environmental Science* 9: 1752–1759.

45 Schueller, E.C., Laurita, G., Fabini, D.H. et al. (2018). Crystal structure evolution and notable thermal expansion in hybrid perovskites formamidinium tin iodide and formamidinium lead bromide. *Inorganic Chemistry* 57: 695–701.

46 Jacobsson, T.J., Schwan, L.J., Ottosson, M. et al. (2015). Determination of thermal expansion coefficients and locating the temperature-induced phase transition in methylammonium lead perovskites using X-ray diffraction. *Inorganic Chemistry* 54: 10678–10685.

47 Steele, J.A. et al. (2019). Thermal unequilibrium of strained black $CsPbI_3$ thin films. *Science* 365: 679–684.

48 Yang, Y. et al. (2020). Suppressing vacancy defects and grain boundaries via ostwald ripening for high-performance and stable perovskite solar cells. *Advanced Materials* 32: 1904347.

49 Rolston, N., Bush, K.A., Printz, A.D. et al. (2018). Engineering stress in perovskite solar cells to improve stability. *Advanced Energy Materials* 8: 1802139.

50 Zhao, J., Deng, Y., Wei, H. et al. (2017). Strained hybrid perovskite thin films and their impact on the intrinsic stability of perovskite solar cells. *Science Advances* 3: eaao5616.

51 Chen, B. et al. (2018). Large electrostrictive response in lead halide perovskites. *Nature Materials* 17: 1020–1026.

52 Li, C., Tscheuschner, S., Paulus, F. et al. (2016). Iodine migration and its effect on hysteresis in perovskite solar cells. *Advanced Materials* 28: 2446–2454.

53 Hoke, E.T., Slotcavage, D.J., Dohner, E.R. et al. (2015). Reversible photo-induced trap formation in mixed-halide hybrid perovskites for photovoltaics. *Chemical Science* 6: 613–617.

54 Kuokstis, E., Yang, J.W., Simin, G. et al. (2002). Two mechanisms of blueshift of edge emission in ingan-based epilayers and multiple quantum wells. *Applied Physics Letters* 80: 977–979.

55 Eliseev, P.G., Perlin, P., Lee, J., and Osiński, M. (1997). "Blue" temperature-induced shift and band-tail emission in ingan-based light sources. *Applied Physics Letters* 71: 569–571.

56 Zhou, W., Chen, S., Zhao, Y. et al. (2019). Constructing $CsPbBr_3$ cluster passivated-triple cation perovskite for highly efficient and operationally stable solar cells. *Advanced Functional Materials* 29: 1809180.

57 Mönch, W. (2001). *Semiconductor Surfaces and Interfaces*, vol. 26. Springer.

58 Marin-Beloqui, J.M., Lanzetta, L., and Palomares, E. (2016). Decreasing charge losses in perovskite solar cells through MP-TiO_2/MAPI interface engineering. *Chemistry of Materials* 28: 207–213.

59 Guerrero, A., Marchesi, L.F., Boix, P.P. et al. (2012). How the charge-neutrality level of interface states controls energy level alignment in cathode contacts of organic bulk-heterojunction solar cells. *ACS Nano* 6: 3453–3460.

60 Seol, D.-J., Lee, J.-W., and Park, N.-G. (2015). On the role of interfaces in planar-structured $HC(NH_2)_2PbI_3$ perovskite solar cells. *ChemSusChem* 8: 2414–2419.

61 Zhou, H., Chen, Q., Li, G. et al. (2014). Interface engineering of highly efficient perovskite solar cells. *Science* 345: 542–546.

62 Zhang, H., Liang, C., Zhao, Y. et al. (2015). Dynamic interface charge governing the current–voltage hysteresis in perovskite solar cells. *Physical Chemistry Chemical Physics* 17: 9613–9618.

63 Zhao, Y. et al. (2015). Anomalously large interface charge in polarity-switchable photovoltaic devices: an indication of mobile ions in organic–inorganic halide perovskites. *Energy & Environmental Science* 8: 1256–1260.

64 Kim, J., Lee, S.-H., Lee, J.H., and Hong, K.-H. (2014). The role of intrinsic defects in methylammonium lead iodide perovskite. *Journal of Physical Chemistry Letters* 5: 1312–1317.

65 Xiao, Z., Yuan, Y., Shao, Y. et al. (2015). Giant switchable photovoltaic effect in organometal trihalide perovskite devices. *Nature Materials* 14: 193–198.

66 Le Corre, V., Duijnstee, E., El Tambouli, O. et al. (2021). Revealing charge carrier mobility and defect densities in metal halide perovskites via space-charge-limited current measurements. *ACS Energy Letters* 6: 1087–1094.

67 Zhao, Y., Zhou, W., Tan, H. et al. (2017). Mobile-ion-induced degradation of organic hole-selective layers in perovskite solar cells. *Journal of Physical Chemistry C* 121: 14517–14523.

68 Li, J., Dong, Q., Li, N., and Wang, L. (2017). Direct evidence of ion diffusion for the silver-electrode-induced thermal degradation of inverted perovskite solar cells. *Advanced Energy Materials* 7: 1602922.

69 Zhao, Y. et al. (2021). Discovery of temperature-induced stability reversal in perovskites using high-throughput robotic learning. *Nature Communications* 12: 2191.

70 Zhao, Y. et al. (2022). A Bilayer conducting polymer structure for planar perovskite solar cells with over 1,400 hours operational stability at elevated temperatures. *Nature Energy* 7: 144–152.

12

Light Emission of Halide Perovskites

David O. Tiede, Juan F. Galisteo-López, and Hernán Míguez

Institute of Materials Science of Seville, Consejo Superior de Investigaciones Científicas (CSIC)-Universidad de Sevilla, C/Américo Vespucio 49, 41092 Sevilla, Spain

12.1 Introduction

The performance of a semiconducting material as a light emitter is given by its internal luminescence quantum yield (QY), η_{int}, which is the fraction of excited charge carriers that recombine radiatively versus the total density of excited carriers. Additionally, one must consider the fraction of emitted photons that are able to exit the semiconductor, associated with an external QY, η_{ext}. The latter is dictated by factors affecting the possibility of internally generated photons exiting the material (total internal reflection, reabsorption, etc.). Furthermore, the ability of a material to behave as a good (external) emitter is also intimately related with its performance as a photovoltaic material, as η_{ext} is related with the deviation of the real open circuit voltage V_{OC}^* in a solar cell from its ideal value V_{OC} [1]:

$$qV_{\text{OC}} - qV_{\text{OC}}^* = k_{\text{B}}T\left|\ln(\eta_{\text{ext}})\right|$$

Thus, understanding the light-emitting properties of a semiconductor, which dictate its performance in an optoelectronic device, involves a deep understanding of the way charge carriers recombine. Over the past decade, a large combined experimental and theoretical effort has been carried out in order to shed some light on the dominating mechanism of charge recombination in lead halide perovskites. These studies have unveiled some peculiarities demonstrating unique properties of this material.

In this chapter, we will consider the different charge carrier recombination mechanisms in lead-halide perovskites (LHP) in order to evaluate their light emitting properties. Understanding these mechanisms will further allow us to decipher recent results on the feasibility of using LHP as an active material for laser devices and, in particular, for room temperature (RT) electrically pumped lasers, a holy grail within solution-processed materials for optoelectronics.

Halide Perovskite Semiconductors: Structures, Characterization, Properties, and Phenomena, First Edition. Edited by Yuanyuan Zhou and Iván Mora-Seró.

Finally, the above-mentioned mechanisms will also be invoked to get an insight into another characteristic of LHP's emitting properties: the slow photo-induced processes reported for this material comprising photo-activation/darkening that take place on a time scale of seconds to hours. These ubiquitous processes, taking place for LHP of all compositions and dimensionalities, are intimately linked with structural changes taking place in this material as a consequence of its "soft nature," such as ion migration or material degradation, that constitute as of today the main bottlenecks for their future commercialization.

12.2 Charge-Carrier Recombination in Lead-Halide Perovskites

Light emission in a semiconductor is dictated by the way charge carriers recombine once they have been promoted to an excited state, mostly through optical or electrical excitation. Hence, to understand light emission in LHP, one must consider how excess charge carriers recombine in this type of materials. In this section, we elaborate on the charge carrier recombination processes involving primarily electrons and holes at the bottom (top) of the conduction (valence) band, as perovskite semiconductors are direct-gap semiconductors and charge carrier cooling mechanisms are expected to happen faster than recombination processes. Recombination can take place not only by radiative means, where a photon is emitted, but also through processes involving trap states associated with structural defects or interactions with other carriers. In order to study the recombination of charge carriers, one usually resorts to differential equations employing rate constants for all recombination processes. While many different models of coupled differential equations have been proposed (refer e.g. to [2]) to describe the recombination dynamics, a number of assumptions valid for *bulk* perovskites help to simplify those equations: primary photogenerated species are free carriers (as expected for perovskites taking into account the reported exciton binding energies [3]), the photogenerated charge carrier density is larger than the intrinsic one, and the densities of photogenerated electrons and holes are similar. Under these assumptions, the rate equations can be reduced to the simplified expression commonly known as the ABC model:

$$\frac{\mathrm{d}n}{\mathrm{d}t} = -An - Bn^2 - Cn^3$$

The above expression, albeit highly simplified, is illustrative of the main recombination processes taking place in halide perovskites and can be used to highlight some of the peculiarities behind the photophysics of this material. The ABC model employs three terms that summarize the probability of monomolecular (one particle, Figure 12.1a), bimolecular (two particle, Figure 12.1b), and trimolecular (three particle, Figure 12.1c) processes happening. In a general picture, electron-hole recombination is a two-particle process including an electron and a hole. However, under certain conditions, this particle dependency can be varied, as elaborated in

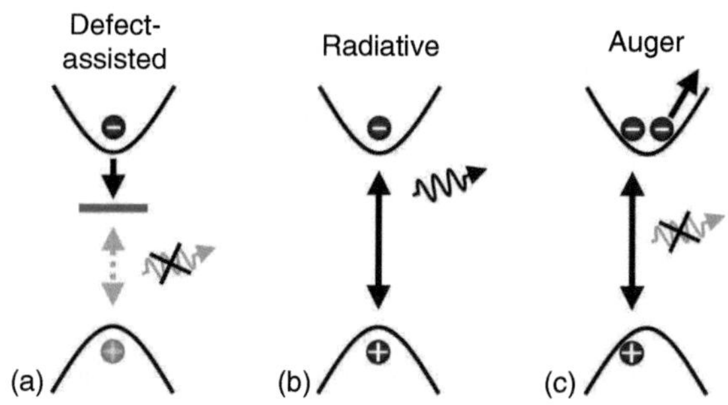

Figure 12.1 Illustration of recombination processes in the ABC model. (a) Defect-induced trapped charge carrier assisted recombination, where the trapping of one charge carrier results in an effective one-particle dependence. (b) Radiative bimolecular recombination, where an electron recombines with a hole and the excess energy is emitted as a photon. (c) Trimolecular Auger process. An electron and a hole recombine in a nonradiative way, and the excess energy is transferred to excite a nearby charge carrier.

the following. We will first focus on bulk-like, free carrier-dominated systems, and in a second step, discuss the variations in confined, exciton-dominated systems.

12.2.1 Monomolecular Recombination

The monomolecular rate constant A includes processes, where the recombination only depends on electron *or* hole populations, which comprise trap-assisted recombination in which localized defect states interacting with either holes or electrons are considered (Figure 12.1a). These processes are particularly important at low excitation regimes commonly attained under operation conditions in photovoltaic devices. In the case of electron traps and considering a fast electron trapping rate k_{trap}, the recombination rate can be described as:

$$R = k_{\text{trap}} n_{\text{e}} N_{\text{T}}(1 - f) = A n_{\text{e}}$$

where n_{e} is the electron density, f is the probability of a trap state to be occupied and k_{trap} the capture coefficient. $k_{\text{trap}} N_{\text{T}}(1-f)$ can be considered as constant, if the transition from electrons to trap states is much faster than the recombination from trap states with holes. In this case, the recombination rate only depends on the density of holes in the material, and the charge carrier density becomes particle-dependent and thus monomolecular. Such charge carrier trapping is found to happen mainly via a phonon-assisted transfer of free carriers to electronic trap states. The energetic position of the involved trap states determines the accessibility of those states for free charge carriers. The charge trapping constant can be described with

$$k_{\text{trap}} = \sigma_{\text{trap}} v_{\text{th}} N_{\text{T}}$$

where σ_{trap} and v_{th} represent the capture cross section and the thermal velocity of the free carriers. In a classical picture, σ_{trap} is dependent on the allowed phonon modes that permit a transition between a free carrier and a trap state energy level. v_{th} is correlated with the thermal energy of the system via

$$E_{th} = \frac{1}{2} m_{eff} v_{th}^2 = \frac{3}{2} k_B T$$

$$v_{th} = \sqrt{\frac{3 k_B T}{m_{eff}}}$$

where m_{eff} represents the effective mass of the corresponding charge carrier. In LHP, m_{eff} has been reported to be in the range $0.2 - 0.5\ m_e$ [3]. Finally, N_T is the (energy dependent) trap state density, where the depth of the individual traps affects the effective σ_{trap}. A transfer to trap states via phonon emission is thus more likely and faster if the energy difference matches that of the available phonon modes in the lattice.

The energetic position of trap states determines their influence on recombination dynamics. Trap states lying deep inside the conduction or valence band will not affect the way charge carriers recombine. Trap states close to the conduction band minimum (CBM) or valence band maximum (VBM) with an energy difference to the CBM (VBM) smaller than the thermal energy $k_B T$ are easily ionized and act as *shallow traps* that do not lead to nonradiative recombination but will induce doping. Trap states with an energy difference larger than $k_B T$ are labeled *deep traps*. Recombination from those sites occurs only in a nonradiative way via vibrational decay paths where electron and electron hole recombine in the trap center. This nonradiative recombination via multiple phonon emission was first proposed by Shockley, Read, and Hall in 1952 and is commonly referred as Shockley–Read–Hall (SRH) recombination [4, 5].

LHP with ABX_3 structure, where A is typically methylammonium (MA), formamidinium (FA), or cesium, B is lead, and X is either iodine, bromine, chlorine, or a mixture of them, are solution-processed materials and possess a soft crystal structure in comparison with conventional inorganic semiconductors, which causes an intrinsically high defect density. In spite of this high defect density, perovskites can behave as good emitters as a result of the prevalence of shallow traps and only a few deep traps within their bandgap. Point defects of different nature, comprising interstitials, vacancies, and antisites of standard components, have been considered, and surprisingly most of them have been identified as shallow traps, with mainly halide-related defects as origin for deep traps (see, for instance, specific examples for $MAPbI_3$ and $MAPbBr_3$ in Figure 12.2) [7]. In particular, the role of halide interstitials and halide vacancies and their presence as Frenkel pairs have been pointed out theoretically to cause electron deep traps, which depend on the atomic number of the employed halide. The intra-band or shallow character of most of the remaining defects explains the high defect tolerance of perovskite semiconductors that turn them into a material with outstanding optoelectronic properties. Much effort has been dedicated to determining the exact nature of surface and bulk defects, a debate that remains open within the perovskite community, as well as

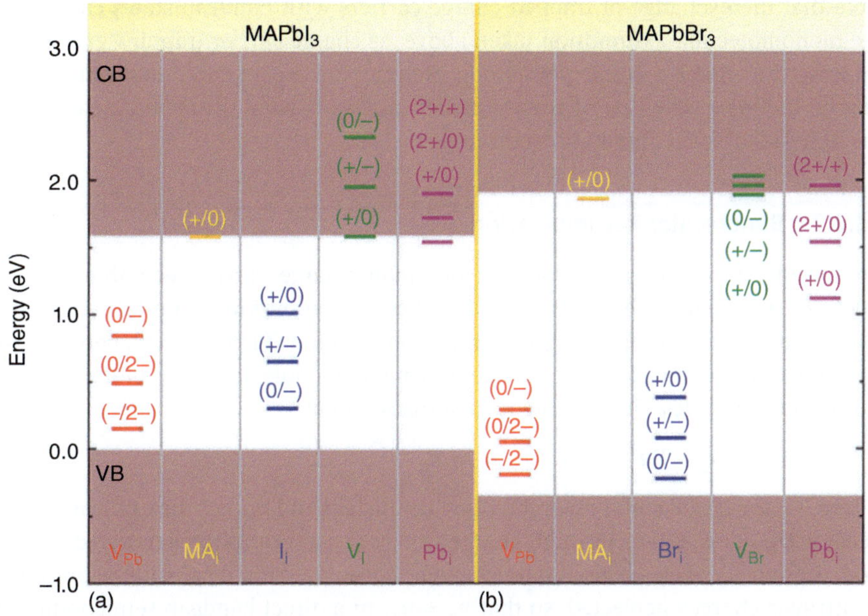

Figure 12.2 Density functional theory (DFT) calculations of different point defects in (a) $MAPbI_3$ and (b) $MAPbBr_3$. Energy levels that are close to the conduction/valence band can be considered shallow traps. Trap states in the band gap are considered deep traps. Source: Motti et al. [6]/Reproduced with permission of John Wiley & Sons.

possible passivation routes or their role in their optical properties via photo-induced processes affecting the defect structure of the material (see Section 12.3).

Measured defect densities for LHP, while depending on the composition and morphology, have been reported to take elevated values ranging from $10^{14}-10^{17}\,cm^{-3}$ in polycrystalline films [8–10] to $10^{9}-10^{12}\,cm^{-3}$ in single crystals [11, 12], slightly higher than in other common semiconductors such as crystalline silicon ($10^{8}\,cm^{-3}$) [13]. Nevertheless, the large defect density does not compromise the optoelectronic properties of this material, something associated with its defect tolerance. Here, we recall that the nonradiative lifetime is not only dependent on the number of defects but also on the ease of exchange of free carriers with defect states given by σ_{trap}. The exact mechanism responsible for the high defect tolerance is still under debate, but has often been assigned to (multiple) interactions between free charge carriers and trap states, which depend on σ_{trap} and can be rationalized in two models. These can be investigated in low temperature experiments, that enable emission from band tail states [14]. In a hopping model, trapped charge carriers can hop between different trap states but are unlikely to contribute to radiative recombination. In a multiple trapping model, after being trapped, charge carriers can escape and contribute to band to band recombination or be recaptured again. The cross section of involved electronic states has been further demonstrated to be dependent on the formation of polarons via a coupling to low frequency phonons [15].

Note that the exchange of trapped charge carriers with bandgap states can break the monomolecular assumption taken above, as charge carrier trapping cycles can increase the effective defect population lifetime and create additional multiexponential lifetimes. The rate of these transfer models is what ultimately determines the non-radiative recombination constant A.

12.2.2 Bimolecular Recombination

Bimolecular processes are those that dominate charge carrier recombination at intermediate excited charge carrier densities. In perovskite materials, they are mainly attributed to direct band-to-band recombination (Figure 12.1b), between an electron at the bottom on the conduction band and a hole at the top of the valence band, and can be described by the recombination rate:

$$R = Bn_\mathrm{h}n_\mathrm{e} = Bn^2 \qquad \text{if} \qquad n_\mathrm{e} = n_\mathrm{h}$$

where B is the bimolecular recombination coefficient and n_e (n_h) the electron (hole) density. Upon excitation through charge carrier separation/photogenerated excitation, the density of electrons and holes can be considered identical and the presence of intrinsic charges neglected, so that $n_\mathrm{e} = n_\mathrm{h}$. In a direct bandgap semiconductor such as LHP, most of the recombination is expected to occur radiatively via emission of a photon. However, some thermal recombination via lattice vibrations can also contribute to band-to-band charge carrier recombination, such that the recombination coefficient B has to be extended to

$$R = Bn^2 = (B_\mathrm{rad} + B_\mathrm{non-rad})n^2$$

In a direct semiconductor like LHP, the radiative term can be expected to be much faster than the nonradiative recombination as the latter would involve several phonons thus reducing its probability. Nevertheless, in LHPs, the experimental recombination lifetimes are much longer than expected from a pure direct bandgap semiconductor with a similar electronic gap, where the recombination lifetime can be estimated from the refractive index through the van Roosbroeck–Shockley relation [16]. Generally speaking, a fluorescence process can be slowed down either if the charges involved in the recombination are trapped or separated spatially or its transition is forbidden momentum- or spinwise. Based on this, several mechanisms have been proposed to account for the prolonged lifetimes observed, which are currently being debated (Figure 12.3). Beyond the role played by the above-mentioned deep-traps, some reports account for shallow trap filling and escaping as mechanisms of extended lifetimes (Figure 12.3a). In shallow trap states, the lifetime of charge carriers is extended to milliseconds, so trap states can act as reservoir for excited charge carriers. Only after an interaction with a lattice phonon can the trapped charge carrier be lifted to its initial energy level and recombine radiatively. Since this process requires a thermal contribution from the lattice, it is often referred to as thermally activated delayed fluorescence [17]. Other mechanisms may affect the bimolecular recombination coefficient B. One approach suggests the formation of polarons, which are local lattice distortions caused by attraction and repulsion

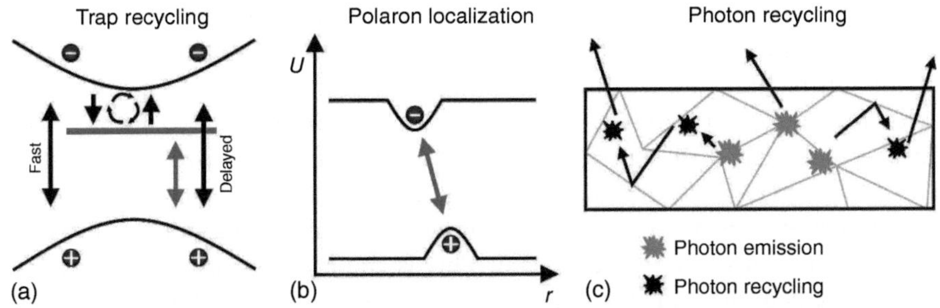

Figure 12.3 Fluorescence delay mechanisms. (a) Energy diagram with possible decay pathways considering defect-induced trap states. A fast decay corresponds to radiative bimolecular recombination. A delayed decay is caused by the charge trapping and detrapping process. The lifetime in trap states is extended as a nonradiative phonon assisted decay process (gray arrow) is less likely. After an eventual detrap, delayed radiative recombination occurs. (b) Illustration of a spatial separation of electron and hole through a polaron formation, which weakens the Coulomb interaction potential. (c) Illustration of photon recycling processes. Upon radiative recombination, the emitted photon can either escape directly or be reabsorbed in the film. Scattering at film impurities can further enhance the path length of emitted photons in the film and thus enhance the probability of a reabsorption event.

of charged ions in the vicinity of the photogenerated carrier. The formation of polaron quasi-particles can localize charge carriers and reduce the recombination rate through spatial separation (Figure 12.3b) [18]. Other reports suggest mainly photon recycling to be responsible for the extended lifetimes (Figure 12.3c) [19]. As photon emission might occur isotropically in the entire volume of the material, reabsorption of the emitted photons by an absorptive material is a likely process, which enhances the effective (measured) radiative emission rate. In thin films, this process is further enhanced, as grain boundaries can act as scattering centers and further disperse photons in the film. In fact, a plethora of studies have demonstrated significant changes in radiative lifetime upon bulk and surface excitation, which can be correlated both to bulk photon reabsorption and distinct defect and/or polaron formation characteristics at the surface and bulk level. Depending on the composition and morphology of the studied perovskite, one or several of the presented radiative recombination retardation processes might dominate the recombination dynamics. Disentangling its contributions in different material configurations will remain a challenging task for the perovskite community in the future.

12.2.3 Trimolecular Recombination

When the carrier density is sufficiently large, the third term in (1) dominates carrier recombination in the semiconductor. This term represents many-body Auger nonradiative recombination processes where excess energy from an excited charge carrier is transferred to another charge carrier at the band edge, which is promoted to a higher energy state. Figure 12.1c shows the situation where energy from an electron-hole recombination is transferred to an electron (eeh process), although the equivalent one involving a hole (ehh process) is also possible. These processes represent a nonradiative recombination mechanism as the relaxation of the third "hot" particle to the band edge takes place via interaction with lattice phonons in order to satisfy momentum conservation that transfers its energy to the lattice.

This nonradiative recombination pathway usually becomes relevant at carrier densities much larger than those generated in LHPs under photovoltaic operation (10^{15}–10^{16} cm^{-3}). While the Auger mechanism is not expected to play a key role in solar devices, it may become critical for light emitting applications where high charge carrier densities are demanded. This is the case for lasing applications, where amplified spontaneous emission (ASE) charge carrier thresholds in the 10^{-17}–10^{-18} cm^{-3} range have been reported (see Section 12.3). These charge carrier densities are similar to those where Auger effects start to dominate carrier recombination and could constitute a burden for the future development of lasers, as experimentally determined Auger coefficients have been suggested to be intrinsic to the band structure of this material [20]. Furthermore, reported experimental values for Auger coefficient C in LHP (3.4×10^{-28}–3.7×10^{-29} $cm^6\ s^{-1}$) [21, 22] are 2–3 orders of magnitude larger than those reported for other semiconductors such as silicon (2.8×10^{-31} $cm^6\ s^{-1}$) [23] or GaAs (7×10^{-30} $cm^6\ s^{-1}$) [24], evidencing a more efficient process in the former.

12.2.4 Recombination Constants in Excitonic Systems

As initially stated, the above-mentioned recombination dynamics only hold true in a system that is dominated by free carriers. Recently, confined systems such as quantum dots or 2D perovskites have become increasingly popular due to their outstanding optoelectronic properties. In these systems, the spatial confinement enhances exciton binding energies and may cause recombination dynamics to become excitonic as the electron and hole cannot fully dissociate and remain bound as an exciton. The charge carrier recombination dynamics now only depend on the density of excitons (n_{ex}) in the system and can be described as one-particle and two-particle processes:

$$\frac{dn_{ex}}{dt} = -An_{ex} - Bn_{ex}^2$$

Here, a one-particle process describes the recombination of an exciton, predominantly via a radiative recombination. Under higher exciton density, two-particle processes are likely to happen, as described via the term $-Bn_{ex}^2$. For the case of an excitonic system, bimolecular processes involve two excitons rather than two charge carriers. In analogy to the afore described Auger process, exciton species can also interact with surrounding excitons and recombine via a non-radiative exciton–exciton annihilation (EEA) process (Figure 12.4a), where the energy of an electron-hole recombination is transferred to a nearby exciton generating a hot exciton that returns to the lattice temperature by a combination of phonon (nonradiative) and photon (radiative) emission processes. EEA is expected to occur at a much faster time scale than exciton recombination (tens of ps, Figure 12.4b) and can be easily separated from monomolecular contributions in time resolved charge carrier recombination dynamics [25].

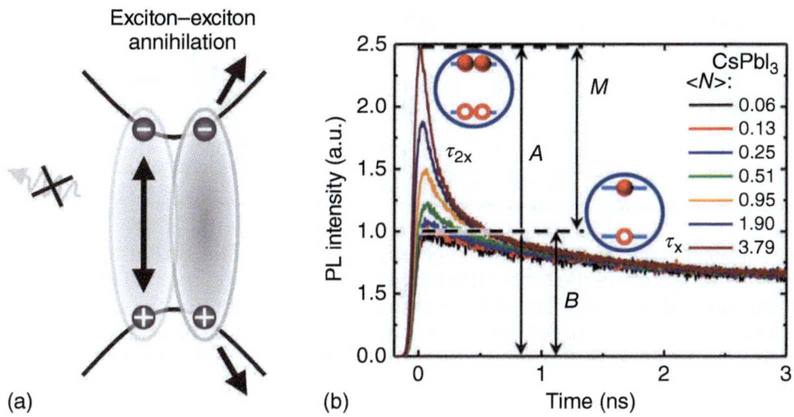

Figure 12.4 Exciton–exciton interaction. (a) Illustration of an exciton–exciton annihilation (EEA) process. The excess energy of an exciton recombination process is transferred to excite a nearby exciton. (b) Charge carrier density- dependent time resolved PL measurements on $CsPbI_3$ nanocrystals. A fast, initial decay is associated with EEA, while at extended timescales the PL decay remains excitation independent. Source: (b) Makarov et al. [25]/Reproduced with permission of American Chemical Society.

In many weakly confined systems and materials with low exciton binding energy, a coexistence of excitonic and free carrier species is expected. This interplay of the two recombination pictures can influence the disentanglement of occurring radiative and nonradiative recombination processes drastically. In that sense, great care must be taken when time-resolved photoluminescence (PL) or absorption measurements are employed to determine the distinct charge carrier recombination dynamics.

12.2.5 Common Recombination Dynamics Measurement Techniques and Experimental Evidence

As we have seen above, a precise determination of the different recombination coefficients that account for the involved processes is essential for studying light emission in any semiconductor. In order to determine each of these coefficients, several experimental techniques can be used. The most appropriate ones are time-resolved approaches that are able to follow the evolution of a material property that can be associated with the population of the different carrier species (free and trapped) involved.

In time-resolved PL measurements, information on the kinetics of electron-hole recombination processes can be extracted. Since the amount of radiative emission can be correlated to the overall charge carrier density, as it is expected to depend as $PL \propto Bn^2$ and B, in turn, is expected to be charge carrier density independent, it remains an indirect technique to analyze the recombination dynamics. Please note that effects like pure nonradiative low charge carrier density recombination can affect its interpretation [2], so care must be taken to not misinterpret time-resolved PL decays. A more direct measurement is offered by transient absorption spectroscopy (TAS). Employing a time resolved pump and probe configuration

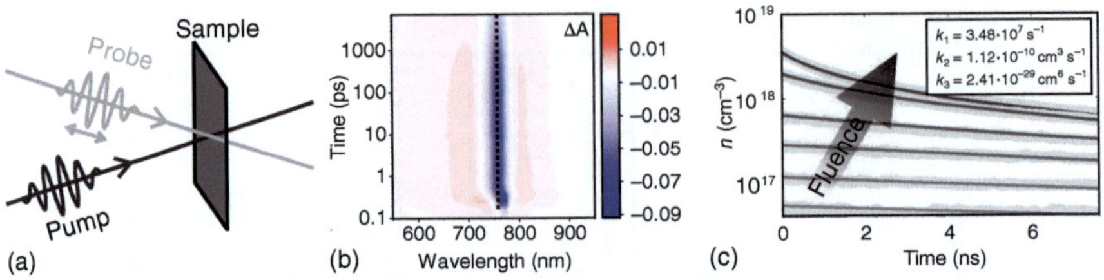

Figure 12.5 Transient absorption measurement of a $MAPbI_3$ thin film. (a) Illustration of a pump and probe setup, where the time delay between the two pulses is varied to probe time-resolved changes in absorption. (b) Color map of changes in absorption ΔA upon a 190 fs, 780 nm pulsed excitation. Dotted line indicates the trace of the ground state bleach (GSB) plotted in (c). (c) Fluence dependent traces of the GSB with a global fit to the ABC-model.

(Figure 12.5a), the changes in absorption ΔA after an ultrafast charge carrier population event can be measured. The ground state bleach (GSB) in ΔA (Figure 12.5b) can be directly correlated to the charge carrier density generated at the valence and conduction band edges. This direct measurement allows to fit the charge carrier density decay to the discussed ABC-model and extract rate constants for every charge carrier recombination process (Figure 12.5c) carrying out a set of measurements for increasing excitation fluences where different processes come into play.

Alternative techniques using a pump and probe configuration have also been employed to study charge carrier recombination in semiconductors in general, and in LHP in particular. For instance, optical pump terahertz probe measurements use a similar configuration as TAS, employing an optical pulse to generate a population of excited carriers and a probe beam comprising a THz pulse. Changes in the THz spectral region can be correlated to the photoinduced conductivity in the perovskite, which can be associated with the carrier density. From the spectral shape of the mobility, which can be derived from the conductivity, one can further extract specific information for free charge carriers, excitons, or polarons, turning this technique into a powerful one to characterize charge recombination in semiconductors. In another pump-probe configuration, time-resolved microwave conductance, is used where changes in micro-wave intensity can be correlated with sample conductance and charge carrier density.

12.3 Photoinduced Effects on Charge Carrier Recombination

According to the above discussion, the internal luminescence quantum yield, η_{int}, of a semiconductor is not a fixed property but strongly depends on the particular regime dominating carrier recombination. We can express the η_{int} as the ratio of the radiative versus the total recombination rates:

$$\eta_{int}(n) = \frac{k_{rad}}{k_{tot}} = \frac{Bn^2}{An + Bn^2 + Cn^3}$$

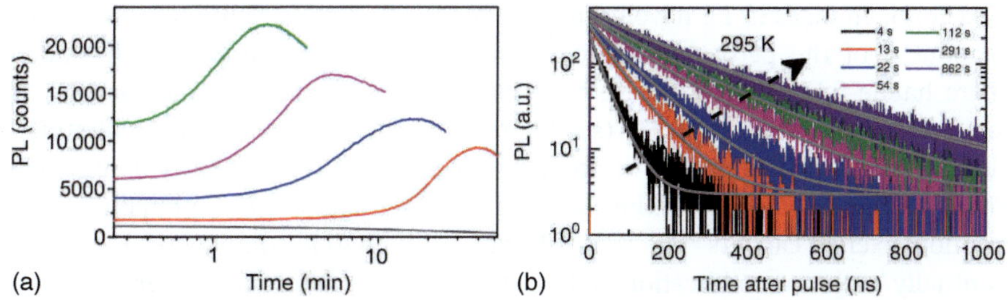

Figure 12.6 (a) Evolution of the PL intensity of a $MAPbBr_3$ thin film during irradiation with a CW laser source under different oxygen pressures. (b) PL decay curves of a $MAPbI_3$ film collected at different times (indicated in the inset) of CW irradiation with a laser beam. Source: (a) Anaya et al. [26]/Reproduced with permission of American Chemical Society. (b) DeQuilettes et al. [27]/Springer Nature/Licensed under CC BY 4.0.

A puzzling observation in LHPs has been the time-dependent evolution of the η_{int} upon optical irradiation with a continuous wave (CW) excitation on time scales (seconds–minutes) orders of magnitude larger than those involved in the charge carrier recombination or trapping at defect states. These changes comprise both an enhancement as well as a drop in the emission of the material (see Figure 12.6a). Such observations, reported already in the early days of the field [28, 29], can be associated with changes in the weight of the nonradiative recombination rate versus the total rate. Since these photoinduced changes are observed for optical excitation conditions where Auger recombination is not relevant, the main source of changes is expected to come from variations in the defect density as B is an intrinsic material property that should not vary if the band structure of the material is not altered. While such variations have not been considered, they cannot be excluded a priori, as some reports have demonstrated spectral changes in the emission of crystals that have been previously irradiated, pointing at structural changes [30].

The connection between photo-induced changes in emission and trap density has been evidenced by a number of observations. When the PL decay dynamics of a LHP film are monitored under monomolecular conditions dominated by trap-assisted recombination [27, 31], upon irradiation with a CW laser source leading to an enhancement in PL intensity, the PL decay dynamics become slower, pointing to a reduction in the A rate coefficient (Figure 12.6b) and a less efficient nonradiative recombination. Such variations in A were related in a quantitative way to the defect density in the material [27]. In a different configuration, resorting to confocal PL microscopy, polycrystalline films presenting an inhomogeneous PL distribution among the different grains were studied [27]. Here, dark (defective) grains underwent slow enhancements of several orders of magnitude in their PL while bright (defectless) grains showed much smaller changes in emission. Thus, upon irradiation, the film emission becomes spatially homogenized.

In an effort to understand the nature of the defects involved in these photo-induced processes, local probe studies have been carried out combining PL confocal microscopy and chemical composition techniques. Using time-of-flight secondary ion-mass spectrometry (ToF-SIMS) [27] and energy dispersive X-ray spectroscopy

(EDS) [32], changes in PL have been correlated with a redistribution in the halide component of the perovskite matrix. These observations coincide with studies where halide-related defects have been associated with the introduction of deep traps within the bandgap of LHP (see Figure 12.2). In particular, halide interstitials within the bulk [6] and halide vacancies at the surface [33] have been identified as potential traps responsible for the introduction of nonradiative recombination paths for photo-excited carriers. The redistribution of the halide component that would eventually lead to a modification of the defect structure within the material is likely connected with the facility of halides to migrate within the perovskite matrix under external stimuli, presenting the smallest activation energies for ionic migration out of all the components [34].

Another ingredient playing a key role in these photo-induced processes is the atmosphere surrounding the material. Early studies demonstrated that under an inert atmosphere (vacuum or N_2), similar to the conditions under which LHP are commonly grown, these materials are poor emitters and further undergo a reversible drop in the PL upon irradiation [35–37]. The latter is likely connected to the light induced formation of Frenkel pairs comprising iodide vacancies (V_I) and interstitials (I_i) leading to an enhanced ionic conductivity [38]. On the contrary, exposure to the combination of light and reactive atmospheres (moisture or oxygen) leads to an initial enhancement of the PL [26, 31, 36, 37] followed by a later degradation [26]. This points to two striking evidences, which are directly related to the performance of these materials in a device. On the one hand, as-synthetized perovskites present low emission, likely due to the presence of a large density of crystalline defects. This situation is improved when the material is exposed to the combined action of external irradiation and ambient air, as evidenced by the enhanced PL. This behavior is now acknowledged to be linked to the so-called "light soaking" phenomenon commonly reported in LHP photovoltaic devices, where the photoconversion efficiencies undergo a variation under continuous illumination. This phenomenon, while observed in other photovoltaic technologies, is particularly pronounced in the case of LHPs. On the other hand, several reports have pointed out that photo-induced enhancement or depletion of the PL are not separate effects but coexist, and the prevalence of one or the other is just a matter of irradiation time [26, 39]. The latter indicates that if photo-induced changes can be frozen at a given time when PL enhancement has reached its maximum, this could be employed as a means to induce "atmospheric light annealing" on the material. In this direction, such light annealing in the presence of humid air was carried out on $CH_3NH_3PbI_3$ films and devices fabricated from them, leading to higher efficiencies and a more stable performance [40]. While this approach represents an exciting avenue to further improve the efficiency of perovskite-based devices, alternative routes that avoid the use of reactive molecules will need to be explored in order to prevent a detrimental effect in the long run.

The presence of molecular oxygen in the surroundings can also modify the defect structure of LHPs even in the absence of irradiation. In particular, O_2 can interact with interstitial iodide in its stable I_i^- charge state responsible for introducing nonradiative recombination paths, leading to a change in the oxidation state of the latter, which lowers its energy level closer to the valence band, turning it into a shallow trap with reduced trapping activity [41]. But molecular oxygen can also have a detrimental role in the optoelectronic properties of LHP when it interacts with the complementary defect: an iodide vacancy V_I [42]. In this situation, O_2 can diffuse through the perovskite lattice and incorporate into an iodide vacancy, where it can absorb a photogenerated electron leading to the formation of a superoxide species O_2^- which then reacts with the oxidized perovskite lattice leading to its decomposition. The combined action of these two processes was observed in a $MAPbI_3$ film [26], where increasing the amount of oxygen in the surrounding atmosphere in a controlled manner allowed for enhanced emission at early times of irradiation, pointing to O_2 passivating traps without the need for light, and leading to degradation at longer times (Figure 12.6a). Surprisingly in this experiment, the amount of oxygen also correlated with the rate of emission enhancement at intermediate times, pointing to a mechanism not described before and associated with the presence of a superficial negatively charged O_2^- layer that could induce the migration of halide-related defects leading to the annihilation of Frenkel pairs. A systematic experimental and theoretical study has also been performed to determine the interaction of different molecules with the surface of $CH_3NH_3PbI_3$, suggesting that O_2 molecules can also passivate surface iodide vacancies, leading to a reduction in the density of trap states and contributing to an enhancement of PL upon irradiation [33]. While the above discussion has been centered on the $CH_3NH_3PbX_3$ formulation of LHPs, photo-induced processes have been reported in other compositions comprising different cations as well as for various morphologies such as nanocrystals and quasi-2D perovskites, pointing to the universal character of these processes in this type of material.

The above discussion highlights the complex scenario that is likely behind these photo-induced changes, as most of these processes can be taking place simultaneously. Thus, a precise prediction involves a large number of variables, from the defect landscape of the material to their charge state. Efforts to unveil the ultimate mechanism will nevertheless pay off, as highly efficient and more stable devices will likely be achieved. In this direction, many passivation strategies targeting surface traps have been implemented that lead to devices with better performance and a higher efficiency. These comprise postsynthetic surface passivation with different organic molecular species depending on the nature of the targeted surface defect, or using inorganic species directly into the precursors in order to neutralize surface vacancies. But, as a certain defect density will always exist within any crystalline lattice at a temperature above 0 K, these photo-induced processes can serve as a means to perform an ultimate light-assisted annealing.

12.4 Lasing in Lead-Halide Perovskites

The above discussion dealing with charge carrier recombination in Section 12.2 assumed that the electron population of the valence band is much larger than that of the conduction band. Therefore, an emitted photon as a result of the radiative recombination of an excited electron with a hole will either exit the semiconductor or be reabsorbed, leading to photon recycling. However, if a sufficiently large population of excited carriers is generated, we can be in the situation where the electron population in the conduction band is larger than that of the valence band. In this scenario, a spontaneously emitted photon can trigger the recombination of an excited electron with a hole, leading to the stimulated emission of an additional photon that will share wavelength, direction, and optical phase with the original one.

The excited carrier population for which stimulated emission outcompetes absorption is known as population inversion, and the semiconductor is said to exhibit optical gain. An additional feedback mechanism like a cavity structure can be introduced so that emitted photons traverse the optical gain medium several times. Above a certain charge carrier population, optical gain overcomes losses (due to scattering or reabsorption) and laser emission is attained. In order to achieve this scenario, one must choose the proper material where nonradiative losses are not dominant for the large density of excited charge carriers needed to achieve population inversion, but also introduce a feedback mechanism with reduced losses. According to the previous discussion, the mono- (A) and trimolecular (C) rates will determine the appropriateness of a given material to be used as active component in a laser device.

LHP have entered the race of electrically pumped lasing solution processed devices where population inversion is achieved by means of electrical injection of charge carriers, critical for its commercial implementation. While this holy grail in the field of solution-processed materials remains elusive, either for LHP or other approaches such as colloidal quantum dots (cQDs) and organic semiconductors, tremendous progress has been achieved over a time span of just a few years. As for the case of photovoltaic and light-emitting devices, advances in LHP-based lasing have profited from the developments in the fields of cQDs and organic semiconductor lasers. The main challenge that has faced solution-processed materials on the path to build a laser device is related to material losses once the high carrier density needed for population inversion is approached. As mentioned, similarities of LHP with other solution-processed materials have permitted great advances over the past years. At the same time, some points about the gain properties of LHP remain to be unveiled that will determine whether this material will finally allow stable electrically pumped lasing required for most applications. Nevertheless, recent results on low- dimensional structures point to this morphology as an exciting avenue for obtaining LHP-based lasers as will be discussed.

After the initial reports in the 1990s, working at cryogenic temperatures [43], the first demonstration of RT ASE in LHP films dates back to 2014 [44] where emission spanning the entire visible was achieved profiting from the bandgap tunability of these materials using the $MAPbX_3$ formulation (where X = I, Br, and Cl or an I–Br or Br–I alloy). While RT emission amplification was observed over several hours, it was achieved using low repetition rate (1 kHz) pulsed (130 fs) optical excitation. Shortly after, a work reporting RT lasing from a $MAPbI_3$ film inserted in a planar optical cavity was demonstrated also under pulsed optical excitation [45]. Over the

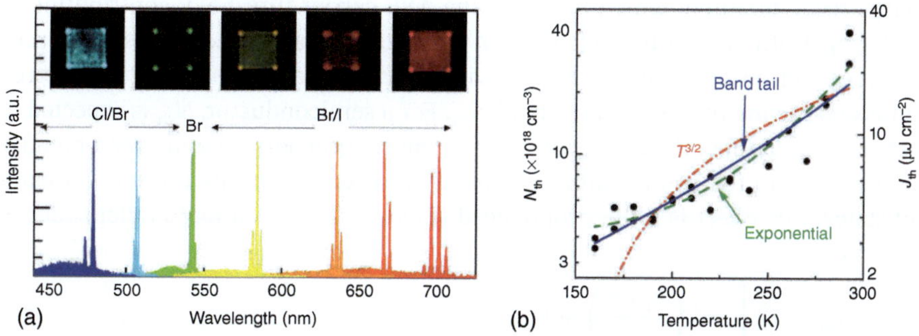

Figure 12.7 Amplified spontaneous emission in perovskites. (a) Wide tunability of ASE wavelength for different perovskite thin film compositions. (b) Temperature dependent ASE threshold for a thin $MAPbI_3$ film fitted to a model, including band tail states as described in the text. Source: (a) Zhang et al. [46]/Reproduced with permission of Wiley. (b) Jia et al. [47]/Reproduced with permission of John Wiley & Sons.

following years, a plethora of articles were reported on RT ASE and lasing exploring different material compositions (employing different cations –MA,FA,Cs- as well as halide anions) and resonant structures (distributed feedback [DFB], planar cavities, etc.) (see Figure 12.7a) [46], but always under pulsed optical excitation. A demonstration of ASE under CW optical excitation was achieved at cryogenic temperatures for mixed cation-mixed anion $Cs_{0.1}(MA_{0.17}\text{-}FA_{0.83})_{0.9}Pb_{0.84}(I_{0.84}Br_{0.16})_{2.68}$ perovskites [48] and for $MAPbI_3$ [49]. For the latter case the gain mechanism involved carrier funneling between two coexisting crystalline phases. While a report of CW lasing at RT was shown in a $MAPbI_3$ film where a nanoimprint procedure was used to turn the film itself into a distributed bragg reflector (DBR), thus providing the resonant mechanism [50], this result was debated [51] and associated with experimental artifacts commonly observed when studying lasing [52]. So far, the only demonstration of RT CW lasing, constituting a milestone in the development of a LHP-based electrically pumped laser, has been presented for low-dimensional quasi-2D perovskites [53] opening an exciting avenue for future developments.

The main issue with lasing and ASE studies using CW pumping with LHP is the short duration of amplified emission (a few hundred ns after its onset), the so-called "lasing death," taking place at RT. Understanding the main limitation that impedes CW light amplification from LHP to take place at RT is a must prior to their incorporation into real life electrically driven devices. A first look at the carrier recombination dynamics in the systems for which CW lasing and ASE have been observed indicates that reported carrier thresholds for ASE (10^{-17}–10^{-18} cm^{-3}) [51, 52] lie in the region where bimolecular recombination dominates [54] and thus losses associated with defects should not interfere. But the role of defect-related recombination has proven to be more complex when one considers the experimental conditions under which light amplification takes place. Due to the low thermal conductivity of LHP [55] the intense pumping needed for obtaining amplified emission leads to a heating of the sample, and this has a deleterious effect on charge recombination, which increases the ASE carrier threshold, contributing to the lasing death [47]. This is related with tail states characteristic of these materials as a result of energetic disorder that will contribute to depopulating the energy bands and raising the carrier threshold N_{th}. For a semiconductor, N_{th} is expected to increase with temperature following $T^{3/2}$ as injected carriers expand over a broader energy range in the bands, limiting population inversion. In the presence of band tail states, this effect is more pronounced (Figure 12.7b) leading to a dependency given by:

$$N_{th} \propto T^3 \left[\cot^2\left(\frac{k_b T\pi}{E_0}\right) + 1\right]$$

where E_0 is the Urbach energy of the material. On top of this effect, CW irradiation at cryogenic temperatures in an inert atmosphere further leads to the introduction of defect states that contribute to the nonradiative recombination of charges. As a result, an increase of one order of magnitude in k_1 has been reported to take place on a timescale similar to that of the T-induced rise in N_{th}, which also ends up contributing to the lasing death. The exact nature of such defects remains to be unveiled.

But the role of temperature under intense pumping also affects the other recombination pathways discussed above, further complicating the picture of lasing death. This has been studied for the mixed formulation $Cs_{0.1}(MA_{0.17}\text{-}FA_{0.83})_{0.9}Pb_{0.84}(I_{0.84}Br_{0.16})_{2.68}$ for which CW ASE was reported [56]. While Auger recombination did not change in the temperature range under consideration (50–290 K), where this perovskite remains in its cubic phase, the bimolecular recombination rate drops according to $T^{-3/2}$ due to a reduction in carrier mobility as a result of increased phonon scattering. This means that as the T rises, the density of injected carriers N_A for which Auger recombination dominates over bimolecular ones ($B = C \cdot N_A$) will decrease and eventually be comparable to N_{th}, bringing ASE to a halt. Additionally, a critical emitted photon density within the ASE band is needed in order to sustain ASE, and this density increases as T rises since the PL spectrum broadens.

Even though the above results were obtained for LHP with different formulations, they provide a general picture of the current limitations on CW and RT lasing in LHP and allow us to point out some improvements that will be needed in order to circumvent or at least minimize them. As low thermal conductivity is a characteristic that spreads across all LHPs [57], efficient device designs to dissipate pump-induced heating will be needed in order to minimize effects related to elevated temperatures. Nevertheless, as some T increase will be present, advances will be needed on two fronts:

- **Defect management:** understanding the precise nature of defects leading to the presence of tail states in LHP as well as photo-induced defects during high intensity pumping needed for light amplification will allow guiding future synthetic approaches to reduce disorder within LHP.
- **Tailoring recombination within LHP:** while the previous point will reduce trap-assisted recombination, acting on the temperature dependence of the bimolecular recombination coefficient could also improve the prospects for CW lasing. As mentioned above, this dependence is dictated by electron–phonon

interaction. Means to act on this could involve introducing strain in the crystalline lattice or looking for material combinations to reduce the soft character of the perovskite lattice while maintaining its band structure, as suggested in [56].

While efforts toward CW and RT lasing in LHP are being actively explored, it should be mentioned that the use of low dimensional structures can constitute a way to circumvent some of the above limitations. As a matter of fact, the use of quasi-2D LHP has led to RT and CW lasing under optical excitation, constituting a landmark in the development of LHP-based lasing devices [54] and setting the pace for future research toward LHP-based lasers. In this system, the main limitation for sustained ASE is singlet-triplet annihilation, which can be overcome by the use of appropriate organic molecules that act as spacers for the quasi-2D structure but also as quenchers for triplet states. While this approach is exciting, it still poses some limitations, such as the difficulty of injecting carriers in the presence of such organic molecules.

12.5 Conclusions

Beyond any consideration for commercial applications, LHP have proven to be a truly fascinating material from the point of view of their emission properties. Their optoelectronic performance is outstanding for a solution-processed material containing organic moieties. While their tolerance to structural disorder is well acknowledged at this stage, future research will be needed to deepen into the precise origin of crystalline defects and their role in LHP emission in order to achieve stable light-emitting diodes as well as electrically pumped lasers. Furthermore, the instability and sensitivity to changes in the environment, characteristic of these materials as a result of their "soft" nature, responsible for the presence of light-induced processes strongly affecting their QY, represent both a burden for their commercialization as well as a route to designing light-assisted postprocessing techniques to improve their performance. Thus, while further efforts to understand the physical mechanisms behind light emission in these materials will be needed, the prospects for their implementation into highly efficient light-emitting devices constitute a motivating goal.

References

1 Miller, O.D., Yablonovitch, E., and Kurtz, S.R. (2012). Strong internal and external luminescence as solar cells approach the Shockley–Queisser limit. *IEEE Journal of Photovoltaics* 2 (3): 303–311.

2 Kiligaridis, A., Frantsuzov, P.A., Yangui, A. et al. (2021). Are Shockley-Read-Hall and ABC models valid for lead halide perovskites? *Nature Communications* 12 (1): 1–3.

3 Baranowski, M. and Plochocka, P. (2020). Excitons in metal-halide perovskites. *Advanced Energy Materials* 10 (26): 1903659.

4 Shockley, W.T. and Read, W.T. Jr., (1952). Statistics of the recombinations of holes and electrons. *Physical Review* 87 (5): 835.

5 Hall, R.N. (1952). Electron-hole recombination in germanium. *Physical Review* 87 (2): 387.

6 Motti, S.G., Meggiolaro, D., Martani, S. et al. (2019). Defect activity in lead halide perovskites. *Advanced Materials* 31 (47): 1901183.

7 Meggiolaro, D. and De Angelis, F. (2018). First-principles modeling of defects in lead halide perovskites: best practices and open issues. *ACS Energy Letters* 3 (9): 2206–2222.

8 Ball, J.M. and Petrozza, A. (2016). Defects in perovskite-halides and their effects in solar cells. *Nature Energy* 1 (11): 1–3.

9 Herz, L.M. (2016). Charge-carrier dynamics in organic-inorganic metal halide perovskites. *Annual Review of Physical Chemistry* 67: 65–89.

10 Manser, J.S., Christians, J.A., and Kamat, P.V. (2016). Intriguing optoelectronic properties of metal halide perovskites. *Chemical Reviews* 116 (21): 12956–13008.

11 Shi, D., Adinolfi, V., Comin, R. et al. (2015). Low trap-state density and long carrier diffusion in organolead trihalide perovskite single crystals. *Science* 347 (6221): 519–522.

12 Dong, Q., Fang, Y., Shao, Y. et al. (2015). Electron-hole diffusion lengths >175 μm in solution-grown $CH_3NH_3PbI_3$ single crystals. *Science* 347 (6225): 967–970.

13 Ayres, J.R. (1993). Characterization of trapping states in polycrystalline-silicon thin film transistors by deep level transient spectroscopy. *Journal of Applied Physics* 74 (3): 1787–1792.

14 Wright, A.D., Milot, R.L., Eperon, G.E. et al. (2017). Band-tail recombination in hybrid lead iodide perovskite. *Advanced Functional Materials* 27 (29): 1700860.

15 Chu, W., Zheng, Q., Prezhdo, O.V. et al. (2020). Low-frequency lattice phonons in halide perovskites explain high defect tolerance toward electron-hole recombination. *Science Advances* 6 (7): eaaw7453.

16 Van Roosbroeck, W. and Shockley, W. (1954). Photon-radiative recombination of electrons and holes in germanium. *Physical Review* 94 (6): 1558.

17 Chirvony, V.S., González-Carrero, S., Suarez, I. et al. (2017). Delayed luminescence in lead halide perovskite nanocrystals. *The Journal of Physical Chemistry C* 121 (24): 13381–13390.

18 Munson, K.T., Kennehan, E.R., Doucette, G.S., and Asbury, J.B. (2018). Dynamic disorder dominates delocalization, transport, and recombination in halide perovskites. *Chemistry* 4 (12): 2826–2843.

19 Pazos-Outón, L.M., Szumilo, M., Lamboll, R. et al. (2016). Photon recycling in lead iodide perovskite solar cells. *Science* 351 (6280): 1430–1433.

20 Shen, J.X., Zhang, X., Das, S. et al. (2018). Unexpectedly strong auger recombination in halide perovskites. *Advanced Energy Materials* 8 (30): 1801027.

21 Wehrenfennig, C., Eperon, G.E., Johnston, M.B. et al. (2014). High charge carrier mobilities and lifetimes in organolead trihalide perovskites. *Advanced Materials* 26 (10): 1584–1589.

22 Yang, Y., Yang, M., Li, Z. et al. (2015). Comparison of recombination dynamics in $CH_3NH_3PbBr_3$ and $CH_3NH_3PbI_3$ perovskite films: influence of exciton binding energy. *The Journal of Physical Chemistry Letters* 6 (23): 4688–4692.

23 Dziewior, J. and Schmid, W. (1977). Auger coefficients for highly doped and highly excited silicon. *Applied Physics Letters* 31 (5): 346–348.

24 Strauss, U., Rühle, W.W., and Köhler, K. (1993). Auger recombination in intrinsic GaAs. *Applied Physics Letters* 62 (1): 55–57.

25 Makarov, N.S., Guo, S., Isaienko, O. et al. (2016). Spectral and dynamical properties of single excitons, biexcitons, and trions in cesium–lead-halide perovskite quantum dots. *Nano Letters* 16 (4): 2349–2362.

26 Anaya, M., Galisteo-López, J.F., Calvo, M.E. et al. (2018). Origin of light-induced photophysical effects in organic metal halide perovskites in the presence of oxygen. *The Journal of Physical Chemistry Letters* 9 (14): 3891–3896.

27 DeQuilettes, D.W., Zhang, W., Burlakov, V.M. et al. (2016). Photo-induced halide redistribution in organic–inorganic perovskite films. *Nature Communications* 7 (1): 1–9.

28 Sanchez, R.S., Gonzalez-Pedro, V., Lee, J.W. et al. (2014). Slow dynamic processes in lead halide perovskite solar cells. Characteristic times and hysteresis. *The Journal of Physical Chemistry Letters* 5 (13): 2357–2363.

29 Stranks, S.D., Burlakov, V.M., Leijtens, T. et al. (2014). Recombination kinetics in organic-inorganic perovskites: excitons, free charge, and subgap states. *Physical Review Applied* 2 (3): 034007.

30 Galisteo-López, J.F., Calvo, M.E., and Míguez, H. (2019). Spatially resolved analysis of defect annihilation and recovery dynamics in metal halide perovskite single crystals. *ACS Applied Energy Materials* 2 (10): 6967–6972.

31 Tian, Y., Peter, M., Unger, E. et al. (2015). Mechanistic insights into perovskite photoluminescence enhancement: light curing with oxygen can boost yield thousandfold. *Physical Chemistry Chemical Physics* 17 (38): 24978–24987.

32 Galisteo-López, J.F., Li, Y., and Míguez, H. (2016). Three-dimensional optical tomography and correlated elemental analysis of hybrid perovskite microstructures: an insight into defect-related lattice distortion and photoinduced ion migration. *The Journal of Physical Chemistry Letters* 7 (24): 5227–5234.

33 Brenes, R., Eames, C., Bulović, V. et al. (2018). The impact of atmosphere on the local luminescence properties of metal halide perovskite grains. *Advanced Materials* 30 (15): 1706208.

34 Eames, C., Frost, J.M., Barnes, P.R. et al. (2015). Ionic transport in hybrid lead iodide perovskite solar cells. *Nature Communications* 6 (1): 1–8.

35 Galisteo-López, J.F., Anaya, M., Calvo, M.E., and Míguez, H. (2015). Environmental effects on the photophysics of organic–inorganic halide perovskites. *The Journal of Physical Chemistry Letters* 6 (12): 2200–2205.

36 Fang, H.H., Adjokatse, S., Wei, H. et al. (2016). Ultrahigh sensitivity of methylammonium lead tribromide perovskite single crystals to environmental gases. *Science Advances* 2 (7): e1600534.

37 Motti, S.G., Gandini, M., Barker, A.J. et al. (2016). Photoinduced emissive trap states in lead halide perovskite semiconductors. *ACS Energy Letters* 1 (4): 726–730.

38 Kim, G.Y., Senocrate, A., Yang, T.Y. et al. (2018). Large tunable photoeffect on ion conduction in halide perovskites and implications for photodecomposition. *Nature Materials* 17 (5): 445–449.

39 Morán-Pedroso, M., Rubino, A., Calvo, M.E. et al. (2021). The role of the atmosphere on the photophysics of ligand-free lead-halide perovskite nanocrystals. *Advanced Optical Materials* 9 (18): 2100605.

40 Brenes, R., Guo, D., Osherov, A. et al. (2017). Metal halide perovskite polycrystalline films exhibiting properties of single crystals. *Joule* 1: 155–167.

41 Meggiolaro, D., Mosconi, E., and De Angelis, F. (2017). Mechanism of reversible trap passivation by molecular oxygen in lead-halide perovskites. *ACS Energy Letters* 2 (12): 2794–2798.

42 Aristidou, N., Eames, C., Sanchez-Molina, I. et al. (2017). Fast oxygen diffusion and iodide defects mediate oxygen-induced degradation of perovskite solar cells. *Nature Communications* 8 (1): 1.

43 Kondo, T., Azuma, T., Yuasa, T., and Ito, R. (1998). Biexciton lasing in the layered perovskite-type material $(C_6H_{13}NH_3)$ $2PbI_4$. *Solid State Communications* 105 (4): 253–255.

44 Xing, G., Mathews, N., Lim, S.S. et al. (2014). Low-temperature solution-processed wavelength-tunable perovskites for lasing. *Nature Materials* 13 (5): 476–480.

45 Deschler, F., Price, M., Pathak, S. et al. (2014). High photoluminescence efficiency and optically pumped lasing in solution-processed mixed halide perovskite semiconductors. *The Journal of Physical Chemistry Letters* 5 (8): 1421–1426.

46 Zhang, Q., Su, R., Liu, X. et al. (2016). High-quality whispering-gallery-mode lasing from cesium lead halide perovskite nanoplatelets. *Advanced Functional Materials* 26 (34): 6238–6245.

47 Jia, Y., Kerner, R.A., Grede, A.J. et al. (2020). Factors that limit continuous-wave lasing in hybrid perovskite semiconductors. *Advanced Optical Materials* 8 (2): 1901514.

48 Brenner, P., Bar-On, O., Jakoby, M. et al. (2019). Continuous wave amplified spontaneous emission in phase-stable lead halide perovskites. *Nature Communications* 10 (1): 1–7.

49 Jia, Y., Kerner, R.A., Grede, A.J. et al. (2017). Continuous-wave lasing in an organic–inorganic lead halide perovskite semiconductor. *Nature Photonics* 11 (12): 784–788.

50 Li, Z., Moon, J., Gharajeh, A. et al. (2018). Room-temperature continuous-wave operation of organometal halide perovskite lasers. *ACS Nano* 12 (11): 10968–10976.

51 Brenner, P., Paetzold, U.W., Turnbull, G.A. et al. (2019). Comment on "Room-temperature continuous-wave operation of organometal halide perovskite lasers". *ACS Nano* 13 (11): 12257–12258.

52 Samuel, I.D., Namdas, E.B., and Turnbull, G.A. (2009). How to recognize lasing. *Nature Photonics* 3 (10): 546–549.

53 Qin, C., Sandanayaka, A.S., Zhao, C. et al. (2020). Stable room-temperature continuous-wave lasing in quasi-2D perovskite films. *Nature* 585 (7823): 53–57.

54 Richter, J.M., Abdi-Jalebi, M., Sadhanala, A. et al. (2016). Enhancing photoluminescence yields in lead halide perovskites by photon recycling and light out-coupling. *Nature Communications* 7 (1): 1–8.

55 Pisoni, A., Jacimovic, J., Barisic, O.S. et al. (2014). Ultra-low thermal conductivity in organic–inorganic hybrid perovskite $CH_3NH_3PbI_3$. *The Journal of Physical Chemistry Letters* 5 (14): 2488–2492.

56 Allegro, I., Li, Y., Richards, B.S. et al. (2021). Bimolecular and auger recombination in phase-stable perovskite thin films from cryogenic to room temperature and their effect on the amplified spontaneous emission threshold. *The Journal of Physical Chemistry Letters* 12 (9): 2293–2298.

57 Haeger, T., Heiderhoff, R., and Riedl, T. (2020). Thermal properties of metal-halide perovskites. *Journal of Materials Chemistry C* 8 (41): 14289–14311.

13

Epitaxy and Strain Engineering of Halide Perovskites

Yang Hu, Jie Jiang, Lifu Zhang, Yunfeng Shi, and Jian Shi

Rensselaer Polytechnic Institute, Department of Materials Science and Engineering, 110 8th St, Troy, NY 12180, USA

13.1 Introduction

Epitaxy is one of the most important processes in fabricating modern microelectronic devices. Epitaxy allows the formation of single crystalline thin film and nano- or phase-matched structured semiconductor materials. Epitaxy could lead to coherent interfaces with high electronic and optoelectronic quality. Epitaxy is a common approach to fabricating semiconducting heterostructures and superlattices. The high performance of epitaxial materials and structures stems from the realization of demanded "ordering" (i.e. epitaxy) of the underlying electronic structures.

Like III–V materials, the material morphology in halide perovskite-based optoelectronic devices (photovoltaics, light-emitting diodes, photodetectors, and phototransistors) is often thin film like. A high-quality thin film should be featured by low dislocation densities, absence of grain boundaries, and the presence of coherent interfaces. Furthermore, in the development of emerging nanostructured halide perovskite devices where interfacial property becomes pronounced, it has been realized that a chemically and electronically periodic interface between the epilayer and the substrate is a prerequisite for receiving designed physical properties [1–4]. To pursue such epilayers and interfaces for halide perovskites, epitaxy strategy has been attempted and developed in past five years [1–16].

Based on the nature of the epilayer–substrate interactions, in general, there are three types of epitaxy methods: chemical epitaxy (Figure 13.1a) where film–substrate interaction is chemical bond-based (e.g. ionic or covalent epitaxy); van der Waals epitaxy/quasi van der Waals epitaxy (Figure 13.1b) where epilayer–substrate interaction is van der Waals bond-based or mixed between van der Waals bond and weak chemical bond; remote epitaxy [17] (Figure 13.1c) where the epilayer–substrate interaction is tuned by a buffer layer (often a two-dimensional [2D] material such as graphene or boron nitride).

In chemical epitaxy, a strong epilayer–substrate interaction could provide a large growth window for tuning the thickness and roughness of the epilayer. However,

Halide Perovskite Semiconductors: Structures, Characterization, Properties, and Phenomena, First Edition. Edited by Yuanyuan Zhou and Iván Mora-Seró.

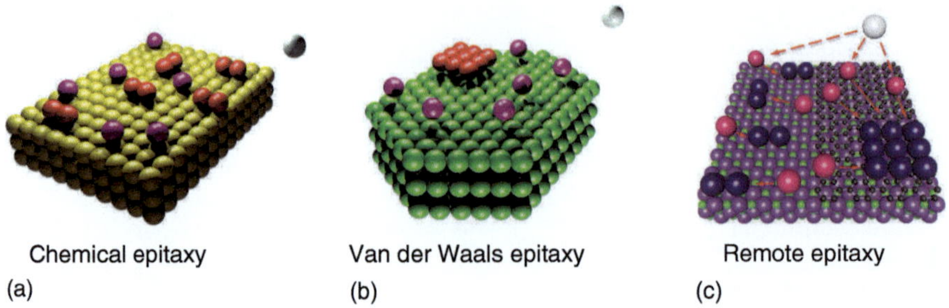

Figure 13.1 Three epitaxy methods have already been demonstrated in halide perovskite growth. (a) Chemical epitaxy; (b) van der Waals epitaxy. Source: (a) and (b) Wang et al. [5]/with permission of American Chemical Society. (c) Remote epitaxy. Source: (c) Jiang et al. [14]/Springer Nature/Licensed under CC BY 4.0.

in chemical epitaxy, the mismatch of lattice symmetry or lattice constant between epilayer and substrate could lead to severe issues in the crystal quality of the epilayer. For example, if a large lattice mismatch exists, one often observes high density misfit and threading dislocations, both of which are harmful for the electronic and optoelectronic properties of the epilayer [14, 18]. In van der Waals/quasi van der Waals epitaxy, and remote epitaxy, one can overcome or alleviate the issues due to lattice mismatch thus leading to better crystal quality [14, 18]. The weak epilayer–substrate interaction also provides opportunities in growing nanostructures such as nanowires or nanosheets [1, 5, 19] though growing ultrathin film becomes challenging.

In current III–V or Si/Ge electronic and optoelectronics or emerging perovskite oxide devices, the reduced dimension of these functional materials allows the realization of elastic strains above 2% [20], far above the elastic strain limit for their bulk counterparts (Figure 13.2). The largely increased elastic strain limit

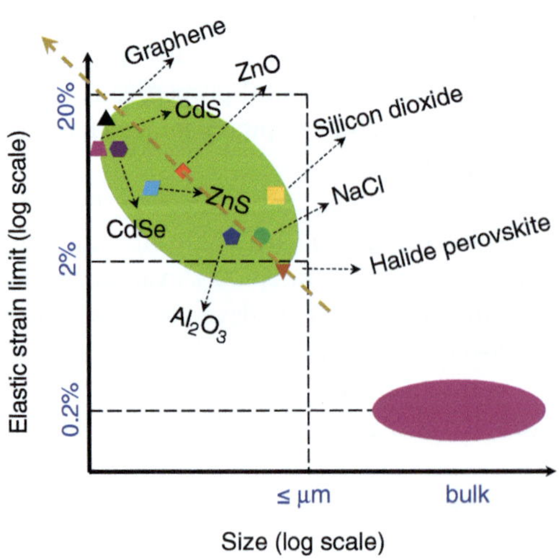

Figure 13.2 Size effect on the elastic strain limit of materials. Source: Data were collected from Refs. [20–22].

for thin film or nanostructure is believed to be a consequence of the switching of the dislocation formation mechanism from the growth-based (of low kinetic barrier) in bulk form to nucleation-based (of high energy barrier) in nanoscale structure. As elastic strains are able to tune and improve the optoelectronic and electronic properties, they have been widely used in many modern electronic and optoelectronic devices [20, 21]. The availability of single-crystal halide perovskites at reduced dimension (i.e. thin film and nanostructures) makes it possible to develop substantial strains possibly with similar values as those demonstrated in other semiconductors [22]. This would further provide opportunities to explore their strain-related phenomena and harness their strain-designed properties.

In conventional semiconductors, a common approach in inducing strains is through the interfacial interaction between thin film/nanostructure and the substrate. For example, in chemical epitaxy, a strong interfacial interaction could provide enough force to strain the epilayer when the epilayer is below a critical thickness. Halide perovskites are mechanically soft and thus a relatively weak interface (e.g. a van der Waals or quasi-van der Waals interface) may be able to stretch or compress them.

In this chapter, we will review the recent progress in the epitaxy and strain engineering of halide perovskites thin film and nanostructures. We will also present our perspectives on the opportunities and challenges in the relevant fields.

13.2 Epitaxy of Thin Film and Nanostructures

13.2.1 Epitaxial Substrates

The success of epitaxy largely relies on the availability of substrates that have close lattice constants and crystal symmetries to the epilayer. 3D non van der Waals halide perovskites have a pseudo-cubic symmetry with lattice constants between 5.5 and 6.5 Å. For the 2D van der Waals halide perovskites, the in-plane facet has similar structure and parameters as the 3D perovskite. A key task in the pursuit of epitaxy of halide perovskite is thus to find substrates with cubic or pseudo-cubic symmetry of close lattice constants.

Figure 13.3 shows a collection of substrates that have been experimentally demonstrated to host epitaxial halide perovskite thin films or nanostructures [1–16, 23–32]. Most of them have the lattice constants close to halide perovskites. Typical ones are NaCl, KCl, $SrTiO_3$, and halide perovskites. They also carry a cubic or pseudo-cubic symmetry. Some of them have non-cubic/non-pseudo-cubic symmetry or very different lattice constants such as muscovite mica, sapphire, and CdS. The epitaxy methods developed so far include chemical epitaxy, van der Waals/quasi-van der Waals epitaxy, and remote epitaxy. A general observation is that: under chemical and remote epitaxy (Figure 13.4), the lattice mismatch between epitaxial perovskites and the substrate has to be relatively small; for van der Waals/quasi-van der Waals epitaxy, the lattice mismatch could be very large and it is not necessary to have similar crystal symmetries between the epitaxial halide perovskites and the substrates.

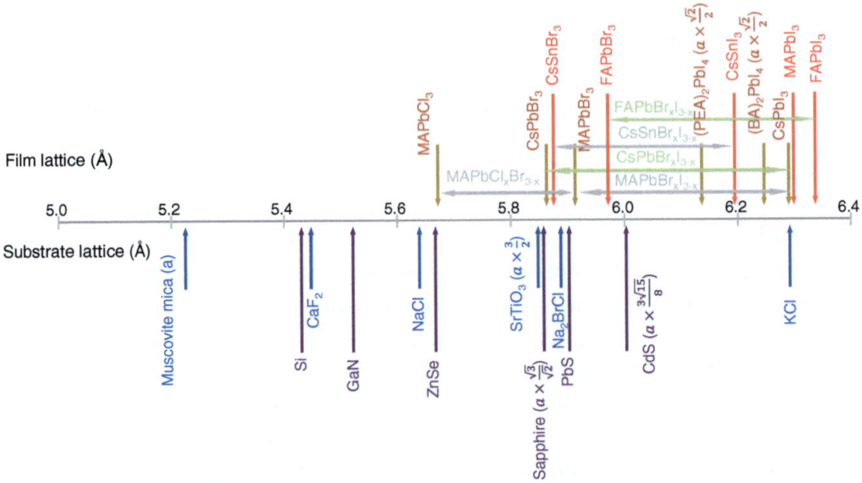

Figure 13.3 Lattice parameters of halide perovskites and experimentally demonstrated epitaxial substrates. Source: Data are collected from Refs. [1–16, 23–32].

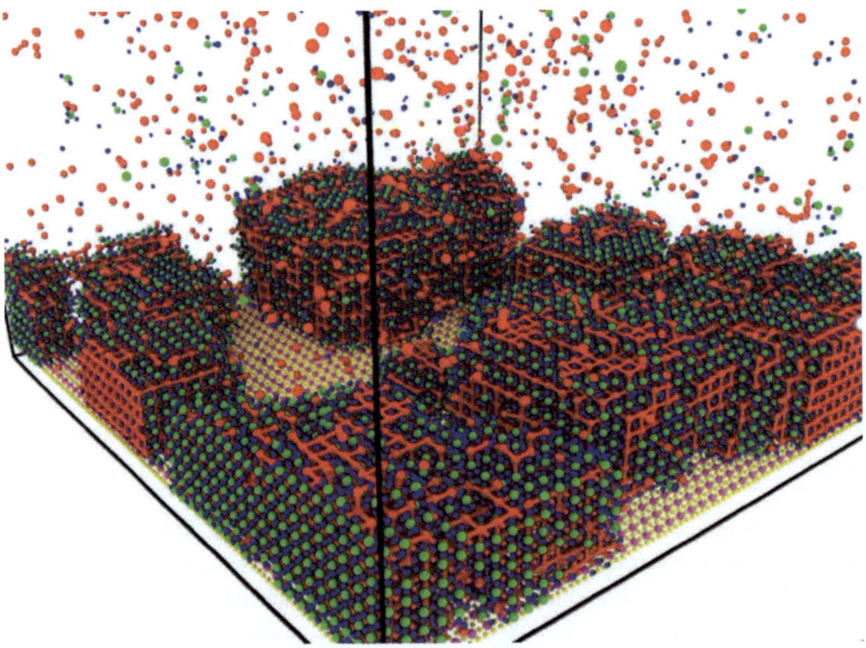

Figure 13.4 Computational simulation of the epitaxy of halide perovskite on a NaCl substrate.

13.2.2 Epitaxial Growth and Defects Formation Mechanisms

Understanding the growth mechanisms and defect formation mechanisms in the epitaxy of halide perovskites has been the central task for designing and controlling their optoelectronic properties. In chemical epitaxy, it has been found that when the lattice mismatch between epitaxial halide perovskite is large (e.g. >1%) and when the epilayer is thick, there exists a large population of threading dislocations, part of which are created to thread the massive misfit dislocations out from the interfaces [14]. This scenario is expected to be consistent with the conventional picture of dislocation formation in epitaxial films: when the film thickness is above the critical thickness, plastic deformation (e.g. misfit dislocation) is introduced to relax the elastic strain and thus strain energy (Figure 13.5a).

For epitaxy at weakly coupled interfaces (e.g. van der Waals/quasi-van der Waals epitaxy and remote epitaxy), it has been found that the density of threading dislocations could be a few orders of magnitude lower than chemical epitaxy (Figure 13.5b), despite the same values of lattice mismatch [14]. For example, the density of the threading dislocation in epitaxial $CsPbBr_3$ on NaCl (through ionic epitaxy) is orders of magnitude higher than on graphene/NaCl (through remote epitaxy), though the lattice mismatch in both cases is identical [14]. Such observations have also been found in conventional III–V materials [18].

What is the reason leading to such a discrepancy in dislocation density? There are at least two factors that may contribute to the low dislocation density in remote and van der Waals epitaxy [14]. First, when two nuclei grow larger and merge into each other, it would create some threading dislocations at the merging interface if a small crystallographic misalignment exists between these two nuclei (or grains). Such misalignment could be created either kinetically or thermodynamically. Kinetic misalignment could be induced due to a fast-growing process in which atoms do not have enough time to reach their equilibrium states/positions. This is a very common reason responsible for the presence of misalignment. Thermodynamically,

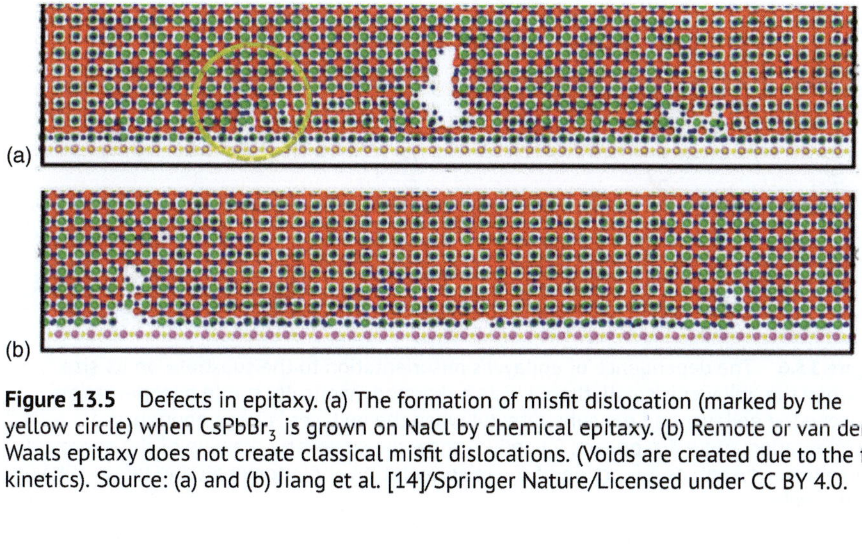

Figure 13.5 Defects in epitaxy. (a) The formation of misfit dislocation (marked by the yellow circle) when $CsPbBr_3$ is grown on NaCl by chemical epitaxy. (b) Remote or van der Waals epitaxy does not create classical misfit dislocations. (Voids are created due to the fast kinetics). Source: (a) and (b) Jiang et al. [14]/Springer Nature/Licensed under CC BY 4.0.

it is known that, when epilayer–substrate mismatch exists and the nuclei are small, the epilayer would have a misorientation (θ) with the substrate [33, 34]. Such misorientation is induced to maximize the percentage of the lowest energy epilayer–substrate bonds. In the presence of lattice mismatch, a larger size of the epilayer leads to a smaller misorientation angle (Figure 13.6a,b where the misorientation angle is exaggerated). This is reasonable since an infinitely large epilayer would support the formation of zero misorientation angle (non-zero angle would make the percentage of non-lowest energy bonds increase significantly). Figure 13.6c,d shows the schematic illustrations of the total energy of an epilayer structure as a function of the misorientation angles at two different sizes (Figure 13.6c: small size; Figure 13.6d: large size). The global minimum of the total energy does not show up at zero misorientation unless the top epilayer is large enough. Since there could be a few equivalent misorientations between epilayer and the substrate (Figure 13.6a,b) when these small nuclei meet each other, a misalignment and thus threading dislocations would be created. Due to the weak epilayer–substrate interaction, with the same growth conditions and lattice epilayer–substrate mismatch, in remote or van der Waals epitaxy, the nucleation rate is much smaller than chemical epitaxy, and thus the corresponding density of threading dislocations created under this mechanism would be smaller.

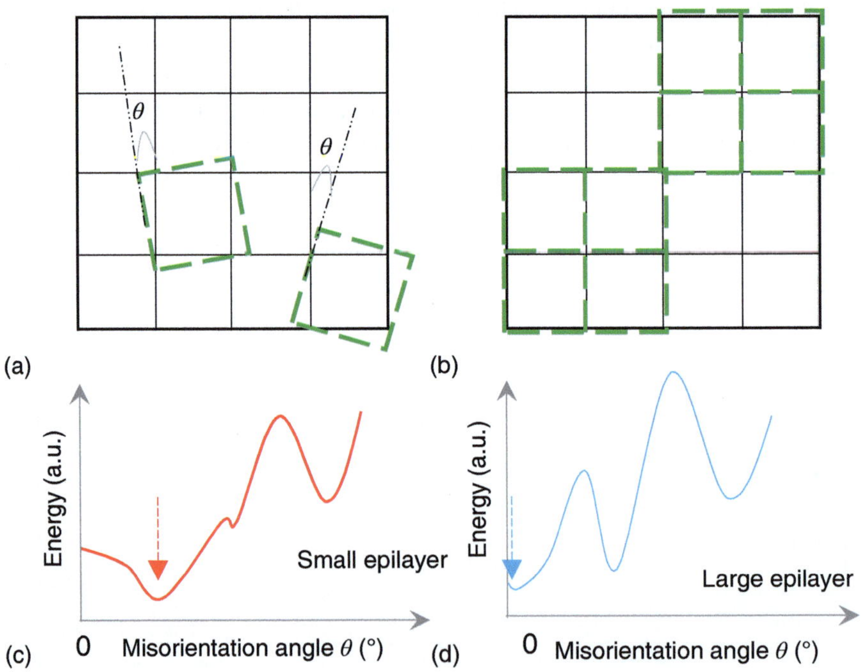

Figure 13.6 The dependence of epilayer's misorientation to the substrate on its size. (a) When the epilayer is small, there exists a thermodynamically stable misorientation between the epilayer and the substrate; (b) when the epilayer is large enough, the misorientation angle disappears; (c) and (d) show the schematic drawing of the energy of the epitaxial system as a function of the misorientation angle for small and large epilayers, respectively.

Second, in the case where there is no such misalignment between two merging grains (assuming they are large enough), in chemical epitaxy, misfit dislocations and thus threading dislocations could still form due to the presence of a lattice mismatch and the need to release the large strain energy when the epilayer is above the critical thickness. However, such issue does not exist in remote or van der Waals epitaxy. Experimental observation shows that the epilayer–substrate relation in remote or van der Waals epitaxy could be incommensurate and misfit dislocation-free [35, 36]. Recent theoretical progress [14] on the epitaxy of halide perovskites shows that one can achieve such scenario by the following mechanistic process: Different from chemical epitaxy, in remote or van der Waals epitaxy, the weak-film interaction could allow the occurrence of random walks for some small grains which receive larger energy gain when merging into the neighboring large grains than when locking them to the local substrate lattice. Figure 13.7a–d shows the gliding (random walk) and rotation ("random rotation") of the small grains during the merging process. By the random walk and "random rotation," remote and van der Waals epitaxy can create an incommensurate and misfit dislocation-free interface. It is noted that when the grains are too large, they stop the random walk and "random rotation" processes, and one still expects to see the formation of dislocations but with lower density.

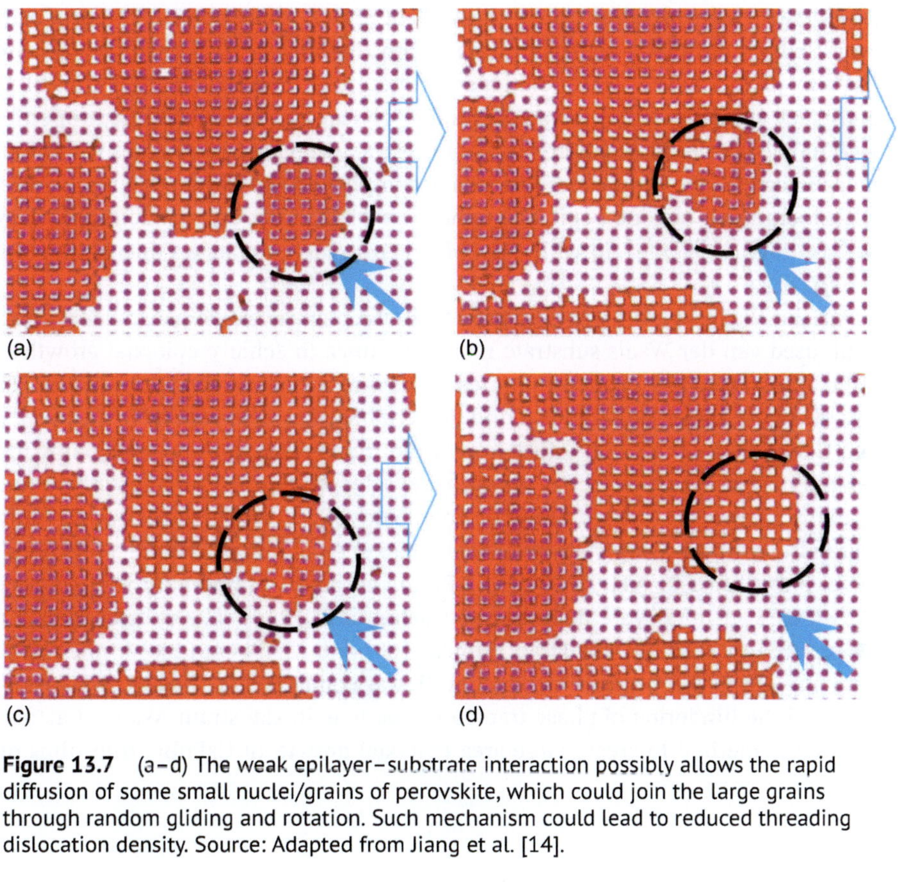

Figure 13.7 (a–d) The weak epilayer–substrate interaction possibly allows the rapid diffusion of some small nuclei/grains of perovskite, which could join the large grains through random gliding and rotation. Such mechanism could lead to reduced threading dislocation density. Source: Adapted from Jiang et al. [14].

13.2.3 Experimental Progresses

In the past several years, there have been many reports [37–77] on the epitaxy of halide perovskites. In this section, some representative experimental progresses in the epitaxy of halide perovskites are introduced. These advancements can be categorized as ionic epitaxy by vapor phase deposition, van der Waals/quasi-van der Waals epitaxy by vapor phase deposition, remote epitaxy by vapor phase deposition, and ionic epitaxy by solution approach.

Ionic epitaxy of halide perovskites has been amply demonstrated with different morphologies on a variety of substrates. By chemical vapor deposition (CVD), Wang et al. successfully realized ionic epitaxial growth of $CsSnBr_3$ and $CsPbBr_3$ thin films (cm by cm) on metal alkali halide substrates NaCl [8], with an epitaxial relation of perovskite [100] parallel to NaCl [100] or perovskite [110] parallel to NaCl [100], which was revealed by pole figure X-ray diffraction (XRD) analysis. As shown in Figure 13.8a, 3D halide perovskites have a lattice constant close to that of NaCl and KCl. With such epitaxial films of high quality, they were able to reveal the hot photoluminescence (PL) and photo-Dember effect in halide perovskite. Ionic epitaxy of halide perovskites has also been achieved on oxide substrates. As shown in Figure 13.8b, Chen et al. realized epitaxial growth of $CsPbBr_3$ on $SrTiO_3$ by CVD [6]. On $SrTiO_3$ (100) substrates, epitaxial relation was revealed to be $CsPbBr_3$ [010] || $SrTiO_3$ [010], where three-unit cells of $SrTiO_3$ matched with two-unit cells of $CsPbBr_3$. On $SrTiO_3$ (110) substrates, both $CsPbBr_3$ [001] || $SrTiO_3$ [100] and $CsPbBr_3$ [010] || $SrTiO_3$ [110] relations were received. Epitaxial $CsPbBr_3$ thin films on $SrTiO_3$ were proved to have comparable optical and optoelectronic properties with halide perovskite single crystals. Ionic epitaxy of halide perovskites with other morphologies, like nanowires, was also accomplished. As shown in Figure 13.8c, Oksenberg et al. obtained epitaxial nanowires of $CsPbBr_3$ on *c*-plane sapphire by surface-guided CVD [2], in which $CsPbBr_3$ $(1\bar{1}0)$ spacing matches that of sapphire $(\bar{1}100)$.

Van der Waals epitaxy was also used to grow epitaxial halide perovskites. Wang et al. used van der Waals substrate muscovite mica to achieve epitaxial growth of $CsPbX_3$ (X = Cl, Br, I, or their mixtures) nanowires (as shown in Figure 13.8d) [1]. By pole figure XRD studies, they found that $CsPbX_3$ (011) was parallel to mica (001), and CsPbX3 (100) had a 30° offset away from mica (020). In those epitaxial wires, unique photon transport behaviors were revealed by temperature-dependent and spatially resolved PL studies. Other morphologies, like epitaxial flakes or thin films, were also realized using van der Waals epitaxy. As shown in Figure 13.8e,f, epitaxial flakes of $(C_4H_9NH_3)_2PbI_4$ ($(BA)_2PbI_4$) and methylammonium lead chloride ($MAPbCl_3$) on mica were demonstrated using CVD [5, 9]. The epitaxial relation was proved to be $(BA)_2PbI_4$ [100] || mica [100] and $MAPbCl_3$ (200) || mica (200) with a 5° offset. Temperature-dependent PL was conducted on the epitaxial system to reveal the hindering of phase transition due to epitaxial strain. Wang et al. also used CVD method to create large-area epitaxial growth of $CsPbBr_3$ thin films on

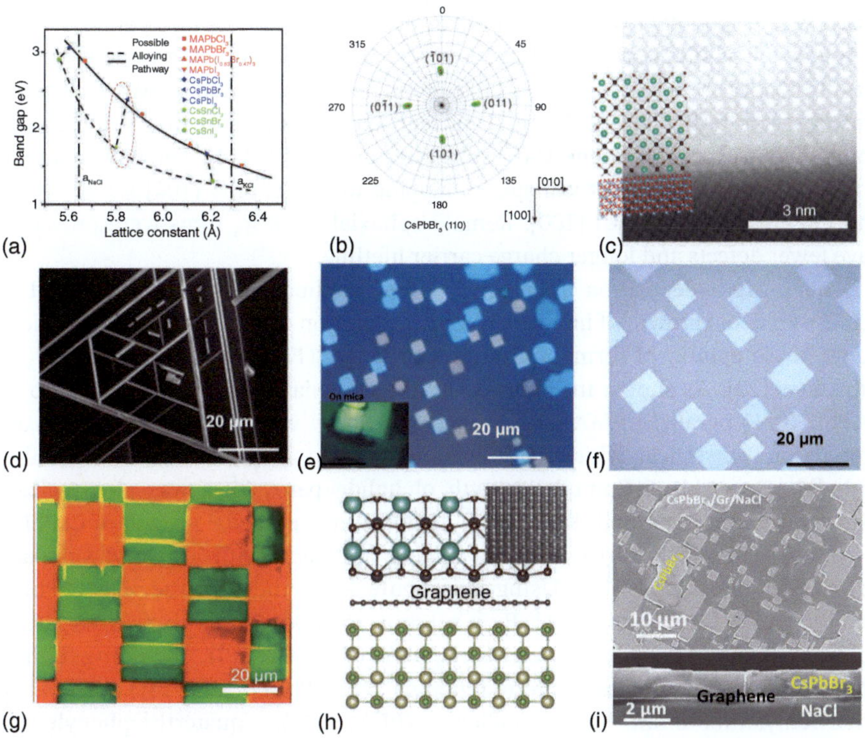

Figure 13.8 Experimentally realized epitaxial growth of halide perovskites. (a) Lattice constants of 3D halide perovskites are similar to those of NaCl, which enables ionic epitaxial growth. Source: Wang et al. [8]/with permission of John Wiley & Sons; (b) ionic epitaxial growth of $CsPbBr_3$ on $SrTiO_3$ was achieved by CVD. Source: Adapted from Chen et al. [6]. Epitaxial relations were revealed using pole figure XRD; (c) ionic epitaxial growth of $CsPbBr_3$ nanowires was realized using c-plane sapphire as substrate. Source: Oksenberg et al. [2]/Springer Nature/CC BY 4.0; (d) By using muscovite mica as substrate, van der Waals epitaxy of $CsPbX_3$ (X = Cl, Br, I, or their mixtures) nanowires was demonstrated. Source: Wang et al. [1], American Chemical Society; (e) Van der Waals epitaxial growth of $(BA)_2PbI_4$ flakes. Source: Wang et al. [5], American Chemical Society; (f) $MAPbCl_3$ nanosheets were epitaxially grown on van der Waals substrate mica. Source: Chen et al. [9], American Chemical Society; (g) Large-area epitaxial thin films of $CsPbBr_3$ could be synthesized on mica by CVD. Source: Wang et al. [16]/with permission of American Chemical Society. The as-grown thin film could then be patterned to form arrays showing spatially resolved PL emission; (h, i) By remote epitaxy, $CsPbBr_3$ was successfully grown on graphene/NaCl and graphene/CaF_2 substrates. Source: Reproduced from Jiang et al. [14]/Springer Nature/CC BY 4.0. Remote epitaxial $CsPbBr_3$ films were proved to have longer charge carrier lifetime.

mica [16]. By transmission electron microscopy (TEM), they showed that in the epitaxial thin film, $CsPbBr_3$ [010] was parallel to mica [010]. As shown in Figure 13.8g, the large-area epitaxial thin film was then patterned and transferred to form device arrays where spatially resolved PL could be observed.

Epitaxy of halide perovskite thin films was also achieved using remote epitaxy (Figure 13.8h,i), where 2D materials like graphene were inserted between substrate and flakes as buffer layers. Jiang et al. realized epitaxial growth of $CsPbBr_3$ on graphene/NaCl and graphene/CaF_2 substrates using CVD [14]. The epitaxial relation was revealed by reciprocal space mapping to be $CsPbBr_3$ [100] || CaF_2 [010] and $CsPbBr_3$ [100] || NaCl [100]. Remote epitaxial $CsPbBr_3$ films were proved to have fewer defects and longer charge carrier lifetime.

Besides vapor deposition approaches, other methods have also been used to achieve epitaxial growth of halide perovskites. Solution method was used to achieve liquid-phase epitaxy of formamidinium lead iodide ($FAPbI_3$) on ($MAPbCl_xBr_{3-x}$) substrates [22]. As shown in Figure 13.8j, the epitaxial relation was proven to be $FAPbI_3$ (001) parallel to $MAPbCl_xBr_{3-x}$ (001). Strain was introduced by epitaxial growth on substrates, which was demonstrated to help increase the stability of halide perovskites. Epitaxial growth of halide perovskites was also realized using spin coating method. Kelso et al. demonstrated epitaxial growth of $CsPbBr_3$ on $SrTiO_3$ (Figure 13.8k) by a spin coating approach with dimethyl sulfoxide (DMSO) as the solvent [78]. Using pole figure XRD, they showed that $CsPbBr_3$ [011] is parallel to $SrTiO_3$ [001]. By a solution method, Shi et al. realized lateral epitaxial halide perovskites heterostructures (Figure 13.8l), including bithiophenylethylammonium lead bromide/iodide ($(2T)_2PbBr_4/(2T)_2PbI_4$), $(BA)_2PbI_4$, phenylethylammonium lead iodide ($(PEA)_2PbI_4$), quaterthiophenylethylammonium tin iodide ($(4Tm)_2SnI_4$), and 7-(thiophen-2-yl)benzothiadiazol-4-yl)-[2,2′-bithiophen]-5-yl)ethylammonium lead iodide ($(BTm)_2PbI_4$) [15]. Sharp interfaces and epitaxial relations were revealed.

13.3 Strain Engineering

Strain engineering has been widely applied in commercially available optoelectronic and electronic devices [79]. The most common approaches in these technologically practical devices are focused on epitaxial and thermal mismatch-induced straining. One can design the strain states through lattice misfit and mismatch of thermal expansion coefficients. In halide perovskites, similar approaches have been adopted [13, 22, 80, 81]. In fact, the shear moduli of halide perovskites (both inorganic and hybrid) are in the range of several GPa (Figure 13.9), which is orders of magnitude lower than many conventional semiconductors [82–88]. This suggests that the energy penalty due to elastic deformation is small, making halide perovskites easily deformable under external stimuli. For example, an interface that has moderate strength could effectively compress or stretch the perovskite epilayer. This chapter focuses on introducing the progress of strain engineering in halide perovskites.

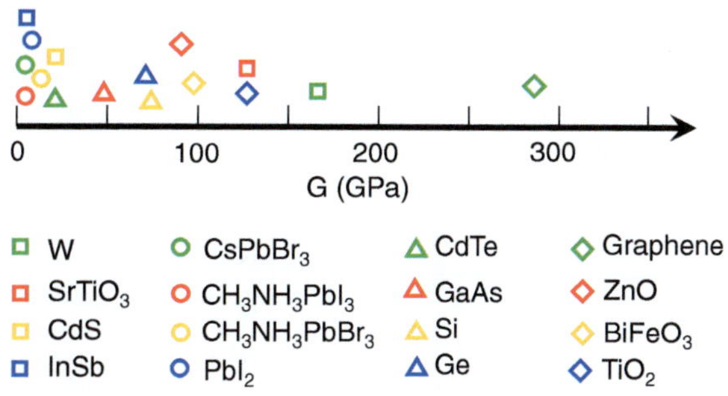

Figure 13.9 Shear moduli of some semiconductors and metals. Source: Data were collected from Refs. [82–88].

13.3.1 Theoretical Progresses

Theoretical and computational efforts have been applied to understand the roles of strain on their physical properties [89–108]. Strain-induced phenomena include band gap modification, effective mass changes, the presence of topological and Rashba phases, the tuning of thermoelectric Seebeck coefficient, and the occurrence of metal-insulator transitions.

As shown in Figure 13.10a, Liu et al. conducted density functional theory (DFT) calculation on halide perovskite $CsGeI_3$ to reveal the effect of strain on its band structure [109]. They showed that, with external strain, bandgap of $CsGeI_3$ could be modified from 0.73 to 2.30 eV. With moderate strain (−1%), $CsGeI_3$ is expected to have a bandgap of around 1.36 eV with enhanced absorption. As shown in Figure 13.10b, DFT calculations were also used to study the strain effect in Cs_2AgInX_6 (X = Cl, Br) [110]. It was revealed that with a compressive strain, $Cs_2AgInBr_6$ would be tuned from indirect to direct bandgap. Also, $Cs_2AgInBr_6$ was predicted to have a higher charge carrier mobility under compressive strain, which makes it a potential candidate for photovoltaic device. Lin et al. studied the effect of biaxial strain on thermal and thermoelectric properties of $CsPbI_{1.5}Br_{1.5}$ [111]. As shown in Figure 13.10c, Seebeck coefficient of $CsPbI_{1.5}Br_{1.5}$ could be increased with external biaxial compressive/tensile strain. Also, thermal conductivity of $CsPbI_{1.5}Br_{1.5}$ would be dramatically reduced with large strain (>6%).

Phase transitions and novel phases were also predicted to exist in strained halide perovskites. Zhang et al. conducted DFT calculations to predict the strain-stabilized phases in $MAPbI_3$ and revealed that new phases would originate from the deformation of inorganic frameworks and rotation of organic molecule [115]. Lee et al. predicted that with hydrostatic pressure, $MAPbI_3$ would go through a metallization process [112]. As shown in Figure 13.10d, with 0.23 GPa pressure, space group of $MAPbI_3$ would change from *Fmmm* to $Im\bar{3}$. With around 100 GPa of pressure, $MAPbI_3$ would undergo a semiconductor-metal phase transition and become a metal. Jin et al. conducted tight binding and first-principle calculations on halide perovskites $CsPbI_3$ and $CsSnI_3$ and predicted a topological insulator (TI) phase enabled by external strain [113]. Based on their results, $CsPbI_3$ and $CsSnI_3$ would become TI with −4.04% and −1.36% strain, respectively (Figure 13.10e). Leppert et al. studied-strain induced Rashba effect in $APbX_3$ (A = MA, $CH(NH_2)_2$, Cs,

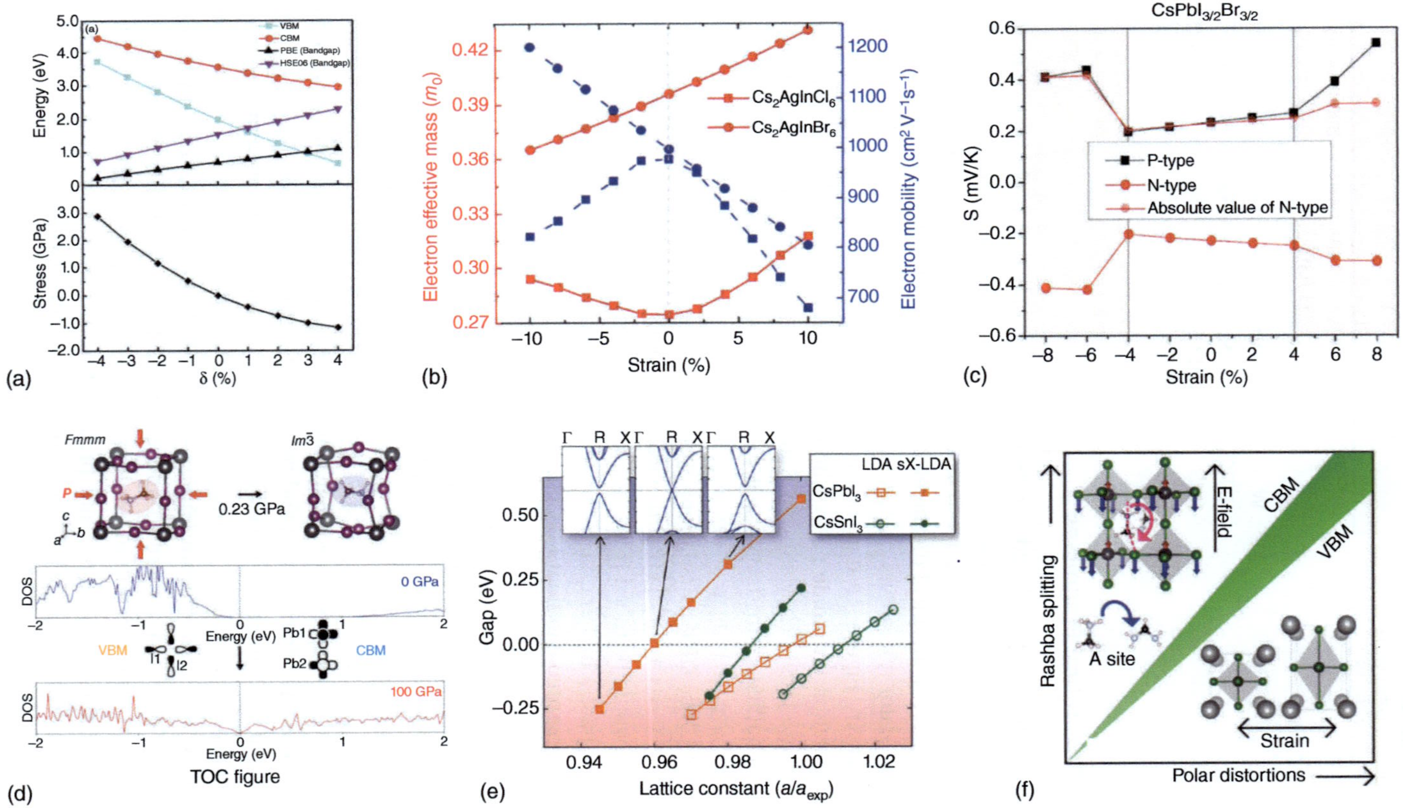

(a)
Energy (eV)
VBM
CBM
PBE (Bandgap)
HSE06 (Bandgap)
Stress (GPa)
δ (%)
(b)
Electron effective mass (m_0)
Electron mobility ($cm^2 V^{-1}s^{-1}$)
$Cs_2AgInCl_6$
$Cs_2AgInBr_6$
Strain (%)
(c)
$CsPbI_{3/2}Br_{3/2}$
S (mV/K)
P-type
N-type
Absolute value of N-type
Strain (%)
(d)
Fmmm
$Im\bar{3}$
P
0.23 GPa
DOS
0 GPa
Energy (eV)
VBM
CBM
Pb1
Pb2
I1
I2
100 GPa
TOC figure
(e)
Γ R X Γ R X Γ R X
LDA sX-LDA
$CsPbI_3$
$CsSnI_3$
Gap (eV)
Lattice constant (a/a_{exp})
(f)
Rashba splitting
E-field
CBM
VBM
A site
Strain
Polar distortions

and X = I, Br) by DFT calculation [114]. They showed that polarization and Rashba effect magnitude of $MAPbI_3$ could be tuned with external strain and predicted a Rashba-active polar phase (with a *R3c* space group) of $MAPbI_3$ and $CsPbI_3$ with ~1% strain (Figure 13.10f).

13.3.2 Experimental Progresses

A plenty of methods have been demonstrated to create strains in halide perovskites [116–125]. These methods include epitaxy, thermal stress, using phase transition substrates, electric field, optical doping, and chemical strains. Epitaxy and thermal stress approaches have been demonstrated to have the potential for wafer-scale strain engineering in halide perovskites.

Finkenauer et al. used mechanical bending to introduce strain on $MAPbI_3$[126] (Figure 13.11a). Lee et al. used an atomic force microscope (AFM) tip to apply local strain on $CsPbBr_xI_{3-x}$ quantum dots [127] (Figure 13.11b). These approaches may be suitable for quick scientific demonstration in the laboratory but cannot be scaled up.

Strain engineering has been demonstrated via epitaxial interface. As shown in Figure 13.11c, Chen et al. used ionic epitaxy to generate strains in halide perovskite films [22], where strain was introduced due to the lattice mismatch of $FAPbI_3$ and $MAPbCl_xBr_{3-x}$. By adjusting x value, they were able to tune the strain value from 0% to 2.4%. The epitaxial strain was found to tune the bandgap and increase hole mobility. By tuning the strain states, they were able to stabilize the preferred halide perovskite phase, which becomes unstable in the absence of the epitaxial interface. This work also demonstrates that such epitaxial approach can be scaled up. Van der Waals epitaxy can also be used to generate strain. Wang et al. showed that the epitaxial interface between $MAPbBr_3$ and van der Waals substrate mica could introduce strains in the perovskite epilayer [12]. By PL study, they revealed that bandgap of $MAPbBr_3$ is modulated by van der Waals epitaxial strain, as shown in Figure 13.11d. It is expected that the van der Waals epitaxial strain is due to the mechanical softness of halide perovskites.

Besides external mechanical force and epitaxy, Zhu et al. were able to take advantage of composition gradients to generate chemical strain in halide perovskite [129].

Figure 13.10 Theoretically predicated physical properties in strained halide perovskites. (a) By DFT, halide perovskite $CsGeI_3$ was predicted to have a reduced bandgap with compressive strain. Source: Liu et al. [109]/Royal Society of Chemistry; (b) Cs_2AgInX_6 (X = Cl, Br) was predicted to have higher charge carrier mobility with compressive strain. Source: Zhang et al. [110]/with permission of American Chemical Society; (c) $CsPbI_{1.5}Br_{1.5}$ was expected to have larger Seebeck coefficient with biaxial compressive/tensile strain. Source: Lin et al. [111]/with permission of American Chemical Society; (d) With ~100 GPa hydrostatic pressure, $MAPbI_3$ was predicted to have a semiconductor-metal phase transition. Source: Lee et al. [112]/with permission of American Chemical Society; (e) Based on tight binding method calculation, $CsPbI_3$ and $CsSnI_3$ were expected to have a TI phase stabilized by external strain. Source: Jin et al. [113]/with permission of American Physical Society; (f) strain-induced Rashba effect and polarization were anticipated in $APbX_3$ (A = MA, $CH(NH_2)_2$, Cs, and X = I, Br) based on DFT calculation. Source: Leppert et al. [114]/with permission of American Chemical Society.

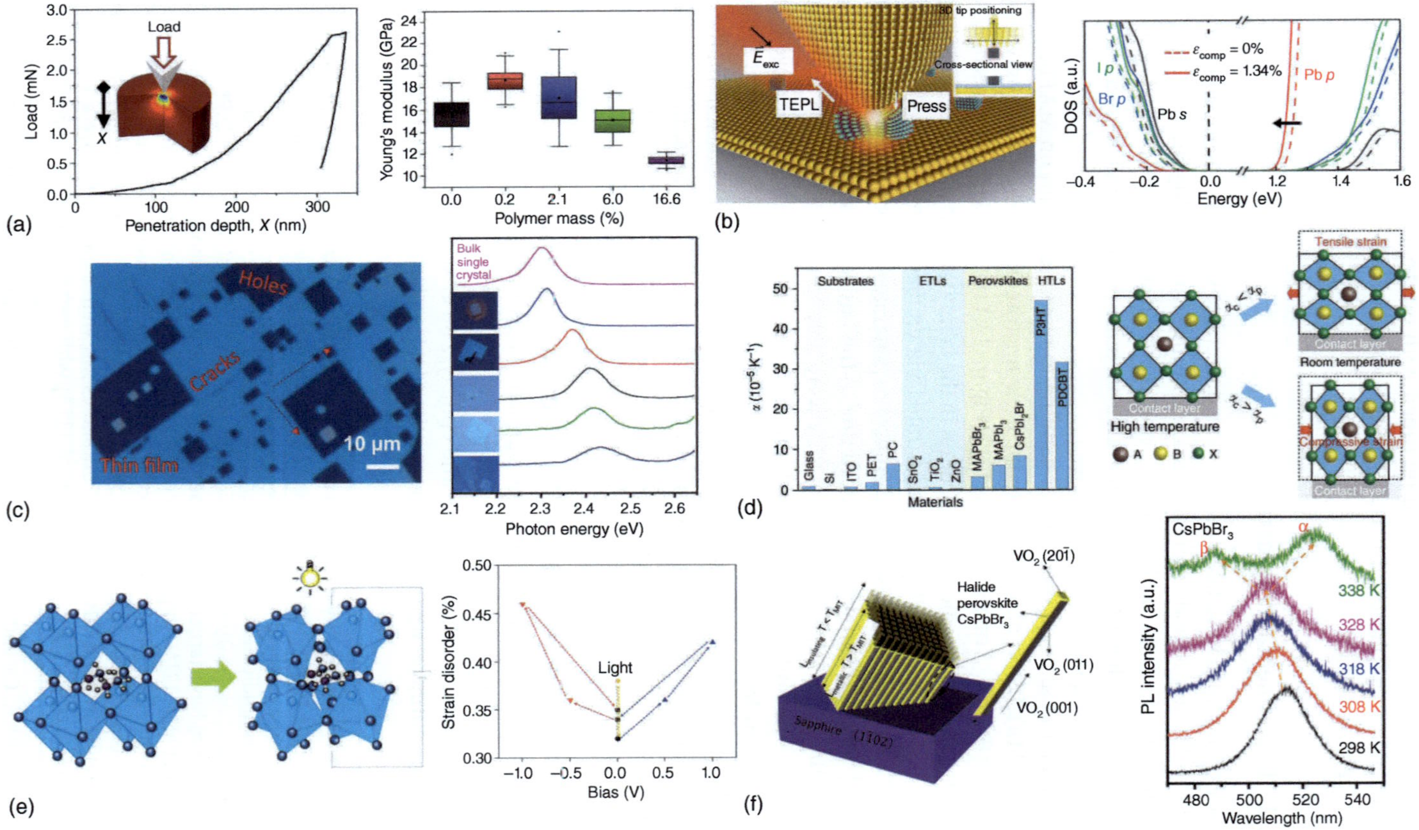
(a)
Load
X
Load (mN)
Penetration depth, X (nm)
Young's modulus (GPa)
Polymer mass (%)
(b)
E_{exc}
TEPL
Press
3D tip positioning
Cross-sectional view
DOS (a.u.)
ε_{comp} = 0%
ε_{comp} = 1.34%
I p
Br p
Pb s
Pb p
Energy (eV)
(c)
Holes
Cracks
Thin film
10 μm
Bulk
single
crystal
Photon energy (eV)
(d)
Substrates
ETLs
Perovskites
HTLs
α (10^{-5} K^{-1})
Glass
Si
ITO
PET
PC
SnO_2
TiO_2
ZnO
$MAPbBr_3$
$MAPbI_3$
$CsPbI_2Br$
P3HT
PDCBT
Materials
Tensile strain
Contact layer
Room temperature
High temperature
Compressive strain
A
B
X
(e)
Strain disorder (%)
Light
Bias (V)
(f)
VO_2 (20$\bar{1}$)
Halide
perovskite
$CsPbBr_3$
VO_2 (011)
VO_2 (001)
Sapphire (1$\bar{1}$02)
$CsPbBr_3$
α
β
338 K
328 K
318 K
308 K
298 K
PL intensity (a.u.)
Wavelength (nm)

Figure 13.11 Experimentally demonstrated strain effects in halide perovskites. (a) Strain was introduced to $MAPbI_3$ by mechanical bending. Source: Finkenauer et al. [126]/with permission of Elsevier. (b) Local strain was achieved on $CsPbBr_xI_{3-x}$ quantum dots using an AFM tip. Source: Lee et al. [127]/with permission of American Chemical Society. Band structure of quantum dots was modified by strain; (c) Van der Waals epitaxy between $MAPbBr_3$ and mica was utilized to introduce strain. Source: Adapted from Wang et al. [12]. Bandgap (PL peak position) of $MAPbBr_3$ was modulated by strain; (d) Strain could be introduced to perovskite films owing to the difference between its thermal expansion coefficient and that of the substrate. Source: Xue et al. [128]/Springer Nature/CC BY 4.0; (e) Light illumination and external bias were shown to generate strain in $(FAPbI_3)_{85}(MAPbBr_3)_{15}$ based solar cell. Source: Zhu et al. [129]/Springer Nature/Licensed under CC BY 4.0; (f) It was revealed that strain could also be introduced by a phase transition material. The metal-insulator phase transition of $VO_{2\pm\delta}$ was used to create ~1% strain in $CsPbBr_3$ flakes. Source: Wang et al. [13]/American Association for the Advancement of Science. This strain initialized another phase transition in $CsPbBr_3$.

In their work, strain gradient was introduced by the phase separation and consequential composition gradient of halide perovskite $(FAPbI_3)_{0.85}(MAPbBr_3)_{0.15}$. The strain was measured using grazing incident X-ray diffraction (GIXRD) and TEM. As shown in Figure 13.11e, they proved that, with external strain, charge carrier mobility of the material was reduced. Further, Zhao et al. showed that strain may exist in perovskite films owing to the difference in their thermal expansion coefficient with the substrate [81]. This strain was proved to decrease the ion migration activation energy (Figure 13.11f) and hence reduce the stability of halide perovskite films.

Another mechanism that can introduce strain in halide perovskite is through light illumination and external electric field (Figure 13.11g). Kim et al. showed that with light or electric field, strain would be created in halide perovskites $(FAPbI_3)_{85}(MAPbBr_3)_{15}$ [130]. They declared that this strain was related to the domain formation of halide perovskites under illumination and bias. Finally, Wang et al. exploited the metal-insulator phase transition of $VO_{2\pm\delta}$ to create ~1% strain in $CsPbBr_3$ flakes [13]. The strain was introduced to the $CsPbBr_3$ flakes when it was grown on $VO_{2\pm\delta}$ which changes from $P2_1/c$ to $P4_2/mnm$ during thermally induced phase transition. As shown in Figure 13.11h, this strain initialized another phase transition in $CsPbBr_3$, which was tracked by temperature-dependent PL study.

13.4 Opportunities and Challenges

The tremendous success of III–V, Si/Ge and perovskite oxide heterostructures and superlattices in the relevant community has inspired us to pursue a similar enterprise in halide perovskite for applications such as lasing, tunneling diodes, photodetectors, spintronic logic transistors, and light-emitting diodes. For example, if one could grow halide perovskite superlattice with precise control on the electronic levels of individual layers, one may have rich opportunities in engineering the optoelectronic performance of halide perovskites-based devices by mimicking the design principles developed in III–V quantum well and superlattice systems. However, a couple of challenges prevent the rapid development of such devices. First, so far most epitaxial substrates are electronically insulating; for example, NaCl or KCl are insulators; halide perovskite substrates themselves are mostly nonconductive due to the lack of effective dopants. Without electronically conductive substrates or epitaxial buffer layers, the development of devices with all interfaces being coherent like the ones in III–V would be challenging. Second, the high volatility of halide perovskites makes it hard to control the growth rate in vapor phase epitaxy, while vapor phase epitaxy is the mostly effective approach in developing quantum well or superlattice structures in semiconductor industry at wafer scale. Resolving these issues would need us to better understand the atomistic scale mechanism of epitaxy in halide perovskite and develop more sophisticated techniques, possibly with revolutionary concepts. To understand the epitaxial mechanism, in-situ characterization of the epitaxial growth process would provide unmatched information on the nucleation and kinetic transformation. New techniques involving the use of low deposition temperatures and low supersaturation growth may be worth exploring.

Epitaxy can be used to scale up the strain engineering strategy. Through the selection of right substrates, one can design the strain state of the halide perovskite films. The strain can be uniaxial or biaxial depending on the demanded magnitudes. In the development of strain engineering focused epitaxial approaches, one must find approximate substrates that often should be electrically conductive or matched to the epilayers. Current pool of epitaxial substrates for halide perovskites is unfortunately very limited. A significant effort in future could be to search for better substrates, expanding the substrate selection space. In fact, the rapid development of strain engineering in oxide perovskites has been largely driven by the progress in developing the relevant substrates. Furthermore, for device design or discovering new materials physics or phenomena under strain engineering, further theoretical/computational prediction and experimental implementation may need to consider not only the strain states but also various interfaces. Such interfaces have been shown to influence the performance of the quantum well devices of III–V substantially through various mechanisms. Halide perovskites carry strong spin-orbit coupling, so any heterogeneous interface would bring large modifications to their band structures, which would add extra complexity to the strain-tuned electronic structure.

Addressing these challenges would unleash unprecedented opportunities in the design and realization of practical and emerging perovskite devices.

Acknowledgments

This work is supported by the US Air Force Office of Scientific Research under grant number FA9550-18-1-0116, the NYSTAR Focus Center at Rensselaer Polytechnic Institute with contract number C C180117, and National Science Foundation under awards of 2024972, 2031692, and 1916652.

References

1 Wang, Y., Sun, X., Shivanna, R. et al. (2016). Photon transport in one-dimensional incommensurately epitaxial $CsPbX_3$ arrays. *Nano Letters* 16 (12): 7974–7981.

2 Oksenberg, E., Merdasa, A., Houben, L. et al. (2020). Large lattice distortions and size-dependent bandgap modulation in epitaxial halide perovskite nanowires. *Nature Communications* 11 (1): 489.

3 Gu, Z., Zhou, Z., Huang, Z. et al. (2020). Controllable growth of high-quality inorganic perovskite microplate arrays for functional optoelectronics. *Advanced Materials* 32 (17): 1908006.

4 Yang, T., Wang, X., Zheng, B. et al. (2019). Ultrahigh-performance optoelectronics demonstrated in ultrathin perovskite-based vertical semiconductor heterostructures. *ACS Nano* 13 (7): 7996–8003.

5 Wang, Y., Shi, Y., Xin, G. et al. (2015). Two-dimensional van der Waals epitaxy kinetics in a three-dimensional perovskite halide. *Crystal Growth & Design* 15 (10): 4741–4749.

6 Chen, J., Morrow, D.J., Fu, Y. et al. (2017). Single-crystal thin films of cesium lead bromide perovskite epitaxially grown on metal oxide perovskite ($SrTiO_3$). *Journal of the American Chemical Society* 139 (38): 13525–13532.

7 Wang, Y., Guan, X., Li, D. et al. (2017). Chemical vapor deposition growth of single-crystalline cesium lead halide microplatelets and heterostructures for optoelectronic applications. *Nano Research* 10 (4): 1223–1233.

8 Wang, Y., Sun, X., Chen, Z. et al. (2017). High-temperature ionic epitaxy of halide perovskite thin film and the hidden carrier dynamics. *Advanced Materials* 29 (35): 1702643.

9 Chen, Z., Wang, Y., Sun, X. et al. (2018). Remote phononic effects in epitaxial Ruddlesden–Popper halide perovskites. *The Journal of Physical Chemistry Letters* 9 (23): 6676–6682.

10 Fan, C., Xu, X., Yang, K. et al. (2018). Controllable epitaxial growth of core–shell PbSe@$CsPbBr_3$ wire heterostructures. *Advanced Materials* 30 (45): 1804707.

11 Ji, L., Hsu, H.-Y., Lee, J.C. et al. (2018). High-performance photodetectors based on solution-processed epitaxial grown hybrid halide perovskites. *Nano Letters* 18 (2): 994–1000.

12 Wang, Y., Gao, L., Yang, Y. et al. (2018). Nontrivial strength of van der Waals epitaxial interaction in soft perovskites. *Physical Review Materials* 2 (7): 076002.

13 Wang, Y., Sun, X., Chen, Z. et al. (2018). Defect-engineered epitaxial $VO_{2\pm\delta}$ in strain engineering of heterogeneous soft crystals. *Science Advances* 4 (5): eaar3679.

14 Jiang, J., Sun, X., Chen, X. et al. (2019). Carrier lifetime enhancement in halide perovskite via remote epitaxy. *Nature Communications* 10 (1): 4145.

15 Shi, E., Yuan, B., Shiring, S.B. et al. (2020). Two-dimensional halide perovskite lateral epitaxial heterostructures. *Nature* 580 (7805): 614–620.

16 Wang, Y., Jia, C., Fan, Z. et al. (2021). Large-area synthesis and patterning of all-inorganic lead halide perovskite thin films and heterostructures. *Nano Letters* 21 (3): 1454–1460.

17 Kim, Y., Cruz, S.S., Lee, K. et al. (2017). Remote epitaxy through graphene enables two-dimensional material-based layer transfer. *Nature* 544 (7650): 340–343.

18 Bae, S.-H., Lu, K., Han, Y. et al. (2020). Graphene-assisted spontaneous relaxation towards dislocation-free heteroepitaxy. *Nature Nanotechnology* 15 (4): 272–276.

19 Chen, J., Fu, Y., Samad, L. et al. (2017). Vapor-phase epitaxial growth of aligned nanowire networks of cesium lead halide perovskites ($CsPbX_3$, X = Cl, Br, I). *Nano Letters* 17 (1): 460–466.

20 Zhu, T. and Li, J. (2010). Ultra-strength materials. *Progress in Materials Science* 55 (7): 710–757.

21 Li, J., Shan, Z., and Ma, E. (2014). Elastic strain engineering for unprecedented materials properties. *MRS Bulletin* 39 (2): 108–114.

22 Chen, Y., Lei, Y., Li, Y. et al. (2020). Strain engineering and epitaxial stabilization of halide perovskites. *Nature* 577 (7789): 209–215.

23 Jin, B., Zuo, N., Hu, Z.-Y. et al. (2020). Excellent excitonic photovoltaic effect in 2D $CsPbBr_3$/CdS heterostructures. *Advanced Functional Materials* 30 (49): 2006166.

24 Wang, L., Chen, P., Thongprong, N. et al. (2017). Unlocking the single-domain epitaxy of halide perovskites. *Advanced Materials Interfaces* 4 (22): 1701003.

25 Jung, Y.-K., Butler, K.T., and Walsh, A. (2017). Halide perovskite heteroepitaxy: bond formation and carrier confinement at the PbS–$CsPbBr_3$ interface. *The Journal of Physical Chemistry C* 121 (49): 27351–27356.

26 Sytnyk, M., Yakunin, S., Schöfberger, W. et al. (2017). Quasi-epitaxial metal-halide perovskite ligand shells on PbS nanocrystals. *ACS Nano* 11 (2): 1246–1256.

27 Liu, Q., Liang, L., Shen, H. et al. (2021). Epitaxial growth of $CsPbBr_3$-PbS vertical and lateral heterostructures for visible to infrared broadband photodetection. *Nano Research* 14 (11): 3879–3885.

28 Shoaib, M., Zhang, X., Wang, X. et al. (2017). Directional growth of ultralong $CsPbBr_3$ perovskite nanowires for high-performance photodetectors. *Journal of the American Chemical Society* 139 (44): 15592–15595.

29 Yang, T., Jin, C., Qu, J. et al. (2021). Solution epitaxy of halide perovskite thin single crystals for stable transistors. *ACS Applied Materials & Interfaces* 13 (31): 37840–37848.

30 Wang, L., Chen, P., Kuttipillai, P.S. et al. (2019). Epitaxial stabilization of tetragonal cesium tin iodide. *ACS Applied Materials & Interfaces* 11 (35): 32076–32083.

31 Chen, J., Luo, Z., Fu, Y. et al. (2019). Tin(IV)-tolerant vapor-phase growth and photophysical properties of aligned cesium tin halide perovskite ($CsSnX_3$; X = Br, I) nanowires. *ACS Energy Letters* 4 (5): 1045–1052.

32 Wang, Y., Yang, F., Li, X. et al. (2019). Epitaxial growth of large-scale orthorhombic $CsPbBr_3$ perovskite thin films with anisotropic photoresponse property. *Advanced Functional Materials* 29 (43): 1904913.

33 Zhu, S., Pochet, P., and Johnson, H.T. (2019). Controlling rotation of two-dimensional material flakes. *ACS Nano* 13 (6): 6925–6931.

34 Bording, J.K., Li, B.Q., Shi, Y.F., and Zuo, J.M. (2003). Size- and shape-dependent energetics of nanocrystal interfaces: experiment and simulation. *Physical Review Letters* 90 (22): 226104.

35 Guo, Y., Sun, X., Jiang, J. et al. (2020). A reconfigurable remotely epitaxial VO_2 electrical heterostructure. *Nano Letters* 20 (1): 33–42.

36 Utama, M.I.B., Belarre, F.J., Magen, C. et al. (2012). Incommensurate van der Waals epitaxy of nanowire arrays: a case study with ZnO on muscovite mica substrates. *Nano Letters* 12 (4): 2146–2152.

37 Zhou, Z., Qiao, H.W., Hou, Y. et al. (2021). Epitaxial halide perovskite-based materials for photoelectric energy conversion. *Energy & Environmental Science* 14 (1): 127–157.

38 Shi, E. and Dou, L. (2020). Halide perovskite epitaxial heterostructures. *Accounts of Materials Research* 1 (3): 213–224.

39 Killilea, N., Wu, M., Sytnyk, M. et al. (2019). Pushing PbS/metal-halide-perovskite core/epitaxial-ligand-shell nanocrystal photodetectors beyond 3 μm wavelength. *Advanced Functional Materials* 29 (14): 1807964.

40 Fu, Y., Zhu, H., Chen, J. et al. (2019). Metal halide perovskite nanostructures for optoelectronic applications and the study of physical properties. *Nature Reviews Materials* 4 (3): 169–188.

41 Wang, L., King, I., Chen, P. et al. (2020). Epitaxial and quasiepitaxial growth of halide perovskites: new routes to high end optoelectronics. *APL Materials* 8 (10): 100904.

42 Wang, X., Li, Y., Xu, Y. et al. (2020). Solution-processed halide perovskite single crystals with intrinsic compositional gradients for X-ray detection. *Chemistry of Materials* 32 (12): 4973–4983.

43 Cao, J., Tang, G., You, P. et al. (2020). Enhanced performance of planar perovskite solar cells induced by Van Der Waals epitaxial growth of mixed perovskite films on WS_2 flakes. *Advanced Functional Materials* 30 (38): 2002358.

44 Chen, J., Zhou, Y., Fu, Y. et al. (2021). Oriented halide perovskite nanostructures and thin films for optoelectronics. *Chemical Reviews* 121 (20): 12112–12180.

45 Sytnyk, M., Yousefi-Amin, A.-A., Freund, T. et al. (2020). Epitaxial metal halide perovskites by inkjet-printing on various substrates. *Advanced Functional Materials* 30 (43): 2004612.

46 Zhang, Q., Su, R., Liu, X. et al. (2016). High-quality whispering-gallery-mode lasing from cesium lead halide perovskite nanoplatelets. *Advanced Functional Materials* 26 (34): 6238–6245.

47 Nasyedkin, K., King, I., Zhang, L. et al. (2021). Extraordinary phase coherence length in epitaxial halide perovskites. *iScience* 24 (8): 102912.

48 Shi, E., Gao, Y., Finkenauer, B.P. et al. (2018). Two-dimensional halide perovskite nanomaterials and heterostructures. *Chemical Society Reviews* 47 (16): 6046–6072.

49 Gong, X. and Sargent, E.H. (2020). Perovskite single-crystal thin film devices using lithography assisted epitaxy. *Matter* 3 (3): 619–620.

50 Ha, S.T., Liu, X., Zhang, Q. et al. (2014). Synthesis of organic–inorganic lead halide perovskite nanoplatelets: towards high-performance perovskite solar cells and optoelectronic devices. *Advanced Optical Materials* 2 (9): 838–844.

51 Mandal, T.N. and Jana, A. (2020). Lateral epitaxial heterostructures of halide perovskites for diode application. *Matter* 3 (3): 617–619.

52 Wang, X.-D., Li, W.-G., Liao, J.-F., and Kuang, D.-B. (2019). Recent advances in halide perovskite single-crystal thin films: fabrication methods and optoelectronic applications. *Solar RRL* 3 (4): 1800294.

53 Dutta, S.K., Bera, S., and Pradhan, N. (2021). Why is making epitaxially grown all inorganic perovskite–chalcogenide nanocrystal heterostructures challenging? some facts and some strategies. *Chemistry of Materials* 33 (11): 3868–3877.

54 Soto-Montero, T., Soltanpoor, W., and Morales-Masis, M. (2020). Pressing challenges of halide perovskite thin film growth. *APL Materials* 8 (11): 110903.

55 Lee, J.-W., Tan, S., Han, T.-H. et al. (2020). Solid-phase hetero epitaxial growth of α-phase formamidinium perovskite. *Nature Communications* 11 (1): 5514.

56 Jia, C., Zhao, X., Lai, Y.-H. et al. (2019). Highly flexible, robust, stable and high efficiency perovskite solar cells enabled by van der Waals epitaxy on mica substrate. *Nano Energy* 60: 476–484.

57 Zhang, X., Zhu, T., Ji, C. et al. (2021). In situ epitaxial growth of centimeter-sized lead-free $(BA)_2CsAgBiBr_7/Cs_2AgBiBr_6$ heterocrystals for self-driven X-ray detection. *Journal of the American Chemical Society* 143 (49): 20802–20810.

58 Shao, C.-R. and Hu, W. (2020). Lateral epitaxial grown of two-dimensional halide perovskite heterostructures. *Rare Metals* 39 (8): 863–864.

59 Zhang, J., Guo, Q., Li, X. et al. (2020). Solution-processed epitaxial growth of arbitrary surface nanopatterns on hybrid perovskite monocrystalline thin films. *ACS Nano* 14 (9): 11029–11039.

60 Banik, A., Bohannan, E.W., and Switzer, J.A. (2020). Epitaxial electrodeposition of BiI_3 and topotactic conversion to highly ordered solar light-absorbing perovskite $(CH_3NH_3)_3Bi_2I_9$. *Chemistry of Materials* 32 (19): 8367–8372.

61 Zhang, X., Chen, S., Wang, X., and Pan, A. (2019). Controlled synthesis and photonics applications of metal halide perovskite nanowires. *Small Methods* 3 (1): 1800294.

62 Imran, M., Peng, L., Pianetti, A. et al. (2021). Halide perovskite–lead chalcohalide nanocrystal heterostructures. *Journal of the American Chemical Society* 143 (3): 1435–1446.

63 Zhao, J., Wang, X., Pan, Y. et al. (2021). Photo-diodes based on $CH_3NH_3PbCl_3$ perovskite single crystals by epitaxial growth for ultraviolet photo-detection. *Frontiers in Physics* 9 (129).

64 Wei, Y., Li, K., Cheng, Z. et al. (2019). Epitaxial growth of $CsPbX_3$ (X = Cl, Br, I) perovskite quantum dots via surface chemical conversion of Cs_2GeF_6 double perovskites: a novel strategy for the formation of leadless hybrid perovskite phosphors with enhanced stability. *Advanced Materials* 31 (16): 1807592.

65 Ricciardulli, A.G., Yang, S., Smet, J.H., and Saliba, M. (2021). Emerging perovskite monolayers. *Nature Materials* 20 (10): 1325–1336.

66 Koza, J.A., Hill, J.C., Demster, A.C., and Switzer, J.A. (2016). Epitaxial electrodeposition of methylammonium lead iodide perovskites. *Chemistry of Materials* 28 (1): 399–405.

67 Lu, C., Sun, S., and Duo, Y. (2021). Inkjet printing epitaxial metal halide perovskites on various substrates? *Advanced Functional Materials* 31 (27): 2100694.

68 Yuan, B., Wei, H., Li, J. et al. (2021). Epitaxial growth of quasi-intrinsic $CsPbBr_3$ film on a $SrTiO_3$ substrate by pulsed laser deposition. *ACS Applied Electronic Materials* 3 (12): 5592–5600.

69 Oksenberg, E., Sanders, E., Popovitz-Biro, R. et al. (2018). Surface-guided $CsPbBr_3$ perovskite nanowires on flat and faceted sapphire with size-dependent photoluminescence and fast photoconductive response. *Nano Letters* 18 (1): 424–433.

70 Huang, L., Gao, Q., Sun, L.-D. et al. (2018). Composition-graded cesium lead halide perovskite nanowires with tunable dual-color lasing performance. *Advanced Materials* 30 (27): 1800596.

71 Afify, H.A., Sytnyk, M., Zhou, S. et al. (2021). Perspectives of solution epitaxially grown defect tolerant lead-halide-perovskites and lead-chalcogenides. *Applied Physics Letters* 119 (23): 230501.

72 Tang, G., You, P., Tai, Q. et al. (2019). Solution-phase epitaxial growth of perovskite films on 2D material flakes for high-performance solar cells. *Advanced Materials* 31 (24): 1807689.

73 Chen, P., Kuttipillai, P.S., Wang, L., and Lunt, R.R. (2017). Homoepitaxial growth of metal halide crystals investigated by reflection high-energy electron diffraction. *Scientific Reports* 7 (1): 40542.

74 Nie, Z., Gao, X., Ren, Y. et al. (2020). Harnessing hot phonon bottleneck in metal halide perovskite nanocrystals via interfacial electron–phonon coupling. *Nano Letters* 20 (6): 4610–4617.

75 Zhang, H., Qin, M., Chen, Z. et al. (2021). Bottom-up quasi-epitaxial growth of hybrid perovskite from solution process—achieving high-efficiency solar cells via template-guided crystallization. *Advanced Materials* 33 (22): 2100009.

76 Kimura, K., Nakamura, Y., Matsushita, T., and Kondo, T. (2019). Epitaxial $CH_3NH_3Pb(Br_{1-x}I_x)_3$ thin films on $CH_3NH_3PbBr_3$ single crystal substrates formed by physical vapor deposition. *Japanese Journal of Applied Physics* 58 (SB): SBBF04.

77 Sanders, E., Soffer, Y., Salzillo, T. et al. (2021). Remanent polarization and strong photoluminescence modulation by an external electric field in epitaxial $CsPbBr_3$ nanowires. *ACS Nano* 15 (10): 16130–16138.

78 Kelso, M.V., Mahenderkar, N.K., Chen, Q. et al. (2019). Spin coating epitaxial films. *Science* 364 (6436): 166–169.

79 Sun, Y., Thompson, S.E., and Nishida, T. (2009). *Strain Effect in Semiconductors: Theory and Device Applications*, 350. Boston, MA: Springer.

80 Shi, J. (2020). *A Structurally Unstable Semiconductor Stabilized and Enhanced by Strain*. Nature Publishing Group.

81 Zhao, J., Deng, Y., Wei, H. et al. (2017). Strained hybrid perovskite thin films and their impact on the intrinsic stability of perovskite solar cells. Science. *Advances* 3 (11): eaao5616.

82 Rakita, Y., Cohen, S.R., Kedem, N.K. et al. (2015). Mechanical properties of $APbX_3$ (A = Cs or CH3NH3; X = I or Br) perovskite single crystals. *MRS Communications* 5 (4): 623–629.

83 Panomsuwan, G., Takai, O., and Saito, N. (2013). Optical and mechanical properties of transparent $SrTiO_3$ thin films deposited by ECR ion beam sputter deposition. *Physica Status Solidi (a)* 210 (2): 311–319.

84 Brazhkin, Y.A., Korobov, A.I., and Lyamov, V.E. (1981). Electric-Field Effect on Electro-Mechanical Properties of Srtio3. *Fiz Tverd Tela+* 23: 1545–1548.

85 Ferralis, N. (2010). Probing mechanical properties of graphene with Raman spectroscopy. *Journal of Materials Science* 45 (19): 5135–5149.

86 Georgantzinos, S., Giannopoulos, G., and Anifantis, N. (2010). Numerical investigation of elastic mechanical properties of graphene structures. *Materials & Design* 31 (10): 4646–4654.

87 Ashby, M.F. (2010). *Materials Selection in Mechanical Design*. Oxford: Butterworth-Heinemann.

88 Kumar, A., Yan, F., Zeng, K., and Lu, L. (2011). Electric, magnetic and mechanical coupling effects on ferroelectric properties and surface potential of $BiFeO_3$ thin film. *Functional Materials Letters* 04 (01): 91–95.

89 Pilania, G. and Uberuaga, B.P. (2015). Cation ordering and effect of biaxial strain in double perovskite CsRbCaZnCl6. *Journal of Applied Physics* 117 (11): 114103.

90 Yu, Z.-L., Ma, Q.-R., Liu, B. et al. (2017). Oriented tuning the photovoltaic properties of γ-$RbGeX_3$ by strain-induced electron effective mass mutation. *Journal of Physics D: Applied Physics* 50 (46): 465101.

91 Zhang, Y., Sahoo, M.P.K., Shimada, T. et al. (2017). Strain-induced improper ferroelectricity in Ruddlesden-Popper perovskite halides. *Physical Review B* 96 (14): 144110.

92 Li, Y., Behtash, M., Wong, J., and Yang, K. (2018). Enhancing ferroelectric dipole ordering in organic–inorganic hybrid perovskite $CH_3NH_3PbI_3$: strain and doping engineering. *The Journal of Physical Chemistry C* 122 (1): 177–184.

93 Jing, H., Sa, R., and Xu, G. (2019). Tuning electronic and optical properties of $CsPbI_3$ by applying strain: a first-principles theoretical study. *Chemical Physics Letters* 732: 136642.

94 Ma, S., Kim, S.H., Jeong, B. et al. (2019). Strain-mediated phase stabilization: a new strategy for ultrastable α-$CsPbI_3$ perovskite by nanoconfined growth. *Small* 15 (21): 1900219.

95 Saffari, M., Soleimani, H.R., and Tagani, M.B. (2020). Quantum confinement and strain effects on the low-dimensional all-inorganic halide Cs2XI2Cl2 (X = Pb, Sn) perovskites: a theoretical approach for modulating electronic and optical properties. *Physica E: Low-dimensional Systems and Nanostructures* 124: 114226.

96 Soni, A., Bhamu, K.C., and Sahariya, J. (2020). Investigating effect of strain on electronic and optical properties of lead free double perovskite $Cs_2AgInCl_6$ solar cell compound: a first principle calculation. *Journal of Alloys and Compounds* 817: 152758.

97 Chen, Z., Brocks, G., Tao, S., and Bobbert, P.A. (2021). Unified theory for light-induced halide segregation in mixed halide perovskites. *Nature Communications* 12 (1): 2687.

98 Liu, D., Luo, D., Iqbal, A.N. et al. (2021). Strain analysis and engineering in halide perovskite photovoltaics. *Nature Materials* 20 (10): 1337–1346.

99 Liu, Y., Sumpter, B.G., Keum, J.K. et al. (2021). Strain in metal halide perovskites: the critical role of A-site cation. *ACS Applied Energy Materials* 4 (3): 2068–2072.

100 Singh, J., Kaur, H., Singh, G., and Tripathi, S.K. (2021). High thermoelectric performance in cubic inorganic halide perovskite material $AgCdX_3$ (X = F and Cl) from first principles. *Materials Today Energy* 21: 100820.

101 Wang, X., Bi, G., Ali, N. et al. (2021). Theoretical study of the strain influence on lead-free bismuth-based halide perovskites. *Journal of Materials Science* 56 (19): 11377–11385.

102 Zarabinia, N. and Rasuli, R. (2021). Electronic and optical properties of halide double-perovskites under strain: a density functional study. *Energy Sources, Part A: Recovery, Utilization, and Environmental Effects* 43 (20): 2443–2455.

103 Zhou, X. and Zhang, Z. (2020). Strain induced Rashba splitting in $CH_3NH_3PbBr_3$ organic–inorganic lead halide perovskite. *AIP Advances* 10 (8): 085210.

104 Teply, C., Tyler, B.A., and Berger, R.F. (2021). Tuning the band gaps of oxide and halide perovskite compounds via biaxial strain in all directions. *The Journal of Physical Chemistry C* 125 (47): 25951–25958.

105 Tsvetkov, N., Khan, M.E., Moon, B.C. et al. (2021). Strain-induced metallization and defect suppression at zipper-like interdigitated atomically thin interfaces enabling high-efficiency halide perovskite solar cells. *ACS Nano* 15 (1): 1805–1816.

106 Moloney, E.G., Yeddu, V., and Saidaminov, M.I. (2020). Strain engineering in halide perovskites. *ACS Materials Letters* 2 (11): 1495–1508.

107 Grote, C. and Berger, R.F. (2015). Strain tuning of tin–halide and lead–halide perovskites: a first-principles atomic and electronic structure study. *The Journal of Physical Chemistry C* 119 (40): 22832–22837.

108 Sa, R., Zha, W., Yuan, R., and Chen, J. (2020). Exploring electronic and optical properties of Ge-based perovskites under strain: Insights from the first-principles calculations. *Spectrochimica Acta Part A: Molecular and Biomolecular Spectroscopy* 229: 118013.

109 Liu, D., Li, Q., Jing, H., and Wu, K. (2019). Pressure-induced effects in the inorganic halide perovskite $CsGeI_3$. *RSC Advances* 9 (6): 3279–3284.

110 Zhang, Z., Su, J., Hou, J. et al. (2019). Potential applications of halide double perovskite Cs_2AgInX_6 (X = Cl, Br) in flexible optoelectronics: unusual effects of uniaxial strains. *The Journal of Physical Chemistry Letters* 10 (5): 1120–1125.

111 Lin, S., Yan, L., Zhao, L. et al. (2021). Strain engineering for tailored carrier transport and thermoelectric performance in mixed halide perovskites $CsPb(I_{1-x}Br_x)_3$. *ACS Applied Energy Materials* 4 (12): 14508–14519.

112 Lee, J.-H., Jaffe, A., Lin, Y. et al. (2020). Origins of the pressure-induced phase transition and metallization in the halide perovskite $(CH3NH_3)PbI_3$. *ACS Energy Letters* 5 (7): 2174–2181.

113 Jin, H., Im, J., and Freeman, A.J. (2012). Topological insulator phase in halide perovskite structures. *Physical Review B* 86 (12): 121102.

114 Leppert, L., Reyes-Lillo, S.E., and Neaton, J.B. (2016). Electric field- and strain-induced rashba effect in hybrid halide perovskites. *The Journal of Physical Chemistry Letters* 7 (18): 3683–3689.

115 Zhang, L., Geng, W., Tong, C.-J. et al. (2018). Strain induced electronic structure variation in methyl-ammonium lead iodide perovskite. *Scientific Reports* 8 (1): 7760.

116 Rolston, N., Bush, K.A., Printz, A.D. et al. (2018). Engineering stress in perovskite solar cells to improve stability. *Advanced Energy Materials* 8 (29): 1802139.

117 Schueller, E.C., Laurita, G., Fabini, D.H. et al. (2018). Crystal structure evolution and notable thermal expansion in hybrid perovskites formamidinium tin iodide and formamidinium lead bromide. *Inorganic Chemistry* 57 (2): 695–701.

118 Zhao, Y., Miao, P., Elia, J. et al. (2020). Strain-activated light-induced halide segregation in mixed-halide perovskite solids. *Nature Communications* 11 (1): 6328.

119 Kim, G., Min, H., Lee Kyoung, S. et al. (2020). Impact of strain relaxation on performance of α-formamidinium lead iodide perovskite solar cells. *Science* 370 (6512): 108–112.

120 Wang, L., Wang, K., Xiao, G. et al. (2016). Pressure-induced structural evolution and band gap shifts of organometal halide perovskite-based methylammonium lead chloride. *The Journal of Physical Chemistry Letters* 7 (24): 5273–5279.

121 Kim, H.-S. and Park, N.-G. (2020). Importance of tailoring lattice strain in halide perovskite crystals. *NPG Asia Materials* 12 (1): 78.

122 Pal, S., Ghorai, A., Goswami, D.K., and Ray, S.K. (2021). Strain mediated light emission using heterojunctions of all-inorganic mixed-halide perovskite nanocrystals via piezo-phototronic effect. *Nano Energy* 87: 106200.

123 Jones, T.W., Osherov, A., Alsari, M. et al. (2019). Lattice strain causes non-radiative losses in halide perovskites. *Energy & Environmental Science* 12 (2): 596–606.

124 Jiao, Y., Yi, S., Wang, H. et al. (2021). Strain engineering of metal halide perovskites on coupling anisotropic behaviors. *Advanced Functional Materials* 31 (4): 2006243.

125 Kim, D.B., Lee, J.W., and Cho, Y.S. (2021). Anisotropic in situ strain-engineered halide perovskites for high mechanical flexibility. *Advanced Functional Materials* 31 (4): 2007131.

126 Finkenauer, B.P., Gao, Y., Wang, X. et al. (2021). Mechanically robust and self-healable perovskite solar cells. *Cell Reports Physical Science* 2 (2): 100320.

127 Lee, H., Woo, J.Y., Park, D.Y. et al. (2021). Tip-induced strain engineering of a single metal halide perovskite quantum dot. *ACS Nano* 15 (5): 9057–9064.

128 Xue, D.J., Hou, Y., Liu, S.C. et al. (2020). Regulating strain in perovskite thin films through charge transport layers. *Nature Communications* 11 (1): 1514.

129 Zhu, C., Niu, X., Fu, Y. et al. (2019). Strain engineering in perovskite solar cells and its impacts on carrier dynamics. *Nature Communications* 10 (1): 815.

130 Kim, D., Yun, J.S., Sharma, P. et al. (2019). Light- and bias-induced structural variations in metal halide perovskites. *Nature Communications* 10 (1): 444.

14

Electron Microscopy of Perovskite Solar Cell Materials

Mathias U. Rothmann[1,2,3], *Wei Li*[1,4], *and Zhiwei Tao*[1,4]

[1] *Xianhu Hydrogen Valley, Foshan Xianhu Laboratory of the Advanced Energy Science and Technology Guangdong 10 Laboratory, Foshan, 528200, P. R. China*
[2] *University of Oxford, Department of Physics, Clarendon Laboratory, Parks Road, Oxford OX1 3PU, UK*
[3] *University of Oxford, Department of Materials, Parks Road, Oxford OX1 3PH, UK*
[4] *Wuhan University of Technology, State Key Laboratory of Advanced Technology for Materials Synthesis and Processing, 122 Luoshi Road, Hongshan District, Wuhan, Hubei Province, 430070, P. R. China*

14.1 Introduction

This chapter is aimed at researchers working with perovskite solar cells who want to use electron microscopes in their research. We have assumed a very basic knowledge of electron microscopes in general, and we therefore spend a bit of time describing their working methods and overall capabilities before discussing results from the current literature. Electron microscopists who want to begin looking at photoactive perovskites might also find this chapter useful, but they can readily skip the sections about electron microscope basics.

We pay particular attention to the practical issues faced when using electron microscopes to study photoactive perovskites. These materials are generally quite unstable under many conditions, and the high vacuum and strong electron radiation in electron microscopes are a very hostile environment for these materials. As such, we have attempted to pass on many of our observations regarding sample preparation, sample transfer, imaging conditions, and how to recognize beam damage. We use this basis to give examples of studies based on various electron microscope techniques in order to provide an overview of the versatility of electron microscopes, hoping to stir the reader's imagination for what other exciting things can be learned with these microscopes.

14.2 Fundamentals of Electron Microscopy

Electron microscopes rely on the wave/particle duality of electrons and their interaction with materials to obtain a wide range of information about a specimen [1–4]. This information can include surface morphology, elemental distribution, crystallographic orientation, crystal quality, and chemical states, to name a few.

Halide Perovskite Semiconductors: Structures, Characterization, Properties, and Phenomena, First Edition. Edited by Yuanyuan Zhou and Iván Mora-Seró.

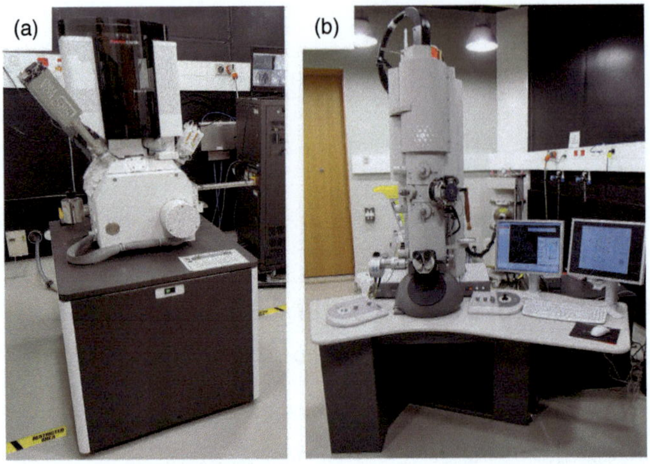

Figure 14.1 Different types of electron microscopes. (a) A Thermo Scientific Verios 5 UC FEG scanning electron microscope. (b) An FEI Tecnai G2 T20 TWIN transmission electron microscope. © Monash University Centre for Electron Microscopy.

The microscopes themselves can roughly be divided into scanning electron microscopes (SEM) and transmission electron microscopes (TEM) (Figure 14.1).

In an SEM, a beam of electrons is extracted from a microscopic tip by a high voltage, typically 0.5–50 kV, and focused to a few picometers in width using electromagnetic lenses. The focused electron beam is scanned across a (usually bulky) specimen placed in a vacuum chamber, and the signals emitted from the specimen are recorded by a variety of detectors found within the chamber. SEM is commonly used to study the morphology of specimens with a resolution of a few nanometers. Additionally, they can also provide highly localized information about the elements found in a specimen through X-ray spectroscopy, as well as information about electronic state of the material through cathodoluminescence (CL), as described below [1, 3, 5].

In its most basic form, a TEM is a long, evacuated tube with an electron source at one end, a sample somewhere in the middle, and various detectors at the other end. Several electromagnets are placed in circular arrangements around the tube, forming electromagnetic lenses, which allow for the focusing of the electron beam passing through the tube. In real life, a TEM is much more complex than this simplified image, with a long list of essential equipment such as vacuum pumps, liquid nitrogen-based cooling systems, and high voltage generators surrounding and interacting with this tube. It is even common to construct specialized rooms dedicated to individual microscopes, to isolate and minimize external vibrations, or simply to house microscopes that can be several meters high. However, from the perspective of visualizing how a TEM works, the evacuated tube analogy works quite well [4].

When electrons are accelerated to high velocities, their associated wavelength becomes very small relative to objects around us, down to just a few picometers when accelerated with a voltage of 100–300 kV. This voltage accelerates electrons to around 70% of the speed of light, and relativistic effects must be taken into account at these velocities when calculating the actual wavelength of electrons, but this has little to no practical impact on the operation of a TEM [4].

The short wavelength of electrons accelerated at high voltage means that the resolution of a TEM is much greater than that of a visible-light microscope. Since the diffraction-limited resolution of a microscope scales with the wavelength of the radiation used, a shorter wavelength results in the distance between two discernible objects also shortening, improving the resolution of the microscope. At acceleration voltage of 100–300 kV, this results in a resolution of a few hundred picometers, and in some cases, well below the distance between two atoms, making atomic resolution imaging possible with a TEM [4].

The main operational difference between SEM and TEM is the transmissive nature of the TEM. Whereas in an SEM, electrons are generally recorded from the surface of a specimen, TEM typically record electrons that have passed through the specimen. The interaction with the specimen can then be interpreted and a wealth of information about the specimen can thus be obtained, ranging from simple surface imaging to atomic-resolution elemental mapping.

TEMs typically operate either in a broad-beam mode (TEM) or a focused, scanning beam mode (scanning TEM or STEM), with each mode having complementary capabilities. In broad-beam TEM, the electron beam is focused in a way that causes it to spread out onto the sample, constantly illuminating a given area. An electron detector (usually CCD, CMOS, or single-electron detectors) records the spatial intensity of the transmitted electrons and forms a contrast image (also known as a micrograph) based on the entire illuminated area that fits within the detector's field of view. The contrast in the electron's image can be caused by several factors, including differences in thickness, composition, and crystallographic orientation throughout the specimen, which can make it difficult to interpret a single micrograph unambiguously. Broad-beam mode also enables the use of selected area electron diffraction (SAED), a very useful technique for obtaining crystallographic information about a desired area of a specimen [4].

14.3 Signal Generation

When energetic electrons impinge on a material, they tend to interact with the atoms and molecules in the material in a variety of ways. This interaction can lead to changes in the electron's properties, such as deflection from the original path or loss of energy, or changes in the sample material's properties, such as excitation of electrons or breaking of bonds. Detecting the changes in properties, or lack thereof, allows us to gain insights into several properties of a range of materials, and this detection is the foundation of electron microscopy [4].

Some of the differences in the uses of SEM and TEM come from the differences in signals they tend to detect and the resolution with which they detect them. Figure 14.2 shows an overview of the main signals generated in electron microscopes as a result of electrons interacting with a material. Broadly speaking, different

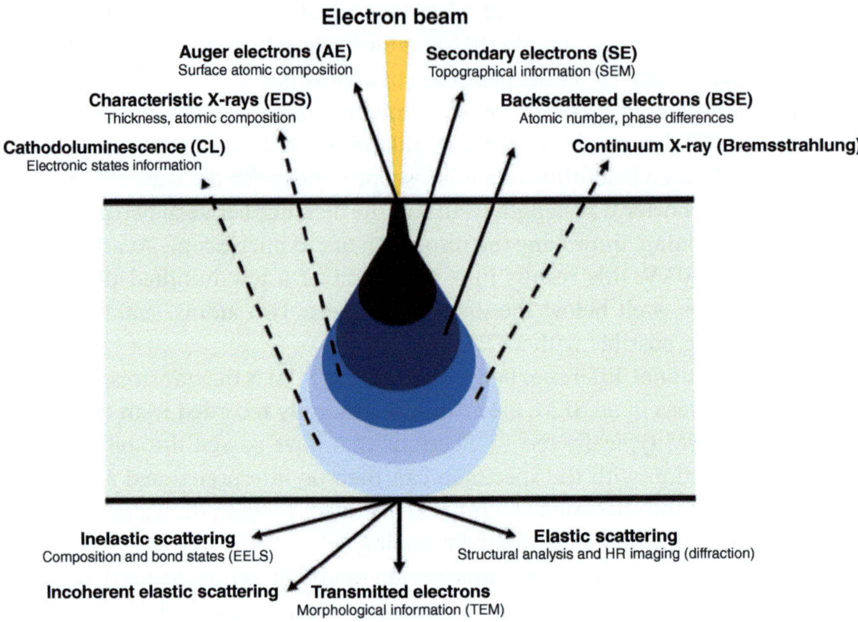

Figure 14.2 Schematic overview of the various signals generated in an electron microscope. Source: Reproduced from Rothmann et al. [6]/with permission of John Wiley & Sons.

signals tend to be used in different types of electron microscopes, with a few signals being used in both SEM and TEM. For example, secondary electrons (SE) and backscattered electrons (BSE) are mostly detected in SEM, whereas elastically and inelastically forward-scattered electrons are more commonly detected in TEM. That being said, some SEM setups are able to detect transmitted electrons, and some TEMs are equipped with secondary electron detectors [6].

Each type of signal is generated in a specific way and is able to provide specific information about the specimen being studied. Below are overviews of the generation and detection of some of the most commonly used EM signals, as well as brief discussions of their limitations when used to study perovskite solar cell materials.

By far the most commonly collected signal in an SEM is the SE emitted by the specimen as a result of inelastic collisions between the incoming electrons and the electrons in the specimen. As the beam is scanned across the surface of the specimen in a raster pattern, the number of SE emitted in each pixel is recorded, and an electron intensity map can be constructed. Assuming a uniform material composition, the emission intensity is largely dependent on the surface morphology of the specimen, with peaks or concave features releasing a larger density of electrons than troughs or convex features. This secondary electron imaging technique is thus able to generate highly magnified images of surfaces, with a typical resolution of a few nanometers.

While most interactions between the electron beam and the specimen are inelastic, where some of the electrons' energy is given off to the specimen, some electrons scatter elastically off the atoms in the specimen, effectively reflecting them back toward the electron source. The intensity of the BSE can be collected, usually by a circular detector sitting near the electron emission pole piece, and an image can be generated simultaneously with the SE image. Since heavier atoms backscatter more electrons than lighter atoms, the BSE signal can be used to map out the relative atomic number distribution of a specimen. Since the backscattered electrons' wave functions interact with the crystal, they form diffraction patterns when they reemerge. These electron backscatter diffraction (EBSD) patterns can be collected and used to determine the crystallographic orientation of a crystalline material, making it possible to generate crystallographic orientation maps of polycrystalline samples [7–9].

Besides crystallographic phenomena, the energetic interaction between incoming electrons and the electrons in the specimen can also provide useful information about the chemical and electronic properties of a material. Electron dispersive X-ray spectroscopy (EDX or EDS), for example, can give highly localized information about the atomic makeup of a specimen by collecting the characteristic X-rays emitted by it under electron irradiation. Each element emits a unique spectrum of X-rays as the energetic incoming electrons eject ground state atomic electrons, and higher orbital electrons relax into the ground state, emitting characteristic X-rays in the process. By correlating this with the pixel position of a scanning beam, elemental maps can be generated and correlated with the microstructure [1]. In a bulk specimen, the interaction volume, and hence the origin of the X-ray signals, can be several tens of nanometers to a few micrometers across, meaning that EDS is a relatively low-resolution technique for bulk specimens. Thinning the specimen, however, improves the resolution, and EDX maps can be obtained with atomic resolution in atomically thin samples.

Visible photons can also be emitted from materials by a process known as cathodoluminescence (CL). SE produced by the electron beam are able to excite electrons in the specimen from the valence band to the conduction band, and when these electrons recombine with holes in the valence band, a visible photon can be emitted. The emission wavelength (or color) is directly related to the bandgap of the material, including the presence of trap states, and CL can thus give very detailed information about the electronic properties of a material. Since the technique relies on secondary electrons, however, it is spatially limited in the same way as EDS, and collecting the emitted photons can be challenging, resulting in large electron doses needed to generate a signal [10].

We will now describe various techniques and applications specific to SEM and TEM studies.

14.4 SEM

14.4.1 Cathodoluminescence

CL can reveal optical and electronic properties of halide perovskite materials on the nanoscale through high-energy electron excitation. Because of its powerful imaging capabilities when associated with an SEM or a TEM, its high lateral and depth resolution make it possible to reveal the distribution of extended defects, including grain boundaries, dislocations in the nature of charge carrier recombination in semiconductors, including perovskites on the nanoscale [11–13].

14.4.1.1 Comparison of CL and Photoluminescence (PL)

CL is similar to PL in working principle but uses energetic electrons with energies ranging from hundreds of eVs to tens of keVs as the excitation source, resulting in some remarkable differences with respect to the excitation with laser of the PL experiments.

In CL, valence electrons are excited into the conduction band by a high energy density electron beam incident onto the perovskite thin film. The excited electrons in the conduction band then recombine radiatively with holes in the valence band, resulting in the emission of photons with an energy equal to the characteristic bandgap (E_g) of the perovskite [14]. The three-dimensional resolution of CL microscopy depends on both the size of the electron beam probe and the extent of electron scattering within the sample. Generally, for CL in the SEM, scattering within the sample will dominate, giving rise to a resolution of some tens to hundreds of nanometers.

In PL, the excitation laser beam is focused with an optical microscope objective. Therefore, the spatial resolution of PL is determined by the size of the laser beam at focus, which is diffraction-limited to the order of the laser wavelength, typically a few hundred nanometers. The volume of excited material in CL can be much smaller than in PL, and the spatial resolution of CL is, therefore, significantly higher than that of PL, which makes it possible to reveal significant spatial heterogeneity in the nature of charge carrier recombination on the nanoscale [12, 14].

Moreover, since the generation volume of electron–hole pairs (EHPs) by CL is controlled by the electron beam accelerating voltage, the penetration depth of the primary electrons can be tuned by varying the acceleration voltage of electron beam [14]. Therefore, CL can also provide specific information about a material's surface or bulk properties. This enables the analysis of the heterogeneity of charge carrier recombination in the depth profile, which is limited to being sensitive to within a fixed depth in the perovskite films by PL [15].

14.4.1.2 Working Principle

Figure 14.3 is a schematic diagram of CL measurements, showing the typical setup used in SEM-based measurements, consisting of the collecting optics, the parts guiding the light to the detector, the spectrometer, and the detector. A parabolic mirror with a large solid angle efficiently collects most of the generated and escaped photons from the sample surface. The microscope column is aligned with a hole in the parabolic mirror so that the electron beam can go through the mirror and strike the sample surface on the focus of the mirror. In this case, the light collected by the mirror is directed parallel to the paraboloid axis and transmitted from the vacuum chamber of electron microscope into a monochromator, where it is then dispersed for spectral detection, a technique known as spectrally resolved CL, or into a photomultiplier tube (PMT) detector to produce a panchromatic CL signal [12].

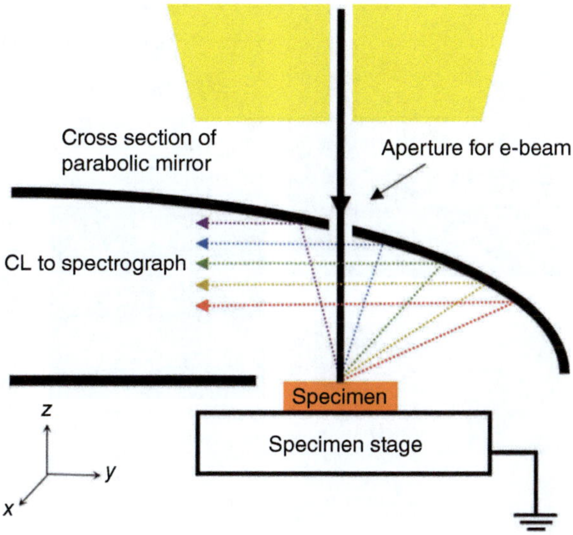

Figure 14.3 Schematic diagram of CL measurements. Source: Zhiwei Tao (co-author).

14.4.1.3 CL for Perovskites

Changes in CL spectra can be associated with trap states at surfaces and grain boundaries, as well as local changes in stoichiometric ratio in halide perovskites [13].

14.4.1.3.1 CL for All Inorganic Perovskites Li et al. observed iodide-rich phases as a result of phase segregation near grain boundaries, as well as the formation of iodide-rich "clusters" in the bulk of $CsPbIBr_2$ film [14]. The panchromatic CL image in Figure 14.4b shows higher emission intensity at grain boundaries relative to grain interiors. To determine the reason for the increased CL emission from grain boundaries, the authors performed CL spectrum imaging over the same region, as shown in Figure 14.4e. In Figure 14.4c, the image intensity is associated with emission between 530 and 590 nm, corresponding to the original $CsPbIBr_2$ phase, whereas in Figure 14.4d, it is associated with emission from 590 to 640 nm, corresponding to a lower-energy emission. It is clear that this lower-energy emission is dominant at the grain boundaries. PL spectra after light-induced phase separation closely resemble the lower-energy spectra in Figure 14.4f; however, the superior spatial resolution of CL confirms that this occurs primarily at the boundaries in this film. The PL spectra and average CL spectra arc dominated by emission from the grain boundary (GB) regions because the lower effective bandgap at these locations facilitates relaxation of carriers in the higher-bandgap grain interiors to these states, where they then recombine radiatively. These phases were previously identified as being iodide-rich [16]. Spectrum imaging revealed cluster-like features in the grain interiors for the lower-energy 630–730 nm emission image shown in Figure 14.4e, as opposed to being concentrated at the boundaries.

Davila et al. reported the presence and spatial distribution of secondary phases in fully inorganic $CsPbBr_3$ deposited by spin coating and co-evaporation by

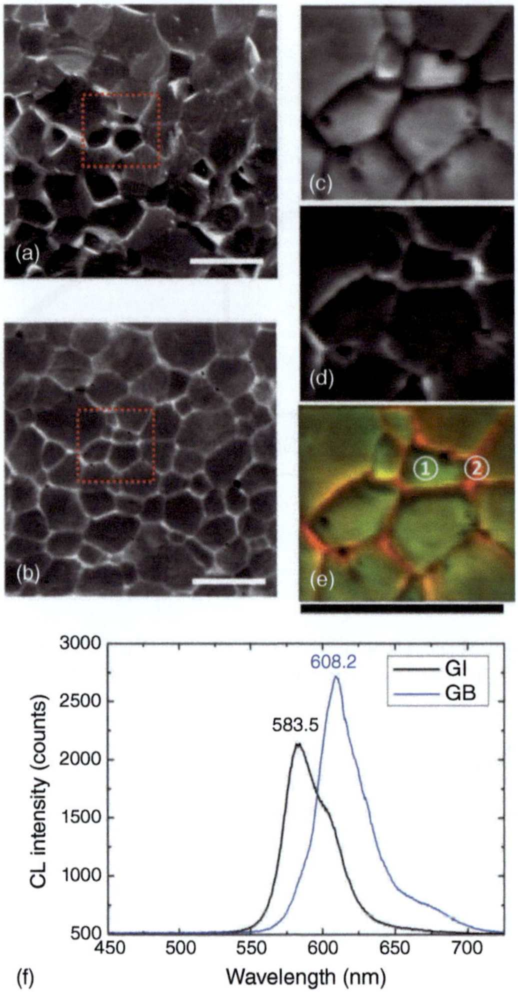

Figure 14.4 Phase segregation on a $CsPbIBr_2$ film surface. (a) Secondary-electron SEM image and (b) panchromatic CL PMT mapping of the $CsPbIBr_2$ film (1024 × 1024 pixels (pixel size of ≈11 nm × ≈11 nm); dwell time of 300 µs per pixel) with e-beam acceleration voltage of 2 kV and current of 440 pA. The square region in (a) is further studied by CL spectrum mapping with different spectral windows of (c) 530–590 nm and (d) 590–640 nm (pixel size of 40 nm × 40 nm; dwell time of 200 ms). (e) A superposition of (c) and (d). The color code in (e) indicates the emitted light wavelength in which orange regions suggest a longer emitted light wavelength than green regions. (f) CL spectra for the area inside a $CsPbIBr_2$ grain (GI) and its grain boundary (GB), corresponding to area 1 and area 2 in (e), respectively. All the scale bars are 3 µm. Source: Reproduced with permission from Li et al. [14], © John Wiley & Sons.

correlating CL spectrum imaging and EDS composition analysis. Figure 14.5a,b shows an area in the center of the bottom with no measurable CL emissions. The emission spectrum from the surrounding region consists of an emission peak (Figure 14.5c) near 2.35 eV, corresponding to $CsPbBr_3$. Information from EDS indicated that the sediment was Cs_4PbBr_6 [17].

14.4.1.3.2 Beam Damage Sufficient CL signals require high generation rates and injection levels, which can lead to degradation of radiation-sensitive materials.

Inorganic perovskites are reasonably stable, but the organic–inorganic hybrid perovskites are extremely fragile under electron beam illumination and can undergo morphological, compositional, and structural degradation [18]. This greatly limits the use of electron microscopy to probe the morphology, microstructure, elemental

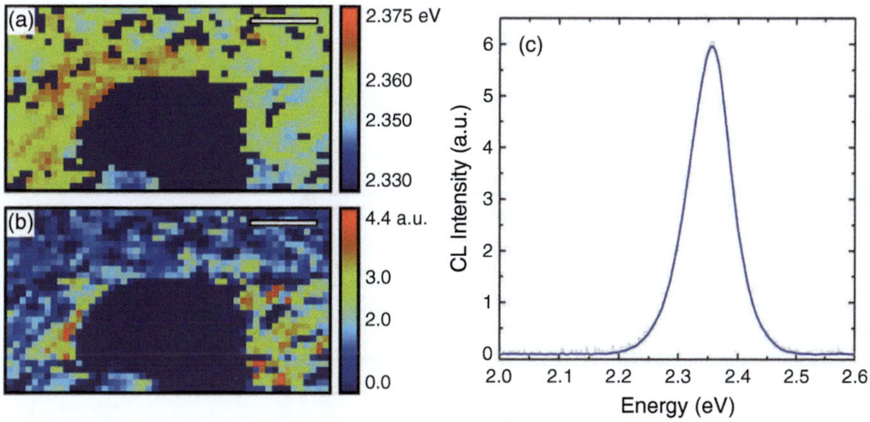

Figure 14.5 Optical properties of a Cs–Pb–Br film spin coated on glass. (a) Map of the emission peaks, (b) CL intensity map, and (c) accumulated CL spectrum. The precipitates show no measurable CL response, whereas the intensity is stronger on the film surrounding them. Note that the CL peak position on the film remains constant. All scale bars are 5 μm and the used exposure time for the CL measurements was 50 ms. Source: Reproduced from Caicedo-Dávila et al. [17]/with permission of American Chemical Society.

composition, crystal structure, and electronic properties of the hybrid materials at the micro- or nanoscale [19, 20].

Xiao et al. studied the changes in emission intensity and emission spectrum due to variable e-beam excitation conditions for different perovskite films in great detail [21]. Figure 14.6 shows the evolution of CL spectra of $CH_3NH_3PbI_3$ polycrystalline films under a series of electron beam conditions and exposure times. For the minimum exposure time of one second, the peak value of photon energy corresponding to 1.6 eV is obvious. As the exposure time increases, the spectral peak shifts to higher energies and broadens slightly, indicating that the incident electron beam induces the formation of a new phase. All these effects are greatly amplified at a higher acceleration voltage.

14.4.1.3.3 Damage Reduction

Reduce Acceleration Voltage Cai et al. recently reported that the knock-on atomic displacement damage in $CH_3NH_3PbI_3$ is significant and highly energy-dependent by first-principles calculations of threshold energies. Through the simulation, thc displaccmcnts of II, C, N, and I begin to occur when the incident energy increases to 2.3, 26.4, 70.4, and 249.4 keV, respectively, but Pb does not displace even at incident energy of 1 MeV. This finding explains the observation that 5 keV or higher energy radiation will cause a significant change in the CL spectrum, while 2 keV will only cause a small change [22]. In other words, reducing the accelerating voltage of the electron beam can effectively reduce the radiation damage to the material.

Reduce Exposure Time As shown in Figure 14.4e, the CL spectrum maps were obtained with a pixel size of 40 nm × 40 nm and a dwell time of 200 ms, which is 667

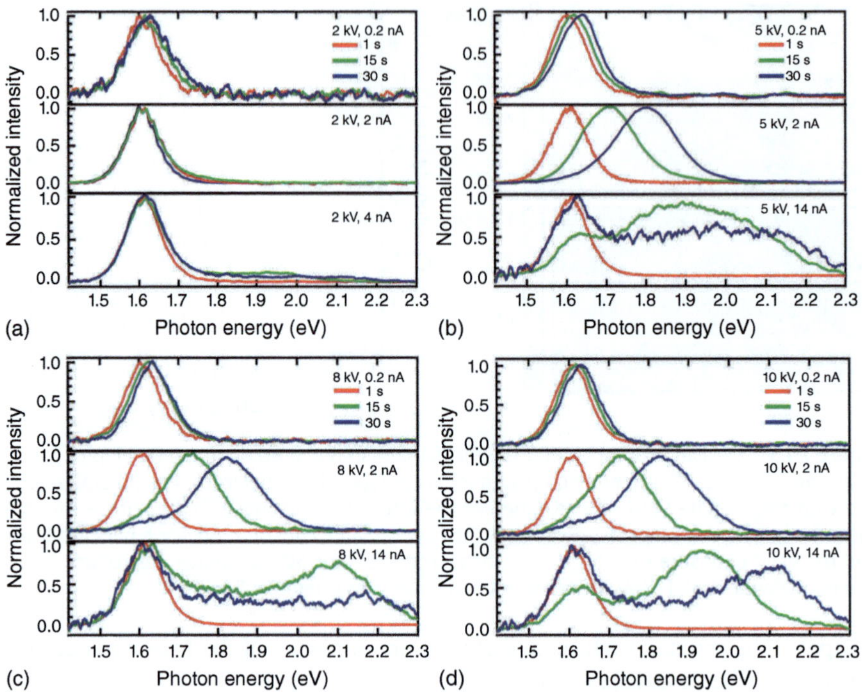

Figure 14.6 CL emission spectra taken after 1 s (red), 15 s (green), and 30 s (blue) of continuous e-beam irradiation. Spectra are shown for accelerating voltages of (a) 2 kV, (b) 5 kV, (c) 8 kV, and (d) 10 kV with beam current varying between 0.2 and 14 nA. Source: Reproduced from Xiao et al. [21]/with permission of American Chemical Society.

times longer than that used for PMT mapping. Spectrum mapping differs from PMT mapping in that not only is the overall CL intensity information collected, but the wavelengths of the emitted photons are resolved through a spectrometer and collected by a spectral camera [14]. This difference results in a longer exposure time and a corresponding increase in electron beam damage. However, combining PMT with a band pass filter may achieve a similar amount of information with slightly reduced spectral resolution while reducing exposure time and beam damage.

Reduce Dose Rate The use of a pulsed beam can reduce the beam current to tens to hundreds of pA [23]. The low injection conditions can be adjusted to match the operando conditions of optoelectronic devices, so that the technique is interesting from the point of view of device characteristics. More importantly, the low-current pulsed beam allows the material to relax electronically and thermally between pulses, which helps to prevent beam damage [24, 25]. In Orri's work, pulsed mode (PM) was obtained by pulsing an electron gun with the third harmonic of an Nd:YAG laser (355 nm) at a pulse width of 7 ps and a frequency of 80.6 MHz (12.41 ns) [26]. PM has been found beneficial for the acquisition of CL on organic hybrid perovskite films in comparison to continuous-wave (CW) mode.

14.4.2 Electron-Beam-Induced Current

Electron-beam-induced current (EBIC) microscopy is an analysis technique for semiconductor materials and devices in SEM or STEM [13]. The combination of EBIC signal and the microscope's scanning capability makes it possible to identify heterogeneous recombination activity and charge carrier collection caused by composition variations, structural defects, or impurities [27–29]. EBIC can also identify the position of charge separation junctions in device cross sections and quantitatively measure the carrier diffusion length of these junctions, thus providing a valuable insight for understanding the working mechanism of the device [13, 30, 31].

14.4.2.1 Working Principle of EBIC

EBIC is based on the generation of electron-hole pairs by electron beam excitation. If the diffusion length of the generated charge carriers is long enough, they will diffuse to the depletion region with a built-in electric field supplied by a Schottky barrier or a p–n junction. The field drifts electrons and holes in opposite directions, and current flows. EBIC measures this current as it joins an external circuit [27]. Normally, this current is on the order of μA, and an amplifier is needed to magnify the current for the EBIC imaging [29]. Generally, there are two forms of EBIC measurement: Cross-sectional and plan-view. Plan-view EBIC can help demystify the role of chemical composition or structure defects in carrier generation and collection.

14.4.2.2 Applications

14.4.2.2.1 Cross-sectional In cross-sectional EBIC, the EBIC signal becomes stronger as the e-beam incidents closer to the depletion region of the p–n junction and becomes weaker as the e-beam scans away, since charge carrier collection efficiency is higher in the depletion region [32]. Therefore, the contrast in cross-sectional EBIC is strongly related to the diffusion length of minority carriers in absorber layer of a photovoltaic device. We can extract the diffusion length from the EBIC data by fitting the EBIC current to a simple exponential decay,

$$I = I_0 + Ae^{[-X/L_{n/p}]} \tag{14.1}$$

where I is the signal intensity, I_0 is the intensity at maximal collection efficiency, A is an arbitrary intensity coefficient, X is a spatial coordinate, and $L_{n/p}$ is the diffusion length for minority carriers. In such a model, where the carrier motion is perpendicular to the line across which carrier separation occurs, the current can be modeled by an exponential decay as described by Eq. (14.1) [33].

Cross-sectional EBIC performed by Edri et al. confirmed the p–i–n operating mechanism in $CH_3NH_3PbI_{3-x}Cl_x$ perovskites [34]. By drawing the variation diagram of charge separation and collection efficiency in the cross-sections (Figure 14.7c), they found that there are two prime high efficiency locations, one at/near the absorber/hole-blocking layer interfaces and the other at/near the absorber/electron-blocking layer interfaces. The former is more obvious. This "twin-peaks" profile is characteristic of p–i–n solar cells.

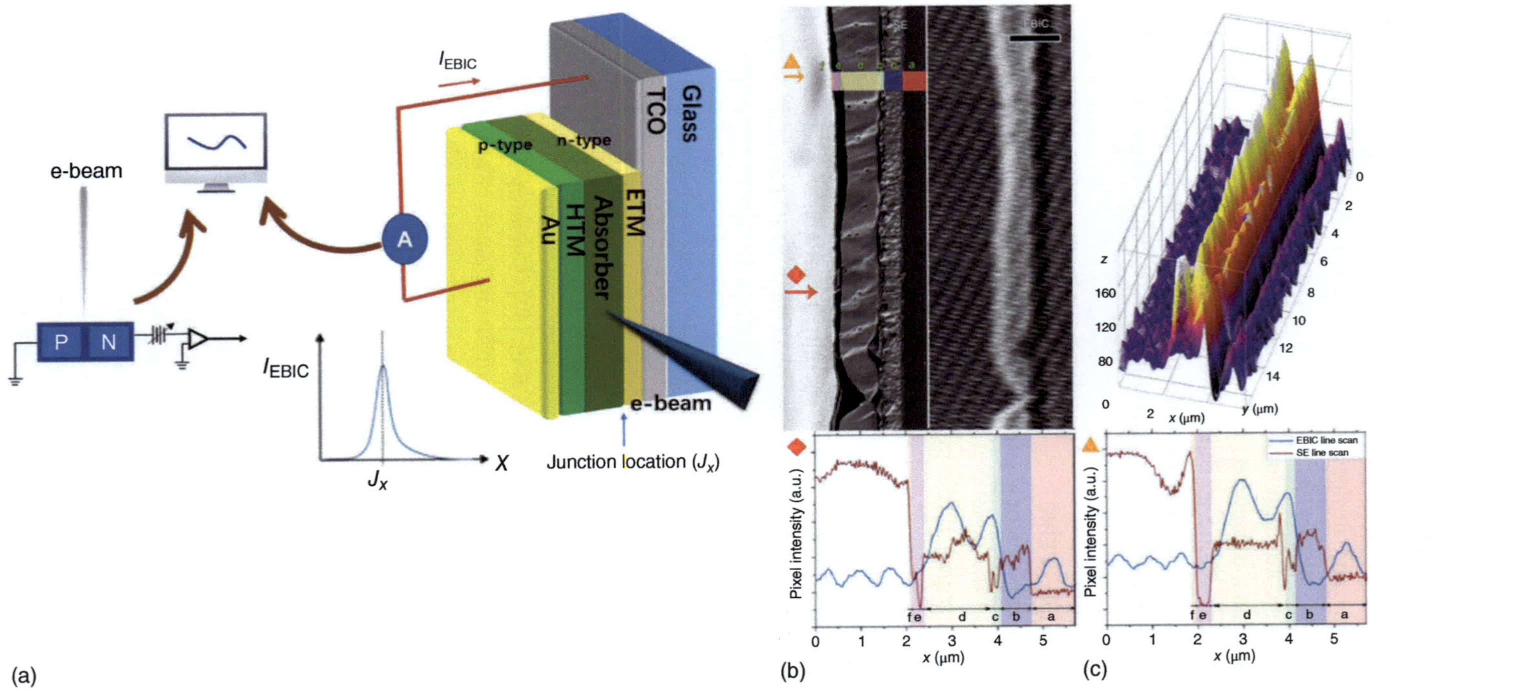

Figure 14.7 EBIC characterization of a flat $CH_3NH_3PbI_{3-x}Cl_x$ device with hole transport material. (a) Schematic diagram of cross-sectional EBIC measurements. (b) Left, SE image; right, EBIC image; scale bar, 2 mm. Letters which correlate with the letters in the line scans (lower panel) designate the layers: a, glass; b, FTO; c, TiO_2; d, $CH_3NH_3PbI_{3-x}Cl_x$; e, HTM; f, Au. (c) 3D surface plot of the EBIC image. Lower panel, line profiles of the SE and EBIC at the indicated regions: red diamond for the lower arrow and orange triangle for the higher arrow position. Overlaying color coding as well as lettering in the graph correlates with the different layers, as shown in (b). Source: Reproduced with permission from Edri et al. [34], Springer Nature.

14.4.2.2.2 Plan-view Planar EBIC is mostly used to image the distribution of nonradiative centers in a semiconductor device. Scanning the electron beam across the solar cell device results in a mapping of the current collected at each scan point. If there are defects such as grain boundaries or dislocations, the recombination at such defects will produce less collection current and darker images on the EBIC plan-view image (corresponding to less collection current collected) [28, 29, 35].

Planar EBIC can reveal the possible origins of current suppression, including low carrier generation, current blocking, and carrier recombination-active sites. Correa-Baena et al. suggest that Rb aggregates are optoelectronically inactive, which leads to current-blocking, nonradiative recombination, and adverse effects on device performance [36]. They showed this by combining EBIC with nanoscale X-ray fluorescence (n-XRF) technology, and were able to correlate weak EBIC signal with Rb clusters [36].

14.4.2.2.3 Electron Beam Induced Damage The degradation caused by electron beam exposure poses a challenge to the study of halide perovskite materials and devices by electron microscopy. Luo et al. determined SEM probe conditions under which dozens of repeated scans on a single region of perovskite solar cells lead to minimum electronic degradation [30]. As shown in Figure 14.8b,c, under higher beam powers, they observed two distinct EBIC responses: at intermediate power, a "quasi-steady" response before decay finally sets in, and finally at "high" power, a direct decay of permanent degradation.

In fact, in the previous work of Edri, it was found that even at a very low energy level (1.5 keV), the electron beam damages the sample over time with increased exposure. Figure 14.8 shows the decrease in EBIC intensity during sequential scanning. After many repeat acquisitions and scannings, the EBIC signal was observed to deteriorate noticeably [34].

One way to avoid damage is to modestly tune modestly the beam voltage up or down, depending upon on the presence of a thicker or thinner hole transport layer (HTL) or back contact, in order to achieve a high-quality EBIC signal while avoiding unnecessary damage. In Luo's work, they recommend starting plan-view EBIC measurement at 7 kV/6.3 pA to avoid rapid beam damage on the samples with typical layer thicknesses of Au (70 nm) and HTL (180 nm) [30].

14.4.3 Electron Backscatter Diffraction

EBSD, when used in an SEM, enables individual grain orientations, local texture, point-to-point orientation correlations, and phase identification and distributions to be determined in polycrystalline materials [8].

As the name implies, EBSD is essentially the analysis of collected backscattered electrons. The EBSD image is based on the Kikuchi diffraction patterns, first observed by Nishikawa and Kikuchi in backscattering mode [9, 13, 37]. EBSD is a very powerful technique for solar cells when used in combination with other techniques such as electron beam induced current (EBIC), atom probe tomography (APT), PL, CL and various scanning probe microscopy (SPM) methods such as

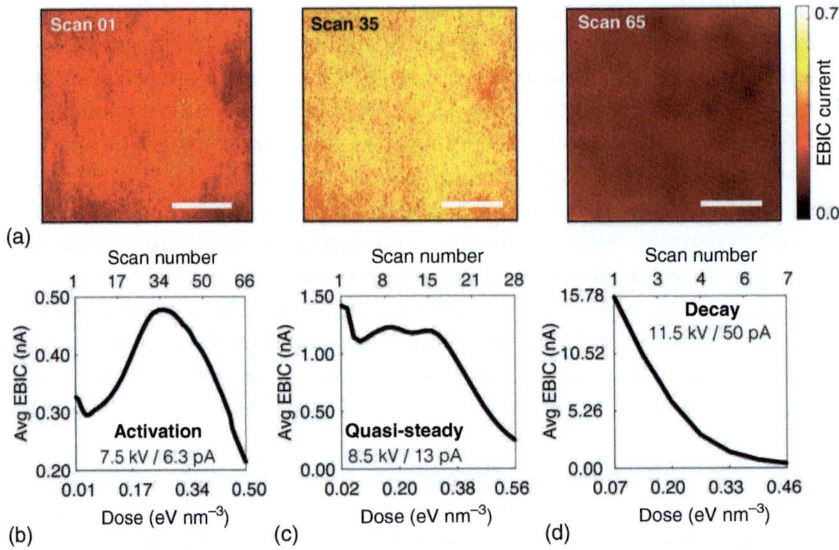

Figure 14.8 (a) Selected scans from a time series of plan-view EBIC maps showing the activation response of a 5% CsI +$(FAPbI_3)_{0.83}(MAPbBr_3)_{0.17}$ device under 7.5 kV/6.3 pA. Scale bar is 10 μm. (b–d) Three characteristic EBIC profiles showing activation, quasi-steady, and decay responses were obtained by plotting the average EBIC signal collected from the EBIC maps as a function of accumulated energy dose in the perovskite. Source: Reproduced with permission from Luo et al. [30], American Chemical Society.

Kelvin probe force (KPFM), scanning spreading resistance (SSRM), or scanning capacitance microscopy (SCM). It makes it possible to reveal structure-property relationships between structural defects, local-orientation distributions and charge carrier separation, chemical defect segregation and nonradiative recombination, respectively [37–40].

14.4.3.1 Differences Between EBSD, XRD, and TEM

TEM and XRD, two common crystallographic techniques, have their own advantages and disadvantages. TEM is a method that has stringent requirements on sample thickness when used to study the atomic structure of thin films and carries the risk of having different sample properties compared to the bulk [41]. Due to the long sample preparation time and the locality of TEM analysis, it is usually only possible to study a small number of locations, thus prohibiting the large-scale acquisition of information about grain size distribution, GB characteristics, or grain orientation effect. The destructive nature of this technique also poses a challenge in directly correlating crystal information with the optoelectronic properties at the other end of the spectrum. X-ray diffraction is usually nondestructive and can easily measure the crystal properties of large areas of thin films, including the average grain size, for example, by peak broadening. However, due to the influence of strain and particle shape, the estimation of grain size becomes complex. In addition, the standard XRD does not provide any local measurements (for example, at a scale of 1 μm or less), which are necessary to understand the grain size distribution and to associate specific grain orientations or special grain boundaries with the resulting optical or electrical properties. Because the local crystal properties require the use of nano-diffraction in a synchrotron, the nanoscale resolution XRD is rarely used to connect local crystal properties with optoelectronic properties [40].

14.4.3.2 Working Principle of EBSD

A direct electron detector or traditional phosphor screen and camera are used in standard EBSD measurements. A polished sample is mounted in the SEM chamber and inclined with approximately 70° tilt relative to normal incidence of the electron beam. At this high tilt angle, incident electrons travel through a longer pathway in the near-surface region, increasing the scattering volume [42]. The traditional detector is a camera equipped with a phosphor screen integrated with a digital frame grabber. A direct detector or phosphor screen is usually placed close to the sample, making it possible to image the BSE escaping from the sample at specific Bragg angles, enabling EBSD to identify local crystal orientation. A diffraction pattern, called a Kikuchi pattern, forms on the screen and is made up of a series of intersecting bands caused by the diffraction of lattice planes called Kikuchi lines (Figure 14.9, right). The Kikuchi lines on the screen are electronically digitized and individually recognized. Just as in TEM electron diffraction, the crossing points of Kikuchi bands correspond to zone axes of the crystal – special high symmetry directions

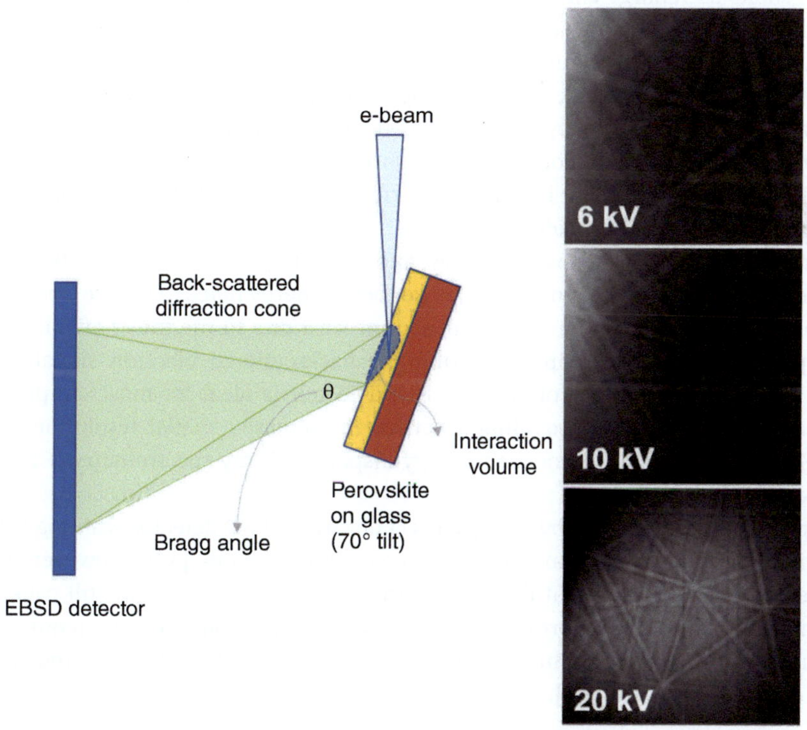

Figure 14.9 Schematic diagram of EBSD and Kikuchi lines taken with EBSD. Source: Reproduced with permission from Jariwala et al. [38], © Elsevier.

[40]. Because the Kikuchi bands correspond to the lattice planes, crystal orientation can be identified by the angles between the bands. Collecting patterns per scan pixel and comparing them with simulated patterns enables indexing. The resulting microstructural map of the sample contains information about phase, strain, crystal orientation, and grain boundaries.

14.4.3.3 EBSD for Perovskites

14.4.3.3.1 Beam Damages and EBSD Detectors EBSD is a useful tool to reveal the grain orientation, texture, and phases of crystalline films [13]. However, this method has been notoriously difficult to apply to the halide perovskites most relevant for photovoltaic applications due to beam-induced damage. Compared to SEM imaging, EBSD requires higher doses (higher current and/or longer integration times) to acquire sufficient contrast in the Kikuchi diffraction lines generated from backscattered electrons. Therefore, the use of traditional EBSD detectors leads to significant beam-induced damage when characterizing halide perovskite solar cell materials, limiting the use of traditional EBSD detectors for characterizing halide perovskite materials [38].

As mentioned above, EBSD currently mainly adopts two kinds of detectors: one is traditional phosphor screen and camera, and the other is a direct electron detector. All commercially available EBSD detectors are based on charge-integrating sensors such as charge-couple device (CCD), complementary metal-oxide-semiconductor (CMOS), phosphorous screen/camera, or a combination thereof [43, 44]. The detector system undergoes two conversion processes from electron-to-photon (phosphorous screen), then back again from photon-to-electron (CCD), along with inevitable efficiency losses (parasitic absorption and scattering) along the way. Those detectors, in combination with fast data acquisition, work well at relatively high doses but show insufficient performance when it comes to imaging sensitive materials such as halide perovskites.

The relationship between beam energy and the successful analysis of beam-sensitive samples is complex. In general, lower beam energy helps to reduce sample damage, but this comes at the cost of speed, not least due to the lower efficiency of the phosphor scintillator in converting the backscattered electron signal to light. An accelerating voltage of between 8 and 15 keV is ideal for most samples, with the lower energy also providing a benefit in increased spatial resolution of the EBSD results (and any associated energy dispersive X-ray spectrometry (EDS) measurements) [45].

This problem can be solved by using a direct electron EBSD detector with higher sensitivity than the traditional phosphor screen and camera [37]. However, it has recently been reported that the use of direct electron detector will still cause electron beam damage. Determining the correct orientation robustly requires high-quality EBSD patterns, which in turn require a high electron dose and induce further damage to the film [46].

14.4.3.3.2 EBSD for Grain Size Adhyaksa et al. used 30 keV accelerating voltage combined with direct detector to obtain high-quality EBSD mapping. The grain size

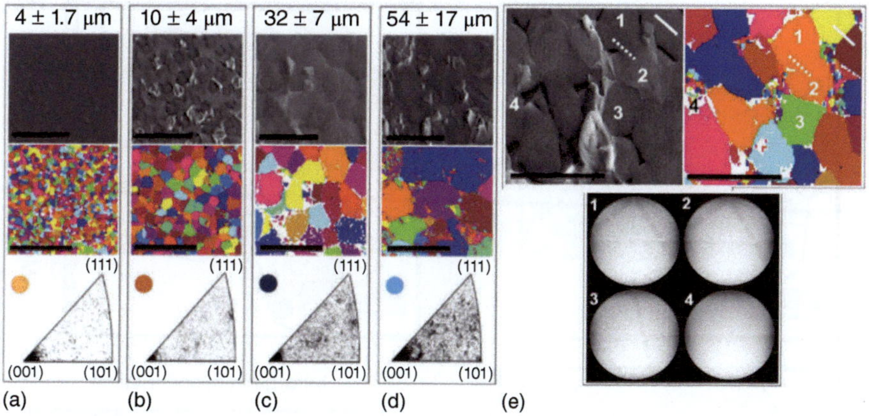

Figure 14.10 Electron backscattered diffraction (EBSD) maps of $CH_3NH_3PbBr_3$ perovskite films with different grain sizes. Collection of scanning electron microscope (SEM) image (top panel) with its corresponding EBSD grain maps (middle panel), and distribution plot of crystallographic orientations (bottom panel) of perovskite films with (a) 4 ± 1.7 μm, (b) 10 ± 4 μm, (c) 32 ± 7 μm, and (d) 54 ± 17 μm grain sizes. The EBSD maps for 32 ± 7 μm grain size show disordered regions near the GB (featured as a white color). The color maps do not represent the actual crystallographic orientation. (e) Deceiving features of grain boundaries showing two contrasting cases where the boundary that is visible under SEM (top left panel; indicated by the white dashed line) is not an actual GB from the EBSD (see the corresponding EBSD maps; top right panel), whereas within the apparent smooth grain under SEM (top left panel; indicated by the white solid line), there is an invisible GB (top right panel; see the different color maps under the EBSD). The bottom panel shows Kikuchi bands obtained by averaging multiple diffraction patterns per pixel within one scanning area (≈200–600 pixels). The scanning areas 1 and 2 show identical Kikuchi bands, and this pattern is different from that of scanning area 3. Scanning area 4 shows no distinct backscattered diffraction. All scale bars are 50 μm. Source: Reproduced with permission from Adhyaksa et al. [37], © John Wiley & Sons.

was directly obtained by EBSD mapping (Figure 14.10a–d). They highlighted specific cases where SEM contrast suggested a GB where there was none and where SEM contrast was smooth even though EBSD showed a clear GB (Figure 14.10e) [37].

14.4.3.3.3 EBSD for Orientation David et al. used a state-of-the-art solid-state EBSD detector and optimized the beam current to be 100 pA at 6 kV accelerating voltage with a pixel integration time of 100 ms, as shown in Figure 14.11. By combining PL with EBSD, they found a negative correlation between grain orientation spread and PL intensity with high statistical significance. In other words, the larger the grain orientation spread, the lower the PL intensity of the local perovskite. This result provides a direct link between local intragrain strain and local nonradiative defect density by showing that grains with higher grain orientation spreads have lower PL intensity and, therefore, higher trap densities contributing to nonradiative recombination [38].

14.4.3.3.4 EBSD for Intragrain Defects Liu et al. studied the crystallographic orientation and ferroic properties of $CH_3NH_3PbI_3$ twin stripes, using electron EBSD and

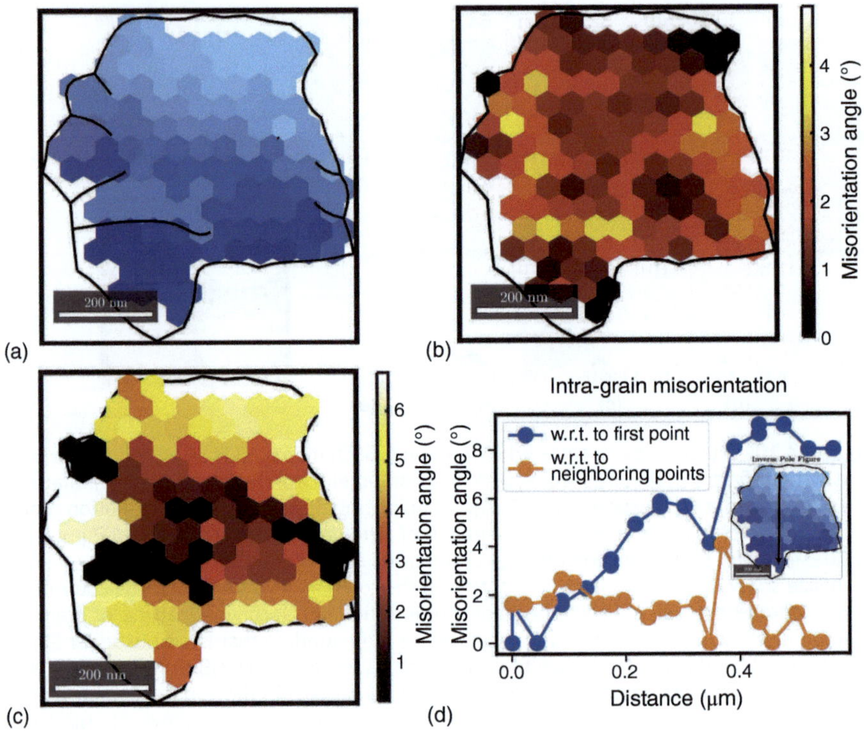

Figure 14.11 Correlating EBSD and PL maps. (a) EBSD map showing coexistence of outer grain boundaries (boundary with other grains), and sub-GB within the individual grain on $CH_3NH_3PbI_3$ thin-film, and (b) its average pixel-to-pixel misorientation demonstrating a local crystallographic heterogeneity in the grain. (c) Confocal PL map overlapped with GB network obtained from EBSD in $CH_3NH_3PbI_3$, and (d) its statistical plot of PL intensity as a function of grain orientation spread showing a negative correlation that indicates local crystallographic heterogeneity may induce a higher nonradiative recombination loss. Source: Reproduced with permission from Jariwala et al. [38], Elsevier.

advanced piezoresponse force microscopy (PFM), as shown in Figure 14.12. Using EBSD, they discovered that the orientation relationship across the twin walls in $CH_3NH_3PbI_3$ is a 90° rotation about $\langle\bar{1}\bar{1}0\rangle$, with the $\langle 030\rangle$ and $\langle 111\rangle$ directions parallel to the direction normal to the surface. This observation of twin stripe structure using EBSD provides essential information regarding crystal lattice for understanding such a twin structure [46].

14.4.3.3.5 EBSD for Crystal Quality Erick et al. used EBSD to explore the relationship between crystal $CsPbBr_3$ quality and image quality (IQ). As shown in Figure 14.13, the higher the IQ_{RT} value, the sharper the Kikuchi line, which represents better crystal quality. In addition, they combined ultrafast pump–probe microscopy and EBSD to directly correlate charge carrier transport with the local diffraction pattern contrast, an indicator of crystal quality. They finally found that even on individual single crystal $CsPbBr_3$ domains, microscopic variability in the crystal quality profoundly impacts the efficiency of charge carrier transport [47].

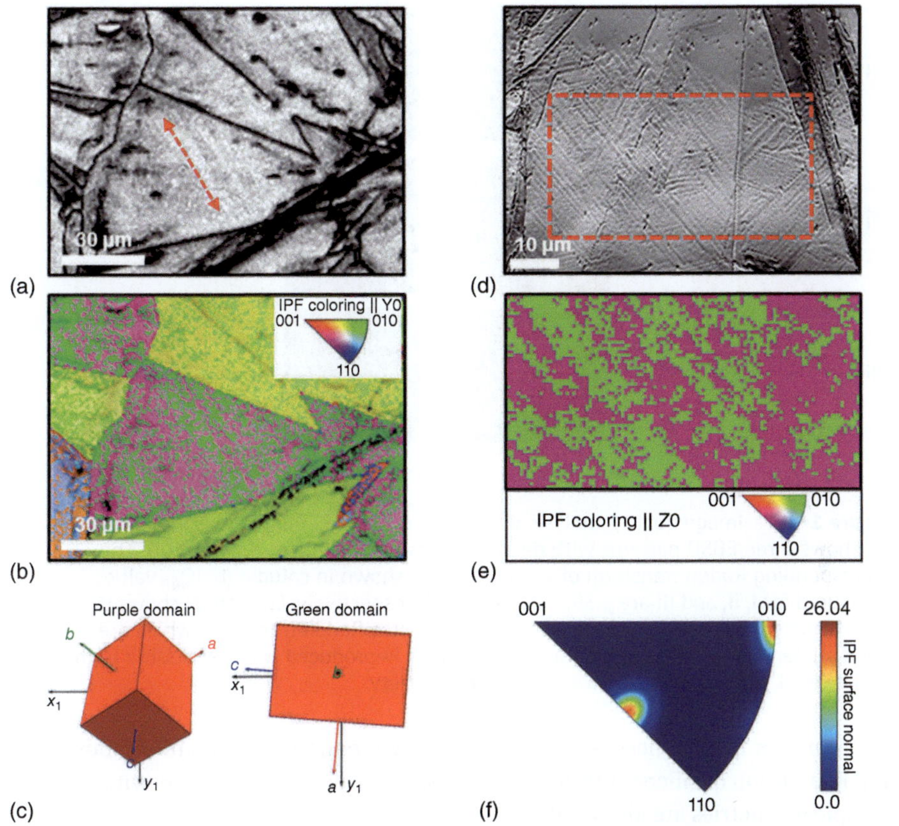

Figure 14.12 (a–c) Initial EBSD results of $CH_3NH_3PbI_3$ thin film using conventional indexing and a measurement step size of 750 nm. (a) EBSD pattern quality map (Kikuchi band contrast) and (b) corresponding EBSD inverse pole figure (IPF)-y orientation map (see color scheme, inset) showing twin domains 90° rotated around the $\langle \bar{1}\,\bar{1}0 \rangle$ orientation; (c) schematic unit cell orientations show the crystallographic orientation of the purple and green domains in (b). (d–f) Higher resolution EBSD results from the $CH_3NH_3PbI_3$ thin film were obtained using pattern matching solution refinement and a measurement step size of 500 nm. (d) Forescatter electron image showing twin-like domain structures within the film. The red box marks the EBSD analysis area. (e) EBSD's IPF-z orientation map clearly shows the domain structures. (f) An inverse pole figure (surface normal direction) showing the clear alignment with the <030> and <111> directions. Source: Reproduced with permission from Liu et al. [46], © 2021, American Chemical Society.

14.4.4 TEM

14.4.4.1 Sample Preparation and Transfer

While the main requirements for an SEM sample are that it is (somewhat) conductive and fits inside the vacuum chamber, the transmissive nature and the geometry of the TEM place some restrictions on which samples can be studied. Most commercial TEM sample holders are designed to hold very thin sample-containing metal grids. These grids are typically around 3 mm in diameter and a few tens of

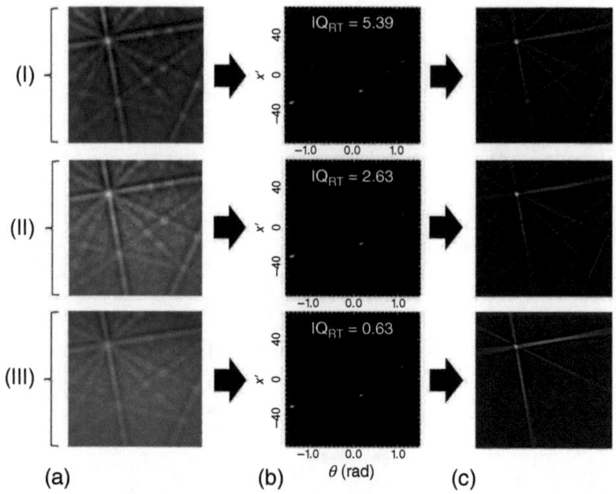

Figure 14.13 Image quality analysis of a single-crystal $CsPbBr_3$ perovskite. Column (a) shows three EBSD patterns with decreasing levels of contrast (I, II, and III). The corresponding Radon transform of each pattern is shown in column (b) IQ_{RT} values measured for I, II, and III are 5.39, 2.23, and 0.63, respectively. Column (c) shows the reconstructed diffraction patterns obtained by inverse Radon transform, which are in good agreement with the original pattern. Source: Reproduced with permission from Hickey et al. [47], © 2020, American Chemical Society.

micrometers in thickness, and covered in a few nanometers of amorphous material like carbon or silicon nitride, although specialized sample holders with different sample geometries are also available.

Since most detectors in a TEM are found below the sample relative to the electron gun, the sample itself must also be thin enough to allow for transmission of at least part of the electron beam. This is typically achieved with samples ranging from a few nanometers to a few hundred nanometers in thickness, with different techniques having different maximum thicknesses. For example, for hybrid perovskite thin films, SAED is readily possible with samples of around 300 nm in thickness, whereas atomic resolution imaging is best done on samples of around 30 nm in thickness. These strict limits on sample thickness make it challenging to produce samples that accurately and precisely reflect the nature of a bulk material, and great care must therefore be taken during sample preparation to avoid preparation-induced artefacts.

Preparing a sample for TEM studies typically involves either reducing a macroscopic sample to an appropriate thickness, or forming a thin enough layer of the material directly on the TEM grid itself. The first approach is typically done using focused ion beam (FIB) milling, scraping a thin film off a substrate, or simply grinding a macroscopic crystal. Each method has advantages and drawbacks, which must be carefully considered when choosing the preparation method. FIB milling uses heavy ions to sputter material off the surface through transfer of momentum from the ions to the surface material, effectively digging into the sample using the ion beam. Ga^+ is typically used, but instruments based on other ions, like helium, are also commercially available. By carefully controlling the beam spread and position, it is possible to cut a very thin lamella of the specimen, down to less than 100 nm in thickness, making it possible to produce TEM-compatible samples from macroscopic specimens. This is excellent from a device perspective, as it enables the study of the internal structure of the device with very high resolution. However, due to the large momentum and charged nature of the gallium ions, the milling process can be highly damaging to the material in the lamella, potentially causing both Ga^{+}-implementation and the introduction of a high density of crystallographic defects. While this effect is often not detrimental to radiation-hardy samples like metals and hard semiconductors, the beam-sensitive nature of hybrid perovskites makes it highly challenging to produce FIB-milled samples that precisely represent the properties of the bulk sample. As such, if one wishes to use FIB milling to prepare a TEM sample, one should always attempt to quantify the damage caused by the preparation method, for example via a SAED pattern, and the damage's potential impact on the information obtained from the sample.

Scraping a film off a substrate has the advantage of directly using a film that could be used in a solar cell device, allowing for a range of device-relevant chemistries to be studied. However, mechanically removing a film is likely to damage the film, alter its morphology, and potentially introduce defects at a crystallographic level. Additionally, relative spatial information, such as relative grain orientation or GB makeup, can be lost in the process since the film has to be re-deposited onto a TEM grid in a more or less damaged state. It is also possible that microscopic pieces of the substrate can be scraped off along with the film, adding potential artifacts that need to be accounted for. As such, the use of this method has been limited.

Crushing a macroscopic crystal is a classic method of preparing TEM samples and can work well for hybrid perovskite single crystals. Using a mortar and pestle (typically agate equipment a few centimeters across), one can reduce a macroscopic single crystal to a fine powder, resulting in particles down to a few hundred nanometers. Typically, a non-dissolving liquid is added to the mortar to prevent the powder from escaping and to ease the transfer of the particles to a TEM grid. It is best to gently press the pestle downwards repeatedly with as little shear motion as possible, resulting in better cleavage of the crystals with a minimum of defects introduced. It can be challenging to get the crystal particles small enough for full electron transparency, but sharp edges can be thin enough for good electron transparency, especially for electron diffraction. The dispersion can then be transferred to the grid with a pipette by locking the grid in self-closing TEM tweezers and placing a single drop onto the grid. The excess liquid can be removed by placing an absorbent wipe between the tweezers and using capillary forces to suck the solvent away. Placing the grid on a hot plate or in a vacuum chamber can help remove the last bits of solvent, an important step since solvent has a tendency to evaporate and explode under an electron beam.

Depositing perovskite thin films directly onto TEM grids is a useful method for studying many of the properties found in native films, such as grain boundaries, relative grain angles, extended defects, and secondary phases. It thus allows for a direct correlation to the properties found in devices, since similar preparation methods can be used, and the films on the grids can be exposed to the same conditions as devices *ex-situ*. One disadvantage, however, is that the films can only be studied top-down, and transverse information relevant to charge carrier properties within a solar cell device can be hidden or difficult to observe. An oblique GB, for example, would appear as two crystals superimposed on top of each other, and the geometry of the interface would not be easily discernible. Despite this, the versatility of the preparation method has led to several important breakthroughs in correlating microscopic and macroscopic properties in halide perovskites. Since it is theoretically possible to use practically any deposition method to cover a TEM grid (with some smart engineering), film-coated TEM samples are a highly useful tool and likely to become a standard characterization method for novel material chemistries and deposition methods.

TEM samples have been prepared using both solution processing and thermal evaporation. Solution-processed films have the advantage of corresponding to the photoactive films found in the (currently) highest efficiency single junction solar cells, but maintaining a complete film with thicknesses below a few hundred nanometers has proven to be difficult. As such, solution-processed TEM samples have mostly been used for lower magnification and diffraction studies. Devices made with thermally evaporated photoactive layers tend to have lower efficiencies but offer much better control of the film thickness, making this preparation method ideal for making TEM samples. Films of around 30 nm have been reported, allowing for excellent electron transparency and resulting in the ability to obtain images of native perovskite thin films with atomic resolution.

Due to their environmental instability, perovskite films are often produced in specific environments, such as nitrogen-filled glove boxes or low-humidity dry boxes. Since TEM samples cannot be encapsulated before being placed within the microscope itself, it is important to minimize environmental exposure after preparation of the sample and during transportation to the microscope. Several methods are possible, ranging from basic to highly technical. The simplest transportation method is simply to enclose the TEM samples in several zip-lock bags while they are in an inert environment. When done carefully and sealed tightly without any solvent in the atmosphere, this can provide excellent inert transportation for short to medium distances. Similarly, enclosing the samples in plastic film and vacuum sealing the film or using rubber-sealed vacuum transfer containers can also prevent exposure to the atmosphere. Commercial vacuum transfer TEM sample holders also exist, allowing samples to be put under vacuum before being placed in the microscope, enabling exposure-free transport from a glove box to the microscope.

14.4.4.2 Imaging Conditions

Since the electromagnetic lenses used in a TEM are not static, unlike glass lenses in a visible light microscope, a certain amount of calibration, or alignment, is often required at the start of each session, which can involve highly concentrated electron irradiation of some parts of the sample. Since each microscope design is different and requires slightly different alignments, we will not go into details on how to align a microscope, but rather focus on the impact the alignment can have on perovskite samples. The radiation associated with alignments can be highly detrimental to beam-sensitive materials and can change the crystallographic nature of a sample. Hybrid perovskites are particularly vulnerable to beam damage, as discussed later, and electron irradiation causes them to degrade rapidly, losing the organic compounds and forming stable lead halide crystals. These fully inorganic crystals are much more beam stable than the perovskite and require a much higher dose before damaging under the electron beam. Additionally, they are crystallographically somewhat similar to perovskites, and distinguishing pristine and damaged perovskites from each other has proven to be difficult. From an operator's perspective, it is thus imperative that different parts of the sample be used for alignment and imaging. The electron-induced beam damage may result in material loss in the specimen, causing it to become thinner and effectively changing the focal height of the sample. This is particularly prominent in very thin (around 30 nm) samples and can make aligning precisely challenging.

A TEM typically operates either in a broad-beam mode (TEM) or a focused, scanning beam mode (STEM or scanning TEM), with each mode having complementary capabilities. In broad-beam TEM, the electron beam is adjusted to be spread out onto the sample, constantly illuminating a given area. An electron detector (usually CCD, CMOS, or single- electron detector) records the spatial intensity of the transmitted electrons and forms a contrast image (or micrograph) based on the entire illuminated area that fits within the detector's field of view. The contrast in the electron's image can be caused by several factors, including differences in thickness, composition, and crystallographic orientation throughout the specimen, which can make it difficult to interpret a single micrograph unambiguously.

Broad beam mode also enables the use of SAED, a very useful technique for obtaining crystallographic information about a desired area of a specimen. Since the electrons behave as waves when they pass through a specimen, they form diffraction patterns upon exit, which can be recorded for crystallographic analysis. These patterns contain information about the plane spacings and angles in the specimen and can be used to evaluate the local crystallography with a resolution of a few hundred nanometers. When placing an aperture in the image plane below the sample, only electrons passing through the aperture are detected. If the aperture is smaller than a single grain, it is possible to obtain single-grain diffraction patterns with very low electron dose.

In scanning mode, a focused electron beam is scanned across the sample and signals are recorded using a raster method, similarly to the operation of an SEM. Since the electron beam can be focused to a width less than an atom's, it is possible to use it to record a wide range of spatially localized information with atomic resolution. Two commonly used STEM techniques are low/high-angle annular dark field (LAADF/HAADF) imaging and scanning electron dispersive X-ray spectroscopy (scanning EDS). LAADF/HAADF imaging is done using a ring detector (hence the *annular* in the name) placed below the specimen, which records electrons that have been scattered by the sample. Since there is a direct relationship between the average atomic number of an atomic column and the angle electrons scatter at when they interact with the column, an annular dark field (ADF) image contains quantifiable information about the relative atomic masses found in an image. When calibrated properly, ADF can even be directly quantitative, and can, for example, tell the number of atoms in each column of a nanoparticle, which can be used to reconstruct a precise model of the particle.

14.4.4.3 Beam Damage

Electron microscope images are formed as a result of the interaction between the incoming accelerated electrons and the atoms contained in the material they impinge on. This interaction can happen either elastically, where the incoming electrons scatter as a result of interaction with the atomic nucleus, or inelastically, where the incoming electrons interact with the electron cloud surrounding the nucleus. Both of these interactions can result in damage to the specimen in a range of ways. Figure 14.14a–d shows various types of beam damage. These are (a) electron charging in SEM, where electrons are not extracted from the specimen, creating a local electric field, deflecting incoming electrons; (b) GB damage in SEM, seen as widening, caused by electron sputtering and radiolysis, damaging the structure; (c) damage to the crystal structure after a single scan of atomic resolution STEM imaging, visible to the left, with a near-pristine structure to the right; and (d) GB splitting observed in STEM.

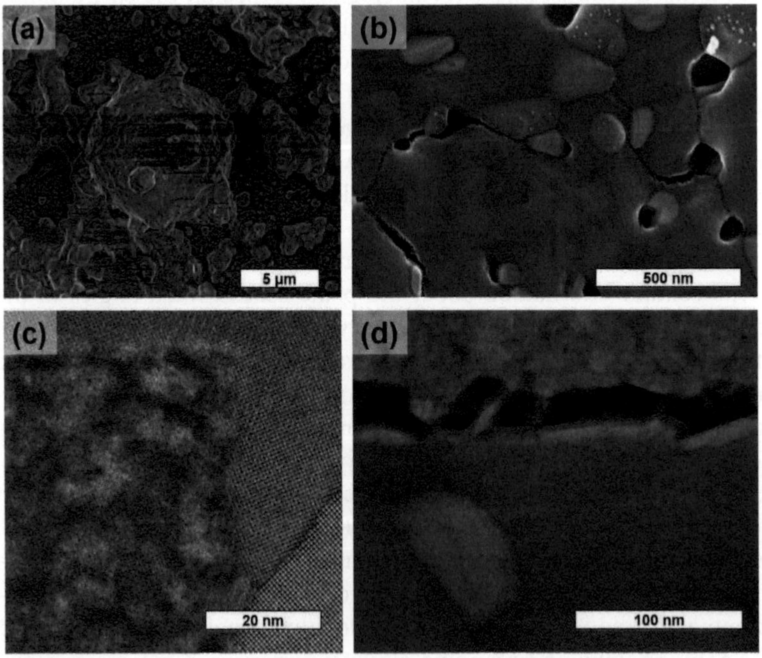

Figure 14.14 Different types of beam-induced damage. (a) Charging is seen as dark stripes on top of a $MAPbI_3$ particle in an SEM. (b) Grain boundaries are seen opening up in a $FAPbI_3$ thin film in an SEM. (c) A single scan has caused serious damage to the perovskite crystal, seen on the left, while the unscanned area remains somewhat pristine in a TEM. (d) A GB opening up due to electron exposure above a pristine crystal in a TEM. Source: Mathias Uller Rothmann (Chapter Author).

Typical methods to lower the impact of beam damage on inorganic materials involve using lower doses of electrons with lower accelerating voltages at lower temperatures. Lowering the electron dose is perhaps the safest way to avoid beam damage, but this also limits the signals generated and can impact the resolution.

In the TEM, the dose can be lowered by lowering the emission current, using a smaller condenser aperture, or spreading the beam. This needs to be balanced by the exposure time of the detector used, since a better signal can be recorded with longer exposure time, but this also increases the total dose. It is very important to remember that the total dose is not simply the dose experienced by the sample during recording, but includes every bit of exposure the sample has been exposed to. As such, it is essential to always image a "fresh" area with no previous exposure for most reliable information. This can be done by aligning the beam in one area of the sample, blanking the beam, and moving to a new area "in the dark," and recording the image as soon as the beam turns on.

In scanning electron microscopes, it is possible to control the resolution and dwell time of the scan, which further determine the total dose along with the beam current. From a microscope perspective, the resolution of a micrograph corresponds to the scan resolution of the coordinates where the beam is parked, and the dwell time is how long the beam rests in each coordinate. As such, increasing the scan resolution and the dwell time can lead to higher spatial resolution images, but also increases the total dose experienced by the sample.

Lowering the accelerating voltage reduces the kinetic energy of the impinging electrons, which can reduce the energy deposited in the sample and thus the beam damage. In organic materials; however, the lower velocity of the electrons can increase the likelihood that electrons interact with the covalent bonds and can thus increase the beam damage. Since many photoactive perovskites contain both ionic and covalent bonds, it is still not clear exactly which voltage is optimal for their study, and it is likely that each material composition has its own optimal voltage, but most (S)TEM studies use 200–300 kV.

SEM is typically used to study bulk specimens, which absorb a large proportion of the electron energy, unlike transmission microscopes, where most electrons pass through the sample. As such, the electrons are more likely to deposit a greater amount of energy in the sample, leading to more damage at lower voltages, meaning that more beam damage can take place, despite the lower voltage. For simple morphological imaging of surface and cross-section specimens, 5 kV accelerating voltage is typically a good balance between resolution, signal intensity, and beam damage, but a higher voltage can be used for imaging fresh areas.

Lowering the temperature of the sample can reduce some of the beam-induced damage caused by heating but may also induce phase transitions in the material studied. Most cooling stages cool the sample to liquid nitrogen temperature (around 60 K), well below the phase transition temperature of many photoactive perovskites. As such, it is likely that the materials will have different properties than those observed at the solar cell operating temperature, and the results may not be representative for real-world applications. Despite this, cooling the sample may be able to provide some protection against beam damage.

14.4.4.4 Examples of Applications of TEM

Below are four examples of studies of photoactive perovskites that have been performed with the use of (scanning) TEMs as a key aspect. Three of the studies have been chosen because they focus on low-dose imaging and diffraction studies of films deposited directly onto TEM grids, as these imaging conditions and sample preparation methods are most likely to introduce artifacts from the sample preparation and imaging. The fourth study shows how high-dose techniques, like FIB preparation and elemental EDX mapping, can be used to study some properties of perovskite solar cell devices despite the introduction of beam damage. Each study represents a different approach to investigating hybrid perovskites in thin films using (S)TEM: atomic resolution imaging, low-dose scanning diffraction, low-dose broad beam diffraction and imaging, and FIB preparation combined with EDX mapping.

14.4.4.4.1 Atomic Resolution Imaging of $FAPbI_3$ and $MAPbI_3$: The ability to resolve individual columns of atoms in a thin sample is one of the most powerful properties of a scanning transmission electron microscope. Using an ADF detector makes it possible to obtain quantitative information about the relative average atomic mass of the sample. This is very useful when studying perovskites of all types, as the crystal structure results in characteristic image intensity patterns, like those shown in Figure 14.15, making it possible to identify accurately the local atomic makeup of the crystal. This can be used to resolve the structure of, for example, grain boundaries, extended crystal defects, interfaces, and secondary material phase inclusions.

Rothmann et al. combined a few of the key aspects needed to study hybrid perovskites reliably and were able to produce the first images of a hybrid perovskite thin film with atomic resolution [48]. The film was prepared via thermal coevaporation and controlled to be around 30 nm in thickness. Using an excess of FAI relative to PbI_2 in the evaporation process allowed for the formation of large grains with few defects and few inclusions of impurities like remnant PbI_2. In order to image the film, the authors used a LAADF setup in an aberration corrected JEOL ARM-200 analytical STEM running at 200 kV acceleration voltage at room temperature. By lowering the emission current and using an annular detector with an appropriate camera length (8 cm camera length with inner and outer collection angles of 33 and 120.77 mrad, respectively), they were able to reduce the total impinging dose per image to around 66 e $Å^{-2}$, while maintaining atomic resolution. Due to the low electron dose, the signal-to-noise ratio was not ideal, so the images were processed using a linear Butterworth filter in the Digital Micrograph software. This filter removes high-frequency information in the image, which is often noise, and leaves the image with an optimized contrast.

14.4.4.4.2 Correlating the Composition of $MA_{1-x}FA_xPbI_3$ with Structure, Defects, and Device Properties for a Range of x Li et al. used SAED to describe the local crystallography of individual grains in $MA_{1-x}FA_xPbI_3$ thin films, where x and $1-x$ describe the relative amounts of FA^+ and MA^+ in the films, respectively, shown in

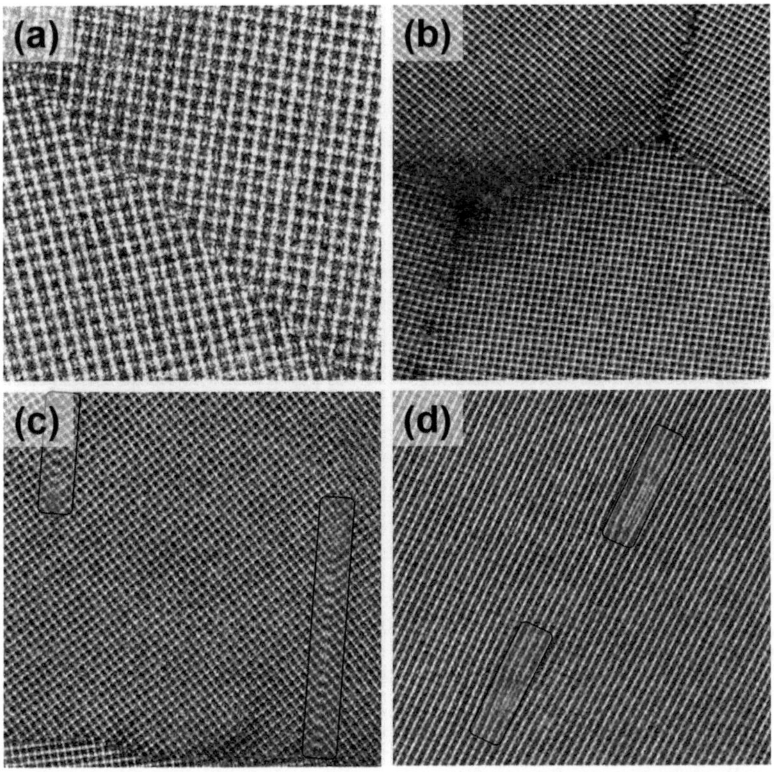

Figure 14.15 Overview of the microstructure of $FAPbI_3$ thin films recorded with atomic resolution. (a) Straight grain boundaries. (b) Two triple junctions. (c) Stacking faults are highlighted in orange. (d) Edge dislocations are highlighted in green. Source: Reproduced with permission from Rothmann et al. [48], American Association for the Advancement of Science - AAAS.

Figure 14.16 [49]. Changing the ratio of FA^+ to MA^+ changes the "average" crystal structure of the perovskite material since the cations have different ionic radii. $MAPbI_3$ is tetragonal at room temperature, but when the A-site cations are made up of 80% MA^+ and 20% FA^+, the structure changes to an overall "average" cubic structure (the term "average" is used since the structure cannot be cubic locally, but the different cations allow for the flexible perovskite structure to accommodate each other better, forming a more relaxed structure that is cubic on average). Increasing the amount of FA^+ induces the presence of crystallographic twinning within the film, and the density of twins increases with the amount of FA^+ until it forms a high density of stacking faults.

These crystallographic changes were observed using low-dose SAED, where only electrons passing through a single-crystal grain are recorded. Samples of the various compositions were prepared from solutions by spinning coating the perovskite thin film directly onto TEM grids while blowing nitrogen gas on the grid, the so-called gas-assisted method. Since the samples were solution-processed, it was possible to prepare solar cells with the same solutions, allowing for a direct

Figure 14.16 Micrographs and diffraction patterns showing the change in microstructure of $MA_{1-x}FA_xPbI_3$ thin films as x changes from (a,d) 0, with crystallographic twinning, (b,e) 0.2, with no extended defects, and (c,f) 1, with a high density of stacking faults. Source: Reproduced with permission from Li et al. [49], Springer Nature.

comparison between the microscopic crystallographic properties of the thin films and the overall performance of perovskite solar cells made with these films.

The authors found that the best-performing solar cells were the ones with the highest symmetry, i.e. the cubic perovskite, with no extended crystal defects, which corresponded to what other studies had found. The use of SAED, however, made it possible to correlate the changes in performance parameters with the local crystallography and the presence of defects, which TEM is uniquely able to do.

14.4.4.4.3 Correlating Charge Carrier Traps with Crystallography Using Scanning Electron Diffraction and Photoemission Electron Microscopy Charge carrier traps can be detrimental to the overall performance of solar cell devices, and understanding how the local crystallography correlates with the presence of traps is key to minimizing any trap-associated performance loss. In this study, Doherty et al. combined scanning electron diffraction with photoemission electron microscopy (PEEM) to correlate the local crystallographic properties of a thin film with the local density of traps, as shown in Figure 14.17 [50]. A full description of PEEM is beyond the scope of this text, but, briefly, it relies on recording the electrons emitted from a surface when a sample is irradiated with UV photons. By illuminating with the right photon energy, electrons can be excited from trap states within the bandgap without the excitation of valence band electrons, and mapping these emitted electrons gives a spatial map of trap states. The authors then used a parallel electron beam of around 4 nm in width, also referred to as a pencil beam, to scan across the same sample in scanning electron diffraction (SED) setup. They had placed unique gold particles

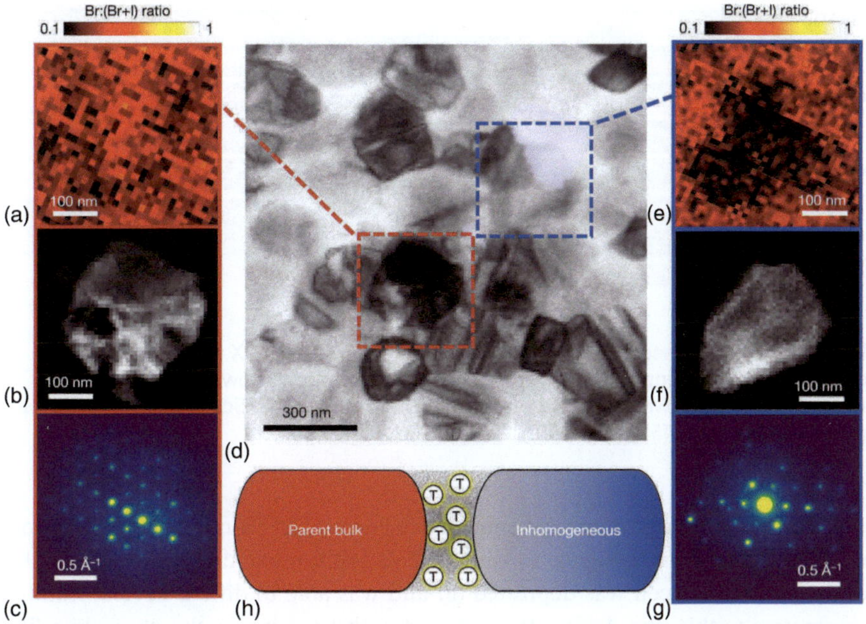

Figure 14.17 Correlating information obtained with SED and PEEM. (d) shows a virtual bright-field electron image overlaid with the PEEM-derived trap clusters, seen in light blue in the top right blue box. (a, e) show differences in bromine and iodine content in different grains. (b, c) show a pristine grain that can be indexed with a cubic perovskite phase, and (f, g) show a grain that cannot be indexed with such a phase. (h) Schematic of how traps are found between cubic and inhomogeneous regions. Source: Reproduced with permission from Doherty et al. [50], Springer Nature.

on the sample, allowing them to correlate the PEEM signals with the SED data. SED enables the recording of diffraction patterns with very high spatial resolution, around the 4 nm width of the beam, and by scanning areas with high and low trap densities, the authors were able to correlate certain interfaces and the presence of a non-cubic phase to the presence of charge carrier traps. This is a great example of how the high resolution and crystallographically interactive nature of electron microscopy can be used to explain important phenomena in perovskite solar cells. For this study, the authors used a JEOL ARM300CF microscope running at 300 kV set up in the nanobeam alignment of a convergence angle of ~1 mrad, electron probe width of ~4 nm, probe current of ~2 pA, scan dwell time of ~1 ms, and a camera length of 15. This resulted in an electron dose of ~6.25 e$Å^2$.

14.4.4.4.4 Observing the Diffusion of Sodium from Glass Substrates into the Perovskite Layer While we have been careful to highlight the potential issues related to using high-dose techniques, like FIB sample preparation and EDX mapping, there are some examples of perovskite solar cell studies using these exact techniques to reliably describe important phenomena. In this study, Kosasih et al. used STEM-EDX, alongside other spectroscopy techniques, to observe the presence of sodium within

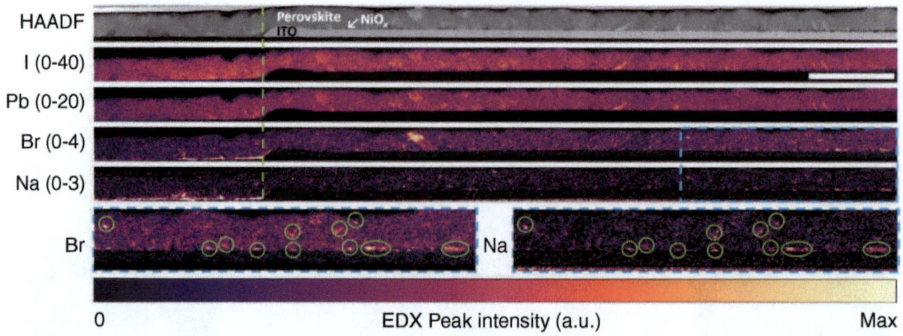

Figure 14.18 Cross-sectional STEM-HAADF image and STEM-EDX peak intensity maps showing the distribution of various elements. The bottom row shows the presence of bromine and sodium, which correlate within the perovskite layer. Scale bar represents 2 μm. Source: Kosasih et al. [51]/John Wiley & Sons.

both the charge transport and the photoactive layers of a FIB-milled perovskite solar cell device, as shown in Figure 14.18 [51]. This sodium had diffused from the glass substrate the solar cell was deposited on and was able to diffuse several hundred nanometers within the device. It combined with bromine in the perovskite film, forming bromine-poor, iodine-rich perovskite, which helps passivate defects and reduces non-radiative recombination. While the effects are beneficial to solar cell performance, the authors caution that the longer-term effects are unknown and can cause unabated sodium diffusion under operating conditions.

In this particular case, having a pristine perovskite crystal is less important since sodium is not found within the perovskite structure and any presence must have come from diffusion from the glass substrate. While it is possible that the FIB preparation can have caused some redeposition of sodium, the spatial distribution of the sodium suggests that this is caused by diffusion and not induced by beam damage. As such, this study is an excellent example of the use of high-dose techniques to describe important phenomena in perovskite solar cell devices with high spatial resolution.

14.5 Conclusions

Electron microscopy is a very powerful family of techniques that can provide immense information about photoactive perovskites with appropriate care. In this chapter, we have gone through the basics of scanning and transmission electron microscopy and their associated techniques, how best to optimize the microscope settings for use on perovskite solar cell materials, what to look out for to avoid beam damage, and given some examples of studies that show the versatility and power of electron microscopes in general. There is clearly much more to learn about perovskite solar cells, and electron microscopes are likely to remain key techniques in their characterization and further development.

Acknowledgments

This work has been financially supported by National Natural Science Foundation of China (NSFC 52202219, 52372222 and 91963209), Nature Science Foundation of Hubei province (2023AFD034) and the Fundamental Research Funds for the Central Universities (WUT: 2019IVB055 and 2019IVA066).

The authors gratefully acknowledge financial support from the Engineering and Physical Sciences Research Council (UK) (EPSRC) through grant number EP/P033229/1 and through the EPSRC CDT for New and Sustainable Photovoltaics.

References

1 Goldstein, J.I., Newbury, D.E., Echlin, P. et al. (2017). *Scanning Electron Microscopy and X-ray Microanalysis*. Springer.

2 Vernon-Parry, K. (2000). Scanning electron microscopy: an introduction. *III-Vs Review* 13: 40–44.

3 Reimer, L. (2013). *Scanning Electron Microscopy: Physics of Image Formation and Microanalysis*, vol. 45. Springer.

4 Williams, D.B. and Carter, C.B. (1996). *Transmission Electron Microscopy*, 3–17. Springer.

5 Pennycook, S.J., Lupini, A.R., Varela, M. et al. (2006). Chapter 6: Scanning transmission electron microscopy for nanostructure characterization, pp. 152–188. In: *Scanning Microscopy for Nanotechnology* (ed. W. Zhou, R. Apkarian, Z.L. Wang, and D. Joy), 1–40. Springer.

6 Rothmann, M.U., Li, W., Etheridge, J., and Cheng, Y.B. (2017). Microstructural characterisations of perovskite solar cells–from grains to interfaces: techniques, features, and challenges. *Advanced Energy Materials* 7.

7 Prior, D.J., Boyle, A.B., Cheadle, F. et al. (1999). The application of electron backscatter diffraction and orientation contrast imaging in the SEM to textural problems in rocks. *American Mineralogist* 84: 1741–1759.

8 Schwarzer, R.A., Field, D.P., Adams, B.L. et al. (2009). Present state of electron backscatter diffraction and prospective developments. In: *Electron Backscatter Diffraction in Materials Science* (ed. A.J. Schwartz, M. Kumar, B.L. Adams, and D.P. Field), 1–20. Boston, MA: Springer US.

9 El Dasher, B. and Deal, A. (2009). Application of electron backscatter diffraction to phase identification. In: *Electron Backscatter Diffraction in Materials Science* (ed. A.J. Schwartz, M. Kumar, B.L. Adams, and D.P. Field), 81–95. Boston, MA: Springer US.

10 Yacobi, B.G. and Holt, D.B. (2013). *Cathodoluminescence Microscopy of Inorganic Solids*. Springer Science & Business Media.

11 Coenen, T. and Haegel, N.M. (2017). Cathodoluminescence for the 21st century: Learning more from light. *Applied Physics Reviews* 4: 031103.

12 Guthrey, H. and Moseley, J. (2020). A review and perspective on cathodoluminescence analysis of halide perovskites. *Adv. Energy Mater.* 10: 1903840.

13 Ran, J., Dyck, O., Wang, X. et al. (2020). Electron-beam-related studies of halide perovskites: challenges and opportunities. *Advanced Energy Materials* 10: 1903191.

14 Li, W., Rothmann, M.U., Liu, A. et al. (2017). Phase segregation enhanced ion movement in efficient inorganic $CsPbIBr_2$ solar cells. *Advanced Energy Materials* 7: 1700946.

15 Bischak, C.G., Sanehira, E.M., Precht, J.T. et al. (2015). Heterogeneous charge carrier dynamics in organic–inorganic hybrid materials: nanoscale lateral and depth-dependent variation of recombination rates in methylammonium lead halide perovskite thin films. *Nano Lett.* 15: 4799–4807.

16 Hoke, E.T., Slotcavage, D.J., Dohner, E.R. et al. (2015). Reversible photo-induced trap formation in mixed-halide hybrid perovskites for photovoltaics. *Chem Sci.* 6: 613–617.

17 Caicedo-Dávila, S., Funk, H., Lovrinčić, R. et al. (2019). Spatial phase distributions in solution-based and evaporated Cs–Pb–Br thin films. *The Journal of Physical Chemistry C* 123: 17666–17677.

18 Milosavljević, A.R., Huang, W., Sadhu, S., and Ptasinska, S. (2016). Low-energy electron-induced transformations in organolead halide perovskite. *Angew Chem Int Ed Engl* 55: 10083–10087.

19 Bischak, C.G., Hetherington, C.L., Wang, Z. et al. (2015). Cathodoluminescence-activated nanoimaging: noninvasive near-field optical microscopy in an electron microscope. *Nano Letters* 15: 3383–3390.

20 Rothmann, M.U., Li, W., Etheridge, J., and Cheng, Y.-B. (2017). Microstructural characterisations of perovskite solar cells – from grains to interfaces: techniques, features, and challenges. *Adv. Energy Mater.* 7: 1700912.

21 Xiao, C., Li, Z., Guthrey, H. et al. (2015). *J. Phys. Chem. C* 119: 26904–26911.

22 Cai, Z., Wu, Y., and Chen, S. (2021). Energy-dependent knock-on damage of organic–inorganic hybrid perovskites under electron beam irradiation: first-principles insights. *Appl. Phys. Lett.* 119: 123901.

23 Merano, M., Collin, S., Renucci, P. et al. (2005). High brightness picosecond electron gun. *Review of Scientific Instruments* 76: 085108.

24 VandenBussche, E.J. and Flannigan, D.J. (2019). Reducing radiation damage in soft matter with femtosecond-timed single-electron packets. *Nano Letters* 19: 6687–6694.

25 VandenBussche, E.J., Clark, C.P., Holmes, R.J., and Flannigan, D.J. (2020). Mitigating damage to hybrid perovskites using pulsed-beam TEM. *ACS Omega* 5: 31867–31871.

26 Ferrer Orri, J., Tennyson, E.M., Kusch, G. et al. (2021). Using pulsed mode scanning electron microscopy for cathodoluminescence studies on hybrid perovskite films. *Nano Express* 2: 024002.

27 Abou-Ras, D. and Kirchartz, T. (2019). Electron-beam-induced current measurements of thin-film solar cells. *ACS Applied Energy Materials* 2: 6127–6139.

28 Hanoka, J.I. (1980). Electron-beam-induced current characterization of polycrystalline silicon solar cells. *Solar Cells* 1: 123–139.

29 Hanoka, J.I. and Bell, R.O. (1981). Electron-beam-induced currents in semiconductors. *Annual Review of Materials Science* 11: 353–380.

30 Luo, Y., Parikh, P., Brenner, T.M. et al. (2020). Quantitative specifications to avoid degradation during E-beam and induced current microscopy of halide perovskite devices. *The Journal of Physical Chemistry C* 124: 18961–18967.

31 Kedem, N., Brenner, T.M., Kulbak, M. et al. (2015). Light-induced increase of electron diffusion length in a p-n junction type $CH_3NH_3PbBr_3$ perovskite solar cell. *J Phys Chem Lett* 6: 2469–2476.

32 Salviati, G., Lazzarini, L., Sekiguch, T., and Chen, B. (2011). *Comprehensive Semiconductor Science and Technology* (ed. P. Bhattacharya, R. Fornari, and H. Kamimura), 308–356. Amsterdam: Elsevier.

33 Park, B.-W., Kedem, N., Kulbak, M. et al. (2018). Understanding how excess lead iodide precursor improves halide perovskite solar cell performance. *Nature Communications* 9: 3301.

34 Edri, E., Kirmayer, S., Mukhopadhyay, S. et al. (2014). Elucidating the charge carrier separation and working mechanism of $CH_3NH_3PbI_{3-x}Cl_x$ perovskite solar cells. *Nature Communications* 5: 3461.

35 Chen, J., Sekiguchi, T., Yang, D. et al. (2004). Electron-beam-induced current study of grain boundaries in multicrystalline silicon. *Journal of Applied Physics* 96: 5490–5495.

36 Correa-Baena, J.-P., Luo, Y., Brenner, T.M. et al. (2019). Homogenized halides and alkali cation segregation in alloyed organic-inorganic perovskites. *Science* 363: 627–631.

37 Adhyaksa, G.W.P., Brittman, S., Āboliņš, H. et al. (2018). Understanding detrimental and beneficial grain boundary effects in halide perovskites. *Advanced Materials* 30: 1804792.

38 Jariwala, S., Sun, H., Adhyaksa, G.W.P. et al. (2019). Local crystal misorientation influences non-radiative recombination in halide perovskites. *Joule* 3: 3048–3060.

39 Abou-Ras, D., Kavalakkatt, J., Nichterwitz, M. et al. (2013). Electron backscatter diffraction: an important tool for analyses of structure–property relationships in thin-film solar cells. *JOM* 65: 1222–1228.

40 Sun, H., Adhyaksa, G.W.P., and Garnett, E.C. (2020). The application of electron backscatter diffraction on halide perovskite materials. *Advanced Energy Materials* 10: 2000364.

41 Rothmann, M.U., Li, W., Zhu, Y. et al. (2018). Structural and chemical changes to $CH_3NH_3PbI_3$ induced by electron and gallium ion beams. *Adv. Mater.* 30: 1800629.

42 Lindl, J. (1995). Development of the indirect-drive approach to inertial confinement fusion and the target physics basis for ignition and gain. *Physics of Plasmas* 2: 3933–4024.

43 Vespucci, S., Winkelmann, A., Naresh-Kumar, G. et al. (2015). Digital direct electron imaging of energy-filtered electron backscatter diffraction patterns. *Physical Review B* 92: 205301.

44 Roberts, A., Svihra, P., Al-Refaie, A. et al. (2019). First demonstration of 3D optical readout of a TPC using a single photon sensitive Timepix3 based camera. *Journal of Instrumentation* 14: P06001–P06001.

45 Clark, S., Elferich, J., Gai, J. et al. (2019). Strategy for compositional analysis of the hair cell mechanotransduction complex using TIRF microscopy. *Microsc Microanal* 25: 1266–1267.

46 Liu, Y., Trimby, P., Collins, L. et al. (2021). Correlating crystallographic orientation and ferroic properties of twin domains in metal halide perovskites. *ACS Nano* 15: 7139–7148.

47 Hickey, C.L. and Grumstrup, E.M. (2020). Direct correlation of charge carrier transport to local crystal quality in lead halide perovskites. *Nano Letters* 20: 5050–5056.

48 Rothmann, M.U., Kim, J.S., Borchert, J. et al. (2020). Atomic-scale microstructure of metal halide perovskite. *Science* 370: eabb5940.

49 Li, W., Rothmann, M.U., Zhu, Y. et al. (2021). The critical role of composition-dependent intragrain planar defects in the performance of $MA_{1-x}FA_xPbI_3$ perovskite solar cells. *Nature Energy* 6: 624–632.

50 Doherty, T.A., Winchester, A.J., Macpherson, S. et al. (2020). Performance-limiting nanoscale trap clusters at grain junctions in halide perovskites. *Nature* 580: 360–366.

51 Kosasih, F.U., Di Giacomo, F., Orri, J.F. et al. (2022). Sodium diffuses from glass substrates through P1 lines and passivates defects in perovskite solar modules. *Energy & Environmental Materials* e12459.

15

In Situ Characterization of Halide Perovskite Synthesis

Maged Abdelsamie[1,2], Tim Kodalle[3], Mriganka Singh[3], and Carolin M. Sutter-Fella[3]

[1] *Lawrence Berkeley National Laboratory, Materials Sciences Division, 1 Cyclotron Rd., Berkeley, 94720 CA, USA*
[2] *King Fahd University of Petroleum and Minerals, Materials Science and Engineering Department, B63 Academic Belt Rd., Dhahran 31261, Saudi Arabia*
[3] *Lawrence Berkeley National Laboratory, Molecular Foundry Division, 1 Cyclotron Rd., Berkeley, 94720 CA, USA*

15.1 Introduction

Material synthesis, including the fabrication of halide perovskites, is a fundamental aspect of discovering and deliberately tailoring new functional materials for energy applications [1, 2]. To this end, the majority of characterizations are focused on post characterization of the perovskite material, often near equilibrium, or via implementation in optoelectronic devices coupled with device characterization. In situ measurements, in contrast to ex situ measurements (also referred to as "black box" synthesis), enable the direct characterization of chemical transformations of the reagent to the product in real time during the reactions (Figure 15.1) [3–7]. Here, in situ characterization refers to observing a process while it is taking place, whereas ex situ characterization refers to measurements after the process is finished. In situ characterization during perovskite thin film fabrication can provide information about the temporal evolution of crystal phases and optical properties [8–10]. In addition, it can reveal thermodynamic and kinetically driven formation pathways as well as the potential formation of metastable phases [4, 7, 11]. The possibly intricate interplay between these driving forces during fabrication can influence the evolution of functional properties and, consequently, device operation. Hence, deliberate design of halide perovskites and materials in general can be facilitated by knowing how a material evolves and then manipulating its formation.

One prerequisite to capturing the fast evolution of features during the solution deposition and consecutive annealing of halide perovskites is fast data collection, i.e. fast detectors with ms or faster integration times and high-flux beam sources (such as X-rays). Technical developments over the last decade or so now allow for the in situ characterization of soft materials, including their growth kinetics, meaning evolution of properties over time, and structural transformations.

Halide Perovskite Semiconductors: Structures, Characterization, Properties, and Phenomena,
First Edition. Edited by Yuanyuan Zhou and Iván Mora-Seró.

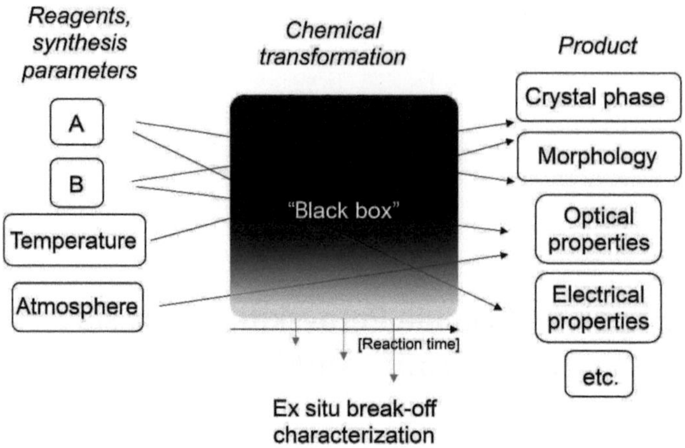

Figure 15.1 Schematic illustration of a "black box" synthesis with known reagents and synthesis parameters. Interrupting the chemical transformations at various points in time can provide snapshots during the transformation of reagents into the final product and are shown as ex situ break-off characterization. Source: Modified from Sutter-Fella [3].

One step further, in situ multimodal characterization is needed to correlate the evolution of functional properties such as bandgap and crystal phase over time during formation [3]. These mechanistic insights will guide the choice of synthesis parameters in order to efficiently design syntheses and significantly increase reproducibility. In the past years, in situ and multimodal characterization of halide perovskites have gained significant interest and provided novel insights to advance the field. Lastly, in an endeavor to close the loop between synthesis, characterization, and computation, in situ characterization is one stepping stone toward synthesis automation and continues to gain momentum and experimental implementation.

In this chapter, the in situ characterization of halide perovskites will be discussed via synchrotron-based X-ray scattering and fluorescence, optical spectroscopy, and the simultaneous combination of different techniques (multimodal characterization). The information one can extract from each technique will be exemplified by recent literature studies. Since in situ characterization is not a common characterization mode, many setups are custom built, which will be described briefly. At the end, probe beam-sample interactions and considerations will be discussed, followed by a summary and outlook.

15.2 Fundamentals of X-Ray Scattering and Fluorescence Techniques

In this section, we provide a short review of the fundamentals of X-ray scattering and X-ray fluorescence techniques used for the in situ characterization of perovskite thin films. Firstly, we cover grazing incidence wide-angle X-ray scattering (GIWAXS) and grazing incidence small-angle X-ray scattering (GISAXS) and their applications to probe the crystallization behavior at the small-length-scale ranging from Angstrom (Å) to a few nanometers (nm) and the perovskite grain formation. Then, we briefly provide fundamentals of X-ray fluorescence (XRF) and its application in probing the compositional evolution during film formation. The major focus, based on its frequent application to halide perovskites, is on GIWAXS. At the small-length-scale, GIWAXS can provide valuable information about the crystal packing structure, the orientation with respect to the device architecture, the relative degree of crystallinity, materials composition, and the crystalline correlation length (i.e. a parameter related to the average size of perovskite crystals) [7, 12, 13]. GISAXS makes use of the contrast between domains of different phases (either amorphous or crystalline) to give information about the average domain size [12, 13]. Both lab-based and synchrotron-based X-ray scattering techniques have been used for perovskite thin film characterization; however, lab-based X-ray measurements are limited to static characterizations due to low photon flux and intensity of the X-ray beam, requiring long measurement times [14]. On the other hand, synchrotrons provide intense and collimated X-ray beams with high photon flux, allowing for the in situ characterization of fast film formation kinetics [14].

15.2.1 Grazing Incidence Wide-Angle X-Ray Scattering (GIWAXS)

Polycrystalline thin films with random crystal orientation may result in low peak intensities due to short X-ray path lengths in the thin film under symmetric diffraction conditions such as in Bragg–Brentano geometry (θ–2θ). To increase the X-ray path length and thus the resulting signal, grazing incidence X-ray diffraction (GIXRD) is used, where the incoming X-ray angle is fixed at a shallow angle while the detector is moved. A portion of the beam is diffracted by the periodic crystallographic planes, and diffraction occurs when Bragg's law is satisfied, as represented in Eq. (15.1) below [15–17]. GIWAXS is focused on analyzing the Bragg diffraction at wide angles using high-energy (or hard) X-rays in the grazing incidence geometry. Figure 15.2 illustrates the X-ray diffraction (XRD) condition and grazing incidence geometry of a WAXS experiment. As shown in Figure 15.2a, the construction of reciprocal space and Ewald's sphere is used for predicting crystallographic scattering patterns [15–17]. Construction of reciprocal space relies on representing the parallel crystallographic planes in real space (green lines) as points in reciprocal space (green dots), where each point corresponds to a set of lattice planes labeled with Miller indices (*hkl*). The magnitude of the reciprocal space vector ($\vec{q}$) is defined by the inter-planar spacing of the lattice planes as $q = 2\pi/d$ and its direction is normal to the lattice planes. A sphere with a radius related to the wavelength of the incident X-ray beam via $2\pi/\lambda$ (equivalent to the magnitude of incident and diffracted beam vectors [$\vec{k}_i$ and $\vec{k}_0$, respectively]) is used to construct the Ewald sphere. Thus, all the possible incident and diffracted vectors are held by the Ewald sphere. The origin of the reciprocal space is the intersection of the incident beam at the edge of the Ewald sphere. The Bragg law is satisfied when reciprocal lattice points intersect the Ewald sphere, so that diffraction occurs (Figure 15.2) and Eqs. (15.1) and (15.3) are fulfilled.

$$2d_{hkl}\sin(\theta) = n\lambda \quad (15.1)$$

$$\vec{q} = \vec{k}_0 - \vec{k}_i \quad (15.2)$$

$$q = \frac{2\pi}{d_{hkl}} = \frac{4\pi\sin(\theta)}{\lambda} \quad (15.3)$$

Here, d_{hkl} is the spacing of the periodic planes *hkl*, 2θ is the diffraction angle, n is the order of reflection [$n = 1, 2, 3, \ldots$], λ is the X-ray wavelength, and $\vec{q}$ and q are the scattering vector and its magnitude, respectively.

Electronic and optoelectronic properties of perovskites are highly affected by their crystal structure; hence, the electronic band structure (or band structure) and its calculations rely on the periodic nature of a crystal lattice. These properties can show anisotropy along different crystallography planes, especially in the case of two-dimensional (2D) perovskite structures. Therefore, determination of the texture or the crystallographic lattice orientation distribution is crucial for understanding the correlation between the microstructure and optoelectronic properties of perovskite thin films. GIWAXS is the most common methodology to investigate the texture orientation, which can be revealed by the shape and polar distribution of the scattering peaks in a reciprocal space map as represented schematically in Figure 15.2c from left to right: (i) Pronounced Bragg diffraction peaks along the q_z direction appear when crystallographic planes are oriented perfectly parallel to the substrate, corresponding to stacking out-of-plane. (ii) Partial arcs appear when the Bragg diffraction broadens, as for crystallites with more evenly distributed texture orientations, where crystallographic planes have angular distribution about the substrate surface. (iii) Full scattering rings similar to powder-like diffraction patterns are obtained for crystals randomly oriented in all directions. These schematics are only for one set of crystallographic planes. In a 3D lattice, there will be multiple peaks from different crystallographic planes. For instance, the diffraction pattern of the 2D perovskite $(PEA)_2PbI_4$ in Figure 15.2b shows two sets of highly oriented diffraction peaks in the out-of-plane (q_z) and in-plane (q_{xy}) directions corresponding to crystallographic planes oriented parallel and normal to the substrate, respectively. Notably, partial arcs appear along the out-of-plane direction, revealing an angular distribution about the substrate surface.

15.2.2 Grazing Incidence Small-Angle X-Ray Scattering (GISAXS)

The GISAXS technique makes use of the difference in the refractive index that arises from inhomogeneities in the electron density of materials/phases depending on their mass density and elemental composition [14, 18]. Large-scale inhomogeneities are expected to be present during film formation due to the evolution of different phases such as the perovskite phase, intermediates, the amorphous phase, or a combination thereof. For instance, in situ GISAXS can be used to quantify the average domain size of emerging phases during film formation [19, 20]. Moreover, SAXS can be applied to liquid samples to investigate the aggregation in solution [7].

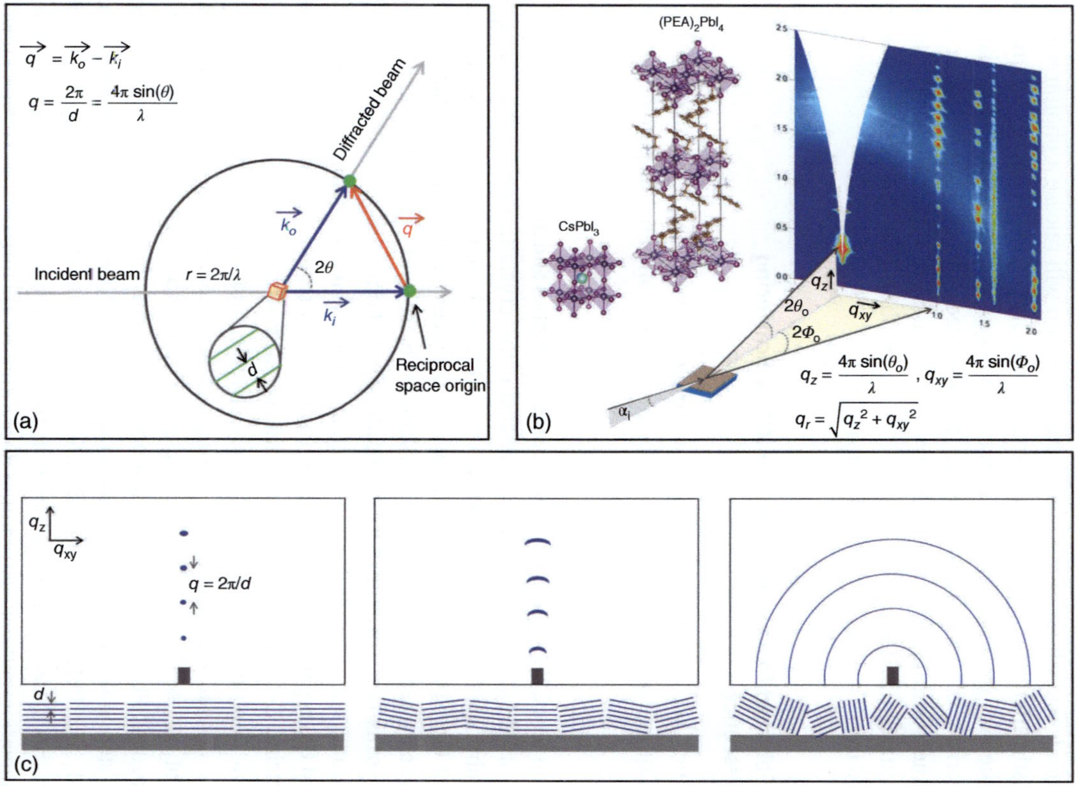

Figure 15.2 (a) Schematic representation of the X-ray diffraction condition. The reciprocal space, calculated from the Fourier transform of the crystallographic planes in Euclidean space (green parallel lines), is represented as green dots. The circle represents the Ewald sphere with radius = $2\pi/\lambda$ and contains all wave-vectors of the incident and diffracted X-ray beams (k_i and k_0, respectively). (b) Schematic illustration of GIWAXS patterns for crystals with different texture orientations. The out-of-plane scattering (q_z) related to its scattering angle ($2\theta_0$) and the in-plane scattering (q_{xy}) related to its scattering angle ($2\Phi_0$) are shown, as well as the calculation of the radial distance (q_r). The diffraction pattern corresponds to GIWAXS of the 2D perovskite $(PEA)_2PbI_4$. (c) Exemplary diffraction patterns correspond to crystallographic planes oriented parallel to the substrate (left pattern), oriented with an angular distribution around the horizontal alignment (middle pattern), and randomly oriented in all directions (right pattern). Source: (a, b) Adapted from Abdelsamie and Toney [16]/with permission of Taylor & Francis. (c) Müller-Buschbaum [17]/with permission of John Wiley & Sons.

15.2.3 X-Ray Fluorescence (XRF)

XRF can provide chemical information via the emission of characteristic X-rays and relies on the interaction and ionization of atoms by X-ray radiation. Ionization of atoms, i.e. the removal of an electron, may occur upon X-ray exposure if the radiation energy is larger than the respective ionization energy [21]. As a consequence, the electronic structure of the atom becomes unstable, and electrons from higher orbitals will fill the inner state. This transition will release a photon with a characteristic energy for the atom involved, corresponding to the energy difference between the two involved orbitals. Since the energy of the fluorescent radiation is characteristic of a transition between specific electron orbitals in a particular element, the resulting fluorescent X-rays can be used to detect the presence of elements during film formation [6]. The relative amounts of different elements can be deduced from their relative XRF intensity [6].

15.2.4 Selected Examples for In Situ X-Ray Scattering and Fluorescence

In this section, we provide a few selected examples of deploying X-ray characterization techniques in an in situ mode to probe the fabrication of halide perovskite thin films.

15.2.4.1 In Situ GIWAXS to Study Crystallization Kinetics and A-Site Doping

Qin et al. used in situ GIWAXS to investigate the crystallization pathways of a two-step-fabricated $FAPbI_3$ perovskite thin film with sequential A-site doping of Cs^+ and GA^+ [4]. Typically, in the two-step process, a PbI_2 layer is fabricated first, followed by its conversion to perovskite in the second step via the deposition of organic salts (e.g. formamidinium iodide, methylammonium bromide, and/or methylammonium chloride dissolved, e.g. in isopropyl alcohol) and interdiffusion of PbI_2 and the organic salts during heat treatment. Qin et al. modified this process by introducing CsI and guanidinium iodide (GAI) into the precursors for the first and second deposition steps, respectively. They found that the addition of CsI altered the crystallization pathway during the first step. In the presence of CsI, the δ-$CsPbI_3$ phase was observed during spin coating together with a weak signal from hexagonal PbI_2, while in the absence of CsI, PbI_2 only appeared during the annealing step. It was concluded that incorporation of Cs^+ leads to the co-existence of δ-$CsPbI_3$ and PbI_2 phases, and that PbI_2 growth is likely a thermally activated process in this system. Then, focusing on the second synthesis step, variations in the crystallization pathway were found depending on the presence of Cs^+ and GA^+. Figure 15.3 shows the integrated scattering intensity, the 2D GIWAXS intensity plots for the evolution of $FAPbI_3$ with Cs^+ and GA^+, and the time evolution of peak areas of δ-$CsPbI_3$, PbI_2, and perovskite phases. For the reference sample without Cs^+ or GA^+, only three stages can be identified, while in the case with Cs^+ and GA^+, four stages were found. Stage I contains the PbI_2 and δ-$CsPbI_3$ phases, Stage II shows the perovskite α-phase formation and a decay of both PbI_2

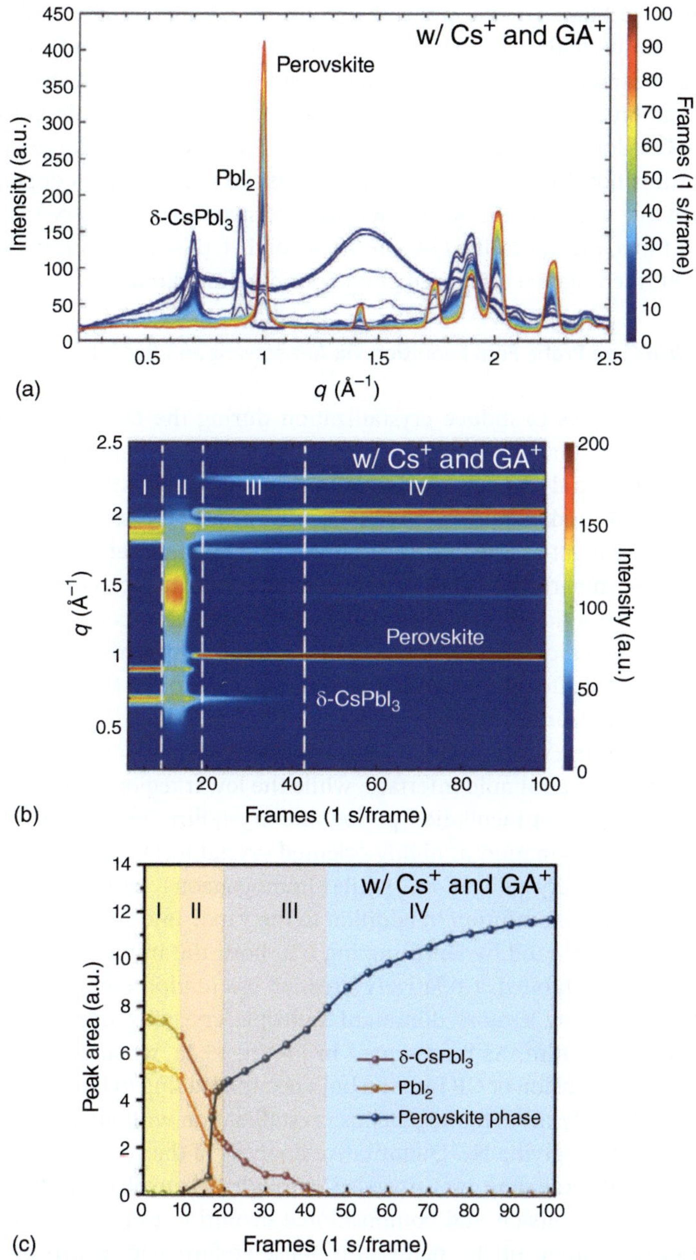

Figure 15.3 (a) GIWAXS integrated scattering intensity and (b) false-color 2D GIWAXS intensity maps showing q versus time (frame number) for the second step of a two-step-fabricated $FAPbI_3$-based perovskite with Cs^+ and GA^+. (c) The time evolution of peak areas of PbI_2, δ-$CsPbI_3$, and perovskite phases. Source: Qin et al. [4]/with permission of John Wiley & Sons.

and δ-$CsPbI_3$ phases. In Stage III, the transition from the δ-$CsPbI_3$ phase to the perovskite phase continues with a much slower transition rate than in Stage II, and it ends with the disappearance of the δ-$CsPbI_3$-phase peak. Stage IV is characterized by further crystallization of pure perovskite α-phase until saturation. While both Cs-doped and Cs + GA-doped syntheses have four similar stages, GA^+ induces an accelerated phase transition from δ-$CsPbI_3$ to the perovskite phase. For Cs-doped films, pinholes form due to the nonuniform nucleation of sparse δ-$CsPbI_3$ crystals. For Cs + GA-doped films, incorporating GA^+ in the second step helps to eliminate pinholes via Ostwald ripening and enhanced grain boundary migration.

15.2.4.2 In Situ GIWAXS to Probe Film Evolution via Antisolvent and Gas Jet Treatments

Typical engineering strategies to induce crystallization during the fabrication of halide perovskites include antisolvent dripping and N_2 gas jet treatment during spin coating [7, 22]. In a study by Abdelsamie et al., in situ GIWAXS was used to provide a fundamental understanding of the crystallization processes to form $Cs_{0.15}FA_{0.85}PbI_3$ via antisolvent-assisted synthesis or N_2 treatment (schematic illustration Figure 15.4a) [7]. In particular, this study extracted phase fractions from the in situ data to reveal compositional fluctuations during synthesis. It was found that when applying N_2, the initial crystallization of the hexagonal non-perovskite (δ_H) $FAPbI_3$ crystallites occurred after an incubation period and were found to be highly oriented in the out-of-plane direction along q_z, followed by crystallization of more randomly oriented crystallites. As depicted in Figure 15.4a, applying N_2 accelerates the solvent removal from the air–liquid interface, while the lower region of the film reaches supersaturation after an incubation period and crystallization starts from the top surface. The initial formation of highly oriented crystallites was attributed to vapor-interface-induced crystallization, while homogeneous crystallization occurs at later times in the bulk solution in addition to the vapor-interface-induced crystallization. On the other hand, when dropping CB, both the initial and later δ_H-$FAPbI_3$ crystallization exhibited a relatively broader orientation compared to the N_2-treated films, revealing a more dominant isotropic crystallization taking place in the bulk of the wet film. As represented in Figure 15.4a, when dropping the antisolvent CB, rapid diffusion of CB into the bulk occurs, leading to immediate supersaturation and relatively more homogeneous crystallization without an incubation period compared to applying N_2. Quantitative analysis of the compositional evolution during thermal annealing was revealed through deconvolution of the contributions of thermal expansion and compositional evolution to the observed shift of the (100) peak position of the perovskite phase (referred to as α-phase, Figure 15.4b). Furthermore, the evolution of the volume fractions of α-phase, δ_H-phase, and amorphous phases was calculated from the intensity evolution of the δ_H- and α-phase peaks during thermal annealing using Eq. (15.4), below [23].

$$I\alpha\left(\frac{I_0\lambda^3}{16\pi d}\right)\left(\frac{e^4}{{m_e}^2c^4}\right)\left(\frac{Vm^{hkl}(F^{hkl})^2}{v^2}\right)(\text{LPF}),\quad \text{where LPF} = \left(\frac{\alpha_{\text{polarization}}}{\sin(\theta)\sin(2\theta)}\right) \tag{15.4}$$

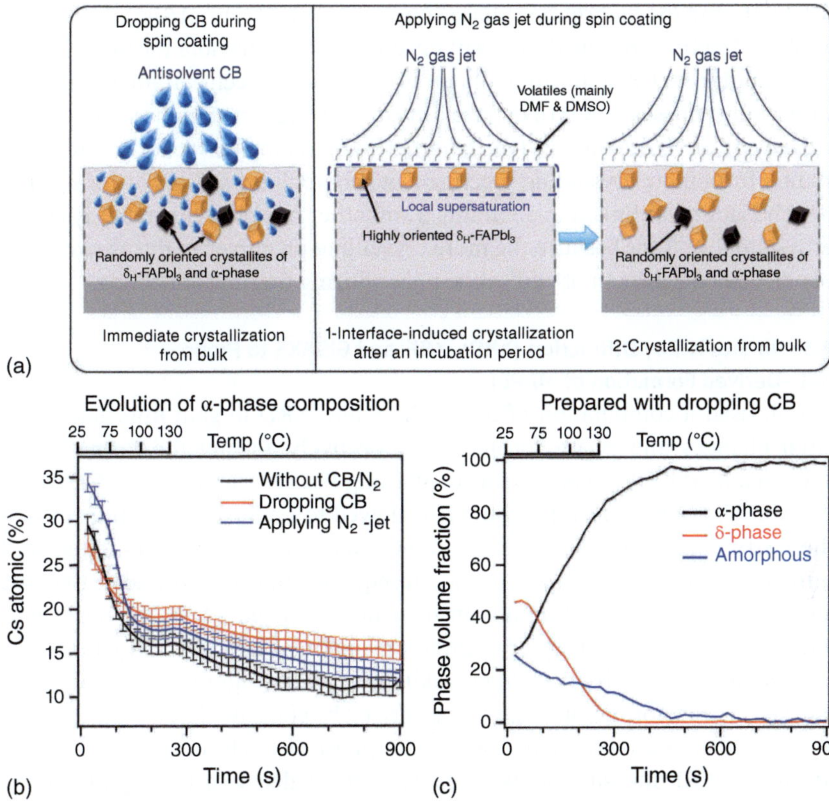

Figure 15.4 (a) Schematic representation of the mechanism of dropping antisolvent (left) and applying N_2 gas jet (right) during the spin coating step. (b) Time evolution of the α-phase composition during thermal annealing of $Cs_{0.15}FA_{0.85}PbI_3$ treated with antisolvent or N_2 and the reference after spin coating. (c) Time evolution of the calculated phase volume fractions of α-phase (black line), δ_H-phase (red line), and amorphous phase (blue line) for films prepared by dropping an antisolvent (chlorobenzene, CB). Source: Abdelsamie et al. [7]/with permission of John Wiley & Sons.

Here, I is the integrated scattering intensity from a reflection hkl, I_0 is the incident intensity of the radiation beam, λ is the X-ray wavelength, d is the sample to detector distance, e is the electron charge, m_e is the electron mass, c is the speed of light, V is the diffracting volume, m^{hkl} is the multiplicity of a reflection hkl, F^{hkl} is the structure factor of a reflection hkl, v is the volume of the unit cell, and LPF is the Lorentz polarization factor, $\alpha_{polarization}$ is the polarization correction factor, and 2θ is the diffraction angle.

Interestingly, the films exhibited an inhomogeneous distribution of Cs and FA in the as-cast films, with the segregation of Cs in the α-phase (with ≈30–35 at% Cs) and FA in the pure δ_H-$FAPbI_3$ (see Figure 15.4b at the start of the annealing). During subsequent annealing, cation (Cs, FA) diffusion occurs, leading to the formation of a more homogenous microstructure with an average stoichiometry of ~15 at% Cs, $Cs_{0.15}FA_{0.85}PbI_3$, while crystallization of the α-phase occurs at the expense of the δ_H-$FAPbI_3$ and the amorphous phase simultaneously (Figure 15.4c). The evolution of phase fractions revealed different crystallization behaviors for the different synthesis conditions. Crystallization of perovskite α-phase occurred predominantly from amorphous phases rather than from a crystalline precursor in the sample without antisolvent or N_2 treatment. On the other hand, crystallization occurred predominantly from more crystalline precursors when applying CB or N_2 during spin coating. The latter was linked to obtaining a smoother morphology and fully covering perovskite thin films because the more crystalline precursor in the as-cast films possibly dictates the growth of perovskite phase during thermal annealing.

15.2.4.3 In Situ X-Ray Diffraction (XRD), XRF, and GISAXS to Probe the $PbCl_2$-Derived Formation of $MAPbI_3$

Stone et al. used a combination of in situ XRF and XRD to gain information on elemental changes and phase formation, respectively, during the formation of $PbCl_2$-derived $MAPbI_3$ from precursors containing $PbCl_2 + 3MAI$ (MA = CH_3NH_3) [6]. This combination of in situ techniques allowed to understand the transformation chemistry from the liquid precursor to the solid perovskite phase (summarized in Figure 15.5c). In situ GIXRD during subsequent annealing revealed that the crystallization pathway goes through an intermediate stage where the peaks corresponding to the Cl-containing crystalline precursor phase (MA_2PbI_3Cl) remain nearly constant with no observable perovskite formation (Figure 15.5a). This stage is followed by a transformation stage characterized by continued loss of MA_2PbI_3Cl concurrent with the growth of the perovskite phase. For all temperatures investigated, in situ XRF measurements revealed a linear decay of Cl (Figure 15.5b). This behavior was attributed to the desorption of MACl from a constant surface area, apparently from the disordered MACl phase in the film. Furthermore, no perovskite crystallization occurs until the removal of 50% of Cl; see the onset of perovskite crystallization in Figure 15.5b. Hence, the precursor transformation to perovskite only happens after the disordered MACl phase evaporated (schematic Figure 15.5c).

Another study relied on small-angle scattering measurements via in situ GISAXS to complement wide-angle characterization down to q values of 10^{-3} Å^{-1} (corresponding to length scales as large as 400 nm in real space) [20]. At thermal annealing at 80 °C, a gradual intensity increase in q value was observed, and it was concluded that as the perovskite film forms, there is an increase in the fraction and number of structures that range from 30 to 400 nm in size.

15.2.4.4 In Situ GIWAXS to Probe the 2D Perovskite Formation on 3D Films

A successful strategy to passivate perovskite thin films is via the incorporation of thin 2D or quasi-2D perovskite layers. The formation of the 2D/3D interface is typically done via solution processing, where the 2D perovskite layer is deposited onto the fully fabricated 3D perovskite film. To provide mechanistic insights into the 2D/3D interface formation, in situ GIWAXS was measured during spin-coating [24]. In this example, a VBABr-based (VBA = vinylbenzylammonium) 2D perovskite was deposited on top of $MAPbI_3$ and $(MAPbBr_3)_{0.05}(FAPbI_3)_{0.95}$ films. In situ GIWAXS

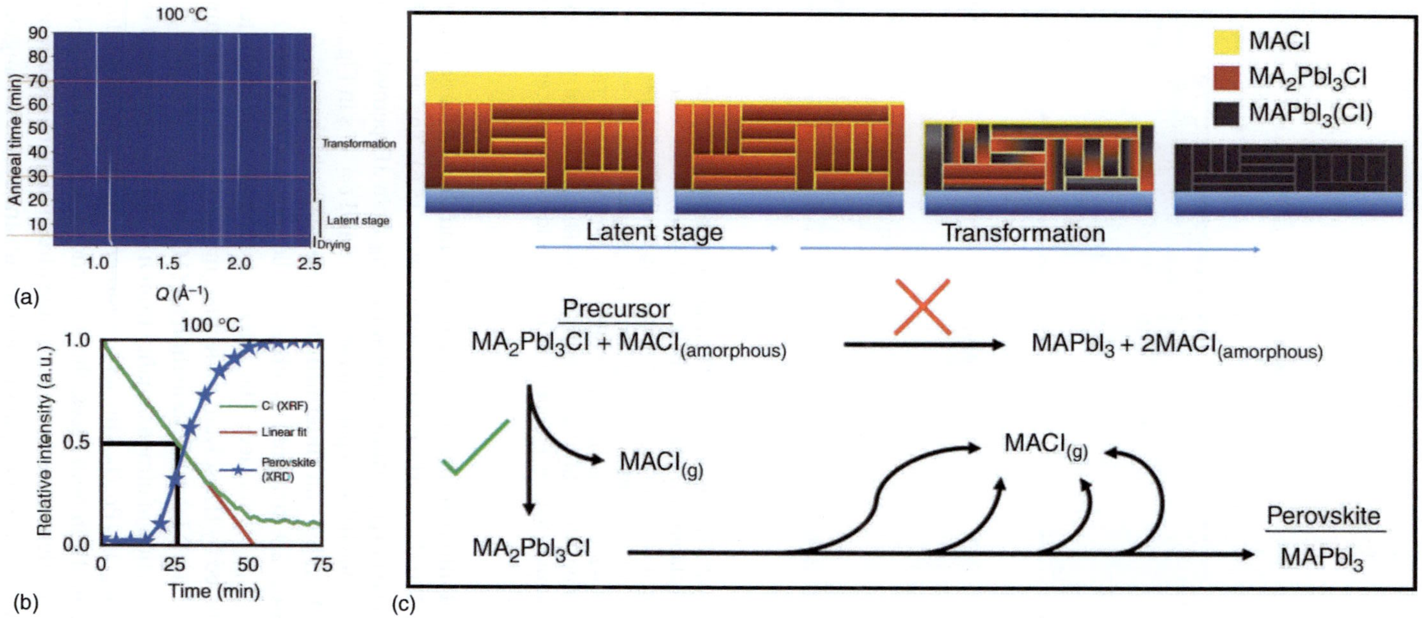

Figure 15.5 (a) In situ XRD of the $PbCl_2$-derived $MAPbI_3$ while annealing at 100 °C. (b) Relative amounts of Cl (from XRF) and crystalline perovskite (from GIXRD) while annealing at 100 °C. (c) Schematic representation of the transformation from a $PbCl_2$-derived precursor to perovskite phase. Source: Stone et al. [6]/Springer Nature/Licensed under CC BY 4.0.

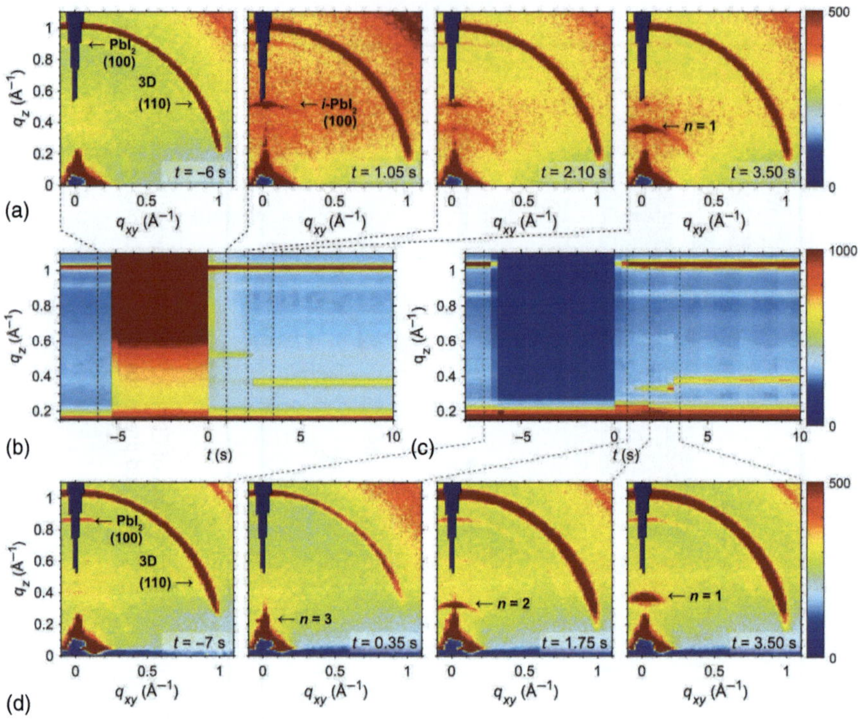

Figure 15.6 (a) 2D GIWAXS patterns for a $MAPbI_3$ film treated with a solution of 5 mg ml^{-1} VBABr in IPA. The intercalated PbI_2 peak is marked as i-PbI_2. (b) False-color 2D GIWAXS intensity maps showing q versus time during spin-coating of VBABr solution on $MAPbI_3$ films. (c) False-color 2D GIWAXS intensity maps showing q versus time during spin-coating of 1 mM VBABr in 3 : 97 IPA:CF solution on $(MAPbBr_3)_{0.05}(FAPbI_3)_{0.95}$ films. Time zero in (b, c) is the start of spin-coating. (d) 2D GIWAXS patterns for $(MAPbBr_3)_{0.05}(FAPbI_3)_{0.95}$ films treated with VBABr solution. Source: Proppe et al. [24]/Springer Nature/Licensed under CC BY 4.0.

has provided insight into both the crystal structure and the orientation of evolving phases during the 2D/3D interface formation (Figure 15.6). The as-cast $MAPbI_3$ and $(MAPbBr_3)_{0.05}(FAPbI_3)_{0.95}$ films show scattering from 3D perovskite and PbI_2 crystals as shown in the GIWAXS patterns at $t = -6$ s and $t = -7$ s in Figure 15.6a,d, respectively. For $MAPbI_3$, an intermediate state with higher q ($\approx$0.50 Å^{-1}) than any VBA-based 2D structure is observed before $n = 1$ of the 2D phase $(VBABr)_2MA_{n-1}Pb_nI_{3n+1}$ is formed. The intermediate state was attributed to 2D PbI_2 sheets intercalated by IPA solvent molecules, marked as i-PbI_2 in Figure 15.6a. On the other hand, 2D/3D interface formation for $(MAPbBr_3)_{0.05}(FAPbI_3)_{0.95}$ occurs via a progressive reduction of dimensionality from $n = 3$ to $n = 1$ of the 2D structure $(VBABr)_2(FA_{0.95}MA_{0.05})_{n-1}Pb_n(I_{0.95}Br_{0.05})_{3n+1}$, rather than the immediate formation of $n = 1$ (Figure 15.6c,d). In both cases, the diffraction peaks of the VBA-based 2D structures are highly oriented in the out-of-plane direction, with arc-like scattering indicating oriented 2D structures parallel to the substrate with an angular distribution around the horizontal alignment. The difference in the 2D/3D interface formation pathways was attributed to differences in the stability between $MAPbI_3$ and $(MAPbBr_3)_{0.05}(FAPbI_3)_{0.95}$ films, where the less stable $MAPbI_3$ is more prone to conversion to PbI_2, thus forming intermediate PbI_2-intercalated phases with the solvent molecules. Moreover, density functional theory (DFT) calculations suggested that the VBA-ligands infiltrate the 3D perovskite layer, leading to splitting of the 3D perovskite phase into lower-dimensional fragments that continuously split till forming $n = 1$, supporting the observations of in situ GIWAXS results. The formation of higher n 2D structures was hypothesized to occur initially, but their (001) diffraction peaks have too low q to be observed in the GIWAXS experiments in the study.

In situ GIWAXS was also used during heating to characterize the evolution of the 2D/3D interface and to reveal the role of 2D perovskites (2-TMAI or PEAI) in inhibiting degradation [10]. The in situ GIWAXS results revealed the transformation of the 2D crystalline structure into a mixed 2D/3D phase, leading to blocking of the degradation of the 3D bulk into lead iodide and keeping the 3D bulk underneath intact. These results provided guidelines for engineering stable device interfaces.

15.3 In Situ Optical Spectroscopy

This section covers UV–vis–NIR optical spectroscopy, discussing the interaction of electromagnetic waves with perovskite materials resulting in absorption, transmission, reflection, and/or emission. More in-depth discussion of the photophysical processes in halide perovskites will be addressed in Chapter 12. Here, the transmission and reflection of light are briefly covered, followed by photoluminescence emission and its application in the in situ characterization of halide perovskite fabrication. Several application examples and information obtained on phase evolution and formation kinetics are discussed.

15.3.1 Fundamentals of Absorption and Emission of Light in Halide Perovskites

Optical spectroscopy plays a key role in the characterization of optical properties and probing of photophysical processes. Figure 15.7a illustrates the photophysical processes occurring in halide perovskites after the absorption of a photon with energy higher or equal to its bandgap energy. The absorbed photon creates free charge carriers or a quasiparticle (electron–hole pair or exciton). It is suggested that free charges are the primary species after photoexcitation in halide perovskites [26]. The excitons in halide perovskites are reported to be Wannier–Mott excitons, where the electron is weakly bound to the hole and excitons are free to move in the material [27]. Many studies report exciton binding energies (E_b) of a few tens of meV and fast dissociation (<1 ps) [26–28]. When applying in situ optical characterization during synthesis, and in particular during the annealing of halide perovskites, the supplied thermal energy (26 meV at room temperature and 32 meV at 100 °C) may be sufficient to dissociate the excitons. Photon-generated free charges are mobile and carrier diffusion lengths between 100 nm to a few micrometers have been found at room temperature (in the

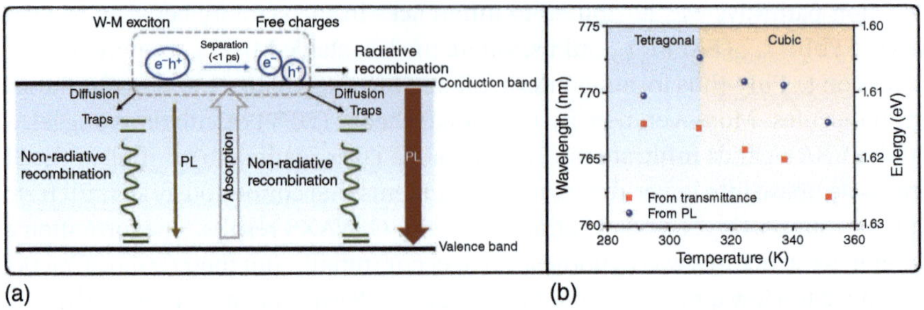

Figure 15.7 (a) Schematic of the photophysical processes in halide perovskites. (b) PL peak energy and absorption band edge as a function of temperature for $MAPbI_3$. Blue = tetragonal phase, orange = cubic phase. Source: (a) Redrawn and inspired by Tian et al. [28]. (b) Data from Milot et al. [25].

low carrier injection regime) [26]. Ultimately, free charges that are not collected at the electrodes will recombine, either nonradiatively or radiatively. Depending on the nature and density of trap states, trapping and de-trapping can occur.

Halide perovskites show strong light absorption with a sharp optical absorption edge and absorption coefficients $>10^4$ cm^{-1} above the bandgap [29]. The light absorption properties can be characterized by wavelength-dependent absorbance (A_λ) measurements along $A_\lambda = -\log_{10}(T_\lambda)$ with the transmitted radiation T_λ. Performing in situ transmission measurements during synthesis is useful to analyze optical features that relate to the band structure, such as the absorption edge and its evolution during fabrication (more in Section 15.3.3). Photoluminescence (PL), or radiative recombination of electrons and holes, is the spontaneous emission of light under optical excitation. For instance, the absorption edge for $MAPbI_3$, calculated from PL and transmittance measurements, shifts monotonically to higher energies with increasing temperature (Figure 15.7b). This behavior is opposite that of other tetrahedrally coordinated semiconductors (such as silicon and gallium nitride) and is explained by a reverse band edge electronic state ordering [30]. In general, with increasing temperature, impurities as well as excitons become ionized. As a consequence, photoexcited free charges increasingly populate the conduction and valence bands, leading to an increasing probability for band-to-band transitions. In first approximation, the maximum of the PL emission peak energy can be interpreted as the bandgap of the sample [31]. In addition, the peak position can be used for compositional approximation via correlation with the bandgap of the perovskite material [32].

Other features of PL spectra include the peak intensity and width (usually characterized by their full-width at half maximum [FWHM]). The PL FWHM characterizes the broadening of PL peaks, which may be caused by thermal effects, bandgap fluctuations, and the population of discrete states. Typically, a high PL intensity is correlated with high-quality materials. During synthesis, and in particular during the thermal annealing stage, the PL intensity exhibits convoluted dependencies. It is influenced by trap states and complicated temperature-dependent recombination dynamics and rates, thermal quenching, which has been explained by thermally activated exciton dissociation, as well as quenching as a result of phonon-assisted nonradiative recombination and thermally activated carrier trapping [27, 33]. Generally, a red shift is observed between the band-edge absorbance and PL spectra, referred to as Stokes shift (Figure 15.7b). As shown for $MAPbI_3$ in Figure 15.7b, the observed shift is nearly constant, and its magnitude is ~4–6 meV and ~11–12 meV in the tetragonal and cubic phase, respectively [25]. As a last comment, fabrication often involves a room-temperature deposition followed by a thermal annealing step, which in the case of the prototype $MAPbI_3$ implies a crystal structure change from tetragonal to cubic around 310 K (Figure 15.7b).

15.3.2 Setup Design for In Situ Optical Spectroscopy

There are a few key components required to run the most basic PL and UV–vis absorbance measurements, namely, a light source (monochromatic in the case of PL) and a spectrometer (Figure 15.8) [31]. More advanced setups include, for example, filtering and focusing optics, objectives, and motorized stages [34]. Application of in situ spectroscopy measurements during halide perovskite fabrication typically involves characterization during deposition and thermal annealing. For in situ PL, since halide perovskites exhibit high luminescence quantum yield, low-power class lasers or light emitting diodes with narrow emission spectrum (e.g. class 3R, output power < 5 mW) can often be sufficient to characterize chemical transformations during fabrication. Consequently, low light source hazards might enable the direct implementation of PL optics into a typical synthesis environment, such as a glovebox (Figure 15.8). Upon excitation, the PL signal is collected by the collection optics, which are commonly fiber-coupled to a spectrometer. In the spectrometer, the signal is spectrally resolved and guided to a detector array. Short acquisition times are required to resolve fast dynamic changes during halide perovskite synthesis, especially during the application of solution engineering strategies such as antisolvent dripping during spin-coating and early crystallization stages. The minimum acquisition time strongly depends on excitation energy and density, quantum yield of the

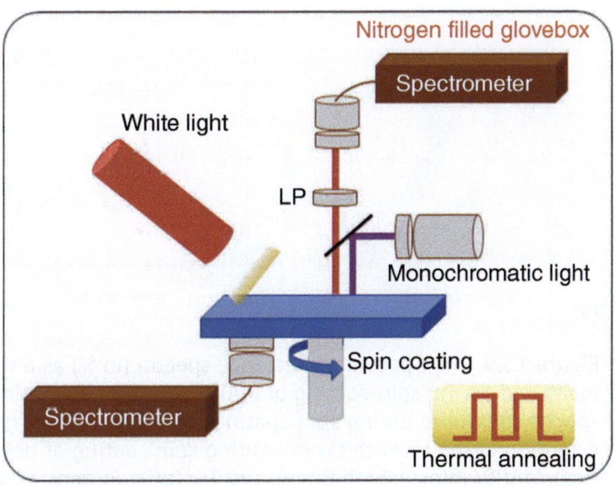

Figure 15.8 Schematic illustration of in situ spectroscopy setups installed in a synthesis glovebox, showing in situ transmission and photoluminescence performed during spin-coating and heating.

sample, as well as setup related specifications. Sub-second acquisition times can be achieved using an incoming photon flux similar to the 1-sun AM 1.5 spectrum. In situ UV–vis transmission and reflection measurements require a white light source, some collection optics, and a spectrometer. The advantage of performing in situ measurements in a real processing environment is the application of findings to typical halide perovskite fabrication processes since the majority of thin films are made via spin coating and annealing in a glovebox with controlled atmosphere.

15.3.3 Selected Examples for In Situ Optical Spectroscopy

In this section, we provide a few selected examples of optical spectroscopy characterization techniques in an in situ mode to probe the fabrication of halide perovskite thin films.

15.3.3.1 Fast In Situ Reflectance Measurements to Characterize the Perovskite Formation

In this study, fast in situ optical reflectance measurements were used to characterize the spin-coating and annealing of different halide perovskite compositions, including ($CH_3NH_3PbI_3$, MAPI), ($CH_3NH_3PbBr_3$, MAPBr), ($HC(NH_2)_2PbI_3$, FAPI), and the triple cation perovskite (TripleCat) $Cs_{0.05}(FA_{0.83}MA_{0.17})_{0.95}Pb(I_{0.83}Br_{0.17})_3$ [35]. Reflectometry measures reflected light and depends on the absorption coefficient, refractive index, and scattering processes. The setup used is inside a glove box with a fiber-coupled tungsten lamp. During spin-coating the light was perpendicular to the film surface but 15° off-axis during annealing to collect diffuse scattered light. Integration times during spin-coating were 0.2–0.5 and 10–40 ms during the annealing, respectively. The interference patterns and fringes from specular reflectance measurements collected during spin-coating are illustrated in Figure 15.9. During the first few seconds of spin coating ($t < 10$ seconds), a fast spectral shift of the interference fringes is observed due to the initial ejection of the solution. At longer times ($t > 10$ seconds), the spectral shift of the fringes gets slower while the spectra appear with a lower number of fringes but with higher amplitude (red color in

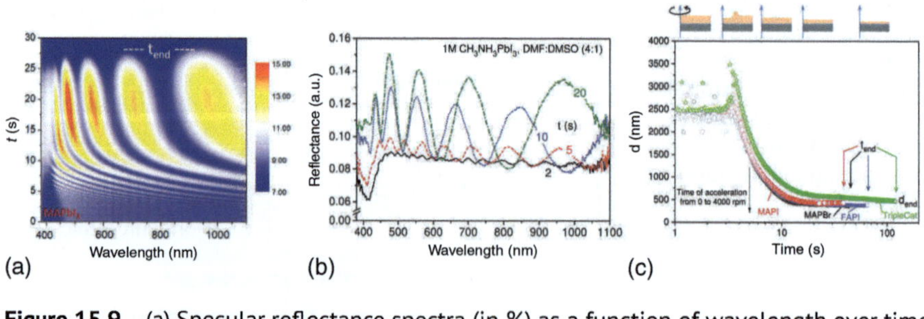

Figure 15.9 (a) Specular reflectance spectra (in %) as a function of wavelength over time measured during spin-coating of $MAPbI_3$ precursor solution. (b) Individual reflectance spectra measured during spin-coating at selected times, given in the figure. (c) Time-evolution of the layer thickness during spin coating of perovskite precursor solution: MAPI (red), MAPBr (black), FAPI (blue), and TripleCat (green). Source: Rappich et al. [35]/John Wiley & Sons/Licensed under CC BY 4.0.

Figure 15.9a). At $t > 28$ s, the surface becomes roughened and specular reflectance disappears, only diffuse light scattering can be measured (indicated as t_{end} in Figure 15.9a). The study found that t_{end} mainly depends on the solvent-ion interaction in solution and concluded that the complex formation energy is the lowest for $MAPbI_3$ and highest for triple cation precursors. The layer thickness (d) during synthesis (Figure 15.9c) was calculated using the interference fringes via [36]:

$$d = \frac{\lambda_1 \cdot \lambda_2}{2 \cdot [\lambda_1 \cdot n(\lambda_2) - \lambda_2 \cdot n(\lambda_1)]}, \ \lambda_1 > \lambda_2 \tag{15.5}$$

For the perovskite precursor solutions, the changes in dispersion of $n(\lambda)$ are small between $500 < \lambda < 1100$ nm and are dominated by the solvent. Thus, Eq. (15.5) can be simplified to Eq. (15.6).

$$d = \frac{\lambda_1 \cdot \lambda_2}{2 \cdot n(\lambda_1 - \lambda_2)}, \ \lambda_1 > \lambda_2 \tag{15.6}$$

Here, λ_1 and λ_2 correspond to wavelengths of two successive maxima or minima, and n is the refractive index.

For all films, the wet film thickness initially decreases rapidly before it levels out after ~10 seconds. The TripleCat wet films exhibited a higher final thickness than the other precursor solutions (Figure 15.9c), which was attributed to a stronger interaction of the precursor ions with the solvent molecules [35]. Furthermore, the in situ reflectance measurements during prolonged spin-coating are used to gain insights into the extent of crystallization of the films. MAPI and FAPI films exhibited no reflectance features associated with perovskite formation, while $FAPbI_3$ still showed interference patterns. In contrast, MAPBr and TripleCat films showed evolving reflectance features, which were interpreted as the development of the perovskite bandgap upon the solvent evaporation. These findings were supported by XRD measurements that have shown perovskite peaks for the as-cast MAPBr and TripleCat films but not for the MAPI and FAPI films. Diffuse in situ reflectance measurements during the annealing were used to further characterize the perovskite formation. They revealed that perovskite formation occurs within ~7 seconds at 100 °C for MAPI but requires much longer time (~60 seconds) at 165 °C for FAPI. For MAPBr and TripleCat films, no significant changes were observed during annealing since the perovskite formation occurred during spin-coating. Notably, since reflectometry cannot provide direct crystal phase information or a precise energetic position of the bandgap, it would be a powerful combination to characterize perovskite synthesis with a combination of reflectance, diffraction, and/or absorbance.

15.3.3.2 In Situ UV–Vis Absorbance Characterization During the Drying Stage

In an effort to scale up the solution process of halide perovskites, Hu et al. used laminar air-knife-assisted meniscus coating at room temperature [5]. To enable controlled drying kinetics during the liquid–solid transformation, in situ UV–vis absorption spectroscopy was performed to establish correlations between supersaturation, nucleation, and growth rate. Two different drying processes, drying with natural air and with a laminar flow N_2 knife in a N_2 glovebox, respectively, were compared. For natural air drying, the in situ UV–vis absorption measurements

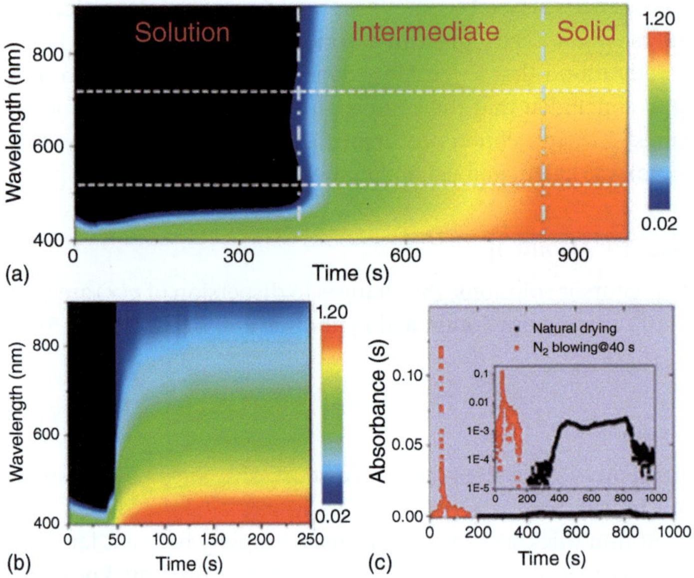

Figure 15.10 (a) Time-resolved contour plot of in situ UV–vis absorbance spectra of natural air-drying hybrid perovskite film. (b) Time resolved in situ UV–vis absorbance spectra of laminar air-knife-assisted meniscus coating nitrogen blowing. (c) The first derivative of the absorbance at 500 nm. Source: Hu et al. [5]/with permission of John Wiley & Sons.

revealed three stages, marked as solution, intermediate, and solid stages in Figure 15.10a. In stage I (solution), only the absorbance from the precursor solution (at $\lambda \leq 440$–450 nm) was observed. In the second stage (intermediate), a gradual red-shifting of the absorbance edge is observed. The red-shifting was correlated to an increased precursor concentration upon solvent evaporation, leading to nucleation and growth of the perovskite crystallites and/or solvent-containing intermediates (DMSO adduct) [37]. In the third stage (solid), stabilized absorbance is observed, revealing the complete transformation of the precursor into solid perovskite. In contrast, when using the laminar N_2 knife, red-shifting in the absorbance spectra occurs 10 times faster, and saturation is reached almost instantly (Figure 15.10b). To quantify the wet film drying rate, the first derivative of the absorbance at 500 nm for the two drying processes was plotted (Figure 15.10c). It was found that using the laminar N_2 knife accelerates the solvent drying rate by two orders of magnitude compared to natural air drying and thus enables controllable and fast perovskite film formation. The use of the laminar N_2 knife enabled rapid nucleation and high-quality films with compact and smooth morphology using a wide processing window (i.e. as long as applying N_2 knife in the solution state).

15.3.3.3 In Situ Photoluminescence Characterization to Investigate the Role of the Precursor

A multimodal in situ approach was used to investigate the role of the lead precursor on the physicochemical evolution of $MAPbI_3$ thin films using three different lead salts (PbI_2, $PbAc_2$, and $PbCl_2$) [38]. Although in situ PL, in situ XRD, and in situ imaging was used, the main focus here is on the in situ PL (Figure 15.11). The benefits of multimodal in situ studies will be discussed in more detail in Section 15.4. Song et al. measured PL by deploying a home-built setup in the fume hood and using a 532 nm laser diode as excitation source. They then fitted each individual PL spectrum using a single Gaussian and investigated both the time-evolution of the intensity and the peak position during thermal annealing (see Figure 15.11a). All three precursors exhibited a similar PL-evolution during annealing; however, the amplitude of the observed shifts of the PL peak positions, intensities, and kinetics strongly depend on the Pb-salt. In all three cases, the initial PL signal occurs at a higher energy than the bulk bandgap energy of $MAPbI_3$, with an energy shift ΔE. The magnitude of the observed shifts in the PL peak positions strongly depends on the Pb-salt. The halide salts (PbI_2 and $PbCl_2$) exhibited an order of magnitude larger energy shift ΔE than $PbAc_2$. Figure 15.11 shows the result for PbI_2 as an example. During annealing, the PL position red-shifts to the position of bulk $MAPbI_3$, while the $PbCl_2$ precursor showed larger delay compared to the other two precursors. The initially blue-shifted PL was attributed to a possible quantum-confined PL emission from small nanocrystallites, while the red-shifting was attributed to the growth of the nanocrystallites [39–43]. Thus, the smaller ΔE for $PbAc_2$-derived precursors was hypothesized to be due to an increased initial size of the $MAPbI_3$ crystallites compared to Pb-halides. Notably, the red-shifting occurred much faster in the acetate case, which was attributed to faster perovskite formation in agreement with other studies [44]. In situ diffraction measurements confirmed the immediate perovskite formation without intermediate phase in the case of $PbAc_2$-derived precursors (agreeing with the small ΔE). In contrast, crystallization of the Pb-halides-derived precursors occurred via the formation of a precursor-solvent complex, leading to slower perovskite formation compared to the $PbAc_2$-derived precursors. Furthermore, all precursors exhibited double maxima in the PL intensity during thermal annealing (see Figure 15.11a for PbI_2 case). Following their interpretation of nanocrystal-nucleation, the authors associate the first increase and drop of the intensity with an increasing volume density of nucleation sites upon supersaturation triggered by solvent evaporation, while the second increase was attributed to perovskite crystal growth, followed by a monotonous intensity decrease due to photo-darkening effects. These findings show high sensitivity of in situ PL to subtle compositional and phase variations.

Another study used in situ PL as a fast tool to optimize the parameter space of annealing temperature and annealing time for the hot-casting of $PbCl_2/PbAc_2$-containing precursors [45]. Moreover, in situ PL was used to characterize the growth dynamics of chlorine-containing $MAPbI_{3-x}Cl_x$ [8]. Relying on the extracted Urbach energy from in situ PL spectra, the defect density of the $MAPbI_{3-x}Cl_x$ was found to be very low during nucleation and initial growth but significantly increased upon the formation of defective crystallite surfaces and grain boundaries.

The abovementioned findings demonstrate the potential of in situ PL as a fast, nondestructive technique to monitor perovskite formation. As discussed

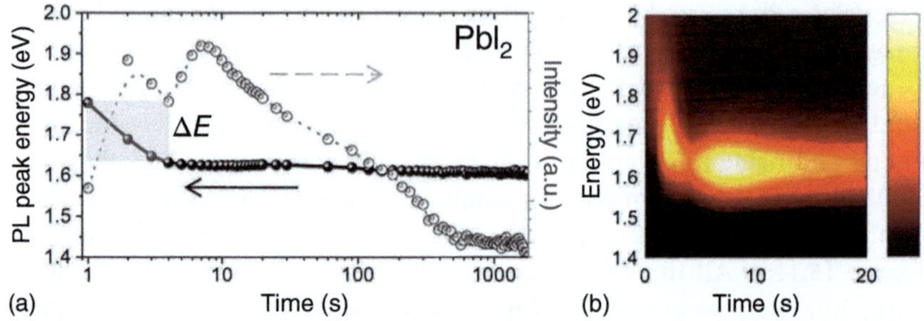

Figure 15.11 (a) Time-evolution of the PL peak position and intensity extracted from the individual PL spectra during thermal annealing of PbI_2-derived $MAPbI_3$ thin films. (b) 2D contour plot of the corresponding PL measurements. Source: Song et al. [38]/with permission of John Wiley & Sons.

later in Section 15.5, there is however a rapid charge carrier funneling that happens upon light excitation that might prevent the PL analysis from gaining full information.

15.4 Examples of In Situ Multimodal Characterization During Solution-Based Fabrication

As discussed in the sections above, in situ measurements can be a very powerful tool for monitoring the evolution of functional properties of perovskite materials during their fabrication process. Combining different, and ideally complementary, in situ techniques, referred to as multimodal characterization, can further improve mechanistic insights of how and when reagents are transformed into the final product. Thus, the additional value of multimodal characterization is the possibility of performing correlative analyses. In this section, two application examples of multimodal in situ characterization will be discussed, including a combination of two in situ optical techniques and combined in situ PL and synchrotron-based XRD to unveil the formation of $MAPbI_{3-x}Cl_x$ and $MAPbI_3$, with particular focus on the antisolvent dropping step, respectively.

Combined in situ reflectometry and PL measurements were used to characterize a one-step deposition routine consisting of the spin-coating of a ($MAI + PbCl_2$): DMF precursor solution, drying of the spin-coated film, and subsequent annealing [46]. While previous studies confirmed the presence of chlorine in the precursor phases (e.g. MA_2PbI_3Cl) after spin-coating as well as a subsequent loss of MACl during annealing, the precise mechanism of the chlorine-loss and the incorporation of chlorine into the $MAPbI_{3-x}Cl_x$ thin film were not well understood. The evolution of the concentration of the chlorine-containing $MAPbI_{3-x}Cl_x$ phase in the thin film was extracted by applying a Beer–Lambert-law-based model to the in situ optical reflectometry data. To do so, the backside of the utilized glass substrate was covered with silver, enabling detection of the reflected light above the sample. Assuming total reflection at the silver layer, an absorption model was used to fit the absorption

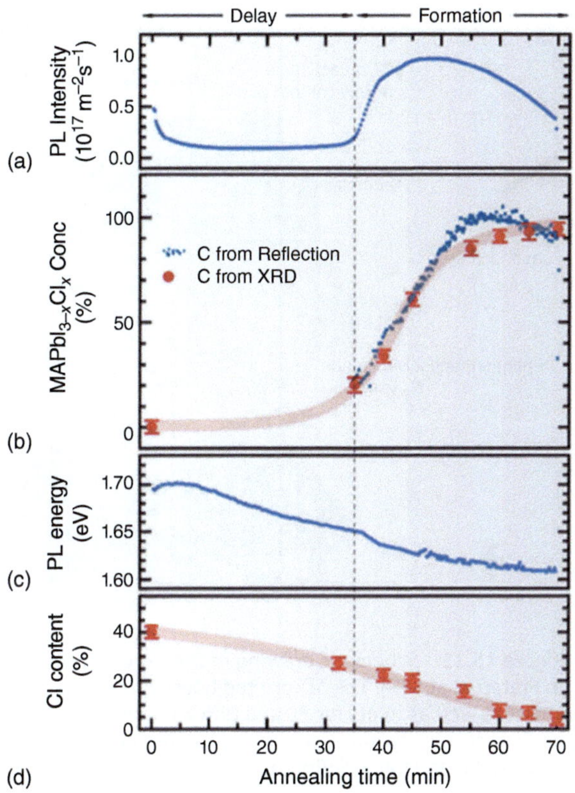

Figure 15.12 Time-evolution of $MAPbI_{3-x}Cl_x$-based halide perovskite properties measured by in situ PL and reflection spectroscopy as well as ex situ XRD and X-ray fluorescence spectroscopy. (a) PL intensity. (b) Different concentrations of $MAPbI_{3-x}Cl_x$ thin film are calculated by fitting the absorption edge of the individual reflection spectra. (c) PL peak energy. (d) Decrement in Cl content in perovskite thin film as measured by XRF. Source: Suchan et al. [46]/Royal Society of Chemistry/ Licensed under CC BY 3.0.

edge with the concentration of chlorine as a free fit parameter. A delay in the formation of the $MAPbI_{3-x}Cl_x$ from the precursor phases was observed (Figure 15.12a,b). At an annealing temperature of 100 °C, the formation of the $MAPbI_{3-x}Cl_x$ phase occurred after a delay of 35 minutes by incorporating the chlorine into the perovskite phase (see Figure 15.12b). In situ PL measurements were performed simultaneously to monitor the evolution of the bandgap energy and PL intensity of the thin film during annealing. The in situ PL measurements supported the hypothesis of delayed chlorine incorporation, revealing that the $MAPbI_{3-x}Cl_x$-formation follows a complex transformation mechanism (Figure 15.12a). The delay in the incorporation of chlorine into $MAPbI_3$ was attributed to the presence of excess chlorine in the drying film, similarly to the case of the latent stage discussed in Section 15.2.4.3. The incorporation of the chlorine into the perovskite phase occurs upon the evaporation of sufficient amount of the excess chlorine and decomposition of the Cl-containing crystalline precursor phase (MA_2PbI_3Cl) (Figure 15.12c,d). Furthermore, the absolute PL intensity was used to calculate the quasi-Fermi level (qFL) splitting versus annealing time following the procedure described in Ref. [47]. The qFL is a measure for the upper limit of the open-circuit voltage of the corresponding solar cell. A maximum of the qFL splitting was observed, indicating an optimal annealing duration for this material, and because of the nature of the multimodal characterization this optimum can be correlated with a certain amount of chlorine-containing

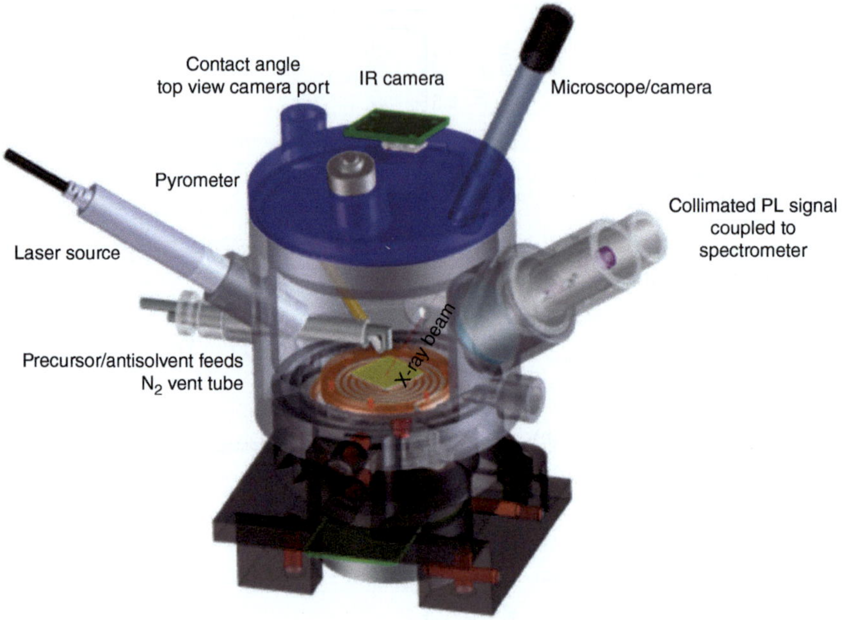

Figure 15.13 Schematic drawing of the experimental setup used for the multimodal study in Pratap et al. [49]. The 3D-printed housing contains an integrated spin-coater/annealing chuck as well as inlets for PL and diffraction measurements, feeds for precursor's solutions and antisolvent syringes, several optical temperature control options, as well as an N_2 flow. Source: Pratap et al. [49]/International Union of Crystallography.

phase and overall chlorine content in the thin film. Correlating these two in situ measurements enabled understanding how the chlorine content of the films influences their fundamental optical and optoelectronic properties. Employing the multimodal approach at different annealing temperatures enabled isolating the kinetic parameters of the $MAPbI_{3-x}Cl_x$ formation reaction. The multimodal measurements revealed that the films at the beginning of the annealing process contain MACl, MA_2PbI_3Cl, and $MAPbI_{3-x}Cl_x$ (with very small x), while MACl evaporates during annealing with an activation energy of ~84 kJ mol^{-1} before MA_2PbI_3Cl starts decomposing and feeding some of the remaining chlorine into the $MAPbI_{3-x}Cl_x$-phase with an activation energy of ~94 kJ mol^{-1}.

Combined in situ PL and synchrotron-based GIWAXS measurements were used to reveal the thin-film formation of $MAPbI_3$ during the spin coating and subsequent annealing [48]. Figure 15.13 shows the experimental setup, which can be attached to a synchrotron diffraction end station and was used for the multimodal characterization of $MAPbI_3$ formation. The complementary insights of this multimodal approach can most clearly be described at the characterization of two crucial points during the synthesis: the antisolvent-dropping during spin-coating and when reaching the final annealing temperature of 100 °C (see Figure 15.14a,b). The GIWAXS data revealed that the investigated material undergoes several stages of transformation during spin-coating and subsequent annealing (Figure 15.14). During spin-coating (stages I and II in Figure 15.14a), the film evolves from a dispersion of the solvent-precursor solution state in stage I (visible as diffuse halos in GIWAXS during the first 24 seconds in Figure 15.14c) to MAPI-DMSO intermediate complexes induced by antisolvent dripping at $t = 25$ seconds. During annealing (stages III and IV in Figure 15.14a), the onset of the crystallization of a perovskite thin film occurs after reaching the final annealing temperature in stage III, inducing perovskite phase formation at the expense of MAPI-DMSO intermediate complexes before crystallization reaches saturation in stage IV.

While the in situ GIWAXS results can already be used to rationalize a growth mechanism, their combination with in situ PL provides additional information. For instance, the nature of abovementioned phase transformations was identified by analyzing the PL peak position and intensity during the different growth stages (identified by GIWAXS). PL results revealed that the initial nucleation upon the antisolvent dripping exhibits a strong PL emission of about 0.1 eV blue-shifted as compared to bulk MAPI, which was attributed to an immediate growth of quantum-confined MAPI nanocrystals upon antisolvent dripping (Figure 15.14b,d). The absence of MAPI diffraction peaks in GIWAXS measurements was attributed to their low volume density in the film compared to the solvent-complex phase, which likely forms the major component of the film at this stage. However, the PL signal is clearly visible because of the high sensitivity of PL to even small amounts of MAPI crystallites due to their high quantum yield. Furthermore, the shape of the corresponding PL peaks, i.e. their FWHM, provided information on the distribution of these nanocrystals upon dripping antisolvent and at later stages. Upon antisolvent dripping, blue-shifted, broad, and asymmetric PL signals were observed, indicating a broad size distribution of nanocrystals. It should be noted, however, that the blue-shifted, broad, and asymmetric PL signals upon antisolvent dripping can also be interpreted as emission from a very luminescent complex composition of precursor and solvent molecules, or these constituents might form a luminescent low-dimensional cluster. A similar argument was presented in a different in situ PL study, and the corresponding discussion is ongoing [50]. At later times, the size of the nanocrystals homogenizes and increases due to cluster coalescence during the rest of the spin-coating, co-existing with the non-luminescent solvent complexes. When reaching the annealing temperature, PL complements the information extracted from GIWAXS data. It shows that the transformation process from both the nanocrystals and the solvent complexes to the solvent-free perovskite thin film occurs via re-dissolution and nucleation processes. From top to bottom of the film, the solvent is removed, inducing growth of highly emissive, quantum confined MAPI-clusters, whose growth and homogenization again become visible via a shift of the PL signal and its FWHM. The observed PL signature is very similar to the first nucleation event observed upon antisolvent dropping.

This study nicely shows how multimodal characterization helps gain a better picture of a complex and fast crystallization process. While GIWAXS was not suited in this study to detect small quantities of early nucleation species, the nonluminescent

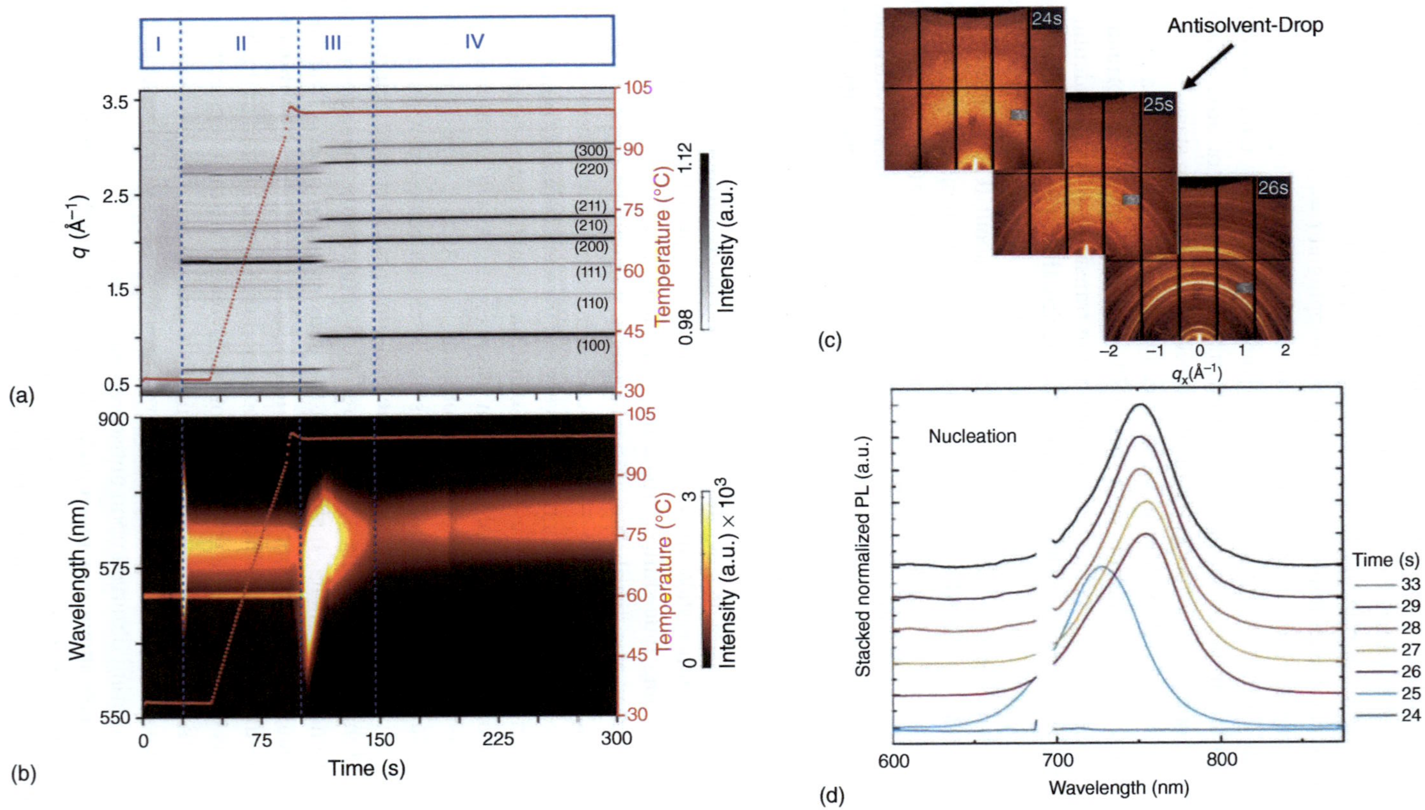

Figure 15.14 Multimodal view of the film evolution of MAPI using in situ GIWAXS (a) and PL (b) as well as selected GIWAXS patterns (c) and PL spectra (d). Source: Pratap et al. [48]/Springer Nature/Licensed under CC BY 4.0.

solvent colloids appearing before the antisolvent dripping could not be detected with PL. The multimodal nature of this study provided novel clues on the dynamic interplay of phases and structures during halide perovskite fabrication. It might also indicate why reproducibility is difficult from lab to lab, including the reproducibility of high-quality devices in the same lab, given the fast changes of thermodynamically versus kinetically driven processes. Multimodal in situ studies can be used to rationalize synthesis parameter choice and film optimization. It should be noted that multimodal in situ studies can further benefit from the combination of ex situ or pseudo in situ measurements. For example, Suchan et al. [46] expanded their study through break-off experiments using time-resolved PL, XRD, and XRF. They, therefore, combine the multimodal in situ study with snapshots of additional information at several points in time during the growth process. However, multimodal in situ measurements are needed to make sure that the findings of such ex situ measurements actually portray the properties of the growing crystals and are not substantially impacted by breaking off the synthesis, e.g. by effects induced by a premature cool-down. Furthermore, the high time resolution of most in situ techniques allows for much better options for optimizing growth procedures and ultimately improving synthesis routines.

15.5 Probing Beam–Sample Interaction

At the end of this chapter, we want to turn our attention to probe beam-sample induced effects, their influence on the sample and the measurement result as well as the mitigation of beam-induced damage. Probe beam–sample interaction involves the transfer of energy (e.g. via photons) to the sample, which can influence the properties of the sample and/or the kinetics of the processes under investigation, for example, due to local heating. The soft nature of halide perovskites and the high volatility of their constituents make them particularly susceptible to beam damage. Furthermore, in situ measurements often require rather high photon fluxes in order to achieve the desired time resolution while maintaining a reasonable signal-to-noise ratio. Therefore, an important requirement for in situ characterization during synthesis is to find experimental conditions that allow capturing transformations at relevant time scales while minimizing the photon flux to avoid damage induced by the characterization. If the latter is not achievable due to the need for high time resolution, probe-induced modifications should be characterized and understood. An extensive overview of possible interactions for different techniques can be found in Hoye et al. [51].

The measurement techniques discussed above are mostly based on interactions of the evolving perovskite film with photons from monochromatic or white light as well as X-ray sources, which both can interfere with the perovskite sample during synthesis. In situ diffraction measurements are mostly performed at synchrotron sources to allow for a high time resolution. Beam damage, due to X ray exposure, can manifest as in reduced diffraction peak intensity as well as peak shifts and can lead to radiation-induced structural changes [51]. In addition, synchrotron X-ray sources, as compared to lab sources, often allow for tunable X-ray photon energies, which can also influence the amount of energy deposited in the material. For example, excitation above the absorption edge of lead (about 13.03 keV) will be much more efficiently absorbed by MAPI than excitation below that energy, therefore, inducing more beam damage [51].

In the case of optical measurements, the interaction with photons from the visible spectrum can have several effects on the properties of the sample under investigation depending on the excitation power, excitation wavelength, sample properties, and measurement atmosphere [9, 52]. In mixed-halide perovskites, for example, illumination with visible light can lead to an elemental redistribution; most widely known is the light-induced halide segregation [33], which can lead to both an increase and a decrease of the PL signal depending on the energy of the used photons [53]. Furthermore, the light dose during a measurement can reduce the trap density of the perovskite [9, 54]. In both cases, the measurement atmosphere plays an important role as well. The presence of oxygen and humidity can influence trap formation and passivation [55, 56] and also contribute to increased and decreased PL signals [57].

PL measurements, in general, are further influenced by charge carrier funneling (typical timescale <1 ns) [58] toward highly emissive low-bandgap regions and regions with low electronic disorder [59]. Therefore, analyzing the PL response (peak position, intensity, and FWHM) but also other optical measurement results requires careful consideration of these effects and ideally complementary measurements to probe the coexistence of compositional or structural fluctuations that might be obscured by carrier funneling.

Additionally, in situ measurements that are taken during film formation can influence the kinetics of the process under investigation. For example, it was shown that the excitation density can alter the timescale of nucleation processes in perovskites by up to two orders of magnitude [38]. However, in that study, it was also found that samples across various conditions were similarly affected without altering the overall trend between samples. Therefore, this is a good example of how one can draw conclusions despite beam-sample interaction by carefully characterizing its effect. Further possible mitigation strategies include, e.g. a reduction of the exposure times by chopping the incident light. Beyond shortened exposure times, the sample spot can be moved during in situ characterization to limit beam interaction. This will, however, lead to an average of the property under study over a certain area and additionally exclude the possibility to laterally resolve the evolution of possibly heterogeneous functional properties.

In summary, there are manyfold experimental parameters that are crucial to monitor and be aware of for a successful and reproducible in situ measurement, including preparation procedures before the actual measurements such as environment (air, N_2, and vacuum), sample handling, and glove box history as well as a wide range of properties of the utilized light and the actual measurement setup. It was reported that there is an interplay of these parameters, i.e. between the properties of the light source and the film properties as well as an amplification of beam-sample interaction in air compared to other environments [51, 53].

15.6 Summary and Outlook

A better understanding of synthesis via in situ (multimodal) characterization of different fabrication steps is powerful for gaining mechanistic insights into how functional properties evolve and how they correlate to synthetic parameters and solution engineering approaches. In addition to guiding the decision on synthetic parameters to improve material performance or stability, real-time characterization can also be integrated into computational machine-learning efforts to autonomously stir experiments as well as synthesis prediction and validation of theoretical frameworks. Advancing the autonomous synthesis of perovskite materials via the development of robots and automated synthesis platforms integrated with in situ characterization techniques can pave the way to accelerating the synthesis process optimization.

The great advantage of high temporal resolution during synthesis comes at the cost of poor or no lateral resolution. This means, in order to get reasonable signal-to-noise data at short acquisition times, large areas are measured, or, in the case of spin-coating, a moving substrate is characterized by averaging over an area. To truly understand synthesis, both in situ characterization during fabrication and in depth post (ex situ) characterization of samples at very high spatial resolution are required. One critical concern is the potential beam damage to the probed materials due to exposure to high-energy X-rays or lasers depending on exposure time, flux, and the energy of the probing beam. Thus, it is of crucial importance to choose the parameters of the probing beam such that the in situ measurements don't significantly interfere with the measurement variable. Since some beam damage is imminent, care must be taken while analyzing the data to decouple the beam-related changes from the actual evolution of the measurement variable.

Acknowledgments

This work was funded by the U.S. Department of Energy (DOE), Office of Science, Office of Basic Energy Sciences, Materials Sciences and Engineering Division under Contract No. DE-AC02-05-CH11231 (D2S2 program KCD2S2). Support from the Molecular Foundry supported by the Office of Science, Office of Basic Energy Sciences, the U.S. DOE under Contract No. DE-AC02-05CH11231 is kindly acknowledged.

References

1 Kim, J.Y., Lee, J.-W., Jung, H.S. et al. (2020). High-efficiency perovskite solar cells. *Chemical Reviews* 120: 7867–7918.

2 Saparov, B. and Mitzi, D.B. (2016). Organic–inorganic perovskites: structural versatility for functional materials design. *Chemical Reviews* 116: 4558–4596.

3 Sutter-Fella, C.M. (2021). The value of watching how materials grow: a multimodal case study on halide perovskites. *Advanced Energy Materials* 11: 2003534.

4 Qin, M., Xue, H., Zhang, H. et al. (2020). Precise control of perovskite crystallization kinetics via sequential A-site doping. *Advanced Materials* 32: 2004630.

5 Hu, H., Ren, Z., Fong, P.W. et al. (2019). Room-temperature meniscus coating of >20% perovskite solar cells: a film formation mechanism investigation. *Advanced Functional Materials* 29: 1900092.

6 Stone, K.H., Gold-Parker, A., Pool, V.L. et al. (2018). Transformation from crystalline precursor to perovskite in $PbCl_2$-derived $MAPbI_3$. *Nature Communications* 9: 3458.

7 Abdelsamie, M., Xu, J., Bruening, K. et al. (2020). Impact of processing on structural and compositional evolution in mixed metal halide perovskites during film formation. *Advanced Functional Materials* 30: 2001752.

8 Mrkyvkova, N., Held, V., Nadazdy, P. et al. (2021). Combined in situ photoluminescence and X-ray scattering reveals defect formation in lead-halide perovskite films. *Journal of Physical Chemistry Letters* 12: 10156–10162.

9 Babbe, F. and Sutter-Fella, C.M. (2020). Optical absorption-based in situ characterization of halide perovskites. *Advanced Energy Materials* 10: 1903587.

10 Sutanto, A.A., Szostak, R., Drigo, N. et al. (2020). In situ analysis reveals the role of 2D perovskite in preventing thermal-induced degradation in 2D/3D perovskite interfaces. *Nano Letters* 20: 3992–3998.

11 Mundt, L.E. and Schelhas, L.T. (2020). Structural evolution during perovskite crystal formation and degradation: in situ and operando X-ray diffraction studies. *Advanced Energy Materials* 10: 1903074.

12 Rivnay, J., Mannsfeld, S.C.B., Miller, C.E. et al. (2012). Quantitative determination of organic semiconductor microstructure from the molecular to device scale. *Chemical Reviews* 112: 5488–5519.

13 Hexemer, A. and Müller-Buschbaum, P. (2015). Advanced grazing-incidence techniques for modern soft-matter materials analysis. *IUCrJ* 2: 106–125.

14 Als-Nielsen, J. and McMorrow, D. (2011). *Elements of Modern X-ray Physics*. Wiley.

15 Hammond, C. (2009). *The Basics of Crystallography and Diffraction*. Oxford University Press.

16 Abdelsamie, M. and Toney, M.F. (2019). Chapter 12: Microstructural characterization of conjugated organic semiconductors by X-ray scattering. In: *Conjugated Polymers: Properties, Processing, and Applications* (ed. J.R. Reynolds, B.C. Thompson, and T.A. Skotheim), 391–425. CRC Press.

17 Müller-Buschbaum, P. (2014). The active layer morphology of organic solar cells probed with grazing incidence scattering techniques. *Advanced Materials* 26: 7692–7709.

18 Müller-Buschbaum, P. (2009). *A Basic Introduction to Grazing Incidence Small-Angle X-ray Scattering*, 61–89. Berlin, Heidelberg: Springer. https://doi.org/10.1007/978-3-540-95968-7_3.

19 Schlipf, J. and Müller-Buschbaum, P. (2017). Structure of organometal halide perovskite films as determined with grazing-incidence X-ray scattering methods. *Advanced Energy Materials* 7: 1700131.

20 Barrows, A.T., Lilliu, S., Pearson, A.J. et al. (2016). Monitoring the formation of a $CH_3NH_3PbI_{3-x}Cl_x$ perovskite during thermal annealing using X-ray scattering. *Advanced Functional Materials* 26: 4934–4942.

21 Young, K.E., Evans, C.A., Hodges, K.V. et al. (2016). A review of the handheld X-ray fluorescence spectrometer as a tool for field geologic investigations on earth and in planetary surface exploration. *Applied Geochemistry* 72: 77–87.

22 Dunlap-Shohl, W.A., Zhou, Y., Padture, N.P., and Mitzi, D.B. (2019). Synthetic approaches for halide perovskite thin films. *Chemical Reviews* 119: 3193–3295.

23 Warren, B.E. (1990). *X-ray Diffraction*. Dover Publications, Inc.

24 Proppe, A.H., Johnston, A., Teale, S. et al. (2021). Multication perovskite 2D/3D interfaces form via progressive dimensional reduction. *Nature Communications* 12: 3472.

25 Milot, R.L., Eperon, G.E., Snaith, H.J. et al. (2015). Temperature-dependent charge-carrier dynamics in $CH_3NH_3PbI_3$ perovskite thin films. *Advanced Functional Materials* 25: 6218–6227.

26 Herz, L.M. (2016). Charge-carrier dynamics in organic-inorganic metal halide perovskites. *Annual Review of Physical Chemistry* 67: 65–89.

27 Baranowski, M. and Plochocka, P. (2020). Excitons in metal-halide perovskites. *Advanced Energy Materials* 10: 1903659.

28 Tian, Y., Peter, M., Unger, E. et al. (2015). Mechanistic insights into perovskite photoluminescence enhancement: light curing with oxygen can boost yield thousandfold. *Physical Chemistry Chemical Physics* 17: 24978–24987.

29 De Wolf, S., Holovsky, J., Moon, S.J. et al. (2014). Organometallic halide perovskites: sharp optical absorption edge and its relation to photovoltaic performance. *Journal of Physical Chemistry Letters* 5: 1035–1039.

30 Even, J., Pedesseau, L., Dupertuis, M.-A. et al. (2012). Electronic model for self-assembled hybrid organic/perovskite semiconductors: reverse band edge electronic states ordering and spin-orbit coupling. *Physical Review B* 86: 205301.

31 Gfroerer, T.H. (2006). Photoluminescence in analysis of surfaces and interfaces. In: *Encyclopedia of Analytical Chemistry* (ed. R.A. Meyers). Wiley. https://doi.org/10.1002/9780470027318.a2510.

32 Noh, J.H., Im, S.H., Heo, J.H. et al. (2013). Chemical management for colorful, efficient, and stable inorganic–organic hybrid nanostructured solar cells. *Nano Letters* 13: 1764–1769.

33 Liu, M., Wan, Q., Wang, H. et al. (2021). Suppression of temperature quenching in perovskite nanocrystals for efficient and thermally stable light-emitting diodes. *Nature Photonics* 15: 379–385.

34 McClelland, A. and Mankin, M. (2018). *Optical Measurements for Scientists and Engineers: A Practical Guide*. Cambridge University Press.

35 Rappich, J., Kaspari, C., Camus, C., and Nickel, N.H. (2021). Fast optical reflectance measurements during spin coating and annealing of organic–inorganic perovskite precursor solutions. *Physica Status Solidi B: Basic Solid State Physics* 258: 2000479.

36 Ghosh, T., Bandyopadhyay, S., Roy, K.K. et al. (2008). Optical and structural properties of lead iodide thin films prepared by vacuum evaporation method. *Journal of Experimental and Industrial Crystallography* 43: 959–963.

37 Li, X., Bi, D., Yi, C. et al. (2016). A vacuum flash–assisted solution process for high-efficiency large-area perovskite solar cells. *Science* 353 (6294): 58–62. https://doi.org/10.1126/science.aaf8060.

38 Song, T.-B., Yuan, Z., Mori, M. et al. (2020). Revealing the dynamics of hybrid metal halide perovskite formation via multimodal in situ probes. *Advanced Functional Materials* 30: 1908337.

39 Wagner, L., Mundt, L.E., Mathiazhagan, G. et al. (2017). Distinguishing crystallization stages and their influence on quantum efficiency during perovskite solar cell formation in real-time. *Scientific Reports* 7: 1–6.

40 Hsieh, T.-Y., Huang, C.-K., Su, T.-S. et al. (2017). Crystal growth and dissolution of methylammonium lead iodide perovskite in sequential deposition: correlation between morphology evolution and photovoltaic performance. *ACS Applied Materials and Interfaces* 9: 8623–8633.

41 Nie, W., Tsai, H., Asadpour, R. et al. (2015). High-efficiency solution-processed perovskite solar cells with millimeter-scale grains. *Science* 347: 522–525.

42 Akkerman, Q.A., Motti, S.G., Srimath Kandada, A.R. et al. (2016). Solution synthesis approach to colloidal cesium lead halide perovskite nanoplatelets with monolayer-level thickness control. *Journal of the American Chemical Society* 138: 1010–1016.

43 Do, M., Kim, I., Kolaczkowski, M.A. et al. (2019). Low-dimensional perovskite nanoplatelet synthesis using in situ photophysical monitoring to establish controlled growth. *Nanoscale* 11: 17262–17269.

44 Zhang, W., Saliba, M., Moore, D.T. et al. (2015). Ultrasmooth organic–inorganic perovskite thin-film formation and crystallization for efficient planar heterojunction solar cells. *Nature Communications* 6: 1–10.

45 van Franeker, J.J., Hendriks, K.H., Bruijnaers, B.J. et al. (2017). Monitoring thermal annealing of perovskite solar cells with in situ photoluminescence. *Advanced Energy Materials* 7: 1601822.

46 Suchan, K., Just, J., Becker, P. et al. (2020). Optical in situ monitoring during the synthesis of halide perovskite solar cells reveals formation kinetics and evolution of optoelectronic properties. *Journal of Materials Chemistry A* 8: 10439–10449.

47 Ross, R.T. (1967). Some thermodynamics of photochemical systems. *The Journal of Chemical Physics* 46: 4590–4593.

48 Pratap, S., Babbe, F., Barchi, N.S. et al. (2021). Out-of-equilibrium processes in crystallization of organic-inorganic perovskites during spin coating. *Nature Communications* 12: 1–9.

49 Pratap, S., Tamura, N., Stan, C. et al. (2019). Probing the in-situ dynamics of structure-property evolution in hybrid perovskite thin films spincoated from complex fluids by a custom designed, beamline compatible multimodal measurement chamber. *Acta Crystallographica Section A: Foundations and Advances* 75: A155–A156.

50 Li, J., Dobrovolsky, A., Merdasa, A. et al. (2018). Luminescent intermediates and humidity-dependent room-temperature conversion of the $MAPbI_3$ perovskite precursor. *ACS Omega* 3: 14494–14502.

51 Hoye, R.L., Schulz, P., Schelhas, L.T. et al. (2017). Perovskite-inspired photovoltaic materials: toward best practices in materials characterization and calculations. *Chemistry of Materials* 29: 1964–1988.

52 Quitsch, W.-A., deQuilettes, D.W., Pfingsten, O. et al. (2018). The role of excitation energy in photobrightening and photodegradation of halide perovskite thin films. *Journal of Physical Chemistry Letters* 9: 2062–2069.

53 Goetz, K.P., Taylor, A.D., Paulus, F., and Vaynzof, Y. (2020). Shining light on the photoluminescence properties of metal halide perovskites. *Advanced Functional Materials* 30: 1910004.

54 DeQuilettes, D.W., Zhang, W., Burlakov, V.M. et al. (2016). Photo-induced halide redistribution in organic–inorganic perovskite films. *Nature Communications* 7: 1–9.

55 Knight, A.J., Wright, A.D., Patel, J.B. et al. (2018). Electronic traps and phase segregation in lead mixed-halide perovskite. *ACS Energy Letters* 4: 75–84.

56 Chen, S., Wen, X., Huang, S. et al. (2017). Light illumination induced photoluminescence enhancement and quenching in lead halide perovskite. *Solar RRL* 1: 1600001.

57 Brenes, R., Guo, D., Osherov, A. et al. (2017). Metal halide perovskite polycrystalline films exhibiting properties of single crystals. *Joule* 1: 155–167.

58 Yuan, M., Quan, L.N., Comin, R. et al. (2016). Perovskite energy funnels for efficient light-emitting diodes. *Nature Nanotechnology* 11: 872–877.

59 Frohna, K., Anaya, M., Macpherson, S. et al. (2021). Nanoscale chemical heterogeneity dominates the optoelectronic response of alloyed perovskite solar cells. *Nature Nanotechnology* 1–7.

16

Multimodal Characterization of Halide Perovskites: From the Macro to the Atomic Scale

Tiarnan A. S. Doherty[1,2,3] *and Samuel D. Stranks*[1,2]

[1] *University of Cambridge, Department of Chemical Engineering and Biotechnology, Cambridge CB3 0AS, UK*
[2] *University of Cambridge, Department of Physics, Cavendish Laboratory, Cambridge CB3 0HE, UK*
[3] *University of Cambridge, Department of Materials Science and Engineering, Cambridge CB3 0HE, UK*

16.1 Introduction

Halide perovskite semiconductors are a versatile family of materials sparking enormous interest [1]. Their tunable bandgaps and dimensionality, facile processability, and exceptional semiconducting properties (which include strong absorption, long carrier diffusion lengths, and high luminescence yields) have led to high device performances when integrated as active layers into optoelectronic devices including photovoltaic (PV) cells [2–4], light-emitting diodes (LEDs) [5], and radiation detectors [6]. For instance, PV device performances have now exceeded 26% in single junction and 33.7% in tandem configurations [7], and X-ray detectors are matching or even exceeding the specifications of commercial analogues. These results are remarkable given perovskites' short development timeline, and, at the time of writing, commercialization is already on the horizon with multiple companies utilizing mature production lines that will deliver commercial perovskite PV products. An exciting paradigm change that is heralded by these materials is the dawning of a new world of tunable and high-performance semiconductors that, in principle, defy what we traditionally teach in semiconductor textbooks – that is, disordered materials that are crudely processed and exhibit high defect densities yet retain excellent device performance. The disorder manifests itself on multiple length scales, from the macroscopic (millimeter) scale down to the atomic scale, and appears in chemical, structural, optoelectronic, and morphological properties [8] – Figure 16.1. The fact that these disordered materials compete on optoelectronic performance metrics with extremely clean semiconductors such as III–Vs (e.g. GaAs) and crystalline silicon – materials, which are highly ordered with minimal tolerance for defects – is truly remarkable. Yet, several challenges still limit the commercialization of halide perovskites. Their ionic nature renders ions mobile, which can lead to changes in the material properties, formation of unwanted ionic defects (e.g. at interfaces in a device), and/or ionic-driven chemical reactions over time under device operation.

Halide Perovskite Semiconductors: Structures, Characterization, Properties, and Phenomena, First Edition. Edited by Yuanyuan Zhou and Iván Mora-Seró.

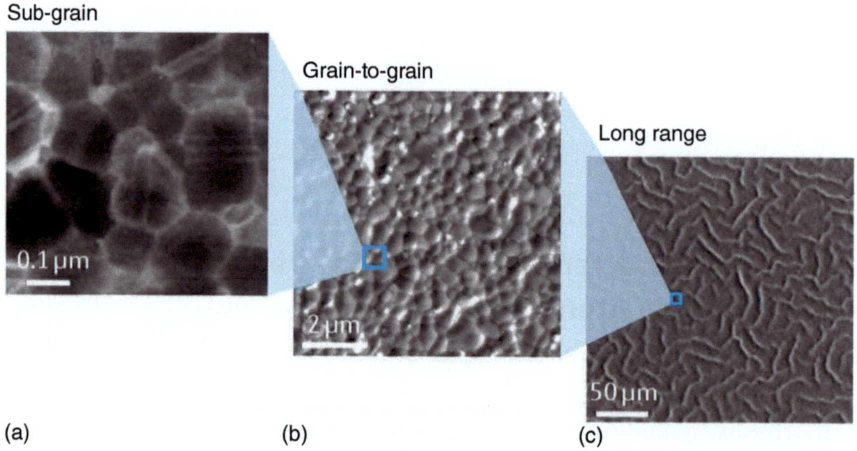

Figure 16.1 (a–c) Electron microscopy images demonstrating the heterogeneity on subgrain (a), grain-to-grain (b), and long range (c) length scales in halide perovskites. Source: Reproduced from [8, 9], Springer Nature.

Ion migration and other material changes resulting from application of external stressors such as light and/or external bias ultimately compromise device longevity. This issue is particularly prolific in LED device configurations and also in X-ray direct detectors, in which strong electric fields are applied across the active layer. Such devices may continue to be hamstrung by these issues. Furthermore, although many point defects introduce either shallow states or states within the bands [10] (in part explaining the apparent performance tolerance to crude processing), not all defects are so benign: there are still energetically deep carrier traps that act as nonradiative recombination centers [11], i.e. sites of substantial power loss. These deep traps appear to arise from complex combinations of structural and chemical imperfections and are particularly problematic in wide bandgap materials (>1.7 eV) required for top cells of tandems [12], as well as in lower bandgap materials [13] (<1.3 eV) for the bottom cells of all-perovskite tandem cells [14]. These traps limit luminescence yields (affecting, for instance, the open-circuit voltage in solar cells) and are likely influenced, or even induced, by the presence of device contacts [15]. When applying external stressors to electrical contacts, e.g. light and/or external bias, even the more "standard" bandgap (1.5–1.6 eV) materials which are close to their radiative efficiency limits when initially fabricated suffer from subsequent instabilities. A global understanding of material chemical, structural, and morphological properties on a range of length scales, how these properties influence carrier recombination (radiative and nonradiative) and transport, and how these relationships evolve over time under device operation, will be paramount for further technology development.

In this chapter, we describe efforts to attain such interproperty relationships and understand the underlying disorder in halide perovskites through use of powerful microscopic techniques utilizing the acquisition of data from multiple measurement modalities, i.e. multimodal measurements – either simultaneously or through sequential correlations on the same scan area. We will review efforts in the field that have led to the current state-of-the-art in our understanding, highlighting in-situ measurements, as well as approaches using ex-situ correlations of different measurements on the same scan areas. For a detailed description of complementary in-situ measurements, we direct the reader to Chapter 15. A key common conclusion is that the length scales that are extremely important – yet least understood – are on the grain-to-grain, sub-grain, and atomic levels. We will identify challenges in utilizing these techniques, including how their very disordered and soft ionic nature makes complete characterization challenging owing to the different length scales and the sensitivity of these materials to experimental probes. Overcoming these challenges to further push the limits of experiment, linked with theory, will further our knowledge, and allow rational approaches to improve performance and stability as these materials move into commercialization. Specifically, these multimodal microscopic techniques will need to move onto the manufacturing line and be coupled with industry standard testing, in turn refining high-throughput processing and device configurations. The approaches will also be adaptable to other technologies based on disordered material families, making such developments in the interests of a much wider community.

Multimodal techniques are powerful approaches to drawing structure-function-performance relationships between key and often complex properties in materials and devices. Although different properties can in principle be inferred from parallel measurements on nominally similar companion samples, multimodal approaches allow direct relationships to be drawn in principle with minimal ambiguity – including those relationships present in a material under device-like environments and operating conditions. Such direct measurements are particularly important in halide perovskite materials, which are especially sensitive to processing conditions such that properties and instabilities may differ locally across the sample, device pixel to device pixel, sample to sample, and batch to batch – thus measuring on the same scan region is crucial to avoid ambiguity. This perspective will focus on microscopic techniques for probing the properties of the as-made films and devices and their evolution under operation; we refer the reader to Chapter 15, which covers bulk multimodal approaches including tracking crystal growth and film formation, as well as other works [16–19].

16.2 Early Multimodal Characterization Work

Nonradiative recombination facilitated by deep trap states in the bandgap of a semiconducting material (see Chapter 12) will directly impact device performance by lowering the quasi-Fermi level splittings and effective charge-carrier densities and, ultimately, the open-circuit voltage – a key PV performance metric in delivering electrical power [11, 20, 21]. Early (2014) theoretical calculations showed that the atomic-scale defects in metal-halide perovskites that are energetically favorable to form create shallow traps, making them benign [22]. This led to a widespread perception of perovskites as "defect-tolerant" materials. This perception is mostly true and in part helps explain why even in the presence of a high density of defects (10^{15}–10^{17} cm^{-3} for polycrystalline perovskite films [23, 24]) perovskite material performance can compare well to that of III–V semiconductors such as GaAs, which have a much lower tolerance to defects and where defect densities as low as 10^7 cm^{-3} can now be achieved [25]. However, experimentally, substantial nonradiative recombination is still observed and prevents perovskite devices from reaching their expected practical performance limits (30% and 35%, respectively, for single-junction and tandem PV devices [26].

Nonradiative recombination can be visualized by spatially resolved photoluminescence (PL) maps. In 2015, de Quilettes et al. in likely the earliest correlated work on perovskites, combined PL of $MAPbI_3$ thin films deposited with Cl-based precursors with scanning electron microscopy (SEM) on the same region to reveal spatial heterogeneity in the luminescence of perovskites, with micron-sized dark regions observed across a high-performing perovskite film [23] (Figure 16.2a,b). These dark regions were dominated by performance-limiting nonradiative recombination, which was particularly aggravated at grain boundaries, though later studies have highlighted the variability in grain boundary properties [28] (i.e. whether they are beneficial or detrimental to performance), and it is highly unlikely that all grain boundaries within even a single sample can be treated as the same, nor that their effect (whether detrimental or otherwise to performance) can easily be generalized across the enormous compositional space that perovskites now inhabit [29]. Nevertheless, the fact that so much heterogeneity in the optoelectronic properties could be present in a material that, at the time, was already achieving PV device efficiencies in excess of 20% was unusual in the context of other thin-film semiconductors for PV, such as GaAs, where high-performance typically necessitates photophysical homogeneity [8, 30]. Another striking observation was that the heterogeneity seemed to be linked to underlying morphology (Figure 16.2c), suggesting that grains within the perovskite differed from each other in either composition or structure and that this impacted performance. Collectively, the results suggested that if methods could be found to eliminate nonradiative recombination and homogenize material properties, substantial improvements in performance could be realized, they also revealed the power of correlated microscopy for understanding emerging materials.

Beyond interrogating local performance, PL, particularly time-resolved PL (TRPL), can be a powerful tool to leverage to understand transient changes in perovskite properties under external stimuli. As halide perovskite is a mixed electronic-ionic conductor, understanding the interplay between ion migration, carrier recombination, and performance is an important topic to understand in order to address stability issues under operating conditions of voltage bias, light, and elevated temperatures. Under these external stimuli, the reactivity of mobile ions has been directly associated with degradation mechanisms of perovskite materials for PV applications [31–33]. One example is the changes induced in the sample's properties, such as PL, over time under illumination [34]. One can observe either PL increases (photobrightening) or PL decreases (photodarkening), and the sign of this change depends sensitively on atmosphere, photon energy, and photon dose [35] – trace amounts of oxygen, for instance, will lead

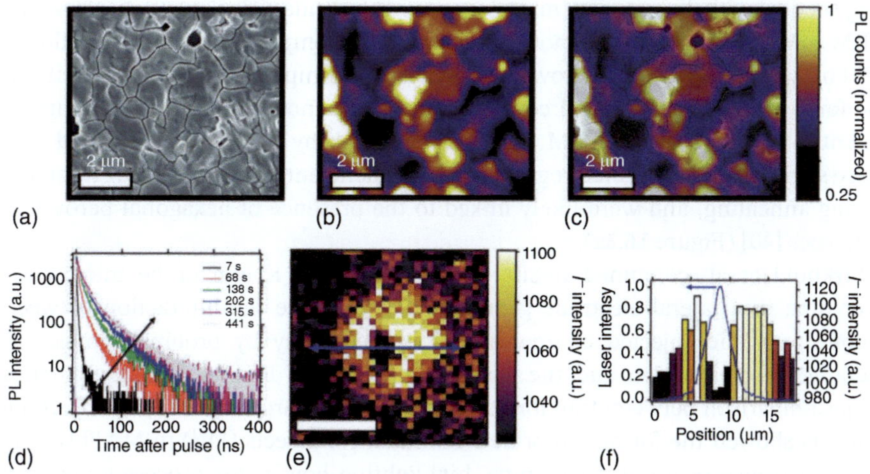

Figure 16.2 (a) Correlated SEM image of a $MAPbI_3$(CL) perovskite film. (b) PL map of the same region as (a) demonstrating substantial heterogeneity in emission. (c) Overlay of PL from (b) and SEM from (a). The heterogeneity in luminescence appears to be dependent on morphology suggesting underlying grain to grain heterogeneity in composition or structure. (d) TRPL decays from a $MAPbI_3$ perovskite film measured over time under illuminance. Increased time under illumination corresponds to increased lifetimes. Source: de Quilettes et al. [27]/Springer Nature/Licensed under CC BY 4.0. (e) Tof-SIMS image of the iodide distribution summed through the film depth, scale bar 10 μm. (f) Line scan of the blue arrow in (e) to demonstrate the iodide distribution (on right axis). The measured spatial profile of the illumination laser (blue) is shown on the left axis. Source: Reproduced from de Quilettes et al. [23], American Association for the Advancement of Science – AAAS.

to passivation of defect-induced electronic trap states and brightening (of the PL) effects [36]. de Quilettes et al. illuminated a local region of a $MAPbI_3$ sample and used TRPL to observe photobrightening and improvements in charge carrier lifetimes (Figure 16.2d). Correlating the same region of the sample mapped with TRPL with the local chemical distribution using time-of-flight secondary ion mass spectrometry (ToF-SIMS) [27] revealed a drop in the local iodide content in the illuminated region and a rise in the adjacent region (Figure 16.2e,f). This is a direct visualization of local light-induced ion migration in halide perovskite films, revealing the key role of such processes in the transient PL changes. Correlative nano-SIMS investigations by Choi et al. further revealed how these effects manifest on a single grain level [37]. These moving ions act as fuel for redox chemistry to occur, which can either passivate or generate traps depending on the sample conditions [33, 35].

Scanning probe measurements have allowed studies linking multiple properties such as those described above to be explored in greater detail and with higher resolution. For example, Xiao et al. fabricated $MAPbI_3$ perovskite on lateral device architectures and utilized PL and Kelvin probe force microscopy (KPFM) to image in real time changes in material properties under continuous application of a field across the perovskite layer [38], showing a switchable photovoltaic effect caused by ions migrating under the field. Gratia et al. further used KPFM in conjunction with helium ion microscopy with secondary ion mass spectroscopy (HIM-SIMS) to show that in a more compositionally complex (than $MAPbI_3$) alloyed $(FAPbI_3)_{0.85}(MAPbBr_3)_{0.15}$ perovskites, nanoscale compositional heterogeneity in the form of halide segregation could be linked to nanoscale variations in surface potential observed from KPFM [39]. Further work by Gratia et al. showed that these same halide-segregated regions were also deficient in Cs^+, grew substantially during annealing, and were likely linked to the presence of hexagonal perovskite polytypes [40] (Figure 16.3a).

Birkhold et al. combined in-situ PL and scanning Kelvin probe microscopy over time in a lateral electrode geometry to investigate the interaction between mobile ions and injected charge-carriers – specifically by probing changes in radiative recombination and the spatial movement of ions and charge carriers [43]. Comparison between trap formation with both charge injecting and blocking contacts showed the formation of new nonradiative defects (Pb^0) in $MAPbI_3$ only in the presence of injected electrons, highlighting how redox processes play key roles in these transient processes. It is important to note that the primary traps in $MAPbI_3$ have been shown to be electron traps [44], and therefore the specific redox chemistry occurring may depend on the nature of the traps and specific perovskite composition. Such device platforms allowing simultaneous access to optical, electronic, and other material properties will be important for enabling further operando measurements.

In further scanning probe work, Garrett et al. employed heterodyne KPFM to spatially (<50 nm) and temporally resolve the open-circuit voltage of $MAPbI_3$ solar cells [42]. They revealed nanoscale sites (even within grains) with voltage variation >0.3 V under 1 sun illumination (Figure 16.3b) and demonstrated transient voltage behavior of these sites both under light and even in retention of photovoltage in the dark for minutes after illumination was removed – observations which are attributed to trap filling under illumination and subsequent slow ion migration.

Leblebici et al. used photoconductive atomic force microscopy on $MAPbI_3$ films to show similar local variations in short-circuit current and open-circuit voltage up to 0.6 V, attributing such findings to facet-dependent densities of trap states [45]. In regions of grains with particular facets, the combination of both high V_{oc} and high I_{sc} could contribute to the local photovoltaic performance approaching theoretical limits; however, in many other regions, the performance was substantially lower. Subsequent studies have shown that the heterogeneity in current from c-AFM measurements can also be influenced by how effectively the AFM tip extracts charge carriers; the heterogeneity can be partially mitigated by using an AFM probe that extracts current well [46]. Nevertheless, there is still underlying heterogeneity in V_{oc} and I_{sc}, and this may be related to specific facet or crystallographic features or structural phase impurities. Thus, it could, in principle, be possible to improve the macroscale performance of halide perovskite devices by eliminating local variations in V_{oc} through engineering materials to grow in preferred orientations or with preferred grain alignments. These results highlighted the importance of the nanoscale in dictating performance and hinted that better understanding the nanostructure of perovskites could result in improvements in device efficiency and stability.

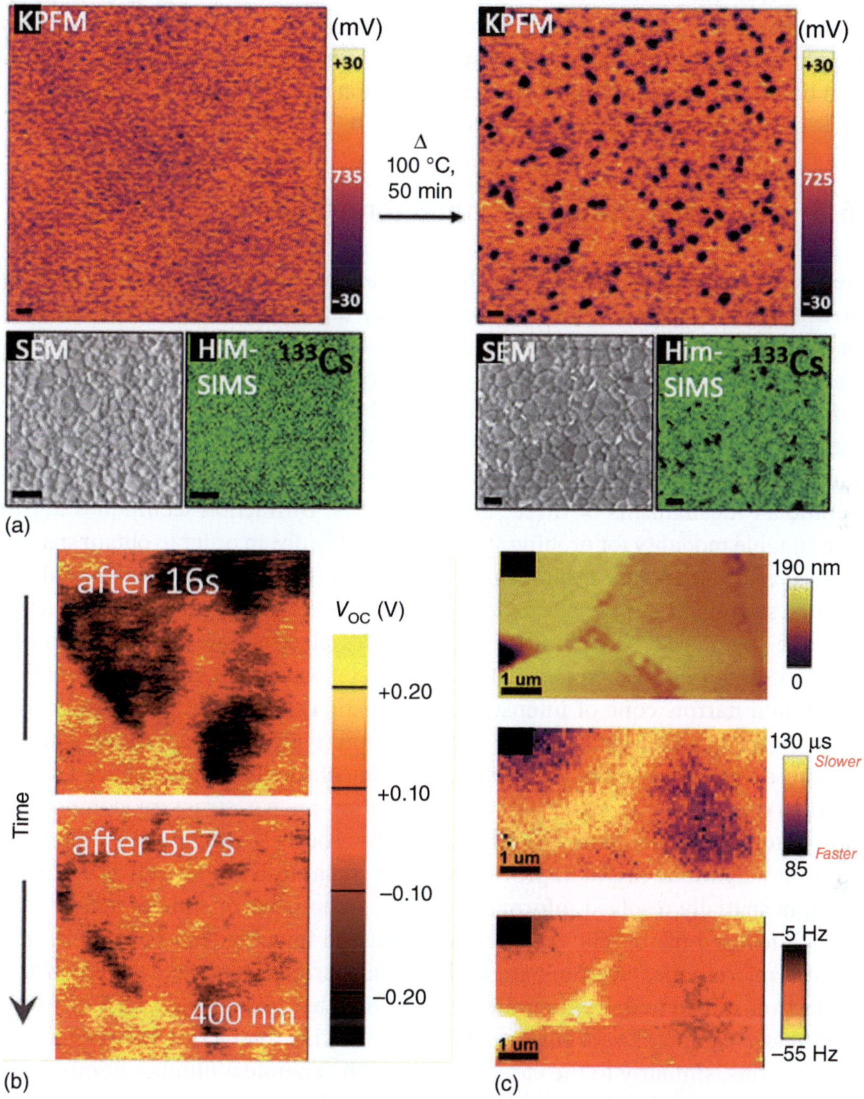

Figure 16.3 (a) KPFM, SEM, and HIM-SIMS maps of a perovskite film fabricated by adding 10% of $CsPbI_3$ to a stock solution of $(FAPbI_3)_{0.85}(MAPbBr_3)_{0.15}$ before and after annealing. After annealing the reduction in the surface potential of certain grains visualized with KPFM corresponds to variations in the *Cs* content of the films, which is visualized with HIM-SIMS. (b) Map of absolute V_{oc} variations in a $MAPbI_3$ perovskite solar cell with nanoscale spatial resolution over time under 1-sun illumination. (c) Fast free, time-resolved electrostatic force microscopy (FF-trEFM) on BAPI (top) topography (middle) FF-trEFM time constant and (bottom) FF-trEFM frequency shift images. In the middle panel brighter regions exhibit slower charging. In bottom panel brighter regions indicate greater total charge. Source: Reproduced from [40–42], American Chemical Society.

Giridharagopal et al. further used time-resolved general mode KPFM and fast free time-resolved electrostatic force microscopy to show slow equilibration effects in 2D perovskites $(C_4H_9NH_3)_2PbI_4$ under illumination, also observing that the surface photovoltage of the 2D grain centers evolves more rapidly in time than at the grain boundaries (Figure 16.3c) – thus highlighting how dimensionality of the metal halide structure influences these properties [41, 47].

16.3 Recent Multimodal Characterization

The multimodal studies discussed to this point are mostly concerned with exploring correlations between photophysics, morphology, and composition. A key outstanding question that arose from this early work, particularly that of de Quilettes et al. relates to the origin of the spatial heterogeneity in local non-radiative recombination events (cf. Figure 16.2). As the deep intraband defect states that cause nonradiative recombination often originate from structural defects, utilizing modalities that can probe structure is important to understand the properties that govern the operational mechanisms of these semiconductors. Diffraction techniques are the most suitable modality for probing structure. Specifically, in order to obtain spatially resolved crystallographic insight into structure on the length scales required (sub 100 nm), synchrotron X-ray diffraction- or electron microscopy-based techniques are most often employed. At synchrotron light sources, electrons are accelerated to close to the speed of light. When these electrons are deflected through a magnetic field, they emit a narrow cone of intense polychromatic radiation in the forward direction, tangent to the electrons orbit. This radiation (typically X-rays) is very intense and can be manipulated (e.g. focused and monochromated) and directed to illuminate a sample of interest to provide information on, e.g. structure, composition, and photophysics. Compared to lab-based X-ray sources, synchrotron sources are orders of magnitude brighter, providing better signal-to-noise as well as the ability to obtain spatially resolved information on the nanometer or smaller length scales through focused nanoprobe or coherent diffraction imaging studies [48, 49]. In an electron microscope, electrons are emitted via thermionic or field emission from an electron source and are then focused and directed to interact with a sample of interest (see Chapter 14 for further information). This interaction between sample and electrons, similarly to the case of X-rays, will generate a number of interesting signals (e.g. cathodoluminescence, diffraction, incoherent scattering, and elemental X-rays), which can describe how the structure, photophysics, and composition of a sample evolve in space and time.

16.3.1 Subgrain Features

In recent years, a particular point that has drawn attention in the perovskite field from multimodal studies incorporating structural measurements is that "morphological" grains in polycrystalline perovskite films as visualized with SEM or AFM can often possess complex subgrain crystallographic features. These hidden crystallographic features can have important implications for performance, as illustrated

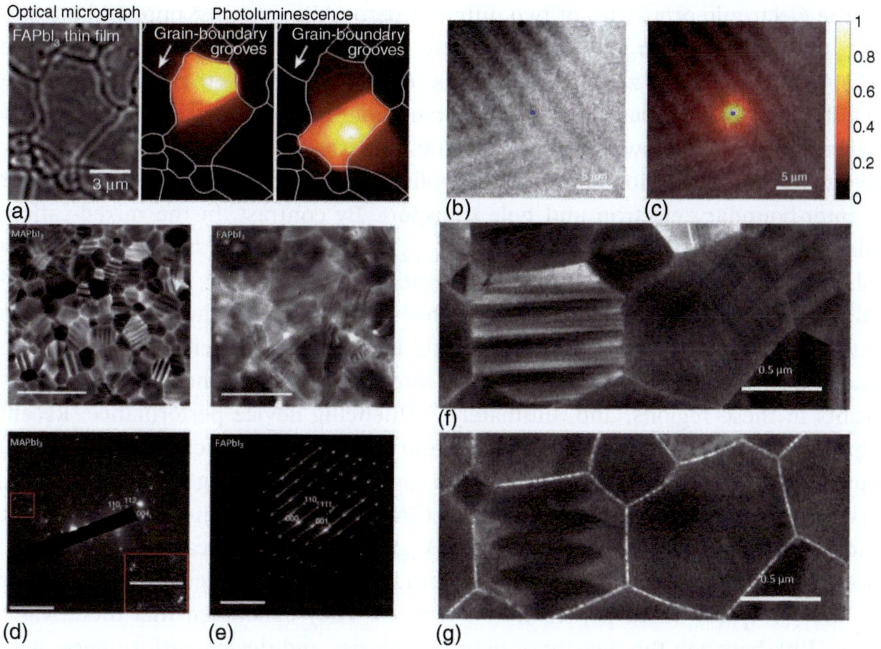

Figure 16.4 (a) Optical micrograph and focused excitation PL intensity distributions illustrating the presence of a subgrain feature not discernible in the optical micrograph map. (b) Polarized optical microscope image of a $MAPbI_3$ single crystal. Striped contrast indicates ferroelastic twin boundaries. (c) Composite image of (b) and a PL intensity image of a $MAPbI_3$ single crystal. The symmetrical intensity distribution across the twin boundaries indicates that charge-carriers are not blocked by the twin boundary. (d) Top: Bright field image of a $MAPbI_3$ thin film. Striped contrast indicates twin boundaries. Bottom: Representative SAED pattern extracted from one of the striped grains (top) revealing the presence of a $\{112\}_t$ twin boundary. (e) Top: Bright field image of a $FAPbI_3$ thin film. Striped contrast indicates twin boundaries. Bottom: Representative SAED pattern extracted from one of the striped grains (top) revealing the presence of $\{111\}_c$ twin boundaries and stacking faults. Bright field TEM images (striped contrast indicates twin boundaries) of a $MAPbI_3$ thin film after (f) brief, low-dose electron exposure, and (g) 1 minutes of exposure using a dose rate of $\sim 2e\,Å^{-2}\,s^{-1}$. After 1 minutes of exposure the twin boundaries have begun to disappear. Source: Reproduced with permission from [50–53], American Chemical Society.

by Li et al. who observed subgrain boundaries with PL and electron backscatter diffraction (EBSD) mapping in $FAPbI_3$ films that were not apparent from simple optical or morphological maps of the same area Figure 16.4a. The subgrain features observed by Li et al. were likely twin boundaries, and recent theoretical and experimental work utilizing diffraction techniques has shed some light on this particular class of subgrain features.

A twinned crystal is an aggregate of two crystals that share some of the same lattice points (the twin boundary) and are, as such, joined together according to a specific symmetry operation. Twins can be produced during growth but can also occur as a result of a phase transformation or structural deformation. Density functional theory calculations were used to investigate the effect of {111} twin boundaries on the electronic properties of two different perovskite systems: pure $FAPbI_3$ and $FA_{0.875}Cs_{0.125}Pb(I_{0.666}Br_{0.333})_3$ [54]. In pure $FAPbI_3$, the formation energy of the {111} twin boundary at zero temperature is very low; therefore, these defects should readily form in real materials at higher temperatures, which was experimentally confirmed [50, 55]. However, the electronic properties of these twins are relatively benign for single-junction photovoltaic applications, presenting only a small barrier to inter-boundary electron and hole transport. By contrast, in the mixed-cation, mixed-halide perovskite $FA_{0.875}Cs_{0.125}Pb(I_{0.666}Br_{0.333})_3$, which is suitable for tandem solar cells, twin boundaries induce segregation of I^- and Cs^+, leading to a local valence band maximum that is 0.2 eV higher in energy than that of the bulk. Such heterogeneous valence band energies could lead to holes preferentially collecting near the twin boundaries, increasing the number of electron–hole recombination events in these regions, and, ultimately, influencing device performance. Recent experimental results were largely consistent with these DFT calculations. By (i) varying the density of twin boundaries via compositional tuning of $MA_{1-x}FA_xPbI_3$, (ii) using selected area electron diffraction (SAED) to visualize the twin boundaries, and (iii) monitoring parameters such as V_{oc}, I_{sc}, and charge-carrier lifetimes, Li et al. were able to elucidate the effect of intragrain planar defects on performance in $MA_{1-x}FA_xPbI_3$-based perovskite PV devices. They concluded that there was a direct link between the density of twin boundaries and device performance, with perovskite compositions that had a lower density of twin boundaries performing better than those that had a higher density. However, deterministically separating the effect that the density of twin boundaries have on performance from the effects that other factors could have on performance, such as composition (variations in MA and FA content were utilized to tune twin density), is challenging and will likely require correlation of, for example, SAED with local V_{oc} and I_{sc} maps.

Twin domains have also been interrogated in the context of ferroelasticity in perovskites. Ferroelasticity is a well-reported phenomenon in oxide perovskites and, similar to ferroelectricity, can be related to nanoscale structural and compositional heterogeneity. This phenomenon is characterized by an elastic hysteresis that results from the mechanical switching between two crystallographic orientations owing to the application of an external stress [56]. Inherent coupling between ferroelasticity and local chemical segregation within $MAPbI_3$ grains has been revealed using several correlated force microscopy and compositional mapping techniques [57]. Piezoresponse force microscopy measurements show striped domains that are assigned as ferroelastic twin regions, which exhibit local ion segregation. It is proposed that this chemical segregation is induced by the release of inherent strain energy from the sample. Furthermore, these periodically ordered, chemically segregated domains may also possess different crystallographic orientations, and these combined properties are a signature of ferroelastic behavior. This chemical heterogeneity on the subgrain scale could have notable positive or negative effects on the electronic properties of halide perovskites. For example, the heterogeneity could either enhance charge-carrier separation near the domain walls owing to ion segregation and thus improve photovoltaic properties or limit device performance owing to the presence of low-conductivity domains [57]. More recent work by Xiao et al. has shed further light on the impact of ferroelastic twin boundaries in $MAPBI_3$ on performance. By correlating scanning photocurrent microscopy with polarized optical microscopy (which is capable of identifying twin domains – Figure 16.4b) and PL measurements (Figure 16.4c), they revealed that ferroelastic twin boundaries are, generally, electronically benign and do not behave as nonradiative recombination center's or impede carrier transport [52].

It is important to note here that the twin boundaries observed in tetragonal $MAPbI_3$ – which have been identified with structural studies by Rothmann et al. to be $\{112\}_t$ type twins in Figure 16.4d – are different than those in FA-rich pseudocubic perovskites – which are $\{111\}_c$ type twins in Figure 16.4e [50, 54, 55]. The $\{112\}_t$ twins in $MAPbI_3$ have been observed to be less stable than those in FA-rich perovskites and disappear quickly under either electron beam irradiation or heating Figure 16.4f,g. This is in part because the $\{111\}_c$ type twins are composed of a series of face-sharing octahedral, which is a very thermodynamically stable structure [50, 53, 58] and in part because $MAPbI_3$ undergoes a phase transition to a cubic structure at ~330 K [59]. Observations such as this mean that it is difficult to generalize the impact of twin boundaries across all compositions of perovskites, as when composition changes, so too does structure. More work is required to understand the impact of this special class of subgrain features, benign or otherwise, on performance and stability in halide perovskites, particularly in the wide-bandgap compositions suitable for tandem PV where twin boundaries are currently under investigation.

16.3.2 Strain and Photophysics

Beyond subgrain features like twin boundaries, other structural properties such as strain can play an important role in modulating a semiconductor's local photophysics. Strain can simply be defined as a material's response to an applied stress (we direct the reader to Chapter 13 for a more rigorous definition) and can be interrogated with techniques such as EBSD, microbeam X-ray Diffraction (μXRD) and nanobeam X-ray diffraction (nXRD), which are capable of mapping fields of view comparable to PL techniques. Jones et al. utilized micro-μXRD and time-resolved confocal PL imaging on the same scan regions of both $MAPbI_3$ thin films to further visualize how strain varies on the microscale and affects the radiative recombination of carriers [60]. By tracking local fluctuations in either the position or broadness of diffraction peaks in μXRD, strain-induced variations in the interatomic spacing of a material could be observed. In a $MAPbI_3$ film, the PL lifetime traces from poorly and brightly emissive regions (Figure 16.5a, left) correspond to regions of higher and lower compressive strain, respectively, in the strain map extracted from μXRD measurements (Figure 16.5a, right; the red and blue circles indicate the regions from which the local PL decays were extracted). As the less strained (brightly emissive) regions have significantly longer lifetimes than the highly strained areas, it indicates that the local strain is associated with trap-related, nonradiative recombination events. Further nXRD experiments on these $MAPbI_3$ samples revealed "super-grain" like structures (which were defined here as regions with common azimuthal scattering angles) as large as 25 μm^2 in lateral area, signifying long-range, uniform regions sharing some form of common orientation that are at least an order of magnitude larger than the average grain sizes identified by SEM [60, 62]. Notably, Jones et al. observed that a film of mixed-cation perovskites, which, on average, has better optoelectronic performance than $MAPbI_3$ [63], contains "super-grains" that are up to ~250 μm in lateral area. In further work, Li et al. used correlations between PL maps and nano-XRD maps of strain in CsPbBr thin-film crystals, revealing PL shifts and instabilities associated with out-of-plane compressive strain [64].

Crystal orientation/relative misorientation may be related to strain variations. Correlating EBSD and SEM with PL allowed Jariwala et al. to examine the effect of crystal orientation, subgrain structure, and strain on nonradiative recombination and efficiency losses. Inverse pole figure maps, which reveal local crystal orientations, again demonstrated that some single morphological grains, as viewed by SEM, were composed of a number of distinct subgrain crystallographic entities, thus revealing boundaries not visible with SEM alone. Interrogating the grain-to-grain misorientation revealed that grain boundaries characterized by a larger degree of intragrain misorientation and thus a larger degree of local strain were linked to higher trap densities as extracted from PL measurements [61] – Figure 16.5b,c.

The above studies clearly highlight the importance of understanding the impact and origin of strain in perovskite materials; however, strain remains relatively poorly understood within the perovskite community. We know that strain can arise globally from, for example, a mismatch between the coefficient of thermal expansion of a perovskite and the substrate on which it is grown [65–67], lattice mismatch between the as-formed perovskite and the substrate, through mechanical bending [68], and hydrostatic pressure. On the local level, strain has also been linked to factors such as atomic defects [69], phase transitions [70], light/bias stimulation [71], and grain boundaries [61]. Interestingly, in addition to impacting performance, as shown by Jariwala et al. and Jones et al. [60, 61] there are hints that strain can be detrimental to material stability, for example by regulating the thermodynamics of defect generation [67] or by decreasing the energy barrier for ionic migration, which in turn will lead to degradation through atomic rearrangement and redox chemistry between migrated ions and other species [65, 66, 72]. This potentially points toward strain as a common link between both material stability and performance [73].

However, many (though not all) of the studies to date that sought to elucidate the impact of strain on perovskite material properties have suffered from a variety of potential drawbacks that may impact their conclusions, such as a lack of an "unstrained" reference state with which to compare experimental observations or lack of understanding in regards to Poisson's effect when applying uniaxial strain to a film and evaluating its effect [65–67]. Possibly related to this, a number of seemingly contradictory general claims have been made about the impact of strain on perovskites. For example, strain is linked to perovskite degradation [65] but also stabilizes the desirable photoactive phase of $FAPbI_3$ [74]; and TRPL measurements of $MAPbI_3$ reveal short carrier lifetimes in regions with greater local compressive strain [60], but in $(FAPbI_3)_{0.85}(MAPbBr_3)_{0.15}$ films, the PL lifetime reportedly

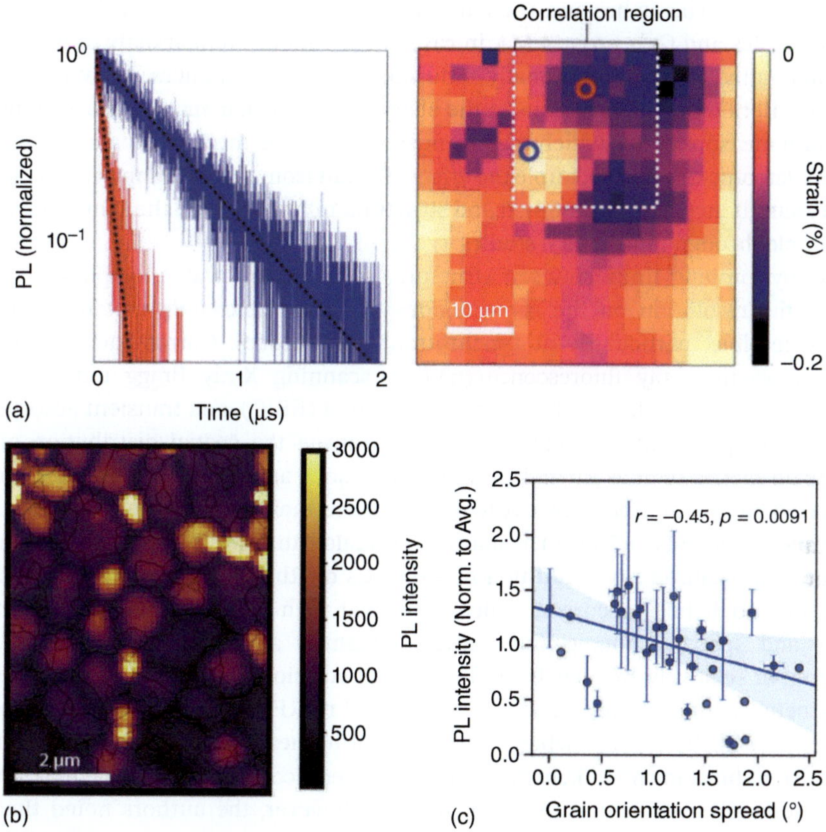

Figure 16.5 (a) Correlated PL and structural measurements. Left: time resolved PL traces extracted from two different regions of a $MAPbI_3$ thin film. The red trace corresponds to a region of larger strain (marked Red on right) while the blue trace (longer lifetime) corresponds to a region of less strain (marked blue on right). Source: Adapted from Jones et al. [60]/Royal Society of Chemistry/Licensed under CC BY 3.0. (b) Confocal image overlayed with a grain boundary map of the same region extracted from EBSD measurements. (c) Plot of PL intensity as a function of grain orientation spread (GOS) in $MAPbI_3$ showing negative correlation, The grains are binned in GOS intervals. The line represents a linear regression fit to the data with the shaded region representing a 95% confidence interval for the regression. Error bars indicate the standard deviation of average PL intensity in a specific GOS interval. Source: Reproduced with permission from Jariwala et al. [61], Elsevier.

increases with compressive strain [75]. In these cases, it is likely that differences in composition, in particular, may play a role; for example, different compositions of perovskites may respond to strain differently. This observation brings us to an important point; when interrogating strain, or any structural feature for that matter, it is important to decouple its effect from that of compositional variations. For example, spatial variations in the halide (or cation) content of a perovskite film in mixed halide and cation perovskite compositions, which readily occur in even high-performing perovskite compositions [76, 77], can locally change the size of the unit cell. This is because the ionic radii of, for example, I^- and Br^- in the case of halides and Cs^+, FA and MA in case of the cations are markedly different. Experimentally, any compositionally induced variation or gradient in local unit cell size will manifest as variations in the observed interplanar spacing observed in structural measurements, which may be misinterpreted as mechanical strain. One way to overcome this and decouple the effect of strain from that of composition is to acquire simultaneous compositional and structural information on the same region of a sample through multimodal studies.

Recently, probing links between structure, composition, and photophysics at high-spatial resolution on the same regions of halide perovskite materials has become feasible, particularly at synchrotron light sources. Correa-Baena et al. used nanobeam X-ray fluorescence (nXRF), scanning X-ray Bragg diffraction microscopy (SXDM), electron beam induced current (EBIC), and transient absorption microscopy (TAM) measurements to investigate the spatial distribution of both alkali metals such as Rb and Cs and halides such as I and Br in mixed cation and mixed halide "alloyed" perovskite materials and their impact on performance. nXRF measurements (~250 nm resolution) revealed that the halide distribution became homogenized upon addition of either Cs or Rb for films that possess an excess of lead halide precursor, which coincided with long-lived charge carrier decays and spatially homogenous carrier dynamics as visualized with TAM. However, at relatively low levels of alkali metal additions (>~1%), second-phase alkali metal-rich aggregates nucleated. SXDM and nXRF revealed that the Rb-rich clusters were $RbPbI_3$, which, when probed with EBIC measurements, were revealed to be current blocking and likely to be charge-carrier recombination active, making them ultimately detrimental for performance. However, the authors noted that this detrimental impact was balanced by the advantageous impact on halide homogenization caused by Rb and Cs additions [78].

Doherty, Winchester et al. pursued higher spatial resolution studies and utilized photoemission electron microscopy (PEEM) correlated with PL to spatially resolve the location of charge-carrier traps on the surface of $(Cs_{0.05}FA_{0.78}MA_{0.17})Pb(I_{0.83}Br_{0.17})_3$ and $(Cs_{0.05}FA_{0.78}MA_{0.17})Pb(I)_3$ thin films [79]. Surprisingly, rather than a uniform distribution of traps within regions of poor PL, they observed discrete nanoscale clusters of traps of photo-excited holes. Correlating these PEEM measurements with AFM, KPFM, and compositional measurements in the form scanning transmission electron microscopy energy dispersive X-ray spectroscopy (STEM-EDX) revealed that the trap clusters were strongly associated with compositional heterogeneity and grain boundaries Figure 16.6a. Trap clusters predominantly manifested at what appeared to be boundaries between grains possessing the compositional signature of the pristine perovskite and "inhomogeneous" grains that were deficient in Br and possessed a slight excess of I. Scanning electron diffraction (SED), whereby an electron probe is scanned across a sample of interest and a diffraction pattern is acquired every 5 nm, was correlated with PEEM and STEM-EDX measurements (Figure 16.6b). The spatially resolved SED results revealed that the compositionally inhomogeneous grains associated with deep trap clusters were also structurally distinct from the pristine perovskite

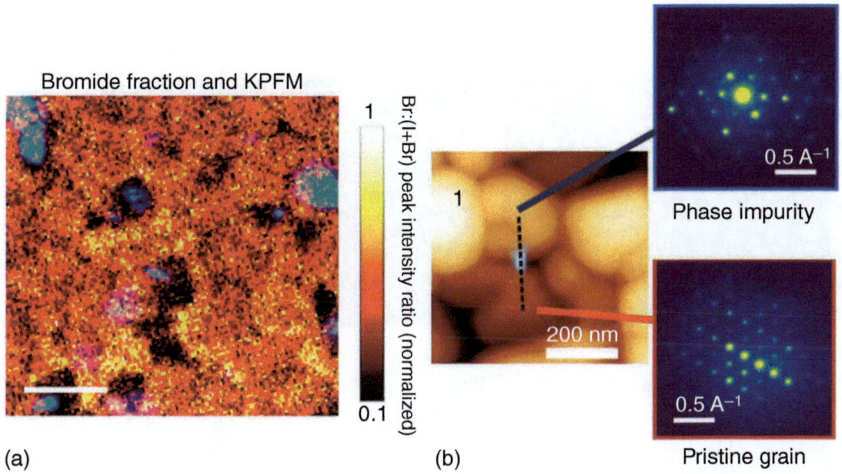

Figure 16.6 (a) KPFM measurements showing the location of trap clusters overlayed on a STEM-EDX map of the fraction of bromide intensity counts out of total halide counts, I(Br Kα)/(I(I Lα) + I(Br Kα)). The majority of trap clusters are associated with grains that are poor in Br. (b) AFM image of a $(Cs_{0.05}FA_{0.78}MA_{0.17})Pb(I_{0.83}Br_{0.17})_3$ film overlaid with a PEEM map denoting a deep trap cluster (blue). Such grain junctions are associated with interfaces between pristine grains and phase impurities, as exemplified by the denoted diffraction patterns. Source: Adapted from Doherty et al. [79], Springer Nature.

structure, although the authors could not, at the time, definitively identify the crystal structure of the inhomogeneous grains. This study helped to reveal the structurally and compositionally disordered landscape that contributed directly to the spatial variations in non-radiative recombination first visualized by de Quilettes et al. (cf Figure 16.2).

Further work by Kosar et al. elucidated the nature of these structurally distinct grains associated with the generation of trap clusters that were observed by Doherty, Winchester et al. [80]. Correlating PEEM (Figure 16.7a,b) with nXRD (Figure 16.7c), Kosar identified three different nanoscale structural entities (defects and phase impurities) associated with trap clusters. The first phase impurity was PbI_2, which was revealed to be relatively benign electronically by time-resolved PEEM (TR-PEEM) as charge carriers photogenerated in the perovskite could not be injected into the PbI_2. The second phase impurity was hexagonal perovskite polytypes, which, in comparison to the PbI_2, did participate in charge-carrier trapping, which was revealed by a decrease in photoemission intensity at hexagonal perovskite phase impurity sites after photoexcitation. The third structural entity are grain boundary type defects, which were found to be the most detrimental for performance as they exhibited the largest reduction in TR-PEEM intensity after photoexcitation, indicating a very significant cross section for hole capture. These grain boundary defects are beyond the resolving power of nXRD techniques employed in this study as they are very likely to be either very small phase impurities or to be atomic scale defects such as vacancies and interstitials that are of a disordered nature making them extremely difficult to directly observe with

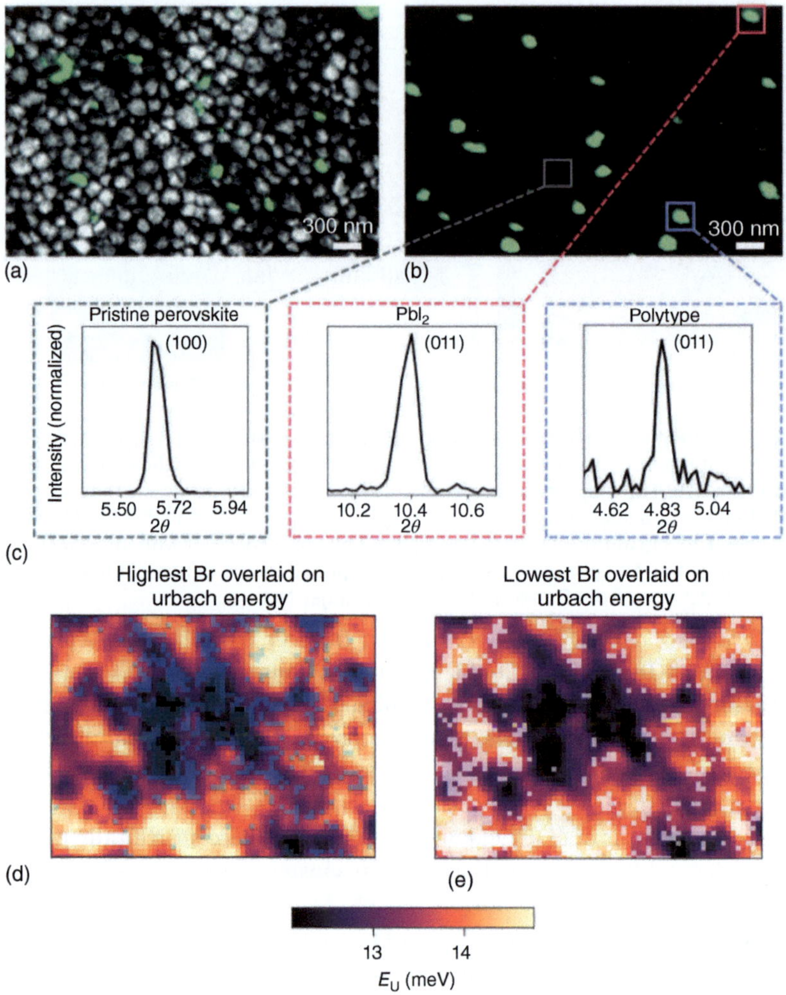

Figure 16.7 (a) PEEM image of the surface morphology (gray contrast) of a $(Cs_{0.05}FA_{0.78}MA_{0.17})Pb(I_{0.83}Br_{0.17})_3$ generated by utilizing 6.2 eV photons as a probe which excite the valence band edge of the perovskite film. Overlaid on the morphology map are trap clusters (green) which are images by utilizing 4.65 eV photons which selectively photoexcite intraband sites. (b) PEEM image correlated with nXRD measurements taken with a 4.65 eV probe showing the spatial location of trap clusters (green). Boxes show nXRD regions of interest. Source: Reproduced from Kosar et al. [80], Royal Society of Chemistry / CC BY 3.0. (c) Local diffraction measurements extracted from the regions indicated in (b) showing from (left to right) pristine perovskite not correlated with any trap clusters. PbI_2 correlated with the presence of trap clusters. Hexagonal polytypes associated with trap clusters. (d and e) Correlations between local luminescence measurements and local composition show that regions of high Urbach energy are anticorrelated with regions of high Br content. Source: Frohna et al. [77]/with permission of Springer Nature.

diffraction techniques. This will motivate further correlated studies with atomic resolution techniques, such as high-resolution STEM in order to resolve the precise nature of these grain boundary defects.

In spite of the presence of the above-identified trap clusters, mixed halide and mixed cation "alloyed" perovskite compositions have achieved high power conversion efficiencies and led most recent PV efficiency tables [7]. Recent studies correlating local structural, compositional, and photophysical measurements have also helped us reconcile high performance in the presence of a relatively high density of what should be performance- limiting deep charge-carrier traps. For instance, work by Feldmann, Macpherson et al. revealed that in FA-rich mixed halide compositions such as $(Cs_{0.05}FA_{0.78}MA_{0.17})Pb(I_{0.83}Br_{0.17})_3$, spatially varying halide content leads to local bandgap variations that act as funnels for charges on local low bandgap sites, which in turn leads to "hot spots" of high radiative recombination at these sites. Frohna et al. further built upon this by correlating local luminescence measurements on thin films with nanobeam X-ray fluorescence and nXRD measurements to spatially map the chemical distribution of $(Cs_{0.05}FA_{0.78}MA_{0.17})Pb(I_{0.83}Br_{0.17})_3$. Local regions of high absolute PL quantum efficiency (PLQE) and low Urbach energy (a parameter that describes the band-edge energetic disorder) were found to be strongly associated with local regions rich in bromide [77] (Figure 16.7d,e). Nevertheless, the local PL spectrum in these highly emissive regions has a slightly red-shifted shoulder with respect to other regions, consistent with emission from sites with more iodide. These results paint a complex picture of these "hot spots": they appear to have an excess of bromide when probing through the bulk region but may have small pockets of lower-bandgap material with slightly more iodide, which may be in the form of a surface layer or small nanoscopic inclusions of iodide-rich material in which strong radiative recombination occurs. Remarkably, this local material is a clean semiconductor with very low Urbach energy. Through further correlations between nXRF and ultrafast TAM, the movement of charges to these local sites was visualized. A key conclusion is that in these alloyed perovskite systems, such compositional disorder dominates the optoelectronic response even over nanoscale strain variations (~1%) in other regions. These shallow energetic gradients funnel carriers and act as a mechanism to draw carrier recombination away from trap clusters associated with electronic disorder. Therefore, this effect can be interpreted as a form of enhanced tolerance to traps that occur serendipitously in these mixed composition materials that have been optimized through empirical efficiency-oriented device developments. Nevertheless, such heterogeneity may become detrimental in devices such as solar cells as they approach the radiative limit, in which deep trap states are at negligible levels where trap avoidance is no longer necessary. Furthermore, there is a fine balance, as larger gradients in which halides segregate more dramatically, which may occur under continued operation under light and/or bias, will be detrimental for performance [78].

Equally important as relating the halide composition to structure, performance, and stability, understands the role of the A-site cation. Though work by Correa-Baena et al. revealed the spatial distribution of Rb and K with nXRF, currently the most technologically relevant halide perovskite materials mix FA,

MA, and Cs^+ on the A-site. This can be challenging to interrogate with nXRF or STEM-EDX as the Cs_L and I_L emission lines overlap and the FA and MA cations are composed of elements that are too light to be effectively mapped with X-ray spectroscopy. However, techniques such as infrared spectroscopy can effectively map the A-site organic cations. Szostak et al. used simultaneous AFM morphological mapping with synchrotron infrared nanospectroscopy (nano-FTIR) to probe the nanoscale organic MA and FA grain compositions in $Cs_{0.05}FA_{0.79}MA_{0.16}Pb(I_{0.83}Br_{0.17})_3$ and $FA_{0.83}MA_{0.17}Pb(I_{0.83}Br_{0.17})_3$ thin films, revealing large variations in both cation species [81]. These large variations in cationic species had important implications. While AFM topography maps revealed morphological grains that appeared isotropic in their physical properties, IR broadband images showed substantial variations in the vibrational activity, independent of image topography. Nano-FTIR spectra extracted from the points indicated in showed variations in the FA characteristic vibrational mods at 1710 cm^{-1} suggesting that the FA content of the film varied spatially, which was likely related to the presence of nanoscale phase impurities (i.e. PbI_2 and hexagonal perovskite phase impurities) – Figure 16.8a,b,c. These are the very same phase impurities that Kosar et al. showed were associated with nanoscale trap clusters. In addition, there is increasing evidence that these phase impurities, which have been observed in a number of FA-rich perovskite compositions [39, 83], are also associated with degradation under operational conditions in perovskite devices, making them particularly sinister inclusions in films [32]. Understanding the origin of these phase impurities, in particular the hexagonal impurities, is thus important to fabricate higher performing and more stable devices.

Szostak et al.'s work hinted that the presence of hexagonal phase impurities is associated with variations in the A-site cation content. Work by Doherty, Nagane et al. provided further information on why and how hexagonal perovskite phase impurities can form in FA-rich perovskite films and in so doing outlined a pathway towards stable $FAPbI_3$-based photovoltaics free of other cationic additives. Generally, $FAPbI_3$ is an excellent candidate for commercial PV applications due to its greatly increased thermal stability compared to MA based perovskite compositions and a near-ideal bandgap. However, $FAPbI_3$ is challenging both to fabricate and stabilize, as the desirable photoactive average cubic phase comprising corner-sharing PbI_6 octahedra is only stable at temperatures greater than 150° C and at room temperature the material rapidly transitions to wide bandgap, face-sharing hexagonal polytypes, such as the 2H δ-phase [39, 84], which are not useful for solar light harvesting. The most promising stabilization strategies to date involve alloying FA on the A-site with Cs^+, MA, or mixtures [63, 85–90]. Utilizing SED, Doherty, Nagane et al. showed that such alloyed, stable, FA-rich perovskites like $(Cs_{0.05}FA_{0.78}MA_{0.17})Pb(I_{0.83}Br_{0.17})_3$ exhibit a minor degree of octahedral tilting at room temperature making them on average non-cubic – Figure 16.8d,e. The octahedral tilting was so minor that it could not be detected with traditional bulk characterization techniques but the local, low-dose nature of SED and the fact that electrons interact strongly with matter meant that the minor structural distortions were readily detected in SED measurements. This is an important

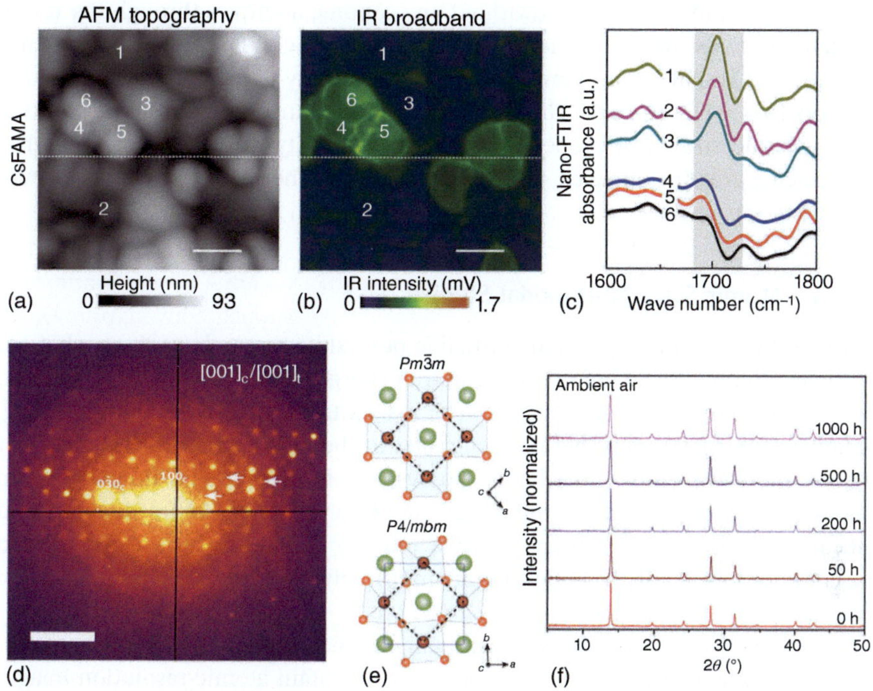

Figure 16.8 (a) AFM topography map of a pristine $Cs_{0.05}FA_{0.79}MA_{0.16}Pb(I_{0.83}Br_{0.17})_3$ (1µm x 1µm region). (b) IR broadband images revealing heterogeneity in vibrational activity. Topography and IR broadband response do not completely correlate, as shown by the profiles in (c) Nano-FTIR spectral from the regions marked by numbers in (a) and (b). Source: Reproduced with permission from Szostak et al. [81], American Association for the Advancement of Science - AAAS. The characteristic vibrational mode of FA shows up only in grains with a weaker IR broadband response. (d) Electron diffraction pattern of a pristine $(Cs_{0.05}FA_{0.78}MA_{0.17})Pb(I_{0.83}Br_{0.17})_3$ perovskite grain oriented near $[001]_c$. Reflections that are forbidden from appearing in a cubic structure are indicated by arrows. These extra reflections come from octahedral tilting. (e) Schematic representation of a cubic *Pm-3m* (top) and an octahedrally tilted tetragonal *P4/mbm* (bottom) perovskite. *P4/mbm* is the actual space group of $(Cs_{0.05}FA_{0.78}MA_{0.17})Pb(I_{0.83}Br_{0.17})_3$. (f) Octahedral tilt stabilized $FAPbI_3$ exhibiting phase stability after exposure to ambient air for 1000 hours. Source: Reproduced with permission from Doherty et al. [82], American Association for the Advancement of Science – AAAS.

result as the previous prevailing wisdom in the perovskite community was that these stable, alloyed, FA-rich compositions possessed the same cubic crystal structure as $FAPbI_3$. Doherty, Nagane et al. further showed that the octahedral tilting frustrated transitions from corner sharing to face-sharing structures, thus providing phase stability to the desirable photoactive corner-sharing structure at room temperature providing a universal mechanism for phase stability in alloyed perovskite compositions. The authors also used AFM-IR measurements to show that heterogeneity in the A-site cation in FA-rich perovskites led to local regions that did not possess octahedral tilting and are thus susceptible to transitioning to hexagonal phase impurities. Finally, the authors showed that octahedral tilting can be induced in $FAPbI_3$ films without any cationic additives through the use of templating agents such as ethylenediaminetetraacetic acid. The author's octahedral tilt stabilized $FAPbI_3$ films were stable against a variety of external environmental stressors [82] – Figure 16.8f. This approach presents further opportunities to tune the nanoscale structural properties and phase stability through exploring other tilt-inducing growth strategies, ultimately enabling higher performance with fewer unwanted trap clusters.

16.3.3 Atomic Scale Multimodal Studies

Many of the remaining questions in halide perovskite research, such as what are the precise atomic-scale defects that can trap charge carriers and see degradation; require, by definition, atomic-scale multimodal studies to further understanding. Obtaining information on these length scales in the most technologically relevant organic–inorganic (thin film) halide perovskites is exceedingly difficult due to a combination of challenges, including sample preparation, beam damage, and data interpretation. Recently, however, a number of studies have demonstrated the communities ability to obtain highly informative information on these length scales.

Rothmann et al. utilized low-angle annular dark field (LAADF) imaging in a scanning transmission electron microscope (TEM) to obtain atomic-resolution images of polycrystalline thin film $FAPbI_3$ and $MAPbI_3$ for the first time. By depositing $FAPbI_3$ and $MAPbI_3$ on carbon TEM grids by thermal evaporation, samples with the required thickness for atomic resolution imaging (~30 nm) were obtained. LAADF images revealed the presence of coherent, defect-free, low-strain interfaces between remnant precursor PbI_2 and $FAPbI_3$ grains, helping to explain why small excess amounts of PbI_2 might not be detrimental to PV performance. In addition, aligned point defects in the form of vacancies on the Pb–I sublattice in the $FAPbI_3$ were observed, which provided direct experimental visualization of defects long predicted by theory. At the same time, stacking faults and edge dislocations, relatively under-explored structural defects in perovskites, were also shown to be common, which could have important implications for performance. Examination of $FAPbI_3/FAPbI_3$ grain boundaries provided further insight: although most triple-point boundaries are crystallographically continuous, some boundaries notably contained amorphous material and aligned point defects [91]. This work by Rothmann et al. is an important step on the pathway to fully correlated atomic-scale studies.

More recently Cai et al. performed high-angle annular dark field (HAADF) STEM measurements on cross-sectional lamellas of $FA_{1-x}Cs_xPbI_3$ perovskite thin films prepared with a focused ion beam (FIB). This was the first atomic resolution study on mixed- cation thin film perovskites. Examining cross-sectional samples such as the ones in this study can provide important information on how contact layers and interfaces in full device stacks interact with the perovskite active layer, a particularly important field of study considering that many remaining device losses are at the contacts. However, preparing lamella using the FIB has been very challenging due to beam-induced damage from the GA ions utilized to mill an electron transparent sample [53]. Following sample preparation, Cai et al. deposited a layer of amorphous carbon on their sample, which they observed dramatically improved the beam damage tolerance of the perovskite by applying a 10 nm-thick layer of amorphous carbon on the perovskite lamella. Acquiring HAADF images revealed an orthorhombic unit cell and heterogeneity in the A-site cation mixing on the atomic scale: In HAADF imaging brighter image contrast corresponds to atoms with a higher Z number. Cai et al. observed nanometer sized regions that they concluded were rich in FA – Figure 16.9a. They also observed a series of intragrain stacking faults and twinning interfaces (along the orthorhombic (011) twin boundary) – Figure 16.9b,c,d. The direct experimental observation of these

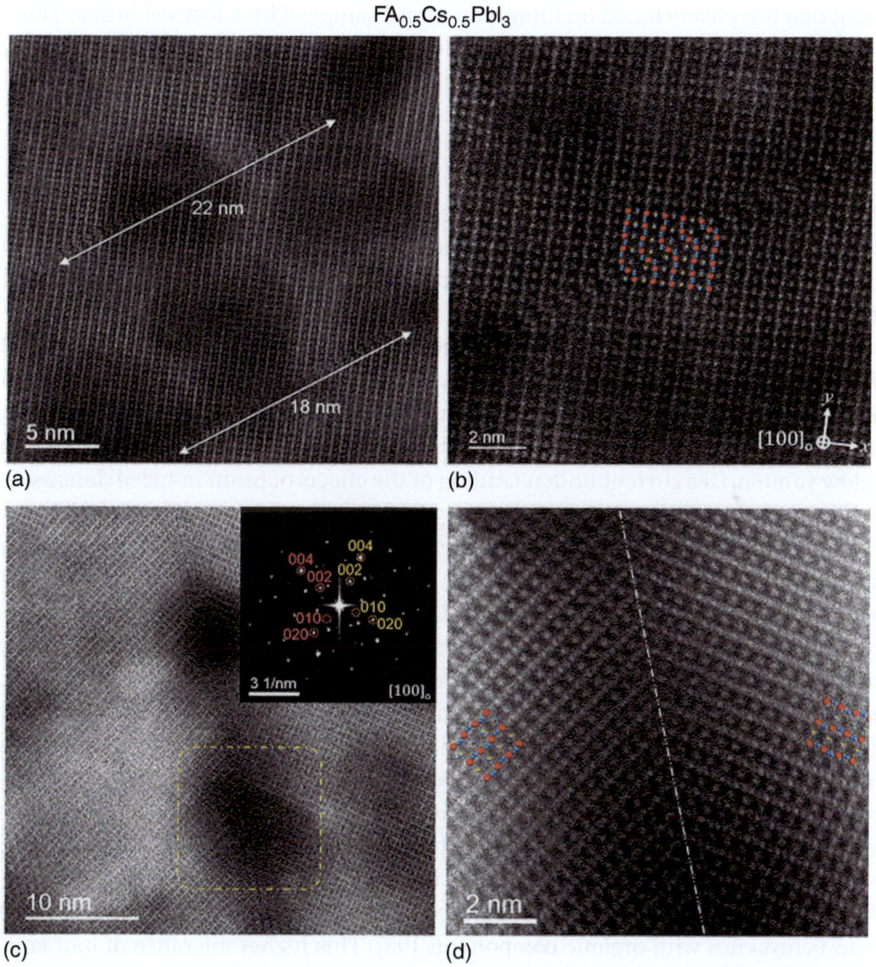

Figure 16.9 HAADF-STEM images of an $FA_{0.5}Cs_{0.5}PbI_3$ thin film perovskite showing (a) compositional heterogeneity on the 10's of nm length scale. Regions of dark contrast are rich in FA. (b) A single stacking fault. (c) A twin boundary. (d) Bragg filtered STEM-HAADF image from the yellow square in (c). Source: Reproduced from Cai et al. [92], American Chemical Society/CC BY 4.0.

defects with high resolution meant that DFT calculations of these exact boundaries could be performed to elucidate their impact on performance. Future work focused on understanding whether the observed variations in FA are present without the sample being preexposed to the FIB will be an informative avenue of research and critical to assess beam induced changes that may occur in these types of studies

16.4 Pressing Challenges and Opportunities

Generally, the probing of properties such as structural, chemical, and optoelectronic properties on different length scales necessitates advanced characterization equipment that is typically based on illumination of a sample with a focused probe. There are several technical and logistical challenges for performing such measurements, especially in multimodal form.

16.4.1 Challenges: Beam Damage

For spatially resolved compositional and structural studies, the most commonly employed probes are composed of X-rays or electrons both of which are forms of ionizing radiation that can modify the properties of the material with which they interact. In the case of halide perovskites, which when compared to other semiconductors are particularly beam sensitive, it is particularly important to understand the impact that ionizing radiation can have on sample properties, particularly in multimodal, correlated samples where multiple techniques and forms of ionizing radiation may be used sequentially on the same region of a sample. The section below summarizes current understanding of the effects of beam induced damage in halide perovskites.

Xiao et al. utilized cathodoluminescence (CL) spectroscopy to examine damage induced by electrons in $MAPbI_3$, $FAPbI_3$, and $CsPbI_3$ perovskite thin films. CL signals were acquired under a number of different experimental operating conditions, varying the voltage from 2 to 10 kV and the current from 0.2 to 14 nA. At lower probe currents and accelerating voltages, excitonic peak broadening and a decrease in CL signal were observed after a relatively short time period of 30 seconds. At higher electron energies, new higher- energy photon peaks appeared in the CL spectrum in addition to the peak broadening. The peak broadening was explained by the generation of defects by the electron beam (as a result of knock-on damage), while the appearance of high energy peaks was attributed to heat- induced damage from the electron beam that led to the formation of intermediary phases with larger bandgaps than $MAPbI_3$. Similar results were observed in $FAPbI_3$, while $CsPbI_3$ was more stable than perovskites with organic components [93]. This higher tolerance of inorganic perovskites compared to organic analogues to radiation damage was echoed by the work of Klein-Kedem et al. who utilized EBIC to investigate the effect of electron beam irradiation on $MAPbI_3$, $MAPbBr_3$, and $CsPbBr_3$. EBIC measurements were carried out with a beam current of 4–5 pA and pixel integration time of 60 μs, resulting in an accumulated dose of $1.5–4.5 \times 10^{16}$ e^- cm^2 per imaging scan. The authors observed that at these imaging conditions, the fully inorganic perovskites such as $CsPbBr_3$ were capable of tolerating electron doses 2 orders of magnitude greater than the organic perovskites [94].

Milosavljevic et al. investigated the effect of low-energy electron-induced transformations in $MAPbI_3$. Energy ranges between 4.5 and 60 eV were utilized. To put this energy in context, most SEM-based studies of perovskites operate between 1 and 30 keV and most TEM studies operate between 100 and 300 keV. Lower-energy electrons, such as those studied in Milosavljevic et al. work, are generated during SEM and TEM characterization as secondary electrons, and so are still highly relevant. They demonstrated that even these low-energy electrons on the order of 10 ev can substantially alter the $MAPbI_3$ crystal structure. As radiation damage progressed, substantial changes were observed in the sample morphology, with an increase in overall roughness due to the appearance of pores and cracks. Notably, the radiation-induced degradation became unobservable at around 4.5 eV, which is quite close in energy to the calculated highest occupied molecular orbital binding energy of MA^+ in defect- free perovskites (4.5–5 eV). The authors thus proposes that electron-induced degradation of $MAPbI_3$ is triggered by ionization of the methylammonium ion [95].

As the field has progressed, more quantitative examinations of the impact of electron radiation on perovskite structure have become possible. Rothmann et al. investigated structural and chemical changes to $MAPbI_3$ induced by electron and gallium ion beams at different accumulated doses and dose rates using selected area electron diffraction. It is important to consider gallium ions as well as electrons in the context of perovskite multimodal studies because many perovskite films for diffraction or other characterization studies are prepared by FIB milling as in work described earlier in this chapter (cf. Figure 16.9) [96, 97]. By continuously exposing a $MAPbI_3$ thin film to an electron dose rate of $2\,e^-\,Å^{-2}\,s^{-1}$, Rothmann et al. illustrated that subtle changes to the structure of a pristine material can occur even at low dose rates and total accumulated dose. The first structural features to disappear are the twin boundaries parallel to the {112} planes, which are intrinsic to tetragonal $MAPbI_3$. Following this, additional forbidden diffraction spots appear, which indicate the formation of a $\sqrt{2} \times \sqrt{2}$ supercell. This confirms Xiao et al.'s proposition that intermediary phases are formed in the material as a result of electron-induced damage. In conjunction with the formation of the supercell and increased electron exposure, the lattice parameters of the (002) and (110) planes continuously decrease. The authors attribute this to a loss of the organic moictics, which is consistent with Milosavljevic et al. proposed degradation mechanism. After the formation of the supercell, the $MAPbI_3$ film eventually degrades into another structure. The I:Pb ratio as a function of total electron dose decreases at similar rates regardless of incident dose rate, indicating that it is the total accumulated dose rather than dose rate that is important to consider with perovskites (at 200 kV). The authors indicated that structural changes occur at accumulated doses as small as $100\,e^-\,Å^{-2}$ at room temperature [53]. Furthermore, the authors were unable to prepare a perovskite sample with FIB milling that was undamaged, regardless of the experimental parameters employed; therefore, FIB prepared

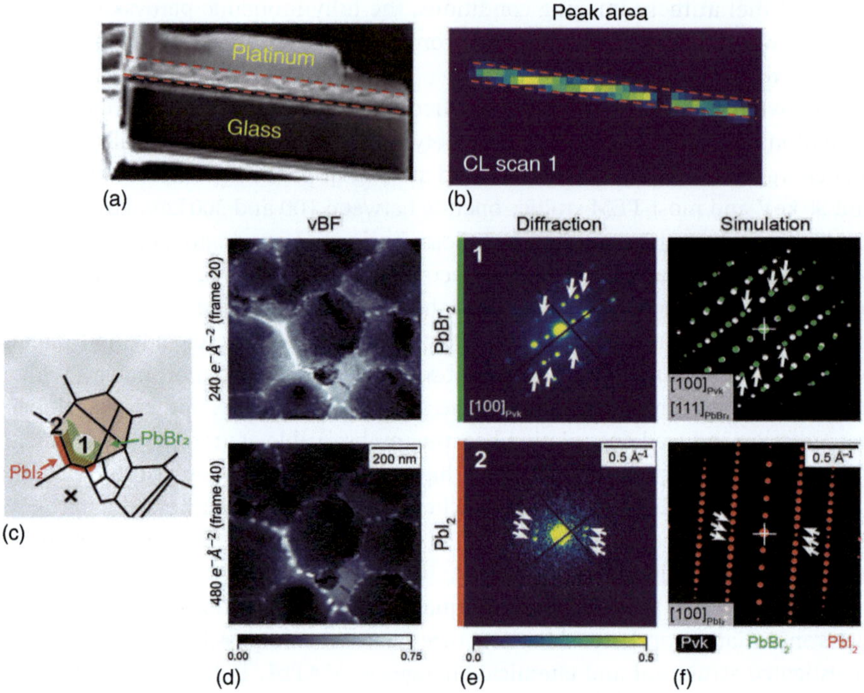

Figure 16.10 (a) Lamella cross section of a triple cation double halide half-solar cell stack consisting of glass/tin-doped indium oxide/NiO and $Cs_{0.05}FA_{0.81}MA_{0.14}Pb(I_{0.9}Br_{0.1})_3$. (b) Cathodoluminescence scan of the lamella in (a) showing emission from the perovskite. (c) Schematic representation of the formation of $PbBr_2$ and PbI_2 due to beam induced damage in a $Cs_{0.05}FA_{0.78}MA_{0.17}Pb(I_{0.83}Br_{0.17})_3$ perovskite. (d) Virtual bright field (vBF) images of the grain schematically represented in (c) following accumulated electron doses of 240 and 480 e^- $Å^{-2}$, respectively. (e) Diffraction patterns extracted from regions indicated in (c). Additional diffraction spots not indexable to perovskite are marked with white arrows. Source: Orri et al. [98], John Wiley & Sons / CC BY 4.0. The additional spots in region 1 can be indexed to $PbBr_2$ and the additional spots in region 2 can be indexed to PbI_2 (f) Simulated diffraction patterns for perovskite, $PbBr_2$ and PbI_2 marked white, green and red, respectively. Source: Reproduced from Kosasih et al. [99], John Wiley & Sons/ CC BY 4.0.

samples are likely not representative of the pristine photoactive material. However, interestingly, Kosasih et al. recently showed that samples prepared with FIB milling still exhibited emission consistent with the presence of perovskite, albeit with a slightly blueshifted luminescence indicating that some of the perovskite structure is preserved following FIB milling (Figure 16.10a,b). The results from Rothmann et al. were confirmed by those of Chen et al. who also utilized SAED to examine the structural instability and decomposition pathways of $MAPbI_3$; Chen et al. also observed a number of structural phases emerging over time with increasing electron dose exposure, which they indexed to an intermediary $MAPbI_{2.5}$ supercell. Following the formation of this unstable phase, the material degraded further into PbI_2, with a critical dose defined as 50–100 e^- $Å^{-2}$. Alberti et al. also observed the emergence of other phases in $MAPbI_3$ due to beam-induced damage [100].

Low-temperature beam damage studies have also been performed. Li et al. applied cryogenic electron microscopy to $MAPbI_3$ and $MAPbBr_3$ in an attempt to improve their resilience to beam-induced damage – a common approach in biological imaging to mitigate radiolysis. While Rothmann et al. and Chen et al. found that low temperatures exacerbate beam damage issues in $MAPbI_3$ [53, 101], the authors here found the converse to be true and claimed a fourfold increase in the critical dose for $MAPbBr_3$. They also inferred a similar improvement for $MAPbI_3$, obtaining atomic resolution images of a pristine $MAPbI_3$ nanowire at an accumulated dose of $12\,e^-\,Å^{-2}$. Notably, this is an order of magnitude less than the claimed critical dose for thin films of $MAPbI_3$ [53], which may indicate that cryogenic treatment makes things worse than better. It is difficult to deconvolute this observation from the different dimensionalities and thicknesses of the materials sampled across different studies (thin films versus nanowires) and the fact that a critical dose for atomic resolution electron microscopy imaging may be different than a critical dose for SAED studies.

This fact that making beam damage comparisons between different techniques, crystal orientations, and samples can be challenging was highlighted by VandenBussche et al. By delivering electrons both continuously via thermionic electron sources and in packets from a pulsed electron source, the authors were able to investigate whether the delivery method could influence the speed at which damage occurs in $MAPbI_3$ at the same total dose rate. Interestingly, it does. At a dose rate of $10\,e^-\,Å^{-2}\,s^{-1}$ they found a 17% reduction in damage from the pulsed electron beam (quantified by tracking the intensities of Bragg peaks) when compared to the continuous, thermionic source [102]. Chen et al. extended the investigation of different experimental conditions on beam damage of $MAPbI_3$ and $MAPbI_3$ single crystals by investigating the effect of accelerating voltage on damage. The authors observed that dose thresholds at 300 kV were 2–3 times larger than those at 80 kV. Facet-dependent electron beam sensitivity is also found for $MAPbI_3$ with dose thresholds 10 times higher for the (100) plane when compared to the (001) plane. This is attributed to a smaller iodine migration energy barrier at the (001) plane. The Br analogue shows dose thresholds twice as high as the I only sample [101].

While most of the previously mentioned studies focus on examination of $MAPbI_3$, $MAPbBr_3$, or $CSPbBr_3$, recent work has focused on more technologically relevant compositions. Rothmann et al. performed atomic resolution scanning-TEM (STEM) LAADF) imaging on samples of $FAPbI_3$ to reveal the decomposition pathways with accumulated electron exposure. Critical dose rates were comparable to $MAPbI_3$ ($\sim 100\,e^-\,Å^{-2}$) and degradation stages were similar (after extensive damage, different structural phases such as PbI_2 emerged) [91]. There have also been reports focused on understanding the effects of beam damage in mixed cation/mixed halide systems. Ferrer-Orri et al. examined the impact of local probes of both electrons and X-rays on the structure of $(Cs_{0.05}FA_{0.78}MA_{0.17})Pb(I_{0.83}Br_{0.17})_3$ using SED and nXRD measurements. In these compositionally complex samples, the authors observed the formation of $PbBr_2$ (within grains) and PbI_2 (at high-angle grain boundaries) after accumulated doses of $\sim 200\,e^-\,Å^{-2}$, followed by the formation of pinholes and a transition of the perovskite structure from the pristine average tetragonal phase to an average cubic phase (Figure 16.10c–f).

Finally, investigations of the impact of X-rays on the degradation of perovskites have largely aligned with electron based observations. Ferrer-Orri et al. defined the critical X-ray dose rate in their nXRD measurements (~20 keV) to be ~4650 photons $Å^{-2}$ [98]. This number compares well to other reports of synchrotron-induced beam damage in the halide perovskite community. Li et al. using a nanoprobe with an energy of 9.0 keV, observed a critical dose of ~10^2 photons $Å^{-2}$ for $MAPbBr_3$ and ~10^4 $Å^{-2}$ for $CsPbBr_3$ [64]. Hoye et al. utilizing a larger area illumination and a photon energy of 10.995 keV, observed substantial material degradation after an accumulated dose of ~46 photons $Å^{-2}$, which is two orders of magnitude lower than the dose threshold observed by Ferrer-Orri et al. and may point to fundamental differences between local and macroscale X-ray irradiation [103]. However, further experimental work is still required to confirm this.

A few common themes emerge from these beam damage studies, and a number more can be surmised that are relevant to understanding multimodal studies discussed in this chapter. (i) Generally, compositions that possess an MA cation have lower critical doses than those that do not. (ii) Compositions with Br may be more resistant to beam-induced damage than compositions with I only. (iii) The critical dose for $MAPbI_3$ thin films with thickness of about 200 nm is ~100 e^- $Å^{-2}$ (determined from SAED). For thinner and low-dimensional $MAPbI_3$ structures, it is approximately an order of magnitude less (determined from HRTEM). For $FAPbI_3$ thin films with a thickness of ~30 nm is ~60–100 e^- $Å^{-2}$ (Determined with HRSTEM). For $(Cs_{0.05}FA_{0.78}MA_{0.17})Pb(I_{0.83}Br_{0.17})_3$ the critical dose is ~200 e^- $Å^{-2}$ (determined with SED). (iv) In addition to thickness and particle dimensionality, critical dose may vary with imaging modality. As such the numbers presented in point (iii) should be taken as a guide only and it is critical to perform internal beam damage checks before characterization of each sample. (v) Generally, it seems that accumulated dose rather than dose rate is the most important parameter to monitor, though it is unlikely that the book is closed on this. (vi) Damage mechanisms in halide perovskites are likely a combination of both radiolysis and knock-on damage. As such, there is no one-size-fits-all solution to avoiding beam-induced changes such as cryogenic freezing to avoid radiolysis [104] (as in biological imaging) or lowering of the accelerating voltage to avoid knock-on damage [105](as in hBN imaging studies). The dominant damage process in a given halide perovskite material is likely related to the composition of the perovskite being examined. (vii) Damage across different perovskite compositions when using either X-rays or electrons to characterize generally follows the same trend: loss of crystallinity/destruction of the pristine perovskite structure followed by the emergence of other crystalline and occasionally amorphous materials/phases. (viii) With points 1–7 in mind, detailed structural measurements of halide perovskites can be completed if care is taken with chosen experimental parameters [106]. Benchmarking pre- and post-beam damage checks using complementary techniques (such as PL) can be one useful strategy. In some instances, doses that exceed the critical dose thresholds may be necessary in order to provide sufficient signal to noise, such as in STEM-EDX measurements, and thus damage may be inevitable. However, such data may still provide useful insight under appropriate measurement conditions particularly when considering any relative changes in composition across a film.

16.4.2 Challenges: Resolution Limits

Another challenge in multimodal microscopy studies of perovskite is – how can we best access the relevant length scales and properties of interest. It is clear that in halide perovskites, there are variations in the properties that govern macroscale device performance and operational stability, even on the atomic and nanometer length scales [8]. PL mapping is a useful way to assess optoelectronic performance, but as it is an optical microscopy technique that mostly utilizes visible light, the diffraction-limited spatial resolution will only approach ~300 nm at typical excitation/emission wavelengths, which is slightly larger than the grain sizes of many of the most technologically relevant perovskite materials. Other luminescence-based techniques that can achieve better resolution are unlikely to be suited for halide perovskites owing to the higher laser powers that must be employed. Similarly to electrons and X-rays, changes to the perovskite material under visible light can lead to rapid changes in the local properties of the perovskite. For example, stimulated emission depletion (STED) microscopy relies on saturating a donut-shaped region around the small region of interest with high excitation power [107], which may lead to rapid photo-induced ion motion that changes even the region of interest. CL is an electron microscopy-based alternative in which electron beams are utilized to excite the sample and can in principle achieve high (even ~1 nm) spatial resolution, but, again, as detailed in the section on radiation damage above, beam damage is problematic. As such, when employing CL, careful control of measurement conditions is required to minimize potential artifacts, meaning that the electron probe typically must be spread beyond its theoretical resolution limit to deposit dose over a larger area [108, 109]. Beyond luminescence, which in principle probes radiative recombination and only gives indirect information about other nonradiative processes, TAM allows tracking of carriers locally with high temporal resolution and spatial resolution beyond the diffraction limit [110, 111]. However, typical carrier densities for these techniques are high, and excitation densities to produce sufficient signal need to be brought down to fluences relevant to carrier recombination in devices, such as carrier densities relevant to maximum power point in solar cells. Photo-emission measurements allow probing of local valence band edge and subgap trap states at the surface and, when coupled into an appropriate electron photo-emission microscope (PEEM), allows extraction of such properties at high resolution (<20 nm) [112]. Time-resolved PEEM measurements allow one to track processes such as carrier trapping in halide perovskites in real time and space [79, 80] providing complementary information to PL and TAM about carrier recombination. Extreme care must be taken to avoid surface artifacts that may appear in such surface-sensitive measurements, and not all perovskite samples will be compatible with these approaches. Scanning probe measurements such as KPFM or photoconductive AFM provide nanoscale information about electrical properties coupled with morphology but are also mostly surface sensitive. Cross-section AFM-based measurements can allow assessment of multilayer device stacks [113], but care must be taken to avoid tip-related artefacts especially, in such soft materials [114]; the tip-sample interaction will vary with the material being probed, thus making cross-section measurements – when scanning across materials with different mechanical and morphological properties – a tricky challenge.

16.4.3 Challenges: Image Registration and Sample Fabrication

It is also essential that both sample drift during measurements and any additional image distortions between modalities be accounted for. This is so that scan regions can be precisely matched between correlative or in-situ measurements. Sharp image registration is critical, and use of fiducial markers may be essential provided they do not themselves influence sample behavior. Some groups have utilized geometrical gold nanoparticles cast on a film surface to be suitable for correlative measurements – forming unique shape clusters that are easily resolvable in optical, chemical, and structural techniques [60, 79], while others have sacrificed regions of a sample to burn in fiducials utilizing techniques such as the FIB [78]. In some instances, careful scratches on the sample surface are appropriate, but this is less useful in highly uniform films given one needs to measure away from the scratch-related debris and still requires a local reference point. The distortion of images during measurement, including nonlinear translations, transformations, and rotations also makes image correlation a challenge. Due to sensitivity or equipment limitations, many measurements need to be performed under conditions or in samples tailored for the measurement technique but which are far from standardized device environments. For example, TEM-based measurements require thin specimens (~30–200 nm depending on modality), thus limiting device layer thicknesses and in general prohibiting use of all contact layers, particularly crystalline ones. These thin films may form in a different manner from the thicker films utilized for devices or those formed on device contacts. Thus without further complementary checks (e.g. PL, current-voltage and/or XRD measurements), it is not obvious whether the sample will be representative of a high-performance device. Furthermore, applying illumination intensities and biases that mimic real devices is not trivial. For instance, the lateral nature of many scanning probe measurements means that only lateral device structures are measurable, which are vastly different from typical vertically stacked devices – leading to different magnitudes of electric field and measurements of different features in lateral (e.g. multiple grains) compared to vertical (single grains extending through the thickness) directions. This also means probing heterogeneity in both vertical and lateral directions is a challenge, where those techniques suited for depth-profiling may not achieve suitable spatial or spectral resolution, and vice versa for techniques utilizing lateral measurements. Tomographic techniques, which could fully reconstruct 3D information through thickness of a film, are typically limited by beam damage considerations. Some characterization techniques are surface sensitive, while others probe the bulk of the material, and it is rare to find techniques that can do both. The same is true when considering the properties being probed. For instance, in chemical mapping, probing the organic cation will typically require orthogonal experimental setups to those probing the inorganic cation. Thus, to achieve a full picture, one needs complementary toolsets that probe the right properties, in the right directions, with the right resolutions.

16.4.4 Challenges: Facility Access and Data Acquisition

Many multimodal techniques that are capable of probing structure and chemical properties with high spatial resolutions require access to large facilities such as synchrotrons, where the highly intense, monochromated light sources make high speed and high spatial resolution characterization possible. Access to these facilities can pose a barrier to progress, in particular when there is the need to iteratively customize specialized equipment and perform long measurement runs in limited and infrequent beamtime allocations. Furthermore, from the perovskite sample perspective, variations between sample batches, individual samples within the same batch, and even pixels on the same device substrate are rife in the halide perovskite field due to these materials sensitivity to local processing conditions. Consequently, multiple measurements must be performed to confidently draw conclusions from multimodal measurements – further stretching the use of limited beamtime. Issues such as these were further exacerbated during the COVID-19 pandemic, where, at best, access became remote, thus prohibiting the more risky, customized innovation that takes hands-on time to develop. A further issue is that many advanced characterization facilities generate data at incredible rates. Modern-day STEM's for instance, can generate up to 200 TB of data per hour, which cannot be analyzed by human inspection alone within a reasonable timeframe [115]. Increasingly, in addition to experimental and materials synthesis expertise, researchers working on multimodal characterization of halide perovskite require deep expertise in data science.

16.5 Outlook and Opportunities

Multimodal characterization studies are driving our understanding of the interrelation between properties in halide perovskites, allowing insights into relationships between, for example, nonradiative recombination and local structure, ion migration and performance, and instability and material changes. However, the need for direct and simultaneous measurement of multiple properties in device-relevant structures and stressors (light, bias, etc.), under continuous device-like operation is critical for monitoring, understanding, and ultimately solving the grand challenge for halide perovskites – long-term operational stability. Achieving these outcomes will require integration of standardized yet modular operando rigs (see Chapter 15), which incorporate application of light, bias, and atmospheric control on multimodal setups in both research laboratories and facilities (beamlines and electron microscopy facilities). Planned or recent beamline upgrades around the world, including ESRF, Diamond II, SPRING-8 and Advanced Photon Source, among others, will enable greater brightness and ever more sensitive detectors, further accelerating developments by allowing faster measurements and new capabilities (while maintaining low dose). Additionally, COVID-19 has shown that many of these advanced experiments can be successfully run remotely with the aid of beamline scientists, which will improve access to these facilities as samples can be routinely posted in.

Currently, a crucially needed capability in the perovskite field is a method to operate and characterize multiple full solar cell devices simultaneously or in rapid succession to ensure that information obtained from multimodal studies is representative and that robust statistics on degradation metrics can be collected in order to inform future fabrication strategies. This brings new challenges in high-throughput experiments and hardware automation, but will be a critical component in the future of self-driving labs, where enormous device fabrication parameter space (composition, device layers, etc.), can be explored through automated fabrication and characterization [116]. This area is a field which the perovskite community is already beginning to explore [117–120] (Figure 16.11a,b). With increased experimental characterization throughput, bottlenecks will move away from measurement and emerge in the storage, access, analysis, and interpretation of large data sets. Artificial-intelligence-driven data analysis techniques such as image classification with neural networks will be essential to exploring the data space in a feasible time span to: extract key features from data [121, 122] (Figure 16.11c,d), reconstruct information from noisy signals in low-dose images [123], enhance image registration accuracy between multiple characterization modalities, and determine when enough data has been acquired to statistically represent a given sample [115]. These data science developments will also play a role in real-time data processing to dynamically guide experiments and to optimize experimental acquisition parameters – including understanding tolerable doses for a given experiment [115]. Key topics to address in the coming years related to this will be how we can leverage innovations in AI, data science, and multimodal characterization, including techniques developed to understand other disordered materials such as batteries [124] and photocatalysts [125], to probe the regions of perovskite device stacks that, to date, have been less accessible, such as buried recombination processes [126], interfaces [127], and device sub-stacks in multijunction cells [128, 129].

More generally, a common theme that has emerged in the perovskite field is that halide perovskite materials are tolerant to a range of additives, which has been observed empirically through high device performance when various passivating additives are added to the perovskite [130, 131]. These additives may reside in the bulk material or at interfaces, are often (though not always) incorporated into the structure, and improve device performance and even stability under feasible

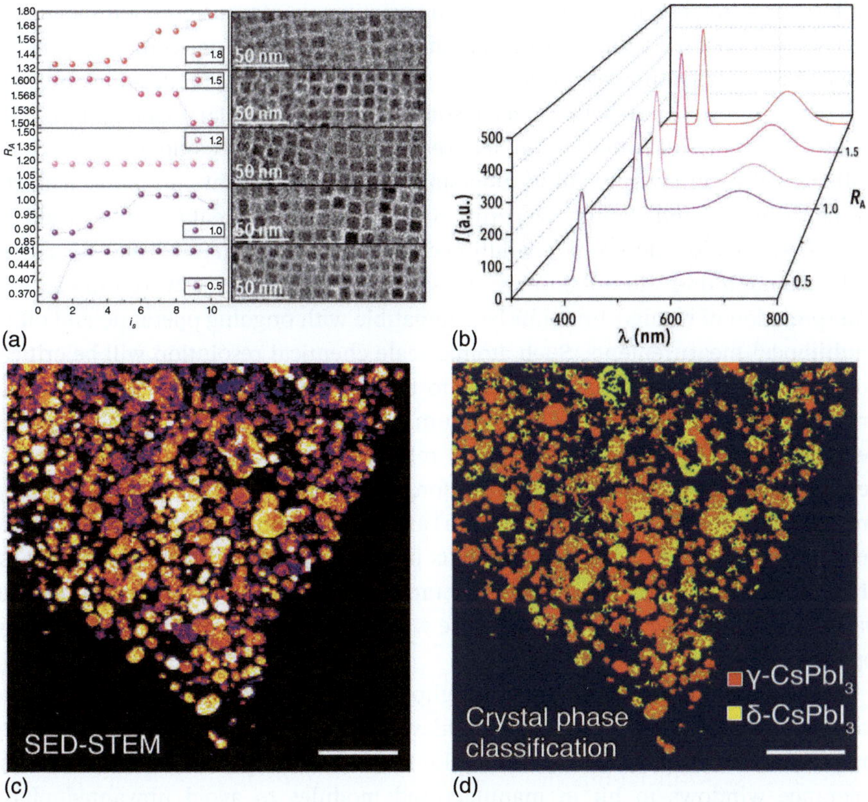

Figure 16.11 (a) Exploitation decision policy (EPLT) autonomous closed loop optimization of perovskite nanocrystals campaign using a self-driving fluidic micro-processor. Representation of best condition-so-far R_A values for five distinct optimization campaigns. TEM images of optimal formulations are shown next to each EPLT plot. Source: Reproduced from Bateni et al. [117], John Wiley & Sons / CC BY 4.0. (b) PL spectra of the optimal compositions of Mn:$CsPb(Cl/Br)_3$. (c) SED map of perovskite ($CsPbI_3$)/metal organic framework composite. (d) Crystal phase map of the perovskite/metal organic framework composite in (c) extracted using convolutional neural networks. Source: Reproduced from Hou et al. [121], American Association for the Advancement of Science – AAAS.

lab-scale testing [132, 133]. In the case of FA-rich perovskites, a number of these additives increasingly appear to work by inducing a slight degree of octahedral tilting that stabilizes the photoactive structure and prevents the formation of local, performance, and stability-limiting phases [32, 63, 82]. However, in spite of the success of some of these approaches, residual phase impurities on the nanoscale still persist [32, 80], such as hexagonal polytypes and lead iodide, which will ultimately act as failure points as they facilitate redox reactions in photovoltaic devices even if only present on the nanoscale [32]. Key questions are: how can we fully eliminate these phases, or is there some length scale or size at which they are no longer problematic? At some point, these impurities will be smaller than the nanoscale resolution of most practical experimental setups, meaning we must push towards atomic scale techniques to resolve them. High-resolution TEM and STEM images of bulk perovskite absorber materials reveal variations in structure on atomic-length scales [91]. Incorporating device-like studies with atomic resolution would be a worthy – and ultimately necessary – pursuit. For example, moving forward, studies that incorporate both atomic resolution imaging and DFT calculations, and ideally photophysical information, focused on different compositions of perovskite will help the community fully understand the impact of atomic scale features on performance and stability in perovskites. A particular breakthrough would be quantitative nanoscale atomic resolution chemical mapping in full device structures at low enough doses to not damage the sample (or at least not overtly impact the interpretation of results), and thus be compatible with ongoing operando and other multimodal measurements. Such atomic-scale chemical resolution will be critical for imaging nanostructured analogues to identify defects and losses in materials more suited to light-emission or quantum applications [134]. Adapting mapping techniques that can systematically and reliably probe strain and stress on these length scales will also be critical for long-term stability assessment (including mechanical effects such as delamination) and fundamental understanding of these materials [66]. In any case, recent results presented here are beginning to suggest that halide perovskites may be less tolerant to additives and local stoichiometric variations than commonly assumed once we start considering extremely long-term operational performance metrics.

As the field moves toward high-throughput manufacturing, understanding these local deformities will be critical, and achieving homogeneity over large area in key properties will be essential. For example, there will likely be local stoichiometry tolerance windows to hit in manufactured modules to avoid unwanted phase impurities and additional mobile ions. These requirements motivate integrating the multimodal techniques discussed in this chapter onto manufacturing lines, allowing real time feedback about unwanted stoichiometry, structural, and optoelectronic variations. Furthermore, ensuring that these multimodal platforms are consistent with industry standard testing – incorporating standardized testing procedures agreed upon by the field [135] – will mean these techniques are directly probing parameters under conditions relevant to field testing, providing useful information for manufacturing. Given the complexity of these nanoscale features, this will require moving beyond more standard luminescence and optoelectronic mapping to identify inhomogeneities in manufactured modules.

The halide perovskite field is on the brink of commercialization, marking an exciting phase in their development. However, a number of aspects are still poorly understood, including power loss and failure mechanisms, and many of these aspects relate to a complex interplay between properties on very small length scales that are in general difficult to probe. As the field matures and experimental capability for multimodal microscopy measurements continues to push the envelope, we will see exciting breakthroughs in this understanding – in turn drive commercialization as well as seed new science, realize unforeseen device platforms, and develop new material design guidelines. In the end, there is very little substitute for direct, operando measurements to monitor a device or material operating in its most relevant environment – and the field must strive to widely adopt, with appropriate care, such techniques.

References

1 Stranks, S.D. and Snaith, H.J. (2015). Metal-halide perovskites for photovoltaic and light-emitting devices. *Nature Nanotechnology* 10: 391–402.

2 Kojima, A., Teshima, K., Shirai, Y., and Miyasaka, T. (2009). Organometal halide perovskites as visible-light sensitizers for photovoltaic cells. *Journal of the American Chemical Society* 131: 6050–6051.

3 Lee, M.M., Teuscher, J., Miyasaka, T. et al. (2012). Efficient hybrid solar cells based on meso-superstructured organometal halide perovskites. *Science* 338: 643–647.

4 Kim, H.-S., Lee, C.-R., Im, J.-H., et al. (2012). Lead iodide perovskite sensitized all-solid-state submicron thin film mesoscopic solar cell with efficiency exceeding 9%. *Scientific Reports* 2: 591.

5 Tan, Z.-K. et al. (2014). Bright light-emitting diodes based on organometal halide perovskite. *Nature Nanotechnology* 9: 687–692.

6 Wei, H. et al. (2016). Sensitive X-ray detectors made of methylammonium lead tribromide perovskite single crystals. *Nature Photonics* 10: 333–339.

7 NREL (2021). Best Research-Cell Efficiency Chart | Photovoltaic Research | NREL. https://www.nrel.gov/pv/cell-efficiency.html.

8 Tennyson, E.M., Doherty, T.A.S., and Stranks, S.D. (2019). Heterogeneity at multiple length scales in halide perovskite semiconductors. *Nature Reviews Materials* 1–15. https://doi.org/10.1038/s41578-019-0125-0.

9 Bercegol, A. et al. (2018). Spatial inhomogeneity analysis of cesium-rich wrinkles in triple-cation perovskite. *The Journal of Physical Chemistry C* 122: 23345–23351.

10 Kovalenko, M.V., Protesescu, L., and Bodnarchuk, M.I. (2017). Properties and potential optoelectronic applications of lead halide perovskite nanocrystals. *Science* 358: 745.

11 Stranks, S. and Nonradiative, D. (2017). Losses in metal halide perovskites. *ACS Energy Letters* 2: 1515–1525.

12 Mahesh, S. et al. (2020). Revealing the origin of voltage loss in mixed-halide perovskite solar cells. *Energy & Environmental Science* 13: 258–267.

13 Dey, K., Roose, B., and Stranks, S.D. Optoelectronic properties of low-bandgap halide perovskites for solar cell applications. *Advanced Materials* 33 (40): 2102300.

14 Zhou, Y., Poli, I., Meggiolaro, D. et al. (2021). Defect activity in metal halide perovskites with wide and narrow bandgap. *Nature Reviews Materials* 6: 986–1002.

15 Stolterfoht, M. et al. (2018). Visualization and suppression of interfacial recombination for high-efficiency large-area pin perovskite solar cells. *Nature Energy* 3: 847–854.

16 Sutter-Fella, C. and The, M. (2021). Value of watching how materials grow: a multimodal case study on halide perovskites. *Advanced Energy Materials* 11: 2003534.

17 Alsari, M. et al. (2018). In situ simultaneous photovoltaic and structural evolution of perovskite solar cells during film formation. *Energy & Environmental Science* 11: 383–393.

18 Suchan, K., Just, J., Becker, P. et al. (2020). Optical in situ monitoring during the synthesis of halide perovskite solar cells reveals formation kinetics and evolution of optoelectronic properties. *Journal of Materials Chemistry A* 8: 10439–10449.

19 Mundt, L.E. and Schelhas, L.T. (2020). Structural evolution during perovskite crystal formation and degradation: in situ and operando X-ray diffraction studies. *Advanced Energy Materials* 10: 1903074.

20 de Quilettes, D.W. et al. (2019). Charge-carrier recombination in halide perovskites: focus review. *Chemical Reviews* https://doi.org/10.1021/acs.chemrev.9b00169.

21 Rau, U. (2007). Reciprocity relation between photovoltaic quantum efficiency and electroluminescent emission of solar cells. *Physical Review B* 76 (February 8): 085303.

22 Yin, W.-J., Shi, T., and Yan, Y. (2014). Unusual defect physics in $CH_3NH_3PbI_3$ perovskite solar cell absorber. *Applied Physics Letters* 104: 063903.

23 de Quilettes, D.W. et al. (2015). Impact of microstructure on local carrier lifetime in perovskite solar cells. *Science* 348: 683–686.

24 de Quilettes, D.W. et al. (2017). Tracking photoexcited carriers in hybrid perovskite semiconductors: trap-dominated spatial heterogeneity and diffusion. *ACS Nano* 11: 11488–11496.

25 Lee, J.W., Shichijo, H., Tsai, H.L., and Matyi, R.J. (1987). Defect reduction by thermal annealing of GaAs layers grown by molecular beam epitaxy on Si substrates. *Applied Physics Letters* 50: 31–33.

26 Hörantner, M.T. et al. (2017). The potential of multijunction perovskite solar cells. *ACS Energy Letters* 2: 2506–2513.

27 de Quilettes, D.W. et al. (2016). Photo-induced halide redistribution in organic-inorganic perovskite films. *Nature Communications* 7: 11683.

28 Castro-Méndez, A.-F., Hidalgo, J., and Correa-Baena, J.-P. (2019). The role of grain boundaries in perovskite solar cells. *Advanced Energy Materials* 9: 1901489.

29 Jacobsson, T.J. et al. (2016). Exploration of the compositional space for mixed lead halogen perovskites for high efficiency solar cells. *Energy & Environmental Science* 9: 1706–1724.

30 Tennyson, E.M. et al. (2015). Nanoimaging of open-circuit voltage in photovoltaic devices. *Advanced Energy Materials* 5: 1501142.

31 Wang, R. et al. (2019). A review of perovskites solar cell stability. *Advanced Functional Materials* 29: 1808843.

32 Macpherson, S. et al. (2021). Local nanoscale defective phase impurities are the sites of degradation in halide perovskite devices. *arXiv:2107.09549 [cond-mat, physics:physics]*.

33 Motti, S.G. et al. (2019). Controlling competing photochemical reactions stabilizes perovskite solar cells. *Nature Photonics* 13: 532–539.

34 Stranks, S.D. et al. (2014). Recombination kinetics in organic-inorganic perovskites: excitons, free charge, and subgap states. *Physical Review Applied* 2: 034007.

35 Andaji-Garmaroudi, Z., Anaya, M., Pearson, A.J., and Stranks, S.D. (2020). Photobrightening in lead halide perovskites: observations, mechanisms, and future potential. *Advanced Energy Materials* 10: 1903109.

36 Anaya, M., Galisteo-López, J.F., Calvo, M.E. et al. (2018). Origin of light induced photophysical effects in organic metal halide perovskites in the presence of oxygen. *The Journal of Physical Chemistry Letters* 9: 3891–3896.

37 Choi, H. et al. (2020). Visualizing the role of photoinduced ion migration on photoluminescence in halide perovskite grains. *Journal of Materials Chemistry C* 8: 7509–7518.

38 Xiao, Z. et al. (2015). Giant switchable photovoltaic effect in organometal trihalide perovskite devices. *Nature Materials* 14: 193–198.

39 Gratia, P. et al. (2016). Intrinsic halide segregation at nanometer scale determines the high efficiency of mixed cation/mixed halide perovskite solar cells. *Journal of the American Chemical Society* 138: 15821–15824.

40 Gratia, P. et al. (2017). The many faces of mixed ion perovskites: unraveling and understanding the crystallization process. *ACS Energy Letters* 2: 2686–2693.

41 Giridharagopal, R. et al. (2019). Time-resolved electrical scanning probe microscopy of layered perovskites reveals spatial variations in photoinduced ionic and electronic carrier motion. *ACS Nano* 13: 2812–2821.

42 Garrett, J.L. et al. (2017). Real-time nanoscale open-circuit voltage dynamics of perovskite solar cells. *Nano Letters* 17: 2554–2560.

43 Birkhold, S.T. et al. (2018). Interplay of mobile ions and injected carriers creates recombination centers in metal halide perovskites under bias. *ACS Energy Letters* 3: 1279–1286.

44 Leijtens, T. et al. (2014). Electronic properties of meso-superstructured and planar organometal halide perovskite films: charge trapping, photodoping, and carrier mobility. *ACS Nano* 8: 7147–7155.

45 Leblebici, S.Y. et al. (2016). Facet-dependent photovoltaic efficiency variations in single grains of hybrid halide perovskite. *Nature Energy* 1: 16093.

46 Eperon, G.E., Moerman, D., and Ginger, D.S. (2016). Anticorrelation between local photoluminescence and photocurrent suggests variability in contact to active layer in perovskite solar cells. *ACS Nano* 10: 10258–10266.

47 Jiang, F. et al. (2021). Scanning Kelvin probe microscopy reveals that ion motion varies with dimensionality in 2D halide perovskites. *ACS Energy Letters* 6: 100–108.

48 Robinson, I. and Harder, R. (2009). Coherent X-ray diffraction imaging of strain at the nanoscale. *Nature Materials* 8: 291–298.

49 Quinn, P., Parker, J., Cacho-Nerin, F. et al. (2018). The hard X-ray nanoprobe beamline at diamond - current status. *Microscopy and Microanalysis* 24: 242–243.

50 Li, W. et al. (2021). The critical role of composition-dependent intragrain planar defects in the performance of $MA_{1-x}FA_xPbI_3$ perovskite solar cells. *Nature Energy* 6: 624–632.

51 Li, W. et al. (2018). Subgrain special boundaries in halide perovskite thin films restrict carrier diffusion. *ACS Energy Letters* 3: 2669–2670.

52 Xiao, X. et al. (2020). Benign ferroelastic twin boundaries in halide perovskites for charge carrier transport and recombination. *Nature Communications* 11: 2215.

53 Rothmann, M.U. et al. (2018). Structural and CHEMICAL Changes to $CH_3NH_3PbI_3$ induced by electron and gallium ion beams. *Advanced Materials* 30: 1800629.

54 McKenna, K.P. (2018). Electronic properties of {111} twin boundaries in a mixed-ion lead halide perovskite solar absorber. *ACS Energy Letters* 2663–2668. https://doi.org/10.1021/acsenergylett.8b01700.

55 Pham, H.T. et al. (2021). Unraveling the influence of CsCl/MACl on the formation of nanotwins, stacking faults and cubic supercell structure in FA-based perovskite solar cells. *Nano Energy* 87: 106226.

56 Salje, E.K. and Ferroelastic, H. (2012). Materials. *Annual Review of Materials Research* 42: 265–283.

57 Liu, Y. et al. (2018). Chemical nature of ferroelastic twin domains in $CH_3NH_3PbI_3$ perovskite. *Nature Materials* https://doi.org/10.1038/s41563-018-0152-z.

58 Rothmann, M.U. et al. (2017). Direct observation of intrinsic twin domains in tetragonal $CH_3NH_3PbI_3$. *Nature Communications* 8: 14547.

59 Whitfield, P.S. et al. (2016). Structures, phase transitions and tricritical behavior of the hybrid perovskite methyl ammonium lead iodide. *Scientific Reports* 6: 35685.

60 Jones, T.W. et al. (2019). Lattice strain causes non-radiative losses in halide perovskites. *Energy & Environmental Science* 12: 596–606.

61 Jariwala, S. et al. (2019). Local crystal misorientation influences non-radiative recombination in halide perovskites. *Joule* 3: 3048–3060.

62 Lilliu, S. et al. (2016). Mapping morphological and structural properties of lead halide perovskites by scanning nanofocus XRD. *Advanced Functional Materials* 26: 8221–8230.

63 Saliba, M. et al. (2016). Cesium-containing triple cation perovskite solar cells: improved stability, reproducibility and high efficiency. *Energy & Environmental Science* 9: 1989–1997.

64 Li, X., Luo, Y., Holt, M.V. et al. (2019). Residual nanoscale strain in cesium lead bromide perovskite reduces stability and shifts local luminescence. *Chemistry of Materials* 31: 2778–2785.

65 Zhao, J. et al. (2017). Strained hybrid perovskite thin films and their impact on the intrinsic stability of perovskite solar cells. *Science Advances* 3.

66 Liu, D. et al. (2021). Strain analysis and engineering in halide perovskite photovoltaics. *Nature Materials* 20: 1337–1346.

67 Xue, D.-J. et al. (2020). Regulating strain in perovskite thin films through charge-transport layers. *Nature Communications* 11: 1514.

68 Rolston, N. et al. (2018). Engineering stress in perovskite solar cells to improve stability. *Advanced Energy Materials* 8: 1802139.

69 Saidaminov, M.I. et al. (2018). Suppression of atomic vacancies via incorporation of isovalent small ions to increase the stability of halide perovskite solar cells in ambient air. *Nature Energy* 3: 648–654.

70 Howard, C.J. and Stokes, H.T. (1998). Group-theoretical analysis of octahedral tilting in perovskites. *Acta Crystallographica, Section B: Structural Science* 54: 782–789.

71 Tsai, H. et al. (2018). Light-induced lattice expansion leads to high-efficiency perovskite solar cells. *Science* 360: 67.

72 Godding, J.S.W. et al. (2019). Oxidative passivation of metal halide perovskites. *Joule* https://doi.org/10.1016/j.joule.2019.08.006.

73 Knight, A.J. et al. (2019). Electronic traps and phase segregation in lead mixed-halide perovskite. *ACS Energy Letters* 4: 75–84.

74 Chen, Y. et al. (2020). Strain engineering and epitaxial stabilization of halide perovskites. *Nature* 577: 209–215.

75 Zhu, C. et al. (2019). Strain engineering in perovskite solar cells and its impacts on carrier dynamics. *Nature Communications* 10: 815.

76 Feldmann, S. et al. (2019). Photodoping through local charge carrier accumulation in alloyed hybrid perovskites for highly efficient luminescence. *Nature Photonics* 14: 123–128.

77 Frohna, K. et al. (2021). Nanoscale Chemical Heterogeneity Dominates the Optoelectronic Response over Local Electronic Disorder and Strain in Alloyed Perovskite Solar Cells.

78 Correa-Baena, J.-P. et al. (2019). Homogenized halides and alkali cation segregation in alloyed organic-inorganic perovskites. *Science* 363: 627–631.

79 Doherty, T.A.S. et al. (2020). Performance-limiting nanoscale trap clusters at grain junctions in halide perovskites. *Nature* 580: 360–366.

80 Kosar, S., Winchester, A.J., Doherty, T.A.S. et al. Unraveling the varied nature and roles of defects in hybrid halide perovskites with time resolved photoemission electron microscopy. *Energy & Environmental Science* 14 (12): 6320–6328.

81 Szostak, R. et al. (2019). Nanoscale mapping of chemical composition in organic inorganic hybrid perovskite films. *Science Advances* 5: eaaw6619.

82 Doherty, T.A.S. et al. (2021). Stabilized tilted-octahedra halide perovskites inhibit local formation of performance-limiting phases. *Science* https://doi.org/10.1126/science.abl4890.

83 Dang, H.X. et al. (2019). Multi-cation synergy suppresses phase segregation in mixed-halide perovskites. *Joule* 3: 1746–1764.

84 Marchezi, P.E. et al. (2020). Degradation mechanisms in mixed-cation and mixed-halide $Cs_xFA_{1-x}Pb(BryI1-y)_3$ perovskite films under ambient conditions. *Journal of Materials Chemistry A* 8: 9302–9312.

85 Yang, W.S. et al. (2017). Iodide management in formamidinium-lead-halide-based perovskite layers for efficient solar cells. *Science* 356: 1376–1379.

86 Yang, W.S. et al. (2015). High-performance photovoltaic perovskite layers fabricated through intramolecular exchange. *Science* 348: 1234–1237.

87 Lu, H. et al. (2020). Vapor-assisted deposition of highly efficient, stable black-phase $FAPbI_3$ perovskite solar cells. *Science* 370.

88 Hui, W. et al. (2021). Stabilizing black-phase formamidinium perovskite formation at room temperature and high humidity. *Science* 371: 1359–1364.

89 Min, H. et al. (2019). Efficient, stable solar cells by using inherent bandgap of α-phase formamidinium lead iodide. *Science* 366: 749–753.

90 Kim, G. et al. (2020). Impact of strain relaxation on performance of α-formamidinium lead iodide perovskite solar cells. *Science* 370: 108–112.

91 Rothmann, M.U. et al. (2020). Atomic-scale microstructure of metal halide perovskite. *Science* 370.

92 Cai, S. et al. (2022). Atomically resolved electrically active intragrain interfaces in perovskite semiconductors. *Journal of the American Chemical Society* https://doi.org/10.1021/jacs.1c12235.

93 Xiao, C. et al. (2015). Mechanisms of electron-beam-induced damage in perovskite thin films revealed by cathodoluminescence spectroscopy. *Journal of Physical Chemistry C* 119: 26904–26911.

94 Klein-Kedem, N., Cahen, D., and Hodes, G. (2016). Effects of light and electron beam irradiation on halide perovskites and their solar cells. *Accounts of Chemical Research* 49: 347–354.

95 Milosavljević, A.R., Huang, W., Sadhu, S., and Ptasinska, S. (2016). Low-energy electron-induced transformations in organolead halide perovskite. *Angewandte Chemie International Edition* 55: 10083–10087.

96 Divitini, G. et al. (2016). In situ observation of heat-induced degradation of perovskite solar cells. *Nature Energy* 1: 15012.

97 Andaji-Garmaroudi, Z. et al. (2020). Elucidating and mitigating degradation processes in perovskite light-emitting diodes. *Advanced Energy Materials* 10: 2002676.

98 Orri, J.F. et al. (2022). Unveiling the interaction mechanisms of electron and X-ray radiation with halide perovskite semiconductors using scanning nano-probe diffraction. *Advanced Materials* 34: 2200383.

99 Kosasih, F.U. et al. (2022). Optical emission from focused ion beam milled halide perovskite device cross-sections. *Microscopy Research and Technique* 85: 2351–2355.

100 Alberti, A. et al. (2015). Similar structural dynamics for the degradation of $CH_3NH_3PbI_3$ in air and in vacuum. *ChemPhysChem* https://doi.org/10.1002/cphc.201500374.

101 Chen, S. et al. (2020). Transmission electron microscopy of organic-inorganic hybrid perovskites: myths and truths. *Science Bulletin* 65: 1643–1649.

102 VandenBussche, E.J., Clark, C.P., Holmes, R.J., and Flannigan, D.J. (2020). Mitigating damage to hybrid perovskites using pulsed-beam TEM. *ACS Omega* 5: 31867–31871.

103 Hoye, R.L.Z. et al. (2017). Perovskite-inspired photovoltaic materials: toward best practices in materials characterization and calculations. *Chemistry of Materials* https://doi.org/10.1021/acs.chemmater.6b03852.

104 Henderson, R. and Clarke, B.C. (1990). Cryo-protection of protein crystals against radiation damage in electron and X-ray diffraction. *Proceedings of the Royal Society of London. Series B: Biological Sciences* 241: 6–8.

105 Egerton, R.F. (2012). Mechanisms of radiation damage in beam-sensitive specimens, for TEM accelerating voltages between 10 and 300 kV. *Microscopy Research and Technique* 75: 1550–1556.

106 Ning, Z. et al. (2021). Reply to: Perovskite decomposition and missing crystal planes in HRTEM. *Nature* 594: E8–E9.

107 Klar, T.A. and Hell, S.W. (1999). Subdiffraction resolution in far-field fluorescence microscopy. *Optics Letters* 24: 954–956.

108 Ferrer Orri, J. et al. (2021). Using pulsed mode scanning electron microscopy for cathodoluminescence studies on hybrid perovskite films. *Nano Express* 2: 024002.

109 Yamamoto, N. (2016). Development of high-resolution cathodoluminescence system for STEM and application to plasmonic nanostructures. *Microscopy* 65: 282–295.

110 Sung, J. et al. (2020). Long-range ballistic propagation of carriers in methylammonium lead iodide perovskite thin films. *Nature Physics* 16: 171–176.

111 Guo, Z. et al. (2017). Long-range hot-carrier transport in hybrid perovskites visualized by ultrafast microscopy. *Science* 356: 59.

112 Man, M.K.L. et al. (2017). Imaging the motion of electrons across semiconductor heterojunctions. *Nature Nanotechnology* 12: 36–40.

113 Bergmann, V.W. et al. (2014). Real-space observation of unbalanced charge distribution inside a perovskite-sensitized solar cell. *Nature Communications* 5.

114 Tennyson, E.M., Howard, J.M., and Leite, M.S. (2017). Mesoscale functional imaging of materials for photovoltaics. *ACS Energy Letters* 2: 1825–1834.

115 Spurgeon, S.R. et al. (2021). Towards data-driven next-generation transmission electron microscopy. *Nature Materials* 20: 274–279.

116 Soldatov, M.A. et al. (2021). Self-driving laboratories for development of new functional materials and optimizing known reactions. *Nanomaterials (Basel)* 11: 619.

117 Bateni, F. et al. (2022). Autonomous nanocrystal doping by self-driving fluidic micro-processors. *Advanced Intelligent Systems* 4: 2200017.

118 Howard, J.M., Tennyson, E.M., Neves, B.R.A., and Leite, M.S. (2019). Machine learning for perovskites' reap-rest-recovery cycle. *Joule* 3: 325–337.

119 Sun, S. et al. (2019). Accelerated development of perovskite-inspired materials via high-throughput synthesis and machine-learning diagnosis. *Joule* 3: 1437–1451.

120 Kumar, R. E. et al. (2021). Opportunities for machine learning to accelerate halide perovskite commercialization and scale-up. *arXiv:2110.03923 [cond-mat]*.

121 Hou, J. et al. (2021). Liquid phase sintering of lead halide perovskites and metal-organic framework glasses. *Science* 374: 621–625.

122 Kaufmann, K. et al. (2020). Crystal symmetry determination in electron diffraction using machine learning. *Science* 367: 564–568.

123 Ede, J.M. (2021). Deep learning in electron microscopy. *Machine Learning: Science and Technology* 2: 011004.

124 Merryweather, A.J., Schnedermann, C., Jacquet, Q. et al. (2021). Operando optical tracking of single-particle ion dynamics in batteries. *Nature* 594: 522–528.

125 Wu, Y.A. et al. (2019). Facet-dependent active sites of a single Cu_2O particle photocatalyst for CO_2 reduction to methanol. *Nature Energy* 4: 957–968.

126 Stavrakas, C. et al. (2018). Probing buried recombination pathways in perovskite structures using 3D photoluminescence tomography. *Energy & Environmental Science* 11: 2846–2852.

127 Yang, X. et al. (2021). Buried interfaces in halide perovskite photovoltaics. *Advanced Materials* 33: 2006435.

128 Tennyson, E.M. et al. (2021). Multimodal microscale imaging of textured perovskite–silicon tandem solar cells. *ACS Energy Letters* 6: 2293–2304.

129 Bowman, A.R. et al. (2021). Relaxed current matching requirements in highly luminescent perovskite tandem solar cells and their fundamental efficiency limits. *ACS Energy Letters* 6: 612–620.

130 Yang, G. et al. (2021). Stable and low-photovoltage-loss perovskite solar cells by multifunctional passivation. *Nature Photonics* 15: 681–689.

131 Jeong, J. et al. (2021). Pseudo-halide anion engineering for α-$FAPbI_3$ perovskite solar cells. *Nature* 592: 381–385.

132 Abdi-Jalebi, M. et al. (2018). Maximizing and stabilizing luminescence from halide perovskites with potassium passivation. *Nature* 555: 497–501.

133 Saliba, M. et al. (2016). Incorporation of rubidium cations into perovskite solar cells improves photovoltaic performance. *Science* 354: 206–209.

134 Shamsi, J., Raino, G., Kovalenko, M.V., and Stranks, S.D. (2021). Halide perovskite light emitters: to nano or not to nano? *Nature Nanotechnology* 16: 1164–1168.

135 Khenkin, M.V. et al. (2020). Consensus statement for stability assessment and reporting for perovskite photovoltaics based on ISOS procedures. *Nature Energy* 5: 35–49.

Index

Halide Perovskite Semiconductors: Structures, Characterization, Properties, and Phenomena, First Edition. Edited by Yuanyuan Zhou and Iván Mora-Seró.

i

n

q

r